C0-ASC-944

Applied Statistics for Business and Economics

Applied Statistics for Business and Economics

Robert M. Leekley

Illinois Wesleyan University
Bloomington, Illinois, U.S.A.

CRC Press is an imprint of the
Taylor & Francis Group an **informa** business
A CHAPMAN & HALL BOOK

CRC Press
Taylor & Francis Group
6000 Broken Sound Parkway NW, Suite 300
Boca Raton, FL 33487-2742

Printed in the United States of America on acid-free paper
10 9 8 7 6 5 4 3 2 1

International Standard Book Number: 978-1-4398-0568-8 (Hardback)

Library of Congress Cataloging-in-Publication Data

Leekley, Robert M.
Applied statistics for business and economics / Robert M. Leekley.
p. cm.
Includes index.
ISBN 978-1-4398-0568-8 (hbk. : alk. paper)
1. Commercial statistics. 2. Economics--Statistical methods. I. Title.

HF1017.L4234 2010
519.5--dc22 2009044113

Visit the Taylor & Francis Web site at
http://www.taylorandfrancis.com

and the CRC Press Web site at
http://www.crcpress.com

Contents

Preface

My aim in writing this textbook is to offer students in business and the social sciences an effective introduction to some of the most basic and powerful techniques available for understanding their world.

- The first requirement to be effective is to engage the students. In my experience, most students are not interested in statistics for its own sake. Nor are they impressed by claims that people in the world of business (or government or academe) actually use these techniques. You will not see pictures of corporate headquarters in this book. You will not read quotes from CEOs affirming the importance of statistics for their enterprises.

There are a few topics that are optional. Regression is introduced in Chapter 3 as a purely descriptive statistic; some instructors may wish to wait with it until it is reintroduced in Chapter 12 as a tool for inference. The discussion in Chapter 12 is written on that assumption. The discussion in Chapter 6 on sampling techniques is not required for later material. Nor are the sections in Chapter 8 on the probability of a type II error. Nor are the last two sections of Chapter 9 on the difference in means with F_Xs unknown and unequal or using paired data. These sections, and the end-of-chapter exercises that rely on them, are all marked with asterisks.

In my experience students are engaged most effectively when they are asked to tackle problems that seem interesting and important to them. To seem interesting and important problems do not need to be cosmic in significance. Rather they need to seem like problems that a real person in a real-life situation would care about answering. And they need to bring to bear data of the sort that this real person might plausibly have available.

It also helps to describe research that slightly more advanced students are actually doing, using statistics as a tool. All such examples in Chapter 1 are real. Doing formal research of their own is a big step for students—one that is beyond the scope of this course. But if I can convince them that such research is something that good students eventually do, they will have a reason (beyond the final exam) to care about this material.

- The second requirement is to focus on what is important. Instructors have "pet" topics—topics that they enjoy covering. In my more than 30 years of teaching this course, I have enjoyed covering such topics as the relationship between the geometric and arithmetic means and the relationship between the hypergeometric and binomial probability distributions. But you will find neither of those topics in this book. For most students, these topics are distractions.

We also have certain "traditional" topics—topics that we cover simply because we always have. Among these are the various "shortcut" formulas—formulas designed to avoid the computational messiness of squared deviations. Once upon a time, students needed to learn these purely mechanical shortcuts because they might really need to calculate standard deviations or regression coefficients by hand. But that time has passed.

Yes, more can be covered. But only so much will be learned. And I want that to be the basic core of applied statistics. By the end of this book, I want students to be able to (1) summarize data in insightful ways (using charts and graphs as well as summary statistics), and (2) make inferences from samples (especially about relationships). Other topics enrich the mix, especially if they reinforce these basics. But these are the basics. The book is not radical in its major topic organization or coverage; indeed, it is fairly traditional. But it attempts to emphasize these basics.

- A related requirement is to bring out the relationships among topics. Many textbooks claim that topics can be skipped or rearranged without loss of continuity. Typically, this means that there is no continuity to be lost. But statistics is not a laundry list of unrelated topics; it is really just a few fundamental ideas that we apply in different ways depending on the data. I have done my best to cover an amount of material that is manageable in a semester, and cover it in a way that highlights the connections as much as possible. Hence, in introducing descriptive statistics I ask students questions about the relationships they think they see. Then in the chapters on hypothesis testing we go back to these same data, to test whether those relationships stand up to formal inference or not. And in introducing contingency tables and analysis of variance, we go back to examples from two-sample tests of proportions and means to see the relationship among them.

One of my student editors cautioned whenever I said "recall" or "turn back" that many students will neither recall nor turn back. I am sure some will not. But I think "recalling" or "turning back" is critical to really mastering this material. This book is not for the lazy student; this book is for the student who wants to master the material enough to "recall" or "turn back."

- The final requirement is, of course, to be clear. I have tried to make good use of examples. I have tried to be direct—not verbose or cute—in my prose. And I have adopted a set of symbols that can be used consistently throughout the book.

Spreadsheets

Spreadsheet software helps meet these requirements in several ways:

- Students know that spreadsheets are ubiquitous in business. They know they are using the tools they will likely use in the real world. This fact, alone, lends exercises more relevance and credibility.
- Moreover, spreadsheets allow the use of larger, richer sets of data. No longer is there any excuse for asking students to find the standard deviation of five numbers. With the **Copy** command, 50 cases involve no more work than five. Nor is it necessary to ask questions that begin: "Suppose a sample of 50 has been taken and the mean and standard deviation have been calculated." Students can be given the raw data and then be faced with the decision about the need to calculate a mean and a standard deviation.

It is worth noting that working from raw data is not easier; indeed it is considerably more difficult. Gone are those artificial clues in the problems that generations of students have learned to exploit, even while realizing that they were not learning what really mattered. Experienced teachers may in fact be disappointed with their students' performance at first. Students seem to learn less. But this is only because they were not really learning as much as it seemed before. My own realization of this fact came the first time I asked my students to work a chi-square contingency table problem from raw data. My students had never had any difficulty with this sort of problem as it is traditionally presented, complete with a two-way frequency table already calculated and laid out for them. But without that completely artificial clue, they did not even recognize the problem.

Happily though working from raw data is more difficult, students do recognize it as much more worthwhile. In the real world, no one is going to ask them for the standard deviation of five numbers. And no one is going to calculate a two-way frequency table and then ask them to finish the problem. However, in the real world, someone may indeed hand them a jumble of data and ask them to make sense of it. Someone may indeed ask them to see whether machine A is really producing more defectives than the other machines. Or someone may indeed accuse them of paying their female employees less than their male employees. These things happen in the real world and students know that. Not all students end up enjoying statistics; however, they do end up understanding its value:

- Until spreadsheets, one could have students calculate standard deviations by hand, using shortcut formulas that provided essentially no insight into the meaning of a standard deviation. Or one could rely on computer programs, black boxes that gave no real insight either. Spreadsheets allow students to calculate standard deviations and similar measures using the definitional formulas. A spreadsheet has no difficulty finding a squared deviation from the mean. Students can (and I think should) be required to go through the steps: find the first squared deviation from the mean, then **Copy** down to get all the others, then sum and divide by the count minus one, and then take the square root. With the **Copy** command this is no longer computationally burdensome for students even with a relatively large sample. And since they are using the definitional formula, there is reason to hope that doing so will provide some insight into the meaning of the standard deviation.
- Spreadsheets are also, in many respects, the ideal medium for accomplishing one of my two basic goals above—that students learn to summarize data in insightful ways. Ironically, descriptive statistics has declined in importance in most statistics books, just as it has become more important in the real world. Most students in our courses will not use inferential statistics extensively, but they will give many presentations with tables and graphs. They ought to learn to do it well. Teaching how to do it well is not easy; arbitrary rules—like formulas for how many class intervals to use in a frequency distribution—are not the answer. But it is worth the effort.
- I have not tried to teach the general use of spreadsheets. At Illinois Wesleyan, students are expected to know something about spreadsheets before taking statistics. Those teaching students without such background will need to invest a little more time on spreadsheets at the beginning of the course. But the spreadsheet skills students need are actually quite modest. They need to know how to enter numbers and labels. They need to be able to create formulas with both relative and absolute references and **Copy** them. They should know how to sort data, create frequency distributions, and create graphs. And of course they should know how to open files, save files, and print. That is about it.

Nor have I tied this book to any particular software package. All the things mentioned above can be done using any of the standard spreadsheet packages. At Illinois Wesleyan we have used both Quattro Pro® and Excel® and everything in this book can be done in Lotus® as well.

These three programs differ very little until one gets to the more advanced features—features that I have not used much anyway. When one gets to the point that more advanced features are necessary, spreadsheets lose their advantage over dedicated statistical packages. I shift to such a package as I get into multiple regression.

Dedicated Statistical Packages

In principle, the whole course could be taught using just spreadsheets. Multiple regression, the most advanced topic presented, is a standard spreadsheet option. But there are advantages to shifting over to a dedicated statistical package:

- Built-in spreadsheet routines for statistics tend to be clumsy and often provide less information than one would like.
- Moreover, by the end of the course, I want students to have begun to imagine doing serious research of their own. And serious research is done using these packages. There is a value to students learning about them as long as the startup cost is not too great.

That qualification is important. This is not the place to spend time covering command languages. I use SPSS in its spreadsheet-like, menu-driven mode of operation. But any other package with a menu-driven interface could be used instead.

Data

This book makes use of a variety of databases, including several from public sources. Internal business data are not so easy to come by. Hence, I have also created a number of hypothetical databases to help with the business examples. All are identified as to source and available for downloading from my Web page: http://iwu.edu/~bleekley.

Acknowledgments

I am grateful to family and colleagues who offered encouragement and a sympathetic ear when I felt overwhelmed or frustrated with my progress.

I am grateful, as well, to the people of Taylor & Francis. David Grubbs, Acquisition Editor for Chapman & Hall/CRC, was very helpful and supportive in the early stages. Kari Budyk, Project Coordinator, and Frances Weeks, Project Editor, helped guide me though the publication process. Katy Smith, Copywriter, was responsible for the promotional efforts. Special thanks go as well to Srikanth Gopaalan of Datapage who oversaw copyediting and layout. I am sure I strained his patience at times, but he refused to show it.

I am especially grateful to a number of special students. The first two people to read major portions of the text were student assistants Katie Hampson and Jeremy Sandford (2001–2002). Both gave it a thorough and thoughtful reading; their suggestions improved the text in style and substance. Nimish Adhia (2002–2003), Lindy Wick (2002–2003), Alexis Manning (2003–2004), and Adrienne Ingrum (2004–2006) all gave careful readings of parts of the revised text. Lindy and Adrien (AJ) Gatesman (2004–2005) contributed data and/or end-of-chapter exercises. Lindy is largely responsible for Appendix B. Adrienne and Jennifer Dawson (2005–2006) worked virtually every end-of-chapter exercise. Amanda Clayton, Lindsey Haines, Kelsey Hample, Anna Konradi, Tian Mao, Teodora Petrova, John Sacranie, Katie Stankiewicz, and Dan Wexler all helped with editing (2008–2010). Finally students in my own statistics classes have given me useful feedback on what works and what does not from the perspective of the most important audience—beginners trying to learn statistics. Thanks to all.

Acknowledgments

Author

Robert M. Leekley is Associate Professor of Economics at Illinois Wesleyan University. He received his BA from Carleton College and his MA and PhD from Michigan State University. He has taught introductory statistics for over 30 years. This text is the result of his attempt, over the last 20 years or so, to incorporate more raw data, using spreadsheets, to promote more meaningful student learning.

In addition to introductory statistics, Professor Leekley teaches a more advanced, applied econometrics course in which students carry out their own empirical research projects. And he has chaired more than 20 successful honors research projects by senior economics majors.

Professor Leekley also teaches intermediate microeconomics, and has taught a variety of applied microeconomic courses such as public finance and environmental and natural resource economics. His own research is in applied microeconomics.

1

Introduction to Statistics

1.1 What Is Statistics Good For?

Imagine you are playing a board game with some friends and a die seems to be coming up "1" too often. Perhaps this is just chance; perhaps the die is unbalanced. How do you decide?

Or imagine you are in a classroom discussion and someone asserts that Democratic politicians support more progressive taxes than Republicans. That could be just rhetoric, of course. Is there any evidence that taxes are actually more progressive when the Democrats are in control? What evidence do you look for?

Or imagine it is your first day on the job. Your boss drops some spreadsheets on your desk and says: "Here's something to start on. See what sense you can make of these." What do you do?

As this range of scenarios suggests, statistics is good for dealing with a wide range of problems from the frivolous to the serious, as a college student, and in the world beyond. Let us look at each of these problems in a bit more detail.

Consider the first example. You would probably roll the die a number of times and keep track of the outcome, much as in Figure 1.1.

You know intuitively that, if the die is *fair*, each of the six possible outcomes should come up *about* the same number of times. But, due to chance, not *exactly* the same number of times. So how close is close enough? Statistics offers a formal framework for making decisions like this.

The second example is more interesting. The notion is that there has been variation in the degree of tax progressivity over time, and that you

Result	Frequency
1	~~||||~~ ~~||||~~ ~~||||~~ |||
2	~~||||~~ ~~||||~~ ||
3	~~||||~~ ||||
4	~~||||~~ ||||
5	~~||||~~ |
6	~~||||~~ |

Figure 1.1 Tally of the results of rolling a die 60 times.

might be able to explain that variation by taking into account the political party in control. Don't worry for now about how you would measure progressivity; assume you have such a measure. A simple approach might be to find the average degree of progressivity for samples of years when each party is in control. If there is really *no* difference between parties, the two sample averages should be *about* the same. Again, due to chance, not *exactly* the same. So how close is close enough? At what point do you conclude that the difference is *not* due just to chance—that the parties really are different? Statistics offers a formal framework for making decisions like this.

Actually, this example merits more consideration. First, there are degrees of party control. You might expect that the degree of progressivity depends on the degree to which a particular party controls. Second, you might expect that other things besides party control also matter. Both these complications can be addressed. Indeed, by the end of this course you will have addressed them. For now, the point is that much of statistics is about trying to explain or account for variation in a dependent variable—in this case tax progressivity—by relating it to variation in other independent or explanatory variables, and then formally assessing the results.

Finally, consider the last example. First, it should be said that your boss's instructions leave much to be desired. It is not exactly clear what he is telling you to do. What are these data about? Is there a particular dependent variable whose variation he would like to understand? Count on getting such vague instructions, though. You will need to look at the data and think for yourself.

Suppose the data are those given in Figure 1.2, for a sample of 50 company employees. The actual data file—Employees1.xls—has five columns of 50; I have displayed it in two blocks of 25 here to save paper. What is there to make sense of here? There is information on a variety of your employees' attributes. Your boss might be interested in knowing something about the distribution of these attributes. You could make up frequency distributions, much like you would have in the die example. Here, though, the purpose would be just to describe. Thus, you might create graphs of these frequency distributions—something like Figure 1.3 for education. It is very common to display information graphically.

Probably, though, your boss also wants to know something about the relationships among these variables. In particular, which employee attributes help explain the variation in salaries? Is the company rewarding its employees for additional years of education and/or work experience? Does it treat male and female employees education and experience of equal equally? By the end of this course you will be able to report to your boss something like the following:

> Based on this sample, I estimate that we are rewarding employees an average of about $2,500 for each additional year of education, and an average of about $500 for each additional year of experience. However, we appear to have a problem with gender equity. I estimate that females are being paid nearly $2,300 less than equally qualified males.

ID	Education (Years)	Experience (Years)	Female	Salary ($1,000)	ID	Education (Years)	Experience (Years)	Female	Salary ($1,000)
19	17	43	0	65.0	149	10	21	1	31.7
20	11	43	0	48.7	156	16	20	0	49.6
25	15	42	1	58.0	164	11	19	0	36.3
30	14	40	1	49.9	168	15	19	0	46.3
34	15	39	1	54.0	173	17	16	1	48.7
42	10	39	0	40.9	182	10	13	1	27.7
46	11	36	0	45.2	191	16	13	0	46.1
52	10	34	0	41.6	194	18	12	0	50.4
59	12	34	0	45.4	199	11	12	0	33.2
69	13	34	1	48.8	208	15	11	1	38.3
74	13	33	0	48.4	217	15	9	0	41.3
82	14	31	0	49.8	226	13	9	0	36.4
87	12	31	0	45.1	234	14	9	0	38.9
92	14	30	0	45.3	239	13	6	1	34.4
99	18	28	1	53.7	240	12	5	0	32.2
105	11	28	0	40.5	246	12	3	1	26.8
107	17	26	1	54.9	249	11	3	1	24.0
110	15	26	0	49.8	258	14	3	1	32.2
115	17	25	0	54.4	267	15	2	1	34.8
117	11	25	0	38.9	270	13	2	0	33.2
121	10	25	0	37.2	277	12	2	0	30.4
130	17	24	0	54.1	286	10	1	1	25.1
139	16	23	0	51.1	290	12	1	0	29.9
140	16	23	1	49.1	291	14	1	1	32.2
141	16	23	0	51.0	292	13	1	0	32.4

Figure 1.2 Information on a sample of 50 employees (Employeesl.xls).

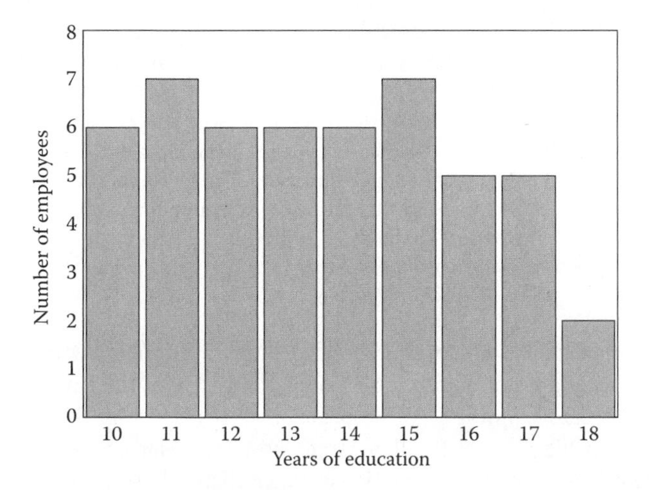

Figure 1.3 Employee education displayed graphically.

1.2 Some Further Applications of Statistics

Some authors of business statistics textbooks try to motivate students with exaggerated claims that you really need to learn this stuff to get and hold a job. Of course that is silly. There are some jobs for which statistics is essential; I will cite some shortly. But learning statistics is certainly less critical than a number of other things—like learning to speak and write effectively—for getting and holding a job.

A far better reason for learning statistics is that it can help you answer questions that you find interesting. And it can help more or less immediately. If you are a sophomore, students just a year or two older than you are undertaking fascinating research projects using statistics—projects that they chose because they were interested. The second example in the last section, on the progressivity of the tax system, is a case in point. Leslie Ayers, a senior economics student at the time, did that study for me a few years ago. I will describe some others as well. But first, a few jobs that really do require statistics.

1.2.1 Statistics in the Real World

1.2.1.1 Quality Assurance

Look at a package of light bulbs. The package makes a number of claims, including how long the bulbs burn on average—perhaps more than 1000 hours. How does the manufacturer know? Hopefully, quality assurance testers do not burn all the bulbs until they go out, or you would have a package of dead light bulbs. Some claims about a product cannot be verified for each unit, because verifying the claim destroys the unit. Yet responsible manufacturers, concerned for their good reputations, want to be right. What do they do?

Quality assurance testers continually examine random samples of their product. Can they say for sure that the sampled items are just like the rest, which were not tested? No. In statistics, nothing is for sure. But they can say something about the probability that the average of a sample will differ from the average of the overall population by some amount. They can say something like the following:

> The probability is only 0.01 that we will get a sample average as high as 1050 hours when the overall population average is 1000 hours or less—that is, when something is wrong and our claim of "more than 1000 hours" is untrue. Thus, if we use 1050 hours as a cutoff, and conclude that our claim is true as long as we get a sample average greater than 1050 hours, we will have only a 1% chance of being wrong.

They can choose whatever probability they want of being wrong (except zero). If they are not comfortable being wrong 1% of the time, they can adjust the cutoff upward, so the probability is even less. But there is a trade-off. A higher cutoff will generate more false alarms.

There is a lot more to quality assurance than this simplified example can suggest. But much of it is based on this same sort of reasoning. Testers look at a sample and try to infer what is true of the population from which it came. There is always a probability that their inference is wrong. But they can set up the test so that this probability is one they are willing to accept.

1.2.1.2 Auditing

Auditing is a sort of quality assurance, too. Auditors assure the quality of financial records using very similar techniques. It would be far too expensive and time-consuming for auditors to look at all the records of even a medium-size firm. They must look at just a sample of a firm's records and infer from this sample what is true of all of them. Again, is there some chance that their sample is misleading? Yes. Again, they must decide on a probability of that with which they are comfortable and choose their sample accordingly.

1.2.1.3 Market Research

It can be very expensive to bring a new product to market. Firms do not want to undergo this expense only to discover that there is no demand. But how can they know? By now you should be able to guess. They use sample surveys and sample test markets, and try to infer from the product's acceptance in these samples how it will fare with the larger population. Again, there is no guarantee that the sample result will carry over to the whole population. Again, though, they can control the probability of being misled.

Advertising a product is expensive, too, and firms want to target that advertising as effectively as possible. Some mistakes would be obvious. No one would advertise a denture adhesive on a kids' show. To get it right, though, firms need to learn about their actual and potential consumers. They might create a product with teenagers in mind, only to discover that it is actually more popular with older young adults. That discovery would be important, since the media that these two groups read or watch are different.

How would they make such a discovery? Again, by taking a sample and trying to infer from that sample what is probably true for the larger population. As always, there is no guarantee that the sample result will actually carry over to the whole population. In this example, it could be that the discovery was merely a quirk of the sample. Again, though, they can control the probability of being misled.

1.2.1.4 Political Polling

Political polling is really just market research in a different context. A political poll is just a sample, from which one tries to infer the political position of the larger electorate. One thing that can make it feel a little different is the way bottom line poll results—who is ahead and by how much—are often reported publicly. Next time you see a political poll result, look at the fine print. There should be a statement something like this:

> This poll was taken in such a way that, 95 times out of 100, the results obtained are within plus or minus 3% of the results that would have been obtained if the whole electorate had been polled.

What does this mean? Well, suppose Senator Smith was found to have 42% of the vote in this sample. There is essentially *no* chance that this result will carry over to the whole population *exactly.* To have any chance of being right, one has to think in terms of ranges. The wider the range, the more often it will be correct, in the sense that the value for the whole population will be somewhere within the range. But of course, the wider the range, the less useful the result. Pollsters seek a balance. In this example, the fact that Senator Smith got 42% of the sample vote allowed them to conclude that the senator actually has somewhere between 39% and 45% of the population vote. And even this conclusion will be wrong 5% of the time. Could this be one of those times? Yes.

1.2.1.5 Social Science Research

Some social scientists specialize in theory, but most do empirical research. (Research in business is really social science research too.) Typically, we develop a theory to explain the variation in something of interest. Then, we put that theory to the test by collecting data and looking for the relationships that the theory implies should be there. Of course there will be random variation in the data as well, and this random variation may suggest a relationship that does not really exist. We can be misled. Again though, statistics gives us a way of controlling the probability that this will happen. We can decide, for example, that we are willing to be misled in this way 5% of the time. Then, we will take seriously only those relationships that are so strong that they would arise randomly only 5% of the time. Why would we be willing to be misled 5% of the time? We may not be; we can adopt any probability we want (except zero). But there is a tradeoff. A more stringent criterion may lead us to miss a relationship that really does exist.

1.2.2 Statistics in the Classroom

I have already suggested that you ought not to think of statistics merely as preparation for a job after graduation, and that statistics offers you a tool for exploring topics of interest to you as a student. The following three studies are real. They are offered as examples of what is possible.

1.2.2.1 Party Control and the Taxation of the Rich

I have already referred to this study several times. Leslie Ayers wondered whether the difference in party ideology and rhetoric concerning the income distribution was actually reflected in the policies that the parties enacted when in control. If so, she expected to see greater progressivity in the income tax when the Democratic Party was in control than when the Republican Party was in control.

First, she needed to devise a measure of tax progressivity. She decided to use the difference between the percentage tax rate paid by the top 10% in terms of family income and the percentage tax rate paid overall. She calculated this difference for every year from 1958 through 1993. She found that it had varied between 4.12 and 10.62 percentage points over those years, so there was indeed some variation to explain.

Next, she had to decide how to measure party control. She created a number of possible measures—each as of the year before, on the grounds that it would take a year for a change in party control to show up in taxes. These measures were

1. Whether or not the president was a Democrat;
2. The percentage of Democrats in the Senate;
3. The percentage of Democrats in the House; and
4. The percentage of Democrats on the House Ways and Means Committee.

Finally, she needed to decide what other factors might also affect progressivity so that she could control for them. She thought that the state of the economy, as measured by the unemployment rate or real average family income, might also affect progressivity.

Leslie's results were interesting. She found that real average family income did indeed affect progressivity. As real average family income rose by $1,000, her measure of progressivity decreased by about 0.1 percentage points, other things equal. This was a small effect, but strong enough that it was unlikely to be due just to chance.

Among the political measures, the percentages of Democrats in the Senate and House did not seem to matter, but the percentage of Democrats on the House Ways and Means Committee mattered a lot. Leslie estimated that a 1% increase in the percentage of Democrats on this committee would increase her measure of progressivity by 0.5876 percentage points, other things equal. Since a one *person* swing on this committee represents a change of several percent, this is quite a large effect. And it was strong enough that it was extremely unlikely to be due just to chance. Finally, Leslie found a surprising result for the presidential variable. Having a Democratic president seemed to *decrease* her measure of progressivity by a little more than 1.0 percentage points, other things equal. And again, this result was strong enough that it was unlikely to be due just to chance.

Leslie's overall conclusion was that differences in party ideology and rhetoric do show up in the policies that the parties actually enact when in control. But party control should not be thought of as controlling the presidency, or even the houses of congress. Party control means having control of the major congressional committees.

1.2.2.2 Racial Discrimination in Major League Baseball: Can It Exist When Productivity Is Crystal Clear?

Racial discrimination could occur in baseball in at least two ways. First, if the team owner is prejudiced, he could pay a minority player less than an equally qualified majority player. However, if there is competition among teams, the prejudiced owner will end up with inferior players at higher salaries. Discrimination is not good for winning; hence, we might expect it to be competed away. Second, if *fans* are prejudiced and are more likely to attend games if the players are of the same race, the owner may feel he needs to take that into account.

Will Irwin, a senior economics major and baseball fan, wanted to know whether there was evidence for either. Evidence for the first would be differences in salary after controlling for ability. Evidence for the second would be differences in team racial makeup that reflected the racial makeup of the cities where they play.

Will limited his salary study to free-agent outfielders who had signed new contracts between 1997 and 2003. He assumed that Salary was a function of offensive ability, defensive ability, experience, and (possibly) race. Baseball is among the most quantified of sports; there are plenty of measures of ability. For offensive ability, he settled on Runs Created (a sort of combination of on-base and slugging percentages) and Steals; for defensive ability, he used Assists; and for experience, he used both

Games Played and Age. For Race, he tried individual variables for Black and Latino; he also tried the two combined as Minority.

Will found strong effects for Runs Created and Games Played. These effects were so large that there was little chance that they were due just to chance. Clearly, team owners were paying for offensive output and experience. However none of the other variables seemed to matter. All, including the race variables, had the wrong sign. And all were so small that they could easily have arisen due to chance. There was no evidence of discrimination.

Will also tested for differences in the racial makeup of the 30 major league teams. Of course, we would not expect them to be exactly the same, even if there were no discrimination. But they should not be too different. And Will found that the differences were so small that they could easily have arisen randomly. Again, there was no evidence of discrimination.

Will did not prove that there was no discrimination. It is not possible to prove a negative. Discrimination might exhibit itself in other ways. Still, his results were hopeful.

1.2.2.3 The Campus Bookstore: Perceptions and Solutions

A colleague, Professor Fred Hoyt, often has his marketing students work on a group project involving market research. One project, which may resonate with students on other campuses, was a market analysis they did for the campus bookstore. As is usual in studies like this, there was less focus on testing some preconceived theory. Rather, the first step was just to find out who does and who does not frequent the bookstore. The next step was to explain these patterns. Why do some students use it and others do not? The final step, then, was to recommend policies that would enhance the experience for those who use it, and attract those who currently do not.

Students created their own market research instruments and surveyed random samples of the student body. These instruments included general demographic questions, such as sex, year in school, and housing arrangements. They included questions about the frequency with which they used the bookstore, and their reasons for doing so when they did. They also included questions concerning their satisfaction with various bookstore qualities such as service, selection, and price.

After collecting their data, the students compared the demographic profile of their sample with the demographic profile of the student body. Did there seem to be biases in who had been surveyed? They decided that there were not. Next, they looked at those students that used the bookstore more often. They found that usage decreased over the four years. Freshmen visited the bookstore roughly three times as often as seniors. And they found that women and students with Greek affiliations visited more often than men and independent students.

What accounted for these patterns? This was less clear. Some of the decline in usage was probably because freshmen tend to be a more captive market, though only 6% of survey respondents listed “no car to go elsewhere” as their reason for using the bookstore. Another explanation appeared to be that some students liked to browse. Some of these students may actually visit *both* the bookstore and its competitors more often. Finally, the bookstore was seen by respondents as having definite strengths and weaknesses that would have attracted some students and not others. The bookstore was seen by respondents as having good service and high quality merchandize. But it was also seen as small, with a relatively limited selection, and pricey.

Finally, the students doing the study recommended marketing strategies to exploit the bookstore's perceived strengths, and to deal with its perceived weaknesses.

1.3 Some Basic Statistical Ideas

I have started this book with examples of the sort of things you can do with statistics because it is a lot easier to learn something when you understand the payoff to learning it. But in this process, I have thrown around some terms and ideas without the benefit of careful definition and discussion. In the discussion that follows, I introduce those ideas more carefully.

1.3.1 Description and Inference

Statistics as a discipline is sometimes divided into two sub-areas—**descriptive** and **inferential** statistics. This is a useful, if somewhat artificial, division.

Sometimes you are not out to test a theory, or estimate something about a larger population. Sometimes your only real interest is to *describe* what a given set of data looks like. You want to inform your intuition or perhaps someone else's. How big are the numbers? How spread out are they? Are they skewed one way or the other?

If this describes your goal, frequency distributions, tables, and graphs are excellent tools. Describing data was at least part of the goal in the example in Section 1.1 on employees, and Figure 1.3 was an example of how one might use a graph to do so. Though I did not reproduce them, the bookstore study I described in Section 1.2 made good use of graphs as well. Chapter 2 will be, devoted to describing data in such ways.

Of course, usually, if you are trying to describe something, it is because you think it is worth describing. It tells a story. There is something there. In that sense, you are trying to *infer* something more fundamental from the data. But those inferences may be intuitive and subjective. Or they may be just a first step, on your way to making more formal inferences.

Most of the examples of the previous two sections involved going beyond description, to formal inference. In the first example, concerning the questionable die, you were not interested in the 60 rolls of the die for their own sake. You were trying to infer something much more fundamental. Is the die fair? In the examples from quality assurance and political polling, the analysts did not care about their little samples, *per se*. They were trying to infer what is probably true of all the other light bulbs, and all the other voters. Leslie Ayers was not interested just in income taxes during the 36 years of her study. She was looking for a tendency for political control to translate into policy that might transcend that particular example. And Will Irwin was not interested just in the 52 baseball players in his salary study. He was trying to infer from them what we should expect more generally, in baseball and perhaps similar labor markets.

Formal inference requires first summarizing what is true of a sample in a way that can be used for further analysis. That means summarizing the sample in quantitative terms. Graphs will no longer do. A number of such quantitative measures—**statistics**—are already familiar to you. You have known for a long time how to calculate an arithmetic mean (a simple average) and a percentage. Others may be new to you. Chapter 3 deals with summary statistics. (Notice that statistics *is* a discipline, whereas statistics *are* quantitative measures you calculate from a sample.)

Chapters 4 through 6 deal with probability, because inference requires knowing something about the probability of certain samples arising from certain populations. In the light bulb example, the quality assurance testers had to know how to choose a cutoff such that there would be only a 0.01 probability of being wrong. In the poll example, the pollster had to know how wide an interval needed to be such that it would be right 95 times out of 100.

Chapter 7 then begins the study of formal inference, which occupies the remainder of the book. You will learn how to decide if that die in the first example was unfair after all. You will learn how to make estimates like in the political poll example and run tests like in the quality assurance example. And you will learn how to estimate relationships between a variable of interest and its possible causes, as in all the other examples.

1.3.2 Explanation and Causation

I have said that much of statistics is about trying to explain or account for variation in a dependent variable; about estimating relationships between a variable of interest and its possible causes. Something needs to be said about what it means to explain things statistically; about the difference between **explanation** and **causation**.

Consider the baseball salary example. Players, on different teams and in different years, have varied in the salaries they have earned. Suppose you had to guess the salary for each. If you had nothing else to go on, your best guess for each would be the overall average. It would be wrong—sometimes too low and sometimes too high—but with nothing else to go on, it would give you the smallest possible errors. A successful explanatory variable helps you tell the cases that are likely to be above or below average, and by how much. It allows you to tailor your guesses so that your errors are smaller. It is in this sense that the explanatory variable "explains." The explanatory variable is a predictor; it need not be the actual cause.

Of course, a lot of social science research hopes to understand causal relationships. But theories about causation need to come from the social science disciplines themselves, not from statistics. Then what statistics can do is tell you whether the variation in the data is consistent with these theories. For example, suppose economic theory suggests that baseball players with greater offensive and defensive skills will earn higher salaries, other things equal. That is, suppose *economic theory* suggests causation. If that theory is correct, then variation in measures of offensive and defensive skill should be able to explain some of the variation in salary. So we would see if it can. If we find that it can, we have not proven the theory. But we *have* found support for the theory. If we find that it cannot, we have *failed* to find support for the theory. In short, statistical explanation cannot establish causation. But it can support or fail to support theories about causation.

1.3.3 The Population and the Sample

I have used the words population and sample repeatedly by now, and I trust that you have figured out from context what each is. In inference, we are interested in knowing something about a whole population. But we do not have access to the whole population. Perhaps it is limitless. Certainly, there is no limit to the number of times a die can be tossed. Perhaps it is finite, but so large that observing it all would be too costly. The electorate in the political polling example is finite but very big. So, we look at a sample taken from that population. But we do not care about the sample for its own sake. We care about what it can tell us about the population.

It is easy, at first, to get confused as to whether you are working with a population or a sample. A simple rule of thumb will usually work. If you can compute it, it is a **sample statistic**. If you are concerned with estimating it, or testing a hypothesis about it, you must not be able to compute it. You are concerned with a **population parameter**.

1.3.4 Variables and Cases

A **variable** is a measure of some quantitative or qualitative outcome. In the example concerning the questionable die, the variable was the number that came up. Since you rolled it 60 times, you had a sample of 60 observations, or **cases**, of this variable, which you organized by grouping all the 1s, all the 2s, and so forth. In the employee example, you were given 50 cases of five variables. (Recall, I displayed them as two sets of 25 just to save paper.) Each of the five columns is a variable; each of the 50 rows is a case. The variables are matched by case. That is, the employee with an ID of 30 is the same person who has 14 years of education, 40 years of experience, and so forth. It is because the variables are matched this way that we can look for relationships among variables.

It is not necessary, on logical grounds, to organize data sets in this format, with columns representing variables and rows representing cases. But it is standard to do so and many statistical programs are written on the assumption that you have.

1.3.5 Types of Variables

1.3.5.1 Numerical and Categorical Variables

Look further at the variables in Figure 1.2. Years of education, years of experience, and salary in thousands of dollars are all **numerical** measures. These values have meaning as numbers. For example, 24 years of experience is twice as much as 12. Female, on the other hand is a **categorical** variable. It categorizes people by sex. Though females are denoted as 1s and males are denoted as 0s, there is no implication that females are more than males. Any other coding, such as F and M would have meant the same thing. In this example there are just two categories but a variable could have several categories. A variable for religion, for example, might be coded Protestant = 1, Catholic = 2, Jewish = 3, and Other = 4. Again, the codes are just alternative names. A Catholic is not twice a Protestant.

A categorical variable like sex, with just two categories, is sometimes called a dichotomous or **dummy** variable. Think of it as a yes–no variable, with 1 being yes and 0 being no. This is the reason for naming the variable Female rather than Sex. If it were named Sex, you would not know whether the 1s were the females or the males.

Why not just use the words "Female" and "Male" instead of numbers? For the presentation of your results, you should. That is, you should not tell your audience that your sample includes 18 1s and 32 0s; you should tell them that it includes 18 females and 32 males. For analysis, though, the numerical codes have advantages, especially for the dichotomous case in which 1 and 0 stand for yes and no.

Why is this distinction between numerical and categorical data interesting? Different sorts of techniques make sense with the different types of data. For example, an average is a sensible way of summarizing a numerical variable. Average years of experience makes sense. An average is generally not a sensible way of summarizing categorical data. In the religion example above, suppose you had a 1 and a 3. You *could* average them, and get a 2. But that would be silly. There is no meaningful way in which a Catholic is the average of a Protestant and a Jew. For categorical data, it makes more sense to look at proportions or percentages. What percentage of the sample is Protestant? What percentage of the sample is Jewish?

1.3.5.2 Discrete and Continuous Numerical Variables

A number of distinctions can be drawn among numerical variables. One is the distinction between discrete and continuous variables. A **discrete** variable takes on a limited number of values. In the example of the suspect die, there were only six possible results of each toss. There is no way to toss a die and get a 3.5. On the other hand, years of experience is **continuous**. Though the numbers in Figure 1.2 are rounded to even years, it makes perfect sense to think of an employee having 3.5 years of experience. Or 3.825 years of experience. Time is infinitely divisible.

Why is this distinction between discrete and continuous data interesting? It can matter for several reasons. It can, for example, affect the way you summarize data. And, more fundamentally, it will affect the way you must think about probabilities.

1.3.6 Sampling Error and Bias

In all the examples of inference, the conclusion reached had some probability of being wrong. It is important to understand that this is not because someone may have made a mistake. If there is a 0.01 probability of being wrong, it is not because people do something wrong 1% of the time. It is because they are unlucky 1% of the time. They get, through no fault of their own, a misleading sample. This is **sampling error**.

Of course it is certainly possible to get a misleading sample by doing something wrong. If you do your political polling in a union hall, you are likely to find the Democratic candidate doing very well indeed. But this will be a misleading sample of the general electorate. This is **bias**, and bias invalidates all the techniques of this course. It is critical that the sample be unbiased—that it be a **random sample** of the population about which you wish to make inferences. This means that it needs to be collected in a manner such that all members of the population have an equal probability of being selected for the sample. You can still be unlucky. But at least you can calculate the probability of that, and take it into account.

1.4 On Studying Statistics

You did not learn to write, ride a bicycle, or play the piano just by watching someone else. You needed to practice, practice, practice. Statistics is really no different. You cannot learn it just by reading this text and watching your instructor. You need to work problems. You need to practice.

For some, it will come relatively easy. For most, it will take more work to become really good. For this second group, consider the bicycle analogy again. You probably started out with training wheels. These kept you from falling. They made you feel safe. But in the end, they also kept you from really learning to ride. It was not until you dared try it without the training wheels—and maybe even fell a few times—that you really learned. And then, all of a sudden, it was easy!

In this course, the training wheels are the worked examples, the answers in the back, the tutor, and the smart kid down the hall. They are great for getting started. Use them. But do not fool yourself. You have not yet mastered the material until you can get by without these aids. That means facing a brand new problem and a blank sheet of paper, and doing it on your own. You may fall a few times. But that is part of how you really learn. And the good news is, for many students, all of a sudden it is easy!

2

Describing Data: Tables and Graphs

This chapter deals with describing data, when your goal is just to see what a variable or set of variables looks like. You want to inform your intuition, or perhaps someone else's. Tables and graphs are the key tools. A well-designed table or graph highlights what is interesting and important, while minimizing distractions in the data. There is a great deal of judgment involved in creating effective tables and graphs. Happily, spreadsheet programs have made the process a lot easier. You can easily create several different versions, and then decide which seems to convey the information most clearly.

2.1 Looking at a Single Variable

2.1.1 Frequency Distributions

2.1.1.1 Ordinary Frequency Distributions

Figure 2.1 gives (hypothetical) data on a sample of 50 undergraduate students. The actual data file—Students1.xls—has four columns of 50; I have displayed it in two blocks here to save paper. Suppose you are asked to describe the data. What do you do?

First, you should examine the data and make sure you understand what you have. There are 50 cases of four variables, though the ID number clearly just numbers the cases. It allows you to refer to a particular case, as in "individual 22 is a male, business major with a very good grade point average (GPA)." The variables that are potentially interesting are Female, Major, and GPA. Of these, Female and Major are categorical variables. Female is a dummy yes–no variable, while Major has six different categories. GPA is a numerical variable.

To tabulate Female and Major you would probably do pretty much what you did in scrutinizing the suspect die in Chapter 1. By hand, you would tally the number of students in each category, much as in Figure 2.2. This is a crude **ordinary frequency distribution**. An ordinary frequency distribution displays each value (or range) of a variable, along with the number of times it appears in the data. It is *ordinary* only to distinguish it from variants that we will get to below.

ID	Female	Major	GPA	ID	Female	Major	GPA
1	1	4	2.537	30	0	1	1.948
2	0	2	3.648	31	1	1	2.717
3	1	3	2.981	32	1	6	1.996
4	0	2	2.683	33	1	3	2.870
5	1	2	3.234	34	0	5	2.986
6	1	3	2.467	35	0	1	3.393
7	1	1	3.384	36	1	2	2.740
8	1	3	3.555	37	1	6	2.499
9	0	2	3.263	38	1	4	3.695
10	1	1	3.711	39	1	4	2.664
11	1	5	1.970	40	1	5	2.306
12	1	4	3.406	41	1	3	3.022
13	0	1	2.523	42	1	6	2.776
14	0	3	1.750	43	1	5	2.175
15	1	6	3.191	44	0	2	3.828
16	1	2	2.795	45	0	6	3.410
17	0	1	2.606	46	0	4	2.330
18	0	5	2.397	47	0	5	3.978
19	1	6	3.791	48	1	3	3.503
20	0	4	3.490	49	1	4	3.253
21	0	3	2.421	50	1	2	2.215
22	0	5	3.937				
23	0	5	2.890				
24	1	6	2.246				
25	1	2	3.371				
26	1	5	3.114				
27	0	1	3.084				
28	0	5	2.703				
29	1	3	3.045				

Codes for Major —
1 — Natural Science
2 — Social Science
3 — Humanities
4 — Fine Arts
5 — Business
6 — Nursing

Figure 2.1 Information on a sample of 50 students (Studentsl.xls).

Sex	Frequency	Major	Frequency
0	𝍸 𝍸 𝍸 𝍸	1	𝍸 |||
1	𝍸 𝍸 𝍸 𝍸 𝍸 𝍸	2	𝍸 ||||
		3	𝍸 ||||
		4	𝍸 ||
		5	𝍸 𝍸
		6	𝍸 ||

Figure 2.2 Tallies for the sex and major of 50 students.

Spreadsheet programs automate the process of creating frequency distributions. They are also better at counting than you and me. There are slight differences among the major programs and even among program vintages. But they all have a menu option that allows you to specify a **values block** and a **bins block**. The values are just the numbers for Sex or Major. The bins are the categories where you want the values organized.

Some programs also require an **output block**; others just put the calculated frequencies next to each bin, as in Figure 2.3.

In most spreadsheet programs, this works only for numerical data—an example of why it pays to code data numerically, even if the variable in question is categorical, like Major. Again, in most spreadsheet programs, the bin number represents the *upper end* of the bin. That is, for Major, everything up to and including 1 will be counted in the 1 bin; everything greater falls through. Everything greater than 1, up to and including 2, will be counted in the 2 bin; everything greater falls through. Everything greater than 2, up to and including 3, will be counted in the 3 bin. In some programs, this convention means that the bins must be in ascending order; if the 6 bin were on top, nothing would fall through to the others. Other programs will reorder the bins. Finally, notice the zero at the bottom of each frequency distribution. This indicates that there were no numbers greater than the upper end of the largest bin.

Of course, these frequency distributions are not yet ready for presentation. Remember, you are doing descriptive statistics. The point is to communicate as clearly as possible. A presentation-quality table will be clearly labeled, with the numeric codes translated back to what they mean in words. While there is certainly room for personal style, Tables 1 and 2 (Figure 2.4) give examples of what the final tables might look like.

	F	G	H	I	J
1	Upper	Ordinary		Upper	Ordinary
2	Bound	Frequency		Bound	Frequency
3	0	20		1	8
4	1	30		2	9
5		0		3	9
6				4	7
7				5	10
8				6	7
9					0

Figure 2.3 Creating frequency distributions for sex and major in a spreadsheet.

Table 1: Sample Breakdown by Sex

Sex	**Number**
Male	20
Female	30
Total	50

Table 2: Sample Breakdown by Major

Major	**Number**
Natural Science	8
Social Science	9
Humanities	9
Fine Arts	7
Business	10
Nursing	7
Total	50

Figure 2.4 Presentation quality frequency distributions.

Both Sex and Major are categorical variables, with relatively few categories, so it was natural just to list each category separately. Notice, though, that Major could have been broken down further. For example, Natural Science is a grouping of Biology, Chemistry, Physics, and so on. Apparently, the person who coded the data decided to group them, in order to avoid having too many very small categories. Too much detail often obscures the more fundamental patterns in data. On the other hand, some information is lost as a result. You often face this sort of trade-off in descriptive statistics. It requires judgment. The goal is to simplify—but not too much.

GPA is a numerical variable with many possible values. For this sort of variable, you *must* group. Otherwise, you end up with 50 different numbers—each with a frequency of 1. So, how do you decide on the number and width of the groupings? Some texts suggest calculating the number of groupings by formula; then dividing the data range by the number of groupings to determine the width of each grouping. However, I have never found such a mechanical approach to be useful. Rather, I suggest the following general guidelines and then trial and error. Again, with spreadsheets, trial and error is really quite painless.

- Make sure your groupings span the entire data range and do not overlap.
- Try to keep all your groupings equal in width.
- Try to use "people-friendly" grouping widths and cutoffs, like 5, 10, 20, 25, and so on.
- Generally, consider between 5 and 10 groupings, especially if you intend to present your frequency distribution in a table. On the other hand, if you have a very large data set, and intend to present your frequency distribution graphically, experiment with a larger number of groupings. When you find successive frequencies jumping around a lot—large, small, large, small, large—you probably have too many groupings.

Consider the GPA variable. The numbers range from 1.75 to just under 4.00. You could cover this range with 10 groupings 0.25 points wide, or with five groupings 0.5 points wide. It would be reasonable to try both. Figure 2.5 shows the results. You would use the one with greater detail if you think the extra detail is interesting. Otherwise, you would not.

2.1.1.2 Relative Frequency Distributions

We found above that nine students in the sample had GPAs over 3.5. Is this a little or a lot? That depends on the size of the sample. Is it 9 out of 10, or 9 out of 50, or 9 out of 100? Generally speaking, it is the proportion or percentage of the total that is really interesting. Thus, it is common to divide all the ordinary frequencies by the sample size, and present a **relative frequency distribution** instead. Figure 2.6 shows the results.

A note on doing this in a spreadsheet program. It may be tempting, at first, to do the calculations by hand and enter them in. Do not. Neither should you enter the formula with numbers (+4/50, +9/50, etc.). Instead, enter the formula with cell locations. Be sure you refer to the location of the four with a **relative reference**, and to the location of the 50 with an

	F	G	H	I	J
14	Upper	Ordinary		Upper	Ordinary
15	Bound	Frequency		Bound	Frequency
16	1.7 5	1		2.00	4
17	2.00	3		2.50	9
18	2.25	3		3.00	14
19	2.50	6		3.50	14
20	2.75	8		4.00	9
21	3.00	6			0
22	3.25	6			
23	3.50	8			
24	3.75	5			
25	4.00	4			
26		0			

Figure 2.5 Deciding on the amount of detail.

	I	J	K	L	M
13				Cumulative	Cumulative
14	Upper	Ordinary	Relative	Ordinary	Relative
15	Bound	Frequency	Frequency	Frequency	Frequency
16	2.00	4	=J16/$J$22 → 0.08	=J16→ 4	0.08
17	2.50	9	0.18	=L16 + J17 → 13	0.26
18	3.00	14	0.28	27	0.54
19	3.50	14	0.28	41	0.82
20	4.00	9	0.18	50	1
21		0			
22	=SUM(J16:J20) → 50		1		

Figure 2.6 Relative and cumulative frequency distributions.

absolute reference. Then you can use the **Copy** command to calculate all the others automatically in one step. There will be a lot of repetitive operations in the material to come. The Copy command will be a godsend. Learn to use it effectively.

2.1.1.3 Cumulative Frequency Distributions

Figure 2.6 also shows **cumulative frequency distributions** for both the ordinary and relative frequency distributions. These do as their name implies—they accumulate, or add up, frequencies. Ordinary frequencies cumulate until the whole sample of 50 is included. Relative frequencies cumulate to 1, or 100%.

Again, these frequency distributions are not yet ready for presentation. For presentation, you would want to clarify the meaning of the bins. They represent ranges, not single numbers, and these ranges include their upper bounds. You would probably also want to change the relative frequencies into percentages, since most people probably find these more intuitive. And you would probably decide that showing all four of these distributions is overkill, since they are really just four different views of the same information. Decide which ones match the story, the way you want to tell it. You might end up with something like Table 3 (Figure 2.7).

A few additional words are in order on specifying the bounds. Recall that groupings must span the entire range and not overlap. This implies that the groupings, as I have listed them in Table 3, must include either their upper or their lower bounds, but not both. For example, 2.00 must be in either the first or second grouping, but not both. Traditionally, class intervals have included the lower bounds. However, this is not how spreadsheets work. Given the way spreadsheet programs work—including everything *up to and including the bin number itself*—2.00 belongs in the first bin, not the second. But how do you communicate that fact?

Table 3 (Figure 2.7) gives one simple approach—a footnote. It is simple and unobtrusive, but it is there for anyone who really cares. But there are other approaches.

One alternative would be to label the ranges as "over 1.50, up to and including 2.00," "over 2.00, up to and including 2.50," and so forth. But this is rather wordy.

A second alternative would be to look back at the data and determine the smallest possible number in each bin. Notice that the original GPA numbers are all rounded to three decimal places. Thus, the smallest possible number that could fall through the first bin—hence the smallest possible number for the second bin—would be 2.001. So you could label the ranges as "1.501–2.000," "2.001–2.500," and so on.

A problem arises with this second alternative, though, if you do not know the smallest possible number. Suppose, for example, that the data set had included variables for "Quality Points," and "Credits Attempted," and that GPA had been calculated in the spreadsheet as the first divided by the second. Now, what is the smallest possible answer that is greater than 2? 2.00001? 2.000000001? 2.0000000000001? There really is no smallest possible number.

Table 3: Sample Distribution of Grade Point Averages

GPA Range*	Frequency	Relative Frequency	Cumulative Relative Frequency
1.50–2.00	4	8%	8%
2.00–2.50	9	18%	26%
2.50–3.00	14	28%	54%
3.00–3.50	14	28%	82%
3.50–4.00	9	18%	100%
	50	100%	

* Each range includes its upper bound.

Figure 2.7 Presentation quality frequency distributions.

2.1.2 Graphs

Tables, like in Figures 2.4 and 2.7, can be effective ways of summarizing what a variable looks like. But often, a graph—in essence, a picture—can be even better.

2.1.2.1 Bar Charts and Pie Charts

Figure 2.8 displays the distribution by Major as a bar chart. It is, in a very real sense, just a picture of Table 2 (Figure 2.4). Most spreadsheet programs will allow you to customize the graph a great deal. You can choose the color and shading of the bars; you can give them 3-D effects; you can rotate them 90 degrees, and so on. Experiment. But remember that your job here is to communicate concerning these data. Too much glitz can get in the way of the message.

Since Table 2 (Figure 2.4) presented an ordinary frequency distribution rather than a relative one, I used the actual numbers on the *y*-axis in this chart as well. You could turn the ordinary frequency distribution into a relative one, though, and use these percentages on the *y*-axis instead. The graph would look exactly the same, except that the numbers along the *y*-axis would be proportions or percentages instead.

However, suppose you want to emphasize the *relative* sizes of these frequencies. In that case, a **pie chart** might be more effective. It emphasizes how much each frequency is of the whole. It also allows you to highlight a particular group, by "exploding" its slice, if there is a group to which you would like to draw attention. And most spreadsheets will allow you to turn a bar chart like Figure 2.8 into a pie chart like Figure 2.9 with just a few clicks of a mouse.

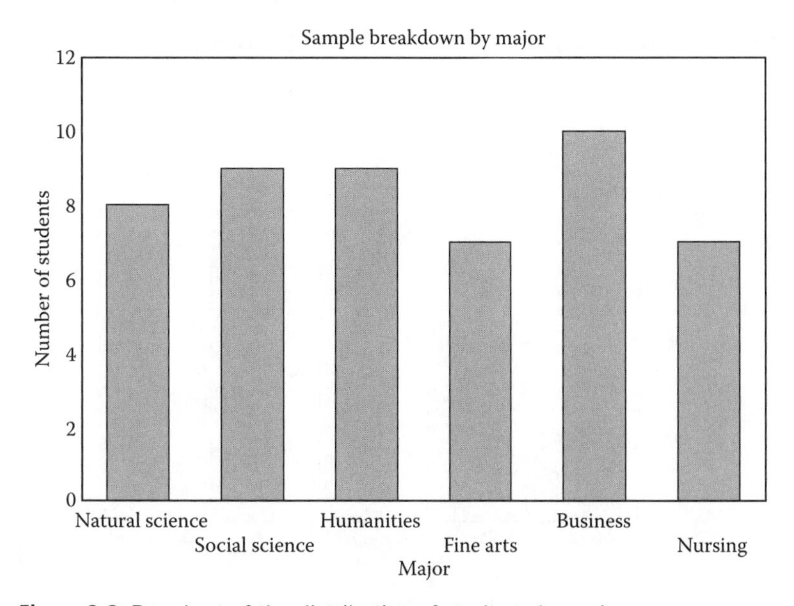

Figure 2.8 Bar chart of the distribution of students by major.

2.1.2.2 Histograms

A **histogram** is essentially a bar chart for continuous numeric data. In the bar chart by Major, large gaps between bars communicate the fact that these majors are distinct. If you were to make a bar chart for discrete numeric data, you would want that as well. With continuous data, though, the groupings are artificial. The values stretch over a range, with the largest value in one grouping being not very different from the smallest number in the next grouping. That fact is indicated by running the bars together. Some spreadsheet programs may not let you run them completely together. And sometimes, the effect of running them together is visually unappealing. In Figure 2.10, I purposely left a tiny gap between bars to prevent them looking like one big blob.

Some graphing programs will let you label the breaks between bars. For a histogram, this is ideal since these breaks are the key individual numbers. If your program will not do that, you need to improvise. In Figure 2.10,

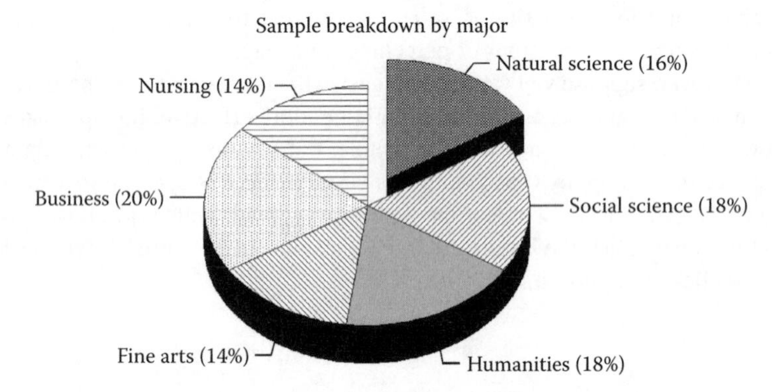

Figure 2.9 Pie chart of the distribution of students by major.

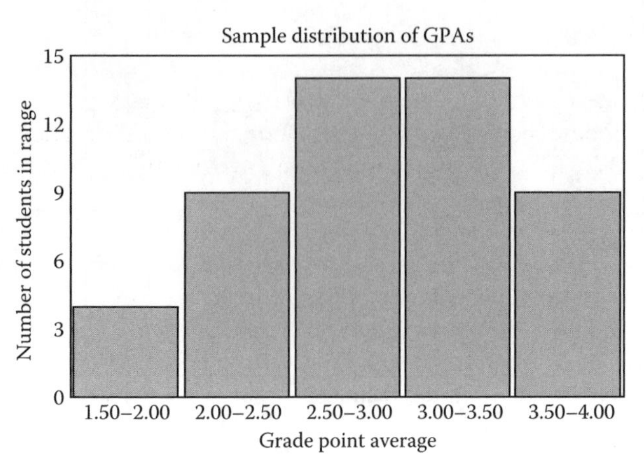

Figure 2.10 Histogram of student GPAs. Each range includes its upper bound.

I listed the whole range that the bar represents. If you have to use a single number, the midpoint of the range would be the best choice.

2.2 Looking for Relationships

As suggested in Chapter 1, very often we are interested in explaining, or accounting for, the variation in something. Suppose you think some other variable might be related and help explain your variable of interest. You might relate the two variables in a table or graph and inspect them visually. However, the most effective approach depends on the sort of explanatory variables you are considering.

2.2.1 Categorical Explanatory Variables

2.2.1.1 Frequency Distributions

If your explanatory variable is categorical, with just a few categories, it can be very useful to create separate frequency distributions for each category. Doing this in a spreadsheet program involves first sorting the data into blocks according to the explanatory variable. Then, you can highlight separate values and bin blocks and get separate frequencies for each category. Finally, you can display these separate frequencies side by side in a table or a graph. If the frequency distributions look different for the different categories of the explanatory variable, you may be able to use that explanatory variable to help explain the value of your variable of interest.

Continuing with the sample of 50 students, perhaps we are interested in explaining, or accounting for, the different choices students make for their major. One possible explanatory variable is sex. That is, if males and females tend systematically toward different majors (for whatever reasons), then taking account of sex may help account for these students' majors.

The first step would be to sort the data by Female. *Beware.* It is critical that you sort all four variables together. Remember, the variables are matched by case. Case 1 is a female, fine arts major, with a 2.537 GPA. Case 2 is a male, social science major, with a 3.648 GPA. These characteristics need to stay with these individuals, whether they get sorted to the top or bottom. There are differences among spreadsheet programs in how you tell them to do this. But they all have a menu option that allows you to specify a **sort block**, several **sort keys**, and whether the sort should be in ascending or descending order. In this case, the sort block should include rows 2–51 *and all four columns*. The sort key should be the Female column. It does not matter whether you sort them in ascending or descending order; the point is just to get all the females in one block and all the males in the other. Figure 2.11 displays possible results of sorting, females first; the order you get within the female and male blocks may differ with the program you use.

Now, you can just create two frequency distributions—one for females and one for males. That is, highlight the first 30 Major values as one values block, and find a frequency distribution of majors for the female students.

ID	Female	Major	GPA	ID	Female	Major	GPA
33	1	3	2.870	9	0	2	3.263
32	1	6	1.996	46	0	4	2.330
37	1	6	2.499	47	0	5	3.978
36	1	2	2.740	2	0	2	3.648
31	1	1	2.717	4	0	2	2.683
25	1	2	3.371	44	0	2	3.828
24	1	6	2.246	45	0	6	3.410
29	1	3	3.045	23	0	5	2.890
26	1	5	3.114	18	0	5	2.397
48	1	3	3.503	17	0	1	2.606
43	1	5	2.175	20	0	4	3.490
50	1	2	2.215	21	0	3	2.421
49	1	4	3.253	22	0	5	3.937
42	1	6	2.776	27	0	1	3.084
39	1	4	2.664	13	0	1	2.523
38	1	4	3.695	34	0	5	2.986
41	1	3	3.022	35	0	1	3.393
40	1	5	2.306	28	0	5	2.703
19	1	6	3.791	30	0	1	1.948
7	1	1	3.384	14	0	3	1.750
12	1	4	3.406				
16	1	2	2.795				
6	1	3	2.467				
5	1	2	3.234				
15	1	6	3.191				
3	1	3	2.981				
10	1	1	3.711				
1	1	4	2.537				
8	1	3	3.555				
11	1	5	1.970				

Codes for Major —
1 — Natural Science
2 — Social Science
3 — Humanities
4 — Fine Arts
5 — Business
6 — Nursing

Figure 2.11 Information on a sample of 50 students (Studentsl.xls) sorted by gender.

Then, highlight the last 20 Major values as a second values block, and find a frequency distribution of majors for the males. Figure 2.12 shows the results.

When looking at a single variable, it did not matter much whether you looked at the ordinary or relative frequency distribution. Their shapes were the same. When comparing frequency distributions, though, it matters. In this example, there will be a tendency for all the ordinary frequencies for females to be greater, just because there are 10 more females in the sample. To compensate, you need to find the female frequencies as a proportion or percentage of 30, and the male frequencies as a proportion or percentage of 20. Notice, for example, that while there are more female social science majors (5 versus 4), a greater proportion of males are social science majors (0.2 versus 0.1667).

Finally, again, you need to prepare your table for presentation. Table 4 (Figure 2.13) gives an example.

	F	G	H	I	J	K
30	Female			Male		
31		Ordinary	Relative		Ordinary	Relative
32	Major	Frequency	Frequency	Major	Frequency	Frequency
33	1	3	0.1	1	5	0.25
34	2	5	0.1667	2	4	0.2
35	3	7	0.2333	3	2	0.1
36	4	5	0.1667	4	2	0.1
37	5	4	0.1333	5	6	0.3
38	6	6	0.2	6	1	0.05
39		0			0	
40		30	1		20	0

Figure 2.12 Comparing frequency distributions.

Table 4: Choice of Major by Sex

Major	Female	Male
Natural Science	10.00%	25.00%
Social Science	16.67%	20.00%
Humanities	23.33%	10.00%
Fine Arts	16.67%	10.00%
Business	13.33%	30.00%
Nursing	20.00%	5.00%
Total	100.00%	100.00%

Figure 2.13 Presentation quality comparison of frequency distributions.

What do the results mean? There do seem to be differences in the patterns between females and males. Are these differences so large that they are unlikely to have arisen just by chance? Eventually, you will learn how to decide. For now, it seems as though they might be.

2.2.1.2 Graphs

As was the case when looking at a single variable, when you are interested in what something looks like, graphs are often even better than tables. Different programs offer different options, but Figures 2.14 through 2.17 offer a number of examples.

For Figure 2.14, which shows both distributions on the same graph, it is critical that relative frequencies be used. If they are, either bar chart does a pretty good job of contrasting the two patterns.

Pie charts, which are so good at showing relative frequencies for a single variable, are not as good at showing how two sets of relative frequencies compare. Slices are hard to compare between pies. The side-by-side columns of Figure 2.17 are probably more effective for that.

The example we have been exploring concerned a categorical explanatory variable—Female—and a categorical dependent variable—Major. But if the dependent variable had been numeric, the same approach would

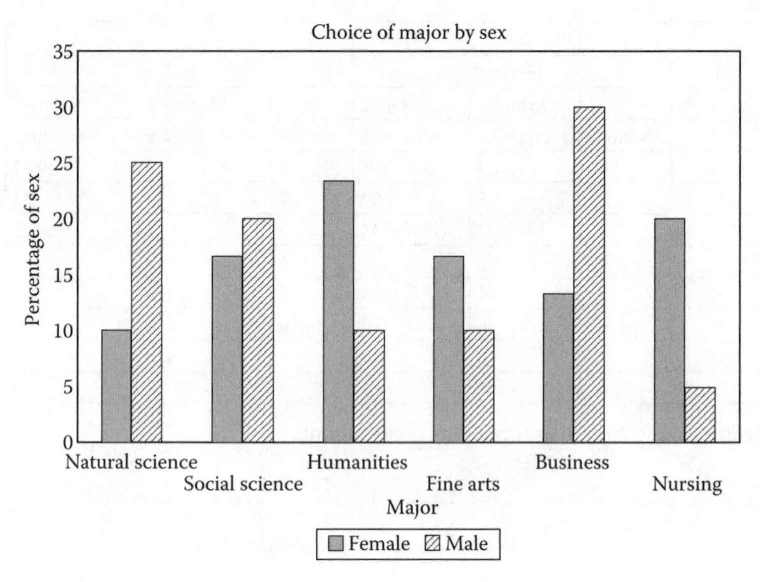

Figure 2.14 Bar chart comparison of majors by sex.

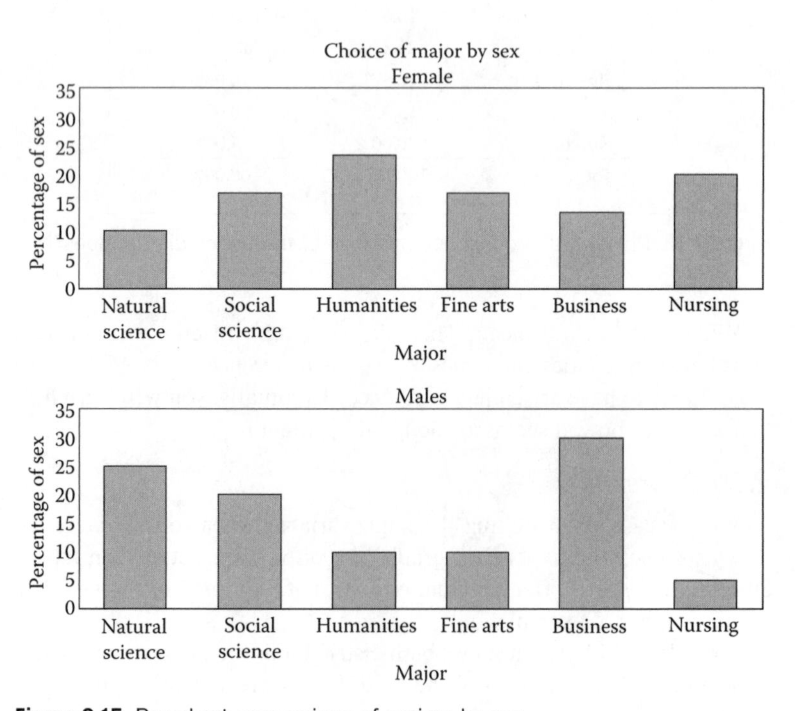

Figure 2.15 Bar chart comparison of majors by sex.

have worked just fine. That is, suppose you are interested in explaining the variation in GPA. Again, it might be that (for whatever reason) females and males have different grade point distributions. You could create separate distributions for females and males again, and see if there are differences in the patterns that seem so large that they are unlikely to have arisen just by chance.

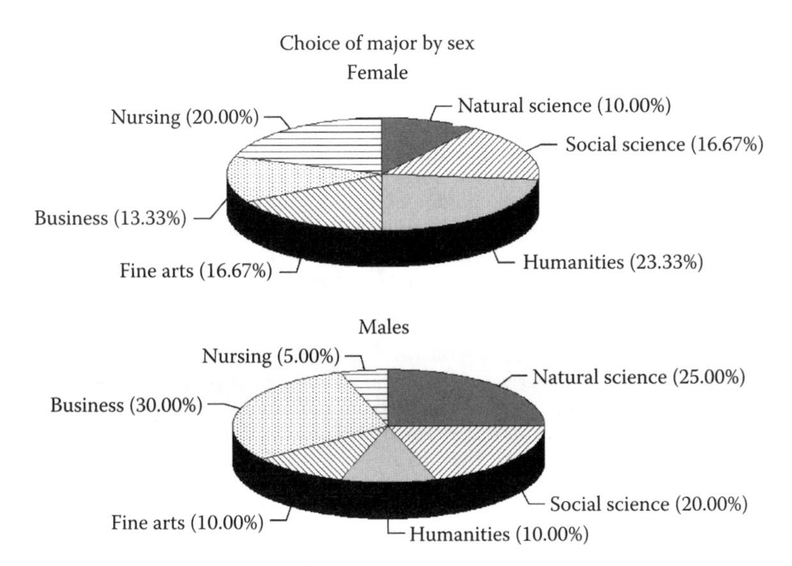

Figure 2.16 Pie chart comparison of majors by sex.

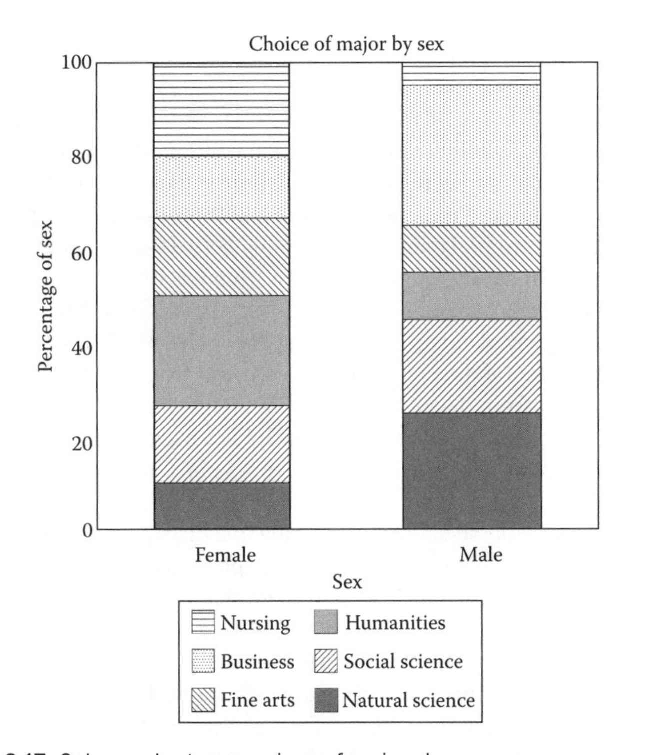

Figure 2.17 Column chart comparison of majors by sex.

2.2.1.3 A More Interesting Example

Consider a new example. Suppose you are examining the weights of young adults, perhaps as part of a larger project concerning health risks facing this age group. You have access to data collected as part of the National Longitudinal Survey of Youth (NLSY). The data, in file NLSY1.

The National Longitudinal Survey of Youth (NLSY) data are real data, collected as part of annual or biennial surveys that have tracked thousands of individuals who were 14 to 22 years old when the survey began in 1979. Overseen by the Center for Human Resource Research at the Ohio State University, with funding from the Department of Labor and other sources, the NLSY is an excellent source of information on this portion of the U.S. population. Variables include demographics, family background and composition, education and training, military and labor market experience, marital status, fertility, health, substance abuse, illegal activity, attitudes and aspirations, and economic well being. And it is downloadable for free at the Bureau of Labor Statistics web site: http://www.bls.gov/nls/nlsy79.htm.

Part of my purpose in using such real data sets is to help you learn about readily available data that you could use to do research of your own. Students of mine (all undergraduates) have undertaken a variety of interesting research projects using the NLSY data. They have tested theories about the determinants of educational attainment. They have asked whether students who take additional high school math and science courses earn higher incomes later on, or whether college majors with higher than average incomes also have higher risks. They have looked for social and economic determinants of women's labor force participation, poverty status, and divorce. If you are a social science major thinking that it would be interesting to do a research project of your own, you should explore these data.

Figure 2.18 A digression on the data.

xls, contain 281 cases of five variables—ID, Female, Age, Height, and Weight. Due to the number of cases, I have not reproduced them here.

Again, the first step is to understand what the data represent. The ID is just a unique number for the individual. Female is again a dummy, yes–no variable. In the original data, it was actually coded 2-1; I changed it to the 1-0, yes–no format. The data were taken from the 1985 survey, when the respondents ranged in age from 21 to 28. In the original data, the actual variable was year of birth; I subtracted that from 1985 to get age. The height, in inches, and weight, in pounds, were from 1985. I took every 20th individual from the approximately 6000 individuals in the representative national sample, which gave me 311 individuals. But 30 of those cases had missing information. Hence, this sample of 281. (For more on the NLSY, see Figure 2.18.)

Again, you could be interested in seeing the distribution of weights. As before, you can do this with a table or a graph. Since weight is a continuous numeric variable, with values in these data ranging from 87 to 260 pounds, you will have to use groupings. With groupings of 10 pounds, it will take 18 groupings—more than you usually want. Groupings of 20 might be better. But this is a fairly large data set, so I decided to try both. As you can see in Figure 2.19, the frequencies for the 10 pound groupings jump around a lot—2 up to 22, down to 16, up to 43, down to 38, etc. Since these fluctuations do not look meaningful, I decided on the wider groupings.

As always, if you intend these results for presentation, they need to be dressed up. Figure 2.20 gives an example. Figure 2.21 shows the same results as a histogram. One thing about these results that you might find interesting is that they are not symmetrical. They are **skewed** to the right. A few individuals are much heavier than most.

	G	H	I	J	K	L	M
1	Upper	Ordinary	Relative		Upper	Ordinary	Relative
2	Bound	Frequency	Frequency		Bound	Frequency	Frequency
3	90	2	0.0071		100	4	0.0142
4	100	2	0.0071		120	38	0.1352
5	110	22	0.0783		140	81	0.2883
6	120	16	0.0569		160	63	0.2242
7	130	43	0.1530		180	55	0.1957
8	140	38	0.1352		200	23	0.0819
9	150	40	0.1423		220	11	0.0391
10	160	23	0.0819		240	4	0.0142
11	170	26	0.0925		260	2	0.0071
12	180	29	0.1032			0	
13	190	13	0.0463			281	
14	200	10	0.0356				
15	210	4	0.0142				
16	220	7	0.0249				
17	230	2	0.0071				
18	240	2	0.0071				
19	250	1	0.0036				
20	260	1	0.0036				
21		0					
22		281					

Figure 2.19 Deciding on the amount of detail.

Table 5: Distribution of Weights

Weight (in pounds)*	Percentage
80–100	1.42%
100–120	13.52%
120–140	28.83%
140–160	22.42%
160–180	19.57%
180–200	8.19%
200–220	3.91%
220–240	1.42%
240–260	0.71%
Total	100.00%

*Each range includes its upper bound.

Figure 2.20 Presentation quality frequency distribution of weights.

So far so good. But probably we are interested in more than just describing the distribution of weights. Probably we are interested in explaining it. Why do some people weigh more than others? Indeed, I included the other variables in the data set—Female, Age, and Height—because they might help. Other things equal, we would probably expect females to weigh less than males. Other things equal, we would probably expect taller

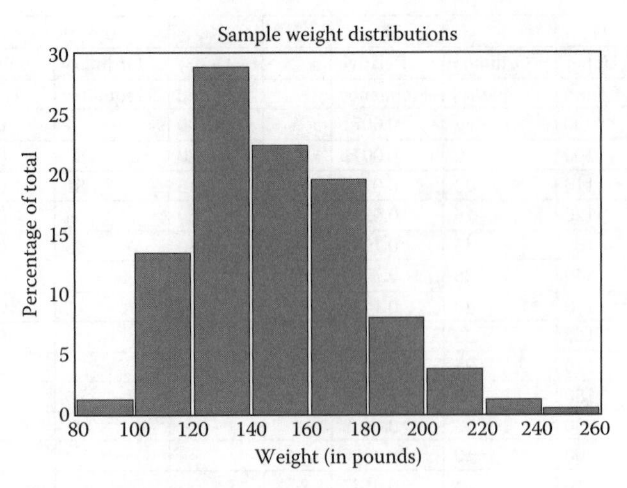

Each range includes its upper bound.

Figure 2.21 Histogram of weights.

individuals to weigh more than shorter ones. And, other things equal, individuals probably gain weight as they age.

Now a warning is in order on the limitations of descriptive statistics. The three previous statements all include the proviso "other things equal." The implication is that each of these variables has its own independent effect. Females do not weigh less just because they are shorter. Females weigh less than males even after controlling for differences in height and age. If all these expectations are correct, we have a four-dimensional relationship here. And it is very hard to show, in a simple, intuitive table or graph, more than two dimensions at a time.

You can and will break down these weights by sex, to see whether sex helps explain some of the variation in weight. But you will not be controlling for height. You can and will break down these weights by height, to see whether height helps explain some of the variation in weight. But you will not be controlling for sex. To see whether each variable actually matters, controlling for the other, you generally must go beyond the tables and graphs that are the stuff of this chapter. But by the end of this course, you will know how to estimate the separate, independent effects of each of these variables, and assess whether these effects are large enough that they are probably not due just to chance.

If you are interested in the possible effect of sex on weight, you can proceed very much as in the example of sex and major. First, sort the data into two blocks, by sex, being careful to include all five variables in the sort. Next, select just the weights in the female block and put them into one set of bins; then select just the weights in the male block and put them into another set of bins. Figure 2.22 shows the results; Table 6 (Figure 2.23) shows how these results might be dressed up, in a presentation-quality table. Clearly, there seems to be a tendency for females to weigh less.

Figures 2.24 through 2.26 show different ways in which these results might be graphed. Again, all show a tendency for females to weigh less. They also show, perhaps more clearly than Table 6, that the overall

	G	H	I	J	K	L
26		Females			Males	
27	Upper	Ordinary	Relative	Upper	Ordinary	Relative
28	Bound	Frequency	Frequency	Bound	Frequency	Frequency
29	100	4	0.0274	100	0	0.0000
30	120	36	0.2466	120	2	0.0148
31	140	64	0.4384	140	17	0.1259
32	160	26	0.1781	160	37	0.2741
33	180	10	0.0685	180	45	0.3333
34	200	4	0.0274	200	19	0.1407
35	220	0	0.0000	220	11	0.0815
36	240	1	0.0068	240	3	0.0222
37	260	1	0.0068	260	1	0.0074
38		0			0	
39		146	1		135	1

Figure 2.22 Comparing frequency distributions.

Table 6: Distribution of Weights by Sex

Weight (in pounds)*	Female	Male
80–100	2.74%	0.00%
100–120	24.66%	1.48%
120–140	43.84%	12.59%
140–160	17.81%	27.41%
160–180	6.85%	33.33%
180–200	2.74%	14.07%
200–220	0.00%	8.15%
220–240	0.68%	2.22%
240–260	0.68%	0.74%
Total	100.00%	100.00%

Each range includes its upper bound.

Figure 2.23 Presentation quality comparison of frequency distributions.

skewness we observed is due to females. The distribution for males is actually fairly symmetric. Figure 2.24 shows the percentages from Table 6 as two interwoven histograms. It is very similar to Figure 2.14, which showed two interwoven sets of bars. The approach is a little less effective for histograms, though, since each of those histograms is really supposed to be continuous. Figure 2.25 is better in that respect, since the histograms can be shown as continuous.

2.2.1.4 Frequency Polygons

Figure 2.26 shows the two frequency distributions as **frequency polygons** instead. A frequency polygon shows a distribution as a segmented line connecting the midpoints of each grouping. Like a histogram, it conveys

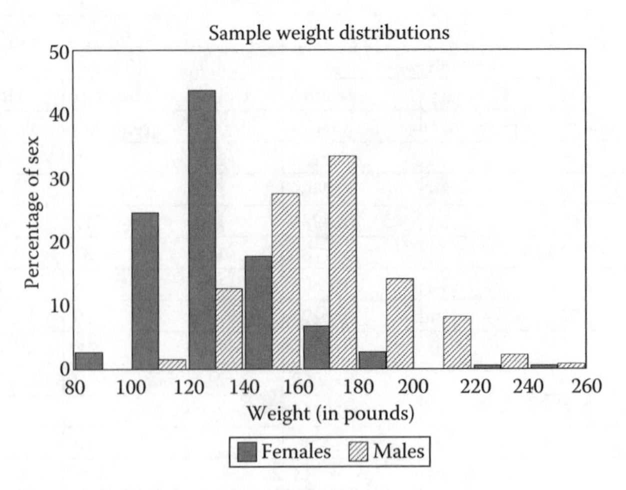

Figure 2.24 Histogram comparison of weights by sex.

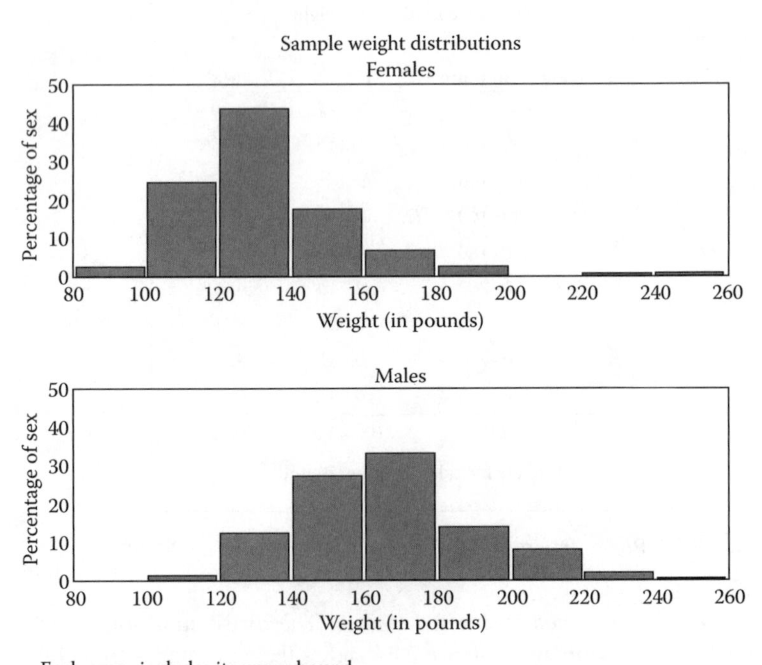

Figure 2.25 Histogram comparison of weights by sex.

the notion that the data are continuous. The advantage of the line, in this context, is that you do not have to worry about one distribution being hidden behind the other. Indeed, I superimposed both on top of the original, overall histogram so that you can see all three.

If you have been reading carefully, you should have noticed something else that is different about Figure 2.26, Figures 2.24, and 2.25 show the percentages from Table 6 (Figure 2.23). They represent the percentages of each

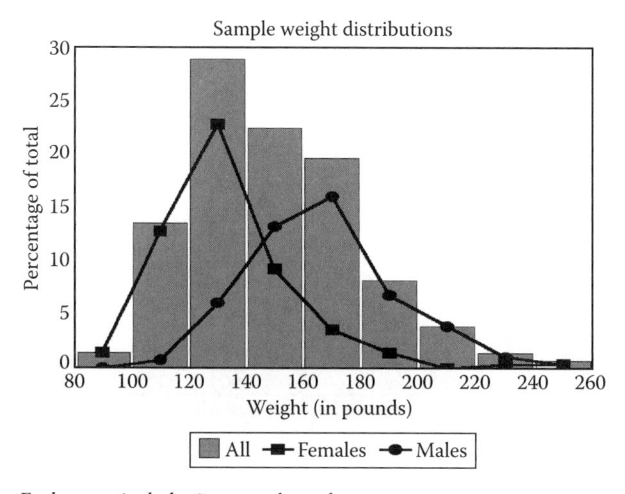

Each range includes its upper bound.

Figure 2.26 Frequency polygon comparison of weights by sex.

sex in each weight class, and add up to 100% for each sex. If your primary interest is in comparing the distributions, this is best. It compensates for the fact that there are different numbers of females and males in the sample.

But for Figure 2.26, when I decided to use the overall histogram as a background, I needed to decide how I wanted to relate the individual sexes to the total. I could again have plotted the percentages from Table 6. In the 120–140 class, for example, the values for females and males would have been 43.84 and 12.59%, respectively, with the overall value in between, at 28.83%. The overall value would have been a weighted average of the individual sexes. I decided, instead, that I wanted the individual sexes to look like subsets of the total. I wanted the values for females and males in each class to add up to the overall value for that class. To do so, I recalculated the percentages for males and females as percentages of the 281 total. Now, the 64 females and 17 males in the 120–140 class represent 22.78 and 6.05% of the total, respectively, and add up to the 81 individuals in that class who represent 28.83% of the total. Either approach is correct, if the graph is properly labeled. But I decided that this second approach would be easier to understand. And ease of understanding is the bottom line in descriptive statistics.

2.2.1.5 Scattergrams

Finally, Figure 2.27 shows a **scattergram** instead. While the histograms and frequency polygons of Figures 2.24 through 2.26 all display frequency distributions based on the data, the scattergram displays the data themselves. It contains 281 points, each one representing the sex and weight combination for one of the 281 individuals in the data. The advantage of this approach is that there is none of the distortion that inevitably occurs when you break a continuous variable into artificial groupings. In this case, it is impossible to identify all 281 cases since the points for two people of the same sex and weight will print over each other. Still, you can see fairly easily where the bulk of the individuals fall and where the outliers are.

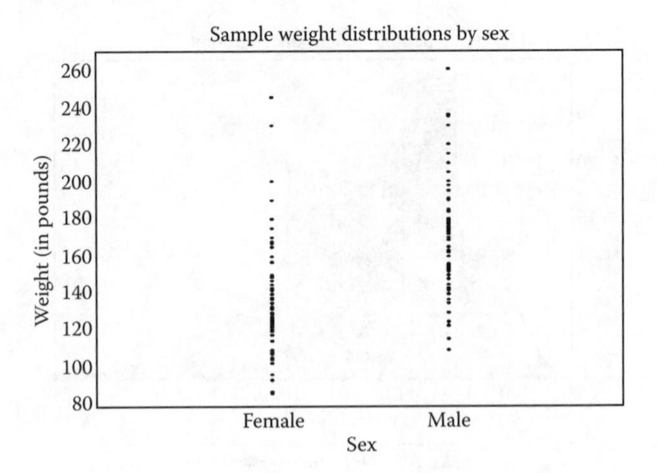

Figure 2.27 Weight distributions by sex.

> Perhaps the qualification—"for whatever reason"—needs a bit more attention. It is easy to think of at least two theories—call them "nature" and "nurture"—that might explain why females end up disproportionately in certain majors, while males end up disproportionately in others. Since both theories predict that we will find a relationship between Sex and Major, and we do, our result is consistent with either. It is also easy to think of several theories as to why females weigh less than males. There is the nature–nurture issue again, if females weigh less than males of the same height. But remember, we have not even controlled for height. It would be consistent with our results so far for sex to be just a proxy for height. Females may weigh less than males for no other reason than that they are shorter. Our results, that sex helped *explain* a student's choice of major and a young adult's weight, are consistent with all sorts of theories about *causation.*

Figure 2.28 A digression on explanation and causation.

So far, we have dealt only with dummy explanatory variables. It has been easy to break down our dependent variable into two groups and compare the distributions of the two groups. If the distributions of the two groups differed, by more than we would expect just by chance, we could say that the dummy variable helps explain the variation in our variable of interest. Thus, in the previous two examples, sex helped explain (for whatever reason) a student's choice of major and a young adult's weight. (see Figure 2.28.)

Suppose an explanatory variable has several categories instead of just two. All the procedures above continue to be appropriate. You will have more than two columns to your frequency tables, more than two sets of histograms, frequency polygons, and so forth. Technically, nothing much changes. Be aware, though, that comparisons become much more difficult, very quickly, as the number of categories rises. Thus, there is a much greater burden on you, putting together the tables and graphs, to exercise judgment. Perhaps several categories appear very similar, and can be collapsed into a single category. Again, the goal is to simplify—but not too much.

2.2.2 Continuous Explanatory Variables

2.2.2.1 Frequency Distributions

The case of a continuous explanatory variable can be thought of as the extreme of the case of an explanatory variable with several categories. To present the relationship in a table, you have no alternative but to group the explanatory variable in the same manner as the dependent variable.

Some spreadsheet programs have advanced features that help automate the creation of two-way tables. However, these are not really necessary. Indeed, when you are just learning, it is probably best to work through the steps yourself.

Continuing with the previous data on the weight of young adults, suppose you are interested in the possible effect of height on weight. You can proceed much as in the example looking at the possible effect of sex. First, you would sort the data by height, being careful to include all five variables in the sort. Now, though, because height is a continuous variable, you would need to group individuals into height ranges. In Figures 2.29 through 2.31, I decided on four inch intervals. I departed from my earlier advice to use ranges like 5 and 10 because there are 12, not 10, inches to a foot. I have seven ranges.

Figure 2.29, then, shows a useful next step in spreadsheets that put frequencies right next to the bins. (It is not necessary in Excel.) I have entered headings that correspond to the seven ranges. I have also entered one column of bins, in the proper place for the shortest height grouping and then copied and pasted it into the six adjacent columns.

The trick, then, is to use the bins from right to left. That is, select the weights of those in the tallest grouping, and use the right-most column of bins. You get the frequencies for this tallest group under the 81–84 heading. Then select the weights of those in the next tallest grouping, and use the next column of bins over. You get the frequencies for this group under the 77–80 heading, writing over the bins that you have already used. Continue across until you get to the shortest grouping, and the left-most bins. Figure 2.30 shows the results.

Table 7 (Figure 2.31) shows how the result might look, dressed up for presentation. In addition to carefully labeling everything, I have changed

	G	H	I	J	K	L	M	N
43	Weight	Height Categories						
44	Upper							
45	Bound	57–60	61–64	65–68	69–72	73–76	77–80	81–84
46	100	100	100	100	100	100	100	
47	120	120	120	120	120	120	120	
48	140	140	140	140	140	140	140	
49	160	160	160	160	160	160	160	
50	180	180	180	180	180	180	180	
51	200	200	200	200	200	200	200	
52	220	220	220	220	220	220	220	
53	240	240	240	240	240	240	240	
54	260	260	260	260	260	260	260	
55								

Figure 2.29 Creating a two-way frequency table in a spreadsheet.

	G	H	I	J	K	L	M	N
43	Weight	Height Categories						
44	Upper							
45	Bound	57–60	61–64	65–68	69–72	73–76	77–80	81–84
46	100	1	3	0	0	0	0	0
47	120	1	22	14	1	0	0	0
48	140	3	27	44	7	0	0	0
49	160	0	11	25	23	4	0	0
50	180	0	4	8	32	11	0	0
51	200	0	1	6	10	6	0	0
52	220	0	0	1	5	4	0	1
53	240	0	1	0	1	2	0	0
54	260	0	0	1	1	0	0	0
55		0	0	0	0	0	0	0

Figure 2.30 A two-way frequency distribution of weight and height.

Table 7: Distribution of Weight by Height

Weight (in pounds)*	Height (in inches)*							
	56–60	60–64	64–68	68–72	72–76	76–80	80–84	Total
80–100	0.36%	1.07%						1.42%
100–120	0.36%	7.83%	4.98%	0.36%				13.52%
120–140	1.07%	9.61%	15.66%	2.49%				28.83%
140–160		3.91%	8.90%	8.19%	1.42%			22.42%
160–180		1.42%	2.85%	11.39%	3.91%			19.57%
180–200		0.36%	2.14%	3.56%	2.14%			8.19%
200–220			0.36%	1.78%	1.42%		0.36%	3.91%
220–240		0.36%		0.36%	0.71%			1.42%
240–260			0.36%	0.36%				0.71%
Total	1.78%	24.56%	35.23%	28.47%	9.61%	0.00%	0.36%	100.00%

* Each range includes its upper bound.

Figure 2.31 Presentation quality two-way frequency distribution.

the ordinary frequencies to relative frequencies. That may or may not have helped. Blanking out the zeros—a simple thing—almost certainly does. It makes the table more "graphical." Clearly, the non-zero numbers tend toward the diagonal, from the upper left to the lower right. Clearly, there seems to be a tendency for taller individuals to weigh more.

2.2.2.2 Scattergrams

The clearest way to show the relationship between height and weight is not with a table at all, but with a scattergram. Figure 2.32 illustrates. Again, there are 281 points representing the height and weight combinations for the 281 individuals in the data. Clearly, the points tend to rise toward the right. Indeed, the line shows an estimate of how the average weight rises

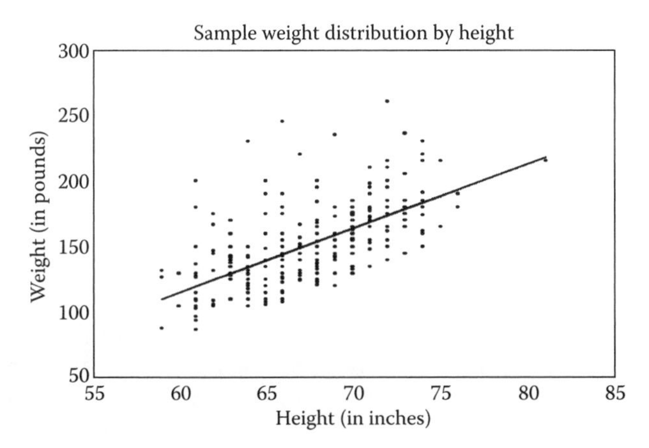

Figure 2.32 Weight distribution by height.

with height. It is called a "regression line," and you will learn how to calculate one in Chapter 3.

2.3 Looking at Variables over Time

So far, in this chapter, all the data sets we have used have been **cross sectional**. That is, the variables represent measures across individuals surveyed at a particular time. But it is also possible to measure something about the same individual across time. Such data sets are called **time series**. And often, with such data, the first thing you would want to know about a variable is how it has actually varied over time. A **line graph**, in which time is taken as the explanatory variable, shows that variation.

Perhaps you have read a lot in recent years about how the U.S. economy is becoming a service economy. Presumably, this means that the service sector of the economy is growing at a greater rate than the economy as a whole. To see, find the *Economic Report of the President;* it is in your library and on the web at http://www.gpoaccess.gov/eop/index.html. You can easily find data like those in Figure 2.33 (Servicesl.xls), measuring overall GDP (which includes the consumer sector), the overall consumer sector (which includes services), and the service sector. Frequency distributions and tables will not help much here. A line graph will. Graphing GDP and Services against year, you get a graph much like in Figure 2.34.

On its face, Figure 2.34 appears to offer little support for the notion that services are growing faster than the overall economy. Indeed, it appears to be the opposite. And you might wonder, as well, why there is so much talk about sluggish economic growth since the 70s. GDP growth appears to have been accelerating.

However, appearances are deceiving. The slope of the two lines in Figure 2.34 represent annual growth in *dollar* terms. And GDP is going to grow faster in dollar terms, just because its values are so much bigger. But, generally, what we really care about is growth in *percentage* terms.

Year	GDP	Consumption	Services	Year	GDP	Consumption	Services
70	1035.6	648.1	291.1	83	3514.5	2283.4	1173.3
71	1125.4	702.5	320.1	84	3902.4	2492.3	1283.6
72	1237.3	770.7	352.3	85	4180.7	2704.8	1416.4
73	1382.6	851.6	384.9	86	4422.2	2892.7	1536.8
74	1496.9	931.2	424.4	87	4692.3	3094.5	1663.8
75	1630.6	1029.1	475.0	88	5049.6	3349.7	1817.6
76	1819.0	1148.8	531.8	89	5438.7	3594.8	1958.1
77	2026.9	1277.1	599.0	90	5743.8	3839.3	2117.5
78	2291.4	1428.8	677.4	91	5916.7	3975.1	2242.3
79	2557.5	1593.5	755.6	92	6244.4	4219.8	2409.4
80	2784.2	1760.4	851.4	93	6553.0	4454.1	2554.6
81	3115.9	1941.3	952.6	94	6935.7	4700.9	2690.3
82	3242.1	2076.8	1050.7	95	7253.8	4924.9	2832.6

Source: *Economic Report of the President*, Government Printing Office, Table B-1, 1997.

Figure 2.33 U.S. nominal GDP, consumption, and services (Servicesl.xls).

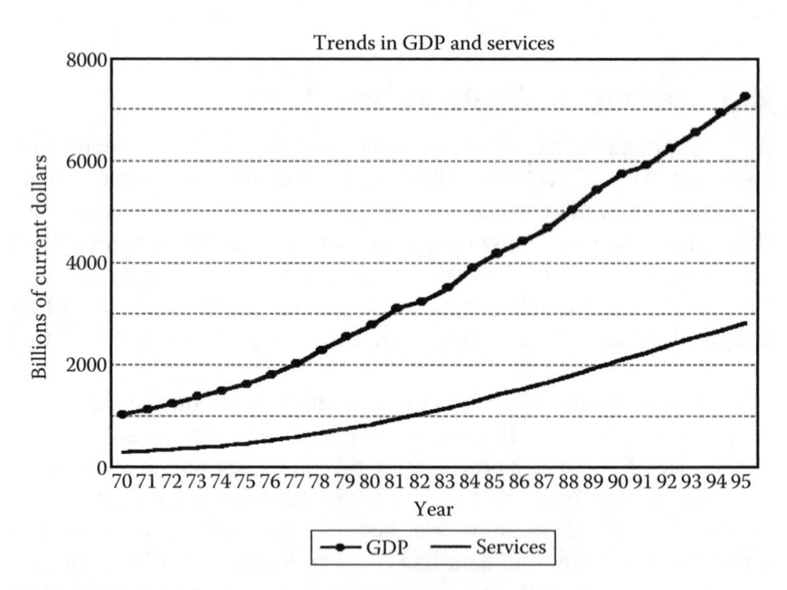

Figure 2.34 Two time trends using an arithmetic scale.

Looking back at the data in Figure 2.33, consider the changes from 1990 to 1991. The GDP grew from $5,743.8 to $5,916.7 billion, a dollar growth of $172.9 billion, but a percentage growth of only about 3%. Services grew from $2,117.5 to $2,242.3 billion, a dollar growth of only $124.8 billion, but a percentage growth of close to 6%. When you want to compare percentage growth rates, as you usually do when looking at time series, a graph like in Figure 2.34 is misleading.

Fortunately, there is a fix. Figure 2.35 shows the same information on a semilog graph. It is called a semilog graph because one of the two axes—the *y*-axis—is scaled according to the logarithms of its values. The

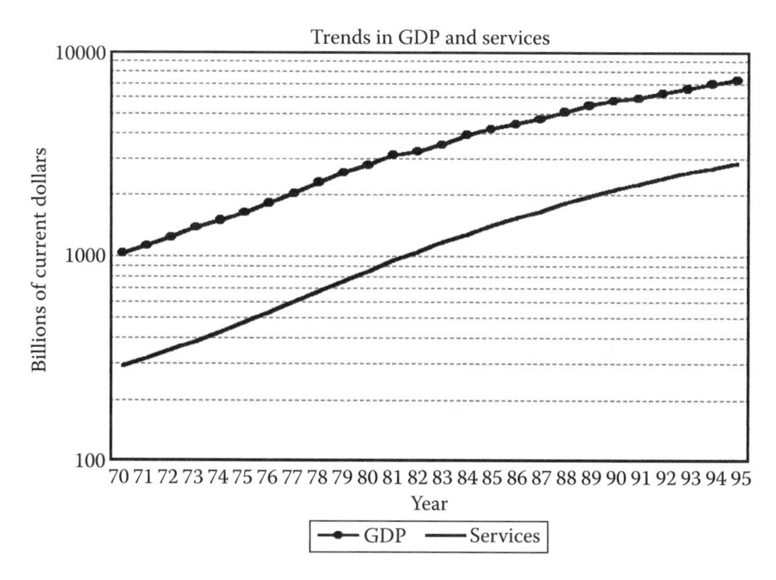

Figure 2.35 Two time trends using a semilog scale.

result is that the movements in the y-direction are *proportional* moves. Measure the distance between $100 and $200 billion—an increase of 100%. Now compare that with the distance between $200 and $400 billion, between $400 and $800 billion, and so on. All increases of 100% look the same.

Now we can look at the graphs of GDP and Services, and tell directly what is growing faster as a percent. The line for Services is a little steeper; Services have indeed been growing at least a little faster. And now we can see that the apparent acceleration in GDP growth was an illusion. Indeed, it appears to be decelerating, though that is largely due to the fact that inflation fell over the period.

2.4 Exercises

2.1 Figure 2.6 shows the calculation of relative and cumulative relative frequency distributions for the student GPAs in file Students1.xls.

a. Read the file into your spreadsheet and, if you have not already done so, reproduce these frequency distributions from the raw data.

b. Consider graphing both relative and cumulative relative frequency distributions in single (i) bar, (ii) line, and (iii) pie charts. One of the three would not make sense. Which one? Explain why it would not make sense.

c. Graph each of those in part b that do make sense. Dress them up. Which one strikes you as most effective? Why?

2.2 Continue with these same data.

a. Break down the data by sex, and create separate frequency distributions of GPA for each sex. Dress these up for formal presentation as a table.
b. In your table for part a, did you present ordinary or relative frequencies? Why?
c. Graph your two frequency distributions in (i) a single histogram and (ii) a single frequency polygon. Dress them up. Which one strikes you as more effective? Why?
d. Graph the original data in a scattergram. Dress it up. How does it compare in effectiveness with the graphs in part c? Why?
e. Do you think you see differences in the distributions of GPA by sex that seem too great to have been due just to chance?

2.3 In Figure 2.28, I said that "females may weigh less than males for no other reason than that they are shorter." But are females actually shorter? Use the NLSY1.xls data to explore that question.

a. Break down the data by sex, and create separate frequency distributions of Height for each sex. Dress these up for formal presentation as a table.
b. In your table for part a, did you present ordinary or relative frequencies? Why?
c. Graph your two frequency distributions in (i) a single histogram and (ii) a single frequency polygon. Dress them up. Which one strikes you as more effective? Why?
d. Graph the original data in a scattergram. Dress it up. How does it compare in effectiveness with the graphs in part c? Why?
e. Do you think you see a difference in the distributions of Height by sex that seems too great to have been due just to chance?

2.4 Continue with these same data. Figures 2.31 and 2.32 attempt to explain individuals' Weights using their Heights, and it looks as though Height does help explain Weight.

a. Can you reverse the variables and use weight to explain height? Try creating a scattergram with the axes reversed.
b. Why would the result in part a be less interesting than the original one?
c. What does all this have to do with the difference between explanation and causation?

2.5 File Employeesl.xls, shown in Figure 1.2 of Chapter 1, contains data on 50 employees of a company. You are interested in explaining the variation in Salary.

a. Create a frequency distribution for Salary, and graph it using any appropriate type of graph.
b. Create frequency distributions for Salary for each sex, and graph them both on a single graph. Do the differences in

the two distributions seem so great that they are unlikely to be due just to chance?

c. Create three scattergrams using (i) education, (ii) experience, and (iii) sex to explain the variation in Salary. Do each of these explanatory variables seem to help explain some of the variation in Salary? For example, is there a tendency for those with more education to earn more?

2.6 File Students2.xls contains (hypothetical) data on a sample of 100 students at a large university. The variables are defined as follows:

ID:	Student identification number;
Female:	Female (1 = yes, 0 = no);
Height:	Height (in inches);
Weight:	Weight (in pounds);
Year:	Year in school (1–5);
Major:	Academic major (1 = mathematics, 2 = economics, 3 = biology, 4 = psychology, 5 = English, 6 = other);
Aid:	Financial aid (in dollars per year);
Job:	Holds a job (1 = yes, 0 = no);
Earnings:	Earnings from job (in dollars per month);
Sports:	Participates in a varsity sport (1 = yes, 0 = no);
Music:	Participates in a music ensemble (1 = yes, 0 = no);
Greek:	Belongs to a fraternity or sorority (1 = yes, 0 = no):
Entertain:	Spending on entertainment (in dollars per week);
Study:	Time spent studying (in hours per week);
HS_GPA:	High school grade point average (5-point scale);
Col_GPA:	College grade point average (4-point scale).

Suppose you are interested in what differences exist between the sexes. For each of the following, create frequency distributions for each sex, and graph them both on a single graph. Do the differences in the two distributions seem so great that they are unlikely to be due just to chance?

a. Height	f. Job	k. Entertain
b. Weight	g. Earnings	l. Study
c. Year	h. Sports	m. HS_GPA
d. Major	i. Music	n. Col_GPA
e. Aid	j. Greek	

2.7 Continue with these same data. Now, though, you are interested in differences between athletes and nonathletes. For each of the following, create frequency distributions for athletes and nonathletes and graph them both on a single graph. Do the differences in the two distributions seem so great that they are unlikely to be due just to chance?

a. Sex	d. Music	g. Study
b. Aid	e. Greek	h. HS_GPA
c. Job	f. Entertain	i. Col_GPA

2.8 Based on the data in file Servicesl.xls, Figure 2.35, based on the data in file Services 1. xls, showed Services growing faster than overall GDP. By now, there should be additional annual data, as well as revised data for the earlier years.

a. Find the *Economic Report of the President*, or some other source, and update the data in file Servicesl.xls.
b. Redo the comparison of GDP and Services. Does it still seem that Services are growing faster?
c. Does it seem that Services are growing faster than overall Consumption?
d. Does it seem that overall Consumption is growing faster than overall *GDP*?

2.9 File Pricesl.xls contains Consumer Price Index (CPI) data for 37 years (1982–1984 = 100). From the U.S. Department of Labor, Bureau of Labor Statistics, *Economic Report of the President*, Table B-60, 2004.

Year: Year (1967–2003);
All Items: All-Item Price Index;
Apparel: Apparel Price Index;
Energy: Energy Price Index;
Food: Food and Beverages Price Index;
Medical: Medical Care Price Index.

a. Create a line graph showing the trends in the overall index and the five selected components.
b. Interpret your results. Which one is growing fastest? Which one is growing slowest? Which one is most volatile?
c. Which form of graph, arithmetic, or semilog did you use? Why?

3

Describing Data: Summary Statistics

3.1 When Pictures Will Not Do

The previous chapter dealt with the description of data, when description is really the only point. When description is really the only point, tables—or better yet graphs–pictures—are by far the most powerful tools available. And indeed, for some readers of this book, these powerful descriptive tools represent the main statistical tools you will use. For some of you, putting together presentation-quality tables and graphs that communicate exactly what *you* want to communicate is the most important thing you will ever need do with statistics. And, indeed, you may need to do it every week of your career.

But when have you ever spent much time and effort describing something that does not matter? The very fact that you are describing something suggests that it matters. Apparently you think there is something there. Indeed, all through Chapter 2, as I presented tables and graphs, I made a point of asking what you saw. And, in the exercises, I asked you what graph you thought most effective. Most effective at what? Most effective at displaying what you thought you saw in the data.

The problem is that, sometimes, what you see in the data is just random, sample to sample variation. In Figures 2.13 through 2.17, females and males seemed to follow different patterns in their choices of majors. But this was in a sample of just 50 students. Do we really know that this difference would show up in any other sample of 50? Do we really know that this difference holds true in the population as a whole?

To answer these questions requires inference. And serious inference generally requires more *quantitative* descriptors of what we have found in our samples. Hence, this chapter deals with such descriptors. You are already familiar with many of them.

3.2 Measures of a Single Numeric Variable

3.2.1 Measures of Central Tendency

Suppose you are given a set of numbers and asked to summarize them with just a single number. What would you do? Probably, you would calculate some sort of average. An average is a measure of the center.

It may be a sort of balance point in the data; it may be some measure of what is typical.

3.2.1.1 The Arithmetic Mean

The **arithmetic mean** is what most people mean by an average. It is just the sum of the numbers divided by their count. It is the arithmetic mean, to distinguish it from other, special purpose means. But in practice, it is often shortened to just the mean, without any modifier.

Figure 3.1 presents the formulas for calculating the mean of a variable, X. Of course, you already know how to calculate it. There is nothing new here except, perhaps, the symbols.

The symbol for the mean of X in a *population* is μ_X ("mu sub X"). Σ_i is th symbol for a sum; X_i represents an individual value of X, and N_X represents the number, or count, of the population. So the formula says simply to sum (Σ) the individual values (X_i) from the first ($i=1$) to the last (N_X) and then divide by the number of individual values (N_X).

The symbol for the mean of X in a *sample* is $\bar{X}$ ("X bar"). Notice, the only difference in the formula is that N_X, the population count, has been replaced with n_X, the sample count. While to calculate μ_X, you add up all the population values and divide by the population count, to calculate $\bar{X}$, you add up all the sample values and divide by the sample count. In other words, whether you have a population or a sample, you do exactly the same thing.

I have written both of the formulas with and without the ranges over which to sum. Usually the range is obvious. If you are working with a population, you are going to want to sum everything in the population. If you are working with a sample, you are going to want to sum everything in the sample. I will not bother to include these ranges in formulas from now on unless there is a danger of ambiguity.

I have written each of the formulas for a variable named X. But, when you are working with a set of variables, you need more than one name. You can add subscripts, and call them X_1, X_2, and so on. Or you can make up names that mean something for the data at hand. For example, in the last chapter's sample of 50 students, ID, Female, Major, and GPA mean a lot more than X_1, X_2, X_3, and X_4. I would use the meaningful names. And then I would refer to the mean GPA of all students in the population as μ_{GPA} and the mean GPA of the 50 students in the sample as $\overline{GPA}$.

The Arithmetic Mean of a Population

$$\mu_X = \frac{\sum_{i=1}^{N_X} X_i}{N_X} = \frac{\sum X_i}{N_X}$$

The Arithmetic Mean of a Sample

$$\bar{X} = \frac{\sum_{i=1}^{N_X} X_i}{n_X} = \frac{\sum X_i}{n_X}$$

Figure 3.1 The arithmetic mean.

	A	B	C	D
1			X-Mean	
2		1	−2.5	← =B2-$B$14
3		1	−2.5	
4		1	−2.5	
5		1	−2.5	
6		1	−2.5	
7		2	−1.5	
8		2	−1.5	
9		2	−1.5	
10		2	−1.5	
11		22	18.5	
12	=SUM(B2:B11)	→ 35	0	← =SUM(C2:C11)
13	=COUNT(B2:B11)	→ 10		
14	=B12/B13	→ 3.5		

Note: While recent versions of Lotus and Quattro Pro understand the Excel format above, their own format uses "@" instead of "=" to indicate a function, and ".." instead of ":" to indicate a range of numbers. The sum, for example, would be "@SUM(B2..B11)."

Figure 3.2 Finding the arithmetic mean: A simple example.

Consider the very simple example in Figure 3.2. Column B contains 10 values of X, five 1s, four 2s, and a 22. As you may know, spreadsheets have special functions to calculate many measures like the mean automatically. I will introduce these eventually. To start, though, I want you to work through the formulas to help you gain a clear understanding of what they mean. Thus, in B12, I have entered the special function=SUM(:) to sum the X-values, which gives us the numerator of the formula for the mean. In B13, I have entered the special function=COUNT(:) to count them, which gives us the denominator. B14, then, divides the numerator by the denominator to get the final answer. In other words, I have done exactly what the formulas say.

Whether you call the answer μ_X or $\bar{X}$ depends on whether the 10 numbers represent the whole population—the whole set of numbers about which you are interested—or just a sample from that population. Usually it is the latter. Recall the whole notion of inference. We are interested in the population, but know only a sample. We are interested in knowing μ_X but—not knowing all the x_i in the population—are unable to use the formula for μ_X. Usually, we will calculate $\bar{X}$, and use that as the basis for estimating μ_X.

Figure 3.2 also illustrates a fundamental characteristic of the mean. In Column C, I have subtracted the mean from each of the original 10 numbers. (Note, I have shown the formula for just the first case, because I entered just that one—a relative reference to the first number, B2 minus an absolute reference to the mean, $B$14—and then used **Copy** for the rest. Make sure you do it that way too.) Then I summed these "deviations from the mean." The deviations sum to zero. This is not a coincidence. The mean is the central number in exactly this sense—that these positive and negative deviations just cancel out. Figure 3.3 illustrates this idea more graphically. The 22 is very much larger than the other nine numbers, and pulls the mean

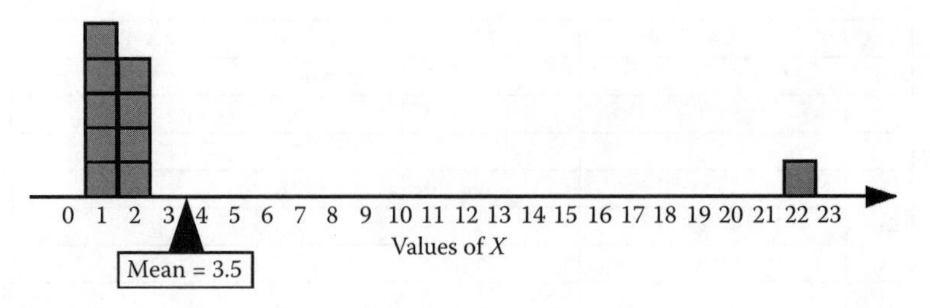

Figure 3.3 The arithmetic mean as a balance point.

in its direction. At 3.5, its 18.5 deviation just balances the sum of the other nine.

3.2.1.2 The Median

The arithmetic mean is the most important measure of central tendency, and the one on which we will focus for most of the rest of this course. But there are others that can be useful too. One is the **median**. The median, Md, is just the middle number or, in the case of an even count, the average of the middle two. In the previous example, with a count of 10, you would average the fifth and sixth numbers. If the numbers were not already in order, you would need to sort them first. Then, simply count from either end. The middle two are a 1 and a 2, so the median is 1.5.

In this example, then, the mean is 3.5 while the median is only 1.5. The difference is due to the way the two measures treat the extreme value of 22. While the 22 pulls the mean up until it is actually greater than all the other numbers in the sample, it does not do this to the median. All that matters for the median is that it is in the larger half. It could be 2 or 12 or 22.

The median is nice because it is so easy to understand. And, for some purposes, it can be more important to know whether a number is in the top or bottom half than whether it is above or below the mean. For example, suppose your instructor hands back an exam and indicates that the class median was 85 while the class mean was 75. Which matters? Well the median is easy to interpret. If you got above an 85, you were in the top half of the class. If you did not, you were not. The fact that the mean is much lower indicates that a couple members of the class did very badly. Perhaps they handed in blank exams. These extremely low scores pulled the mean down well below the median. But the fact that a couple of your classmates did *very* badly instead of just badly should probably not be of any comfort to you.

The difference in the mean and median also provides a simple measure of skewness. Recall, from Chapter 2, that one of the things we might be interested in, as we create histograms and frequency polygons, is whether their shapes are symmetrical or skewed. If they are symmetrical, the pull of extreme values will be about equal in each direction. Hence, the mean will remain close to the median. However, if they are skewed, the extreme values on one side will be more extreme and will pull the mean away from the median in their direction.

In the example of Figures 3.2 and 3.3, the long tail is toward the right—the high numbers—and it pulls the mean up above the median; the data are skewed right. We would know that the data had this shape, even if we did not know the actual numbers, as long as we knew that the mean was much greater than the median. Indeed, in the example of the exams with a median of 85 and a mean of 75, I did not tell you the actual grades. But you know that only a very long tail toward the left—the low numbers—could have pulled the mean this far below the median.

3.2.1.3 The Mode

Still another measure of the middle is the **mode**. The mode, Mo, is the most common number—the one that occurs most frequently in the data. In the previous example, 1 is the most common number, so the mode is 1. The advantage of the mode, like the median, is that it is easy to understand. However, it has serious limitations. A data set could easily have one of each number, in which case it would have no mode. Or it could have several modes.

3.2.1.4 A More Interesting Example

Return to the example in Chapter 2 (page 27) concerning the sample of 281 young adults. One of the variables was Age. Suppose you wanted to calculate its mean, median, and mode. The mean would really not be any more tedious than it was in the previous example with just 10 cases. See Figure 3.4. The sum and count span many more numbers, but that is the only difference. For the median, you could sort the data by Age, and count to the 141st case from either end. But there is an easier way and it will help you find the mode as well. Remember frequency distributions.

3.2.1.5 From a Frequency Distribution

Figure 3.5 shows a frequency distribution for Age. All of those in the sample are in the range 21 to 28, so I created bins for each of those ages and created an ordinary frequency distribution. Then, since the mode is just the number with the largest frequency, it can be found by inspection. The mode is 23.

	A	B	C	D	E
1	ID	Female	Age	Height	Weight
2	6	0	25	65	200
3	18	0	27	66	155
4	77	1	21	63	135
:	:	:	:	:	:
:	:	:	:	:	:
279	12081	0	24	72	155
280	12104	1	28	65	160
281	12118	1	23	67	145
282	12124	0	27	70	176
283			6838	← =SUM(C2:C282)	
284			281	← =COUNT(C2:C282)	
285			24.3345	← =C283/C284	

Figure 3.4 Finding the arithmetic mean: A more interesting example.

	G	H	I	J	K
58	Age	F	← (For the Mode)	Cumulative F	← (For the median)
59	21	32		32	
60	22	40		72	
61	Mo → 23	43	← (Largest F)	115	← (Have not yet reached 141)
62	Md → 24	39		154	← (Have now reached 141)
63	25	33		187	
64	26	32		219	
65	27	32		251	
66	28	30		281	
67		0			

Figure 3.5 Finding the median and mode from a frequency distribution.

	G	H	I	J
71	Age	F	F * Age	
72	21	32	672	← = H72 * G72
73	22	40	880	
74	23	43	989	
75	24	39	936	
76	25	33	825	
77	26	32	832	
78	27	32	864	
79	28	30	840	
80		0		
81	=SUM(H72:H79)	→ 281	6838	← =SUM(I72:I79)
82			24.3345	← =I81/H81

Figure 3.6 Finding the mean from a frequency distribution.

The cumulative frequency distribution is useful for finding the middle number. With 281 individuals, the 141st from either end is the middle one. Cumulating from 21 through 23, we have only 115 individuals, which means we have not yet reached the 141st. Adding in the 24 year olds gets us over 141, so the 141st individual must be in the 24-year-old bin. The median is 24.

Since data are often displayed in frequency distributions, it will be useful to be able to calculate the arithmetic mean from a frequency distribution as well. Figure 3.6 shows how. It is not difficult, though I have found that many students become confused. The key, really, is to remind yourself of what the frequency distribution shows. While it may no longer look like it, it is still a list of 281 numbers. We still have the ages of 281 young adults; the mean is still the sum of those 281 ages divided by their count.

How do we get the sum of the 281 ages? We have 32 individuals who are 21, so we need to add 21 in 32 times—which we can do by just adding in $(32 \times 21) = 672$. We have 40 individuals who are 22, so we need to add 22 in 40 times—which we can do by just adding in $(40 \times 22) = 880$. So, to

	A	B	C	D	E	F
1	ID	Female	Age	Height	Weight	
2	3837	1	22	61	87	
3	2257	1	23	59	88	
:	:	:	:	:	:	
141	4617	0	26	70	145	
142	3009	1	28	62	147	← Median
143	1526	0	27	71	148	
:	:	:	:	:	:	
:	:	:	:	:	150	← Mode
:	:	:	:	:	:	
281	3728	1	26	66	245	
282	5303	0	24	72	260	
283					42457	
284					281	
285					151.0925	← Mean

Figure 3.7 Finding the mean, median, and mode from the actual data.

get the sum of all 281 ages, just multiply each age by its frequency, and then sum. Notice that we get 6838, just as we did in Figure 3.4 working with the actual data.

How would we know we had 281 ages if we had not done a count on the actual data earlier? Again, we have 32 individuals who are 21, 40 individuals who are 22, and so on. If we sum these ordinary frequencies we get the total count. Again, notice that we get 281 just as we did in Figure 3.4 working with the actual data.

In the end, the mean is the sum of the ages (6838) divided by their count (281), just like before.

So far, frequency distributions have posed no difficulties. Indeed, creating a frequency distribution for Age was probably the easiest way (short of special functions) for finding its mean, median, and mode. But suppose, instead of Age, you are interested in Weight. First, we will look at the actual data, so we know the actual mean, median, and mode. Then we will consider finding them from a frequency distribution.

In Figure 3.7 I have sorted the data by Weight. The mean is found in the usual way, by summing and counting the weights and then dividing the sum by the count. The median is just the 141st number that—since the data start in line 2—is in line 142. The mode is harder to find. You can try scrolling through and counting the occurrences of each number. But that is tedious and you are likely to miscount. Or you can create a frequency distribution with 174 bins—one for every number from 87 to 260. In fact, this is what I did. And 150 was the most common number, coming up 20 times.

3.2.1.6 Working from Grouped Data

However, now imagine trying to find these numbers if you did not have the original data—if all you had was a presentation-quality frequency distribution such as the one we produced in the last chapter. Figure 3.8 reproduces it. Of course, it does not have 174 bins (one for every number).

Table 5: Distribution of Weights

Weight (in pounds)*	Percentage
80–100	1.42%
100–120	13.52%
120–140	28.83%
140–160	22.42%
160–180	19.57%
180–200	8.19%
200–220	3.91%
220–240	1.42%
240–260	0.71%
Total	100.00%

*Each range includes its upper bound.

Figure 3.8 Presentation quality frequency distribution of weights.

Numbers are purposely grouped. So, if this is the only information we have, we have a problem in finding the mean, median, or mode.

Consider the median. We can cumulate the percentages, starting with the smallest weights and working up. From 80 up through 140 we have 43.77%. We are not to 50% yet. Adding in the 140–160 grouping takes us over 50%. So the median must be somewhere in the 140–160 range. But we do not know where. Or consider the mode. We know that the most common grouping is from 120 to 140. But that does not mean that the most common number is within that grouping. Indeed, in this case, we know it is not. There are ways of approximating the median and mode in cases like this. But the median and mode are not going to be important enough, in the inference to follow, to warrant additional fuss.

The mean will be important, though, so we will want at least to approximate it. And the approximation is really quite straightforward. We will assume that all values in a grouping are at the midpoint for that grouping. Clearly, this is not correct. Ordinarily, most of the actual numbers will not be at the midpoints. But since, by assumption, we do not know the original data, all we can do is hope that these errors tend to balance out. Figure 3.9 shows these midpoints in column G, and it shows the actual work done in three different ways.

First, suppose you knew the information in Figure 3.8 *and* you knew the sample count was 281. You could turn the relative frequencies back into ordinary frequencies for each grouping. For example, 1.42% of 281 (0.0142×281) is 3.9902. But, since ordinary frequencies must be integers, the actual number must have been 4. The ordinary frequencies in column H in Figure 3.9 were all figured this way.

Once you have the midpoints to use as assumed weights and you have the frequencies for each, the problem is no different from the example in Figure 3.6 for the mean age. Indeed, columns G through I in Figure 3.9 show exactly the same calculations for the assumed weights as those columns in Figure 3.6 showed for the actual ages. You have four individuals

	G	H	I	J	K	L	M
85	Midpoint	F	F * midpoint	F	F * midpoint	F	F * midpoint
86	90	4	360	1.42%	128	0.0142	1.28
87	110	38	4180	13.52%	1488	0.1352	14.88
88	130	81	10530	28.83%	3747	0.2883	37.47
89	150	63	9450	22.42%	3363	0.2242	33.63
90	170	55	9350	19.57%	3327	0.1957	33.27
91	190	23	4370	8.19%	1555	0.0819	15.55
92	210	11	2310	3.91%	822	0.0391	8.22
93	230	4	920	1.42%	327	0.0142	3.27
94	250	2	500	0.71%	178	0.0071	1.78
95		281	41970	100.00%	14936	1.0000	149.36
96		=I95/H95	→ 149.36	=K95/J95	→ 149.36	=M95/L95	→ 149.36

Figure 3.9 Estimating the mean from grouped data.

who are assumed to weigh 90, so you need to add 90 in four times—which you would do by adding in (4 × 90) = 360. You have 38 individuals who are assumed to weigh 110, so you need to add 110 in 38 times—which you would do by adding in (38 × 110) = 4180. To get the sum of all 281 weights, you would just multiply each assumed weight by its frequency and sum. As always, summing the frequencies gives you the count. And, as always, the mean is just the sum of all the values divided by their count.

Suppose you knew the information in Figure 3.8 but *not* that the count was 281. In that case, you could not recreate the ordinary frequencies (4, 38, etc.). But, as it turns out, this does not matter. You can use the *relative* frequencies from Figure 3.8 directly. Columns J and K in Figure 3.9 show the same calculations as Columns H and I, but use the percentages instead of counts and divide by the sum of the percentages—which of course is 100%. Columns L and M show the same calculations using proportions instead and dividing by the sum of the proportions—which of course is just 1. These answers all agree except perhaps due to rounding.

Of course the actual mean, which we calculated in Figure 3.7, was 151.0925. Not surprisingly, our answer using midpoints instead of the actual data is not exactly right. But it is reasonably close.

3.2.2 Measures of Variation

The arithmetic mean (and sometimes the median or mode) will be important in measuring central tendency—the middle of a set of numbers. But, of course, actual numbers are scattered about the middle. And we will also need a measure of how widely they are scattered.

3.2.2.1 The Range

One possible measure of variation is the **range**. It is just the largest number minus the smallest number. Its advantage is that it is easy. Unfortunately, it is not a very good measure.

To see the problem with the range, consider Figure 3.10, which repeats the simple example of Figure 3.2. The range, calculated in B16, is 21. Notice, though, that it is this big only because of a single outlier. The rest of the

	A	B		C	D		E
1			X	X-mean		$(X\text{-mean})^2$	
2			1	−2.5		6.25	← =D2^2
3			1	−2.5		6.25	
4			1	−2.5		6.25	
5			1	−2.5		6.25	
6			1	−2.5		6.25	
7			2	−1.5		2.25	
8			2	−1.5		2.25	
9			2	−1.5		2.25	
10			2	−1.5		2.25	
11			22	18.5		342.25	
12	=SUM(B2:B11)	→	35			382.5	← =SUM(D2:D11)
13	=COUNT(B2:B11)	→	10			9	← =B13-1
14	=B12/B13	→	3.5	Variance	→	42.5	← =D12/D13
15				Standard Deviation	→	6.5192	← =SQRT(D14)
16	=B11–B2	→	21	← Range			
17				CV	→	186.2629%	← =D15/B14

Figure 3.10 Measuring variation: A simple example.

numbers are clustered very tightly—within 2.5 of the mean. A measure of variation should reflect that clustering as well as the fact that there is an outlier. For comparison, suppose the four 2s had been 22s instead. We would have had five 1s and five 22s, the mean would have been 11.5, and not a single point would have been within 10 of the mean. Surely that would have represented greater variation. Yet, since the largest and smallest numbers would have been unchanged, the range would still have been 21.

3.2.2.2 The Variance

The explanation of the problem with the range suggests what we really want—a measure of variation based on how far each of the individual points is from the mean. Indeed, a first intuition might be to find an average deviation from the mean. Recall, though, one of the fundamental characteristics of the mean is that deviations from the mean always sum to zero. The positives and negatives always just cancel out, so the average deviation from the mean is always exactly zero. Somehow, we need to keep the positives and negatives from canceling out.

By far the most important approach is to *square* the deviations from the mean before averaging. This average of the squared deviations from the mean is called the **variance.** Figure 3.11 presents the formulas for calculating it.

The symbol for the variance of a *population* is σ_X^2 ("sigma sub X squared"). It is the sum of the squared deviations from the mean, divided by the count, which makes it just the average of the squared deviations of the population. The symbol for the variance of a *sample* is s_X^2 ("s sub X squared"). It is the sum of the squared deviations from the mean, divided by the count minus one. Since we will have many occasions to work with it, Figure 3.11 also introduces SSD_X as shorthand (for the Sum of Squared Deviations from the mean) for the variable X.

The variance of a population

$$\sigma_X^2 = \frac{\sum (X_i - \mu_X)^2}{N_X} = \frac{\text{SSD}_X}{N_X}$$

The variance of a sample

$$s_X^2 = \frac{\sum (X_i - \bar{X})^2}{n_X - 1} = \frac{\text{SSD}_X}{n_X - 1}$$

Figure 3.11 The variance.

Notice that—unlike the population and sample formulas for the arithmetic mean that really tell you to do exactly the same thing—the two formulas for the variance do not. If you regard a set of numbers as a population, you will divide by the count and get one answer. If you regard it as a sample from a larger population, you will divide by the count minus one and get a slightly larger answer. The difference is not intuitive; it requires some explanation.

First, recall again the whole notion of inference. We are interested in the population, but know only a sample. We are interested in knowing σ_X^2 but—not knowing all the X_i in the population—are unable to use the formula for σ_X^2. Usually we will calculate s_X^2 and use that as the basis for estimating σ_X^2. That means that we ought to calculate s_X^2 in whatever way makes it the best estimator of σ_X^2. And using $n_X - 1$ instead of just n_X does that.

To see why, notice the other substitution in the sample formula. Since we are not going to know μ_X the population mean either, we have substituted $\bar{X}$ instead. We are estimating μ_X with $\bar{X}$ as a first step in estimating σ_X^2 with s_X^2. And we cannot pile estimate upon estimate without cost. In a sense, we have used up one bit of information in the first estimation.

To see this more clearly, consider Figure 3.12. Column B shows a sample of five, taken from a larger population. Suppose the unknown mean of that population is 20. If we somehow knew that mean, we could use it as the basis for our deviations from the mean, as shown in column C. Notice that these sample deviations from the population mean do not sum to zero. This means that even the last one is providing independent information. We have n_X independent deviations from the population mean; hence, we have n_X independent squared deviations. It would make sense to find their average by dividing by n_X.

Now compare that with what we must do, given that we do not know the population mean. In this case, we must use the sample mean as the basis for our deviations from the mean, as shown in column F. Notice that these sample deviations from the sample mean do need to sum to zero. Hence, as soon as we know the first four deviations, we really know all five; the fifth deviation does not provide any independent information. We have only $n_X - 1$ independent deviations from the sample mean. Hence, we have $n_X - 1$ independent squared deviations. It makes sense to find their average by dividing by $n_X - 1$.

	A		B	C	D	E		F	G
1			X	$X - \mu_X$	$(X - \mu_X)^2$			$X - \bar{X}$	$(X - \bar{X})^2$
2			20	0	0			−2	4
3			25	5	25			3	9
4			30	10	100			8	64
5			20	0	0			−2	4
6			15	−5	25	must = −7 to make the	→	−7	49
7	Sum	→	110	10	150	sum of deviations = zero	→	0	130
8	Count	→	5						↑
9	$\bar{X}$	→	22			sum of squared deviations tends to be too small			

Figure 3.12 Squared deviations from μ_X versus $\bar{X}$.

Finally, note that—because the sample deviations around the *sample* mean sum to zero, the sum of their squares will be as small as possible. Since typically, the sample deviations around the *population* mean would not sum to zero, the sum of their squares would be larger. In effect, the numerator using the sample mean will tend to be too small. Reducing the denominator by one offsets this tendency and makes s_X^2, with $n_X - 1$ in the denominator, an unbiased estimator of the true σ_X^2. That is, the expected value of s_X^2, calculated with $n_X - 1$ in the denominator, is σ_X^2.

Returning to the simple example of Figure 3.10, then, columns C and D show the deviations and squared deviations from the mean and cell D14 shows the sample variance. In addition, it shows two new measures, the standard deviation and coefficient of variation, which are based on the variance and will be discussed next.

3.2.2.3 The Standard Deviation

One problem with the variance is that, because the deviations from the mean are squared, the answer is hard to interpret. Suppose the data are in dollars, years, or pounds. Then the variance is in dollars^2, years^2, or pounds^2. These are rather nonsensical units. For that reason, we will ordinarily work with the **standard deviation** instead. As shown in Figure 3.13, the standard deviation is nothing but the square root of the variance. Taking the square root turns the answer back into a measure with the same units—dollars, years, or pounds—as the original data.

3.2.2.4 The Coefficient of Variation

The standard deviation is the measure we will use almost always. However, occasionally, it will be useful to have a *relative* measure. That is, whether a particular standard deviation seems large or small may depend on the size of the numbers with which we are working. As shown in Figure 3.14, the **coefficient of variation** measures the standard deviation as a percentage of the mean. Since the standard deviation and mean have the same units—the dollars, years, or pounds of the original data—the units cancel out. The coefficient of variation is a pure number. Like the median, mode, and range, it is not something we will use

The standard deviation of a population

$$\sigma_X = \sqrt{\frac{\sum (X_i - \mu_X)^2}{N_X}} = \sqrt{\frac{\text{SSD}_X}{N_X}}$$

The standard deviation of a sample

$$s_X = \sqrt{\frac{\sum (X_i - \bar{X})^2}{n_X}} = \sqrt{\frac{\text{SSD}_X}{n_X - 1}}$$

Figure 3.13 The standard deviation.

The coefficient of variation of a population

$$CV_X = \frac{s_X}{\mu_X} \times 100$$

The coefficient of variation of a sample

$$CV_X = \frac{s_X}{\bar{X}} \times 100$$

Figure 3.14 The coefficient of variation.

in inference. Thus, we do not need separate symbols for the population and sample values.

3.2.2.5 A More Interesting Example

Return again to the sample of 281 young adults. Earlier, you found the means of Age and Weight. Suppose you wanted to calculate the standard deviations of these variables as well. Figure 3.15 shows the calculations for Age. Column F shows the squared deviations from the mean. Rather than calculate the deviations in one column and then square them in the next, I did it in just one step. Again, be sure to type the formula for the first one right. The reference to the first age must be relative; the reference to the mean must be absolute. Be sure to include the difference in parentheses, so it is calculated before the square is taken. Then, when you have it right, just **Copy** to get the other 280 cases in a single step. You should be able to confirm the numbers in Figure 3.15. And you should be able to confirm that the standard deviation for Weight is 30.5442.

3.2.2.6 Making Sense of the Standard Deviation: The Empirical Rule

By now, you should be able to calculate a standard deviation. It should make sense, at least in the limited sense that you can see that data that are more spread out will have larger squared deviations from the mean and so will yield larger answers. But what does it really mean to say the standard

	A	B	C	D	E	F	G
1	ID	Female	Age	Height	Weight	$(\text{Age} - \overline{\text{Age}})^2$	
2	6	0	25	65	200	0.44289	← =(C2−$C$285)^2
3	18	0	27	66	155	7.10489	
4	77	1	21	63	135	11.11889	
:	:	:	:	:	:	:	
:	:	:	:	:	:	:	
279	12081	0	24	72	155	0.11189	
280	12104	1	28	65	160	13.43589	
281	12118	1	23	67	145	1.78089	
282	12124	0	27	70	176	7.10489	
283			6838	=SUM(F2:F282)		→ 1388.5552	
284			281	=C284−1		→ 280	
285		Mean	→ 24.3345	=F283/F284		→ 4.9591	← Variance
286				=SQRT(F285)		→ 2.2269	← Standard Deviation
287				=F286/C285 × 100		→ 9.1512	← C of Variation

Figure 3.15 Finding the standard deviation: A more interesting example.

deviation of Age is 2.2269 or the standard deviation of Weight is 30.5442? If the data are at least approximately bell-shaped, the **empirical rule** in Figure 3.16 applies.

We saw in Figure 3.6 that Age is actually fairly flat between ages 21 and 28. That is, the frequencies do not really rise toward the middle and then fall off again. Therefore, the rule will not work all that well for Age. As Figure 3.17 shows, it is at best a crude approximation. The vertical line at 24.3345 marks the mean, and the dotted lines mark off intervals of 1, 2, and 3 standard deviations. In this case, only about 52% of the individuals' ages are within ±1 s_X of the mean, and all are within ±2 s_X.

The frequency distribution for Weight is more clearly bell-shaped and, as Figure 3.18 shows, comes a lot closer to the empirical rule. In this case, about 71% of the individuals' weights are within ±1 s_X, about 95% are within ±2 s_X, and over 99% are within ±3 s_X of the mean.

Clearly, the empirical rule is just a rule of thumb. But it gives your intuition something with which to work. Once you know the standard deviation of a variable, you know something about the values that are likely to occur. For most variables, the majority of values will be within 1 s_X of the mean. And for most variables, a value more than 3 s_X from the mean would be quite unusual.

For data that are at least approximately bell-shaped, roughly:

- 68% of the values will be within 1 standard deviation of the mean;
- 95% of the values will be within 2 standard deviations of the mean;
- Virtually all of the values will be within 3 standard deviations of the mean.

Figure 3.16 The empirical rule.

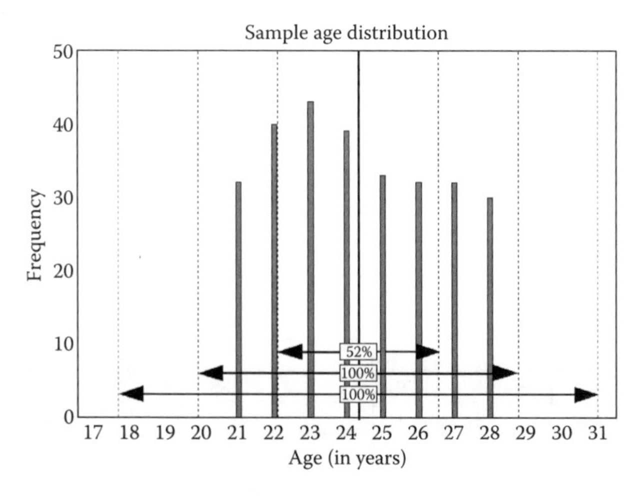

Figure 3.17 The empirical rule: An example (a).

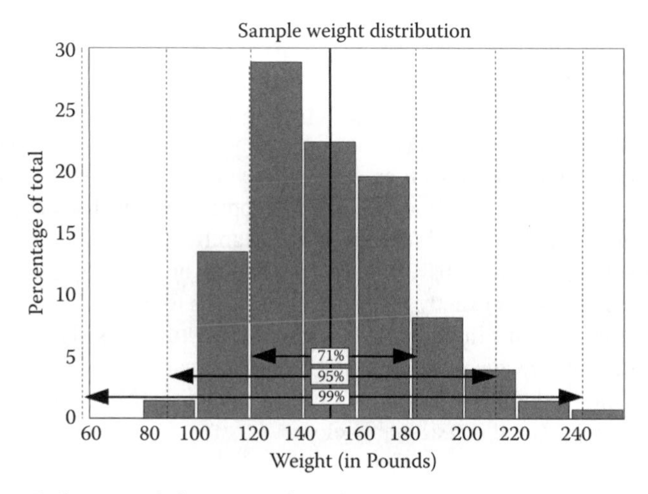

Figure 3.18 The empirical rule: An example (b).

3.2.2.7 From a Frequency Distribution

As with the mean, it will be useful to be able to calculate the standard deviation from a frequency distribution. Columns G, H, and I of Figure 3.19 repeat Figure 3.6, showing the calculation of the mean Age. Column J shows squared deviations for the eight different values in the data. For example, 11.11889 is the squared deviation of 21 from the mean. But we cannot just add this column. Again, it is critical to remember that we still have 281 ages. There are 32 individuals who are 21; thus the squared deviation of 21 from the mean must be added in 32 times—which we do by just adding $(32\times 11.11889)=355.80448$. The sum of column K, then, is the sum of 281 squared deviations from the mean. Dividing that by $(n_X - 1)$ gives the variance. (If you do it in one step, do not forget the parentheses.) The square root gives the standard deviation. Note that all these numbers are exactly the same as the numbers we found in Figure 3.15.

	G	H	I	J		K	L
71	Age	F	F * Age	(Age – $\overline{Age}$)²		F * (Age – $\overline{Age}$)²	
72	21	32	672	11.11889		355.80448	← =H72 * J72
73	22	40	880	5.44989		217.9956	
74	23	43	989	1.78089		76.57827	
75	24	39	936	0.11189		4.36371	
76	25	33	825	0.44289		14.61537	
77	26	32	832	2.77389		88.76448	
78	27	32	864	7.10489		227.35648	
79	28	30	840	13.43589		403.0767	
80		0					
81		281	6838	=SUM(K72:K79)	→	1388.5552	
82		Mean	→ 24.3345	=K81/(H81-1)	→	4.9591	← Variance
83				=SQRT(K82)	→	2.2269	← Standard Deviation

Figure 3.19 Finding the standard deviation from a frequency distribution.

3.2.2.8 Working from Grouped Data

Again, if the only data we have are grouped, as in the example of the frequency distribution of weights (Figure 3.8), we will not be able to find the standard deviation exactly. But, as with the mean, the standard deviation will be important and we will want at least to approximate it. The approximation works exactly as it did for the mean. Columns G, H, and I in Figure 3.20 repeat the same columns of Figure 3.9 based on ordinary frequencies. Columns J and K show exactly the same calculations for midpoint weights as those same columns in Figure 3.19 showed for actual ages. The actual standard deviation, which you were given back on page 57 was 30.5442. Not surprisingly, our answer using midpoints instead of the actual data is not exactly right. But it is reasonably close.

	G	H	I	J		K	L
85	Midpoint	F	F * Midpoint	(Midpoint – Mean)²		F * (Midpoint – Mean)²	
86	90	4	360	3523.54		14094.17	← =H86 * K86
87	110	38	4180	1549.16		58868.26	
88	130	81	10530	374.79		30357.79	
89	150	63	9450	0.41		25.85	
90	170	55	9350	426.03		23431.82	
91	190	23	4370	1651.66		37988.09	
92	210	11	2310	3677.28		40450.07	
93	230	4	920	6502.90		26011.61	
94	250	2	500	10128.52		20257.05	
95		281	41970	=SUM(K86:K94)	→	251484.70	
96	=I95/H95		→ 149.36	=K95/(H95–1)	→	898.16	← Variance
97				=SQRT(K96)	→	29.97	← Standard Deviation

Figure 3.20 Estimating the standard deviation from grouped data.

3.2.3 Spreadsheet Statistical Functions

Until now, I have had you "exercise the formulas" for summary measures, as a way of helping to assure that you understand what they mean, what would make them big or small, and just generally why they make sense. Once you understand them, though, there is really no point in continuing to go through all the steps when spreadsheets have built-in statistical functions that will calculate them for you automatically.

Most modern spreadsheets actually support a variety of approaches, including menu options that will give you a variety of summary statistics in one step. But the specifics vary from spreadsheet to spreadsheet. The formulas in Figure 3.21 are fairly standard, though older releases may not support the functions for the median and mode. And functions for the mode may report just the first mode when, in fact, there may be several. You are probably better off using a frequency distribution to find the mode.

What happens if you ask for the average over a range of data and that range contains nonnumeric data? If a cell is truly blank, it will be treated as a missing value and ignored. This is as it should be. If a cell contains text, though, the answer varies from spreadsheet to spreadsheet. While the standard Excel functions will ignore it, the standard Lotus/Quattro Pro functions will treat it as a zero and average it in. (Later releases of each program have alternative sets of commands that mimic the other.) You should know what your spreadsheet does.

Better yet, avoid mixing numeric and nonnumeric data. If you have a number, enter it; if you do not, leave the cell blank. And you should avoid doing things that make a cell *look* numeric or blank, if it is not.

		Excel*	Lotus/Quattro Pro*
Population or sample (The computations are the same for both)			
Count	N_x or n_x	=COUNT(:)	@COUNT(..)
Sum	Σ	=SUM(:)	@SUM(..)
Mean	μ_X or $\overline{X}$	=AVERAGE(:)	@AVG(..)
Median	Md	=MEDIAN(:)	@MEDIAN(..)
** Mode	Mo	=MODE(:)	@MODE(..)
Maximum		=MAX(:)	@MAX(..)
Minimum		=MIN(:)	@MIN(..)
Population (Assuming the data set is a population, these divide by the count)			
Variance	σ^2	=VARP(:)	@VAR(..)
Standard deviation	σ	=STDEVP(:)	@STD(..)
Sample (Assuming the data set is a sample, these divide by the count −1)			
Variance	S^2	=VAR(:)	@VARS(..)
Standard deviation	S	=STDEV(:)	@STDS(..)
Other useful functions			
Square root	$\sqrt{}$	=SQRT(.)	@SQRT(.)
Natural log	ln(X)	=LN(.)	@LN(.)
Exponential	e^x	=EXP(.)	@EXP(.)

*Lotus/Quattro Pro functions treat text cells as zeros, and include them in the computations. The corresponding Excel functions leave them out. **If there is more than one mode, most spreadsheets report only the first one.

Figure 3.21 Some common spreadsheet statistical functions.

	A	B	C	D	E	F
1		X		X	Did not display zero and thus → forgot it in typing formulas below	X
2	Actually a zero →	0		0		
3		5		5		5
4	Actually a missing value; the cell → should be blank	10	Entered a space which was counted → as a zero instead of a blank below	10		10
5						
6		15		15		15
7		20		20		20
8						
9	=SUM(B2:B7)	→ 50	@SUM(D2:D7)	→ 50	=SUM(F3:F7)	→ 50
10	=COUNT(B2:B7)	→ 5	@COUNT(D2:D7)	→ 6	=COUNT(F3:F7)	→ 4
11	=B9/B10	→ 10	=D9/D10	→ 8.333	=F9/F10	→ 12.5

Figure 3.22 Cautionary spreadsheet tables.

For example, never type numbers in as text. If you type in "6, it will look like the number 6, but the leading quotation mark makes it text. Excel will ignore it and Lotus and Quattro Pro will treat it as a zero. Never delete material from a cell with the space bar. A cell with nothing but a space *looks* blank but it is not. Use your spreadsheet's **Clear** command, or the **Delete** button. Also, avoid turning off the display of zeros, except perhaps as a temporary step in the creation of a presentation-quality table. Many spreadsheets allow you to do this and sometimes it is easier to see the important values in a table without all the zeros. (Recall, I actually did this in Figure 2.31 of the last chapter.) But be very careful.

Figure 3.22 displays some of the strange results you can get if you are careless. Cells B2:B7 contain five numeric values that average 10 and one missing case. Cells D2:D7 look identical, but D5 has a space, which some spreadsheets will treat as an additional zero. Cells F2:F7 look to have only four numbers instead of five because the zero is not displayed.

Finally, a note on when spreadsheet statistical functions will not help. All the functions in Figure 3.21 require that you provide a range that includes *all the individual data values*. Hence, they will not work on data arranged in a frequency distribution. If you need to find summary statistics from a frequency distribution, you will need to do it yourself.

3.2.4 Summing Up

We have dealt with a lot of details in this section; a short recap may help assure that you have not lost the big picture. In describing a single numeric variable, we will be interested in its average (or central tendency), its scatter (or variability), and perhaps its skewness. The mean, median, and mode all represent some sort of average, with the mean being the most important for inference. The range and standard deviation (along with its related values, the variance, and coefficient of variation) measure variability, with the standard deviation being the more important for inference. And along the way, we saw that the difference between the mean and median represents a measure of skewness, though we will not be doing inference with regard to skewness.

Generally we will be interested in the population values of the mean and standard deviation, μ_X and σ_X. However, generally, we will have only a sample from the population; we will be able to calculate only the sample mean and standard deviation, $\bar{X}$ and s_X. And we will do so in the way that makes them the best possible estimates of μ_X and σ_X.

3.3 Measures of a Single Categorical Variable

In the last section, we calculated means and standard deviations, among other things, for the Age and Weight of our sample of 281 young adults. We did not calculate these measures for Female. Since the two sexes are represented with 0 and 1, we could have. But these numbers are just arbitrary category names, with no more numeric meaning than M and F. Hence their mean and standard deviation would have had no numeric meaning either.

With categorical data, the best we can do to summarize a variable is to calculate the **proportions** in the various categories. Figure 3.23 presents the formulas. But, of course, this is exactly what we did in calculating relative frequencies in the last chapter. There is nothing new here except the symbols.

The symbol for the *population* proportion is π_{X_0} ("pi sub *X* naught"). The X_0 represents a particular value (category) of *X*, and f_{X_0} represents the frequency of that value in the population. And, as always, N_X is the number, or count, of the population. The symbol for the *sample* proportion is p_{X_0}. The X_0 again represents a particular value (category) of *X*, and f_{X_0} represents the frequency of that value in the sample. As always, n_X is the number, or count, of the sample.

Figure 3.24 shows the calculations for the proportion of males and females. It is straightforward. However, a couple of minor points may be worth making. In Chapter 2, every time I turned a relative frequency distribution into a presentation-quality table or graph, I converted the proportions (which sum to 1) to percentages (which sum to 100). That was because we were concerned at that point with communicating as clearly and intuitively as possible to someone else. It is my sense that most consumers of statistical information are more comfortable with percentages. However when, for the moment, presentation of our results is not our main concern, we will work with proportions.

The proportion of a population

$$\pi_{X_0} = \frac{f_{X_0}}{N_X}$$

The proportion of a sample

$$p_{X_0} = \frac{f_{X_0}}{n_X}$$

Figure 3.23 The proportion.

	G	H	I	J
100		Ordinary	Relative	
101	Female	Frequency	Frequency	
102	0	135	0.480427	← =H102/$H$105
103	1	146	0.519573	
104		0		
105		281	← =SUM(H102:H103)	
106				

Figure 3.24 Finding a proportion.

Finally, you may have discovered that using the =AVERAGE(:) function on female actually gives you the correct proportion of females in the sample. This works only for a dummy variable, coded 1 for yes and 0 for no. In this case, females each add one to the sum while males do not. So, in this case, the sum, which is the numerator of the mean, is the same as the frequency of females, which is the numerator of the proportion. This convenient result represents another reason for coding a variable this way. Go ahead and make use of it. But do not think of your answer as a mean, even though you used =AVERAGE(:) to get it. It is really a proportion.

3.4 Measures of a Relationship

So far, we have described variables—their average, variability, and the like—in isolation. However, as emphasized repeatedly in the first two chapters, typically we are interested in explaining, or accounting for, the variation in something. And that means looking for relationships. The following section mirrors the material in Section 2.2, in which we were also looking for relationships. Indeed, I will use some of the same examples. This time, though, we are not looking to create intuitive tables and graphs. This time we are looking to create quantitative measures.

3.4.1 Categorical Explanatory Variables

3.4.1.1 Comparing Proportions

Return to the data in file Students1.xls on a sample of 50 undergraduate students. In Chapter 2, we tried to use sex to help explain major. To do so, we sorted students by sex, and made a frequency distribution of major for each. Figure 3.25 reproduces the results. There are actually two sorts of questions in which we might be interested. We might be interested in explaining the choice of a particular major—Natural Science (1), for example. Or we might be interested in explaining the whole distribution of majors.

Addressing the first question is quite straightforward. Sex cannot help explain the choice of a Natural Science major unless the population proportions of each sex choosing Natural Science are different. That is, $\pi_{NS\text{-Female}} - \pi_{NS\text{-Male}}$ cannot equal zero. Now, we do not know the population proportions. However, looking at our sample, the sample proportion of females choosing Natural Science is $p_{NS\text{-Female}} = 3/30 = 0.10$, while

	F	G	H	I	J	K
30		Females			Males	
31		Ordinary	Relative		Ordinary	Relative
32	Major	Frequency	Frequency	Major	Frequency	Frequency
33	→1	3	→0.1	→1	5	→0.25
34	2	5	0.1667	2	4	0.2
35	3	7	0.2333	3	2	0.1
36	4	5	0.1667	4	2	0.1
37	5	4	0.1333	5	6	0.3
38	6	6	0.2	6	1	0.05
39		0			0	
40		30	1		20	0

Figure 3.25 Comparison of proportions of majors by sex (a).

the sample proportion of males choosing Natural Science is $p_{NS\text{-Male}} = 5/20 = 0.25$. The sample proportions are different. Indeed, $p_{NS\text{-Female}} - p_{NS\text{-Male}} = -0.15$ is our best estimate of $\pi_{NS\text{-Female}} - \pi_{NS\text{-Male}}$.

Unfortunately, the fact that our best *estimate* of $\pi_{NS\text{-Female}} - \pi_{NS\text{-Male}}$, is –0.15 does not mean that it is correct. The true population value could still be zero. So the question becomes, how likely are we to get a *sample* difference as big as –0.15, if the true *population* difference is zero? You will learn, when we get to inference, how to make this calculation. Then, if you find it sufficiently unlikely, you will infer that the true value is not zero, that the two true proportions are different, and that sex does help explain the choice of a Natural Science major.

Now suppose, instead of being interested in just one major, we are interested in the whole distribution of majors. The basic idea is the same, though the calculations are different. Again, sex cannot help explain the choice of major unless the population distributions across majors are different between the sexes. Now, again, we do not know the true population distributions. However, looking at our sample, we do see differences in the sample distributions between the sexes.

How do we begin to represent these differences quantitatively? We need, first, to figure out what each frequency would have been if there were no differences. Figure 3.26 illustrates. Notice that 60% of the sample is female. If there were no differences between the sexes, then, 60% of the Natural Science students would be female, 60% of the Social Science students would be female, and so on. With eight Natural Science students, $(0.6 \times 8) = 4.8$ would be female and $(0.4 \times 8) = 3.2$ would be male. With nine Social Science students, $(0.6 \times 9) = 5.4$ would be female, $(0.4 \times 9) = 3.6$ would be male, and so on.

Our actual frequencies differ from these expected frequencies. Of course, some deviation was inevitable; actual frequencies have to be integers. The true population distributions could still be the same. So, again, the question becomes, how likely are we to get actual *sample* frequencies this different from the expected ones if the true *population* distributions are the same. When we get to inference, you will learn how to make this calculation. Then, again, if you find it sufficiently unlikely, you will infer that the true distributions are different and sex does help explain the choice of major.

	F	G	H	I	J	K	L	M
45		Actual Frequencies				Expected Frequencies		
46		Female	Male	Total		Female	Male	Total
47	Natural Science	3	5	8	=G$54 × $I47	→ 4.8	3.2	8
48	Social Science	5	4	9		5.4	3.6	9
49	Humanities	7	2	9		5.4	3.6	9
50	Fine Arts	5	2	7		4.2	2.8	7
51	Business	4	6	10		6.0	4.0	10
52	Nursing	6	1	7		4.2	2.8	7
53	Total	30	20	50		30.0	20.0	50
54	=G53/$I$53	→ 0.6	0.4					

Figure 3.26 Comparison of proportions of majors by sex (b).

3.4.1.2 Comparing Means

In Chapter 2, we also tried to use sex to help explain weight, using our sample of 281 young adults. To do so, we sorted the individuals by sex, and made frequency distributions of weight for each. Using tables, histograms, frequency polygons, and scattergrams, we found visual support for the proposition that males tend to weigh more. To actually test this proposition, though, we need to compare their mean weights.

The basic idea remains the same as in the previous examples on proportions. Sex cannot help explain an individual's weight unless the distribution of weights in the population differ between the sexes—a difference that should be reflected as a difference in their means. That is, $\mu_{\text{Weight–Male}} - \mu_{\text{Weight–Female}}$ should not equal zero. Again, we do not know the population means. However, Figure 3.27 gives the values for our sample. The sample mean weight for males is $\bar{X}_{\text{Weight–Male}} = 169.0148$, while the sample mean weight for females is $\bar{X}_{\text{Weight–Female}} = 134.5205$. The sample means are different. Indeed, $\bar{X}_{\text{Weight–Male}} - \bar{X}_{\text{Weight–Female}} = 34.4943$ is our best estimate of $\mu_{\text{Weight–Male}} - \mu_{\text{Weight–Female}}$.

Again, though, the fact that our best estimate of $\mu_{\text{Weight–Male}} - \mu_{\text{Weight–Female}}$. is 34.4943 does not mean that it is correct. The true population value could still be zero. So the question becomes, how likely are we to get a *sample* difference as big as 34.4943 if the true *population* difference is zero? You will learn, when we get to inference, how to make this calculation. Then, if you find it sufficiently unlikely, you will infer that the true value is not zero, that the two true means are different, and that sex does help explain some of the variation in weights.

Perhaps this example provides a good opportunity to look a little further at what it means to say that sex helps explains some of the variation in weights. Suppose you needed to predict an individual's weight without knowing anything about him or her. Your best guess would be the sample mean, 151.0925. You can think of the sample standard deviation, 30.5442, as a measure of how far off you could be. Now, suppose you found out that the individual was male. You would revise your prediction up to 169.0148. The sample standard deviation, 26.2150, would be smaller as a result. Some of the variation in weights has been eliminated.

	G	H	I
101		Mean	Standard Deviation
102	Total Sample	151.0925	30.5442
103	Males	169.0148	26.2150
104	Females	134.5205	24.3081
105	Difference	34.4943	

Figure 3.27 Comparison of mean weights by sex.

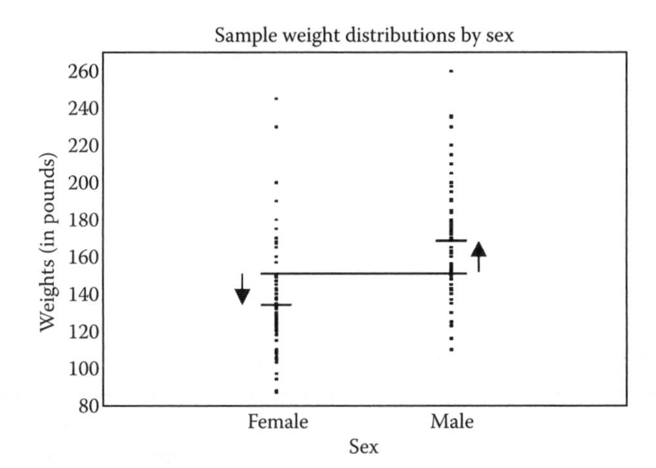

Figure 3.28 Weight distributions by sex.

Figure 3.28 shows the idea graphically. It is just the scattergram from Figure 2.27, with the means drawn in. If you do not know an individual's sex, you must combine the scatters for females and males into one, ranging from 87 for the lightest female to 260 for the heaviest male, with a mean of 151.0925. As soon as you know the individual's sex, the relevant mean shifts—down for females and up for males—and the relevant scatter shrinks.

3.4.2 Continuous Explanatory Variables*

As we saw in the last chapter, when the explanatory variable is continuous, we no longer have the natural groupings that we have with a categorical explanatory variable. While we could group values of the explanatory variable, doing so distorts the variable unnecessarily. It is better to stick with the original values. Graphically, this means creating a scattergram. The question we face now is how to turn the relationship we see in a scattergram into a quantitative measure.

Figure 3.29 shows a scattergram for a small sample of two variables. If you did not know the value of X, you could do no better than predict $\overline{Y}$ for each case. This is the horizontal line in the figure. It is equivalent, in the last example, to not knowing the individual's sex and thus having to predict the overall average weight of 151.0925.

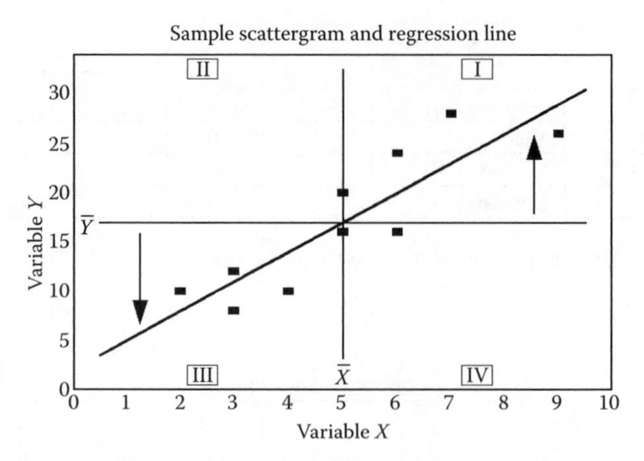

Figure 3.29 The scattergram and the sample regression line.

Now, though, suppose you know the value of X. It certainly appears that higher values of X should lead you to predict higher values of Y, and lower values of X should lead you to predict lower values of Y. This would be equivalent to what we did in the last example, when we adjusted our prediction of weight up or down once we knew the individual's sex. The upward sloping **sample regression line** in the figure gives the best prediction for Y, given X.

How do we find the equation for this line? Any straight line can be represented as $\hat{Y}=a+bX$, where a is the Y intercept, and b is the slope. $\hat{Y}$ ("y hat") is the value of Y predicted by the equation, as opposed to the actual data value, Y. Clearly, we want the prediction errors—the $Y-\hat{Y}$—s to be small. But we cannot simply minimize the sum of these distances since, as always, the positive and negative values would cancel out. So, as you should expect by now, we square these errors and minimize the sum of the squared errors, $\Sigma(Y-\hat{Y})^2$. Figure 3.30 gives the formulas and Figure 3.31 shows the spreadsheet calculations.

Columns B and C in Figure 3.31 give the data. That is, the first point is three across and eight up. And there are 10 such points. The means are calculated in both directions. The mean in the X direction—$\bar{X}$—is 5; the mean in the Y direction—$\bar{Y}$—is 17. On the graph (Figure 3.29) these are the vertical and horizontal lines. Columns D and E find the deviations from the mean in each direction. Of course, as always, these deviations in each direction sum to zero, because positive and negative deviations from the mean always just cancel out. To prevent this we do exactly what we did for the variance—*square* the deviations. The calculations shown in columns F and G are exactly like those for the variance. Indeed, if I had divided SSD_X by $(n-1)$, I would have found the variance of X. If I had divided SSD_Y by $(n-1)$, I would have found the variance of Y.

But now we want to tie together the variation in the two variables. We can do this by multiplying $(X-\bar{X})$ by $(Y-\bar{Y})$ for each point. This is done in column H. The sum of this column is "the Sum of Cross Deviations from the means" for the pair of variables, X and Y. The slope of the line, then, is just $SCD_{XY}/SSD_X = 120/40 = 3$.

$$\hat{Y} = a + bX$$

where the slope is $$b = \frac{\sum (X - \bar{X})(Y - \bar{Y})}{\sum (X - \bar{X})^2} = \frac{SCD_{XY}}{SSD_X}$$

and the intercept is $$a = \bar{Y} - b\bar{X}$$

Figure 3.30 The Formulas for the slope and intercept.

	A	B	C	D	E	F	G	H
1	Case	X	Y	$X-\bar{X}$	$(Y-\bar{Y})$	$(X-\bar{X})^2$	$(Y-\bar{Y})^2$	$(X-\bar{X})(Y-\bar{Y})$
2	1	3	8	−2	−9	4	81	18
3	2	5	20	0	3	0	9	0
4	3	9	26	4	9	16	81	36
5	4	6	24	1	7	1	49	7
6	5	3	12	−2	−5	4	25	10
7	6	2	10	−3	−7	9	49	21
8	7	5	16	0	−1	0	1	0
9	8	4	10	−1	−7	1	49	7
10	9	7	28	2	11	4	121	22
11	10	6	16	1	−1	1	1	-1
12								
13	Means	→ 5	17		Sums →	40	466	120
14						↑	↑	↑
15	Slope	→ 3	← =H12/F12			SSD_X	SSD_Y	SCD_{XY}
16	Intercept	→ 2	← =C12–B14 × B12					

Figure 3.31 Calculating the sample regression line.

Unlike the numbers in columns F and G, the numbers in column H are not squares and do not need to be positive. In this example, most of the points are in quadrants I and III of the graph. That is, most of the points that are above $\bar{X}$ in the X direction are also above $\bar{Y}$ in the Y direction, and most of the points that are below $\bar{X}$ in the X direction are also below $\bar{Y}$ in the Y direction. This causes the cross-products to be positive. Their sum, SCD_{XY}, is positive, and when we calculate the slope, SCD_{XY}/SSD_X, it is positive as well. We have an upward sloping line.

But imagine a different example in which the points had been in quadrants II and IV. The positive deviations would have been paired with negative deviations and the cross-products would have been negative. Their sum, SCD_{XY}, would have been negative. And when we calculated the slope, SCD_{XY}/SSD_X, it would have been negative as well. We would have had a downward sloping line.

Notice on the graph (Figure 3.29) that the line goes through the point $(\bar{X}, \bar{Y})$. This is not a coincidence; the point $(\bar{X}, \bar{Y})$ is always on the line. Thus, we can write $\bar{Y} = a + b\bar{X}$ and, since we know $\bar{X}$, $\bar{Y}$, and b, we can solve for a. In this example, $a=\overline{Y}-b\overline{X}=17-3\times5=2$ So, in this example, the sample regression equation is $\hat{Y}=2+3X$ For an x_i of 2, we would predict a y_i of $2+3\times2=8$. For an x_i of 8, we would predict a y_i of $2+3\times8=26$.

The basic idea of inference remains the same here as in all the previous examples. The X cannot really help explain Y unless the population slope, β, is nonzero. And we do not know the population slope. We have calculated the sample slope, b, in such a way as to make it the best possible estimate of β. However, the fact that it is our best possible estimate does not mean that it is correct. The true population value could still be zero. So the question becomes, how likely are we to get a *sample* slope as big as 3 if the true *population* slope is zero. You will learn, when we get to inference, how to make this calculation. Then, if you find it sufficiently unlikely, you will infer that the true value is not zero, and that X does help explain some of the variation in Y.

3.5 Exercises

3.1 Return to Studentsl.xls. In Exercise 2.1, you calculated frequency distributions for the sample of 50 student GPAs.

a. Working from the original data, and using no special functions except = SUM, = COUNT, and = SQRT, calculate the mean, median, mode (if it exists), range, variance, standard deviation, and coefficient of variation for GPAs.

b. Check your answers in part a making full use of special functions.

c. Working from your frequency distribution, estimate the mean and standard deviation and compare your answers with those in parts a and b.

3.2 Continue with these same data.

a. Using the empirical rule and your answers from the previous question, what interval should contain 68% of the GPAs? 95% of the GPAs? 100% of the GPAs?

b. Evaluate the empirical rule in this case. What proportion of the GPAs actually do fall in each of these intervals?

3.3 Continue with these same data. In Exercise 2.2, you broke down the GPA data by sex, to see if there seemed to be any difference in their distributions.

a. Based on those frequency distributions, which sex do you think has the greater mean GPA? The greater standard deviation in GPA?

b. Working from the original data, and using no special functions except = SUM, = COUNT, and = SQRT, calculate the mean, median, mode (if it exists), range, variance, standard deviation, and coefficient of variation for the GPA of each sex separately.

c. Check your answers in part b making full use of special functions.

d. Working from your frequency distributions, estimate the mean and standard deviation and compare your answers with those in parts b and c.

e. Do your results in parts b through d agree with your expectations in part a?

3.4 Return to NLSY1.xls.

a. Find the following statistics for each variable: n, sum, $\bar{x}$, maximum, Md, minimum, s^2, s, cv, and range. Make full use of special functions.

b. For each variable, calculate the range that should include 95% of all values, according to the empirical rule for bell-shaped data. For each, find the percentage of values that actually do fall in this range. How good is the rule? Is it really better for the variables with bell-shaped distributions?

3.5 Continue with these same data. In Exercise 2.3, using NLSY1.xls, you broke down the Height data by sex, to see whether females were actually shorter.

a. Break down the data by sex and find the mean and standard deviation for the Height of each sex separately.

b. Does taking sex into account help explain any of the variation in Height? Explain.

3.6 Return to Employees1.xls.

a. Find the following statistics for each variable: n, sum, $\bar{x}$, maximum, Md, minimum, s^2, s, cv, and range. Make full use of special functions.

b. For each variable, calculate the range that should include 95% of all values, according to the empirical rule for bell-shaped data. For each, find the percentage of values that actually do fall in this range. How good is the rule? Is it really better for the variables with bell-shaped distributions?

3.7 Continue with these same data. In Exercise 2.5, you used graphs to try to explain variations in Salary.

a. Break down the data by sex and find the mean and standard deviation for the Salary of each sex separately.

b. Does taking sex into account help explain any of the variation in Salary? Explain.

3.8 Continue with these same data.

a. Calculate the regression line relating Salary to Education. Do individuals with more education seem to earn more?

b. Calculate the regression line relating Salary to Experience. Do individuals with more experience seem to earn more?

3.9 Return to Students2.xls.

a. Working from the original data, and using no special functions except =SUM, =COUNT, and =SQRT, calculate the mean, median, mode (if it exists), range, variance, standard deviation, and coefficient of variation for Height.

b. Check your answers in part a making full use of special functions.

c. Calculate the range that should include 95% of all values, according to the empirical rule for bell-shaped data. Find the percentage of values that actually do fall in this range. How good is the rule?

3.10 Continue with these same data.
a. Repeat Exercise 3.9 for Weight.
b. Repeat Exercise 3.9 for Entertain.
c. Repeat Exercise 3.9 for Study.
d. Repeat Exercise 3.9 for Col_GPA.

3.11 Continue with these same data. In Exercise 2.6, you broke down Height by sex, to see if there seemed to be a difference in their distributions.
a. Based on those frequency distributions, which sex do you think has the greater mean Height? The greater standard deviation in Height?
b. Find the following statistics for Height for each group: n, sum, $\bar{x}$, maximum, Md, minimum, s^2, s, cv, and range. Make full use of special functions.
c. Working from your frequency distributions, estimate the mean and standard deviation and compare your answers with those in part b.
d. Do your results in parts b and c agree with your expectations in part a?

3.12 Continue with these same data.
a. Repeat Exercise 3.11 for Weight.
b. Repeat Exercise 3.11 for Entertain.
c. Repeat Exercise 3.11 for Study.
d. Repeat Exercise 3.11 for Col_GPA.

3.13 Continue with these same data. In Exercise 2.7, you broke down Aid by Sports, to see if there seemed to be a difference in the financial aid distributions for athletes and others.
a. Based on those frequency distributions, would you expect athletes to have the higher or lower mean Aid? The higher or lower standard deviation in Aid?
b. Find the following statistics for Aid for each group: n, sum, x–, maximum, Md, minimum, s2, s, cv, and range. Make full use of special functions.
c. Working from your frequency distributions, estimate the mean and standard deviation and compare your answers with those in part b.
d. Do your results in parts b and c agree with your expectations in part a?

3.14 Continue with these same data.
a. Repeat Exercise 3.13 for Entertain.
b. Repeat Exercise 3.13 for Study.
c. Repeat Exercise 3.13 for Col_GPA.

4

Basic Probability

4.1 Why Probability?

In the last chapter, we calculated a variety of sample statistics with an eye toward inferring something about unknown population parameters. For example, we used our sample of 50 students to explore whether sex might help explain the choice of a Natural Science major. For sex to help explain the choice of a Natural Science major, the underlying population proportions of females and males choosing such a major $-\pi_{NS\text{-Female}}$ and $\pi_{NS\text{-Male}}-$would need to differ. That is, the difference $\pi_{NS\text{-Female}}-\pi_{NS\text{-Male}}$ could not be zero.

Of course, we could not find the underlying population difference; we had just a sample. But using our sample, we found a difference in sample proportions of $p_{NS-\text{Female}}-p_{NS-\text{Male}}=-0.15$. The key question, which we had to defer, was how likely we were to have found a sample difference this large if the underlying population difference were actually zero. This likelihood—this probability—is critical. It is the probability of being wrong if we conclude, based on our sample, that the underlying population difference is not zero. If this probability of being wrong is low enough, we will take the chance. We will conclude that the population difference is not zero and that sex can help explain the choice of a Natural Science major. If this probability of being wrong is not low enough—if a sample difference this large could easily come from a population for which the difference were zero—we will be unwilling to make this leap.

Probability, then, is critical to moving from description to inference. The next three chapters will explore probability. This chapter introduces the subject. Chapter 5 introduces two important probability distributions—the binomial and the normal. Chapter 6 applies these probability distributions to questions like the one above.

4.2 The Basics

4.2.1 Experiments and Events

We start with some basic definitions. An **experiment** is a process that can lead to more than one possible outcome or **simple event**. A **compound event** is a combination of simple events. The complete set of possible simple events is the **sample space**.

The roll of a single die is an example of an experiment; the possible simple events making up the sample space are {1, 2, 3, 4, 5, 6}. Getting an even number would be a compound event, since it is made up of the simple events {2, 4, 6}.

4.2.2 Discrete versus Continuous Probabilities

The roll of a single die is an example of an experiment for which the possible simple events are **discrete** and countable. There are only six possible simple events. It is not possible, for example, to roll a 3.5. A different experiment might be to select someone at random and measure his or her height. In this example, the possible events are **continuous**. Indeed, there are an infinite number of possible heights. In this chapter, we will deal just with discrete probabilities; we will begin our consideration of continuous ones in Chapter 5.

4.2.3 The Probability of an Event

The probability of an event is a quantitative measure of the event's likeliness. It has the following characteristics:

- An event that cannot occur—that is not part of the sample space—has a probability of zero.
- An event that can, but need not, occur has a probability between zero and one—with numbers closer to one indicating greater likeliness.
- An event that must occur—that includes the whole sample space—has a probability of one.

All probabilities, then, must be between zero and one, inclusive.

Where do these probabilities come from? There are several ways to think about this.

One notion is that probabilities are **relative frequencies** that apply to the population. Of course we cannot observe the whole population. But imagine rolling a balanced die over and over again, and recording the proportion of the time it comes up 1. We would expect, eventually, that the proportion of 1s would stabilize around 1/6th. Figure 4.1 shows a computer simulation of 1000 rolls of a fair die. The proportion of 1s gradually settles down in the vicinity of 1/6th.

I just said that "we would expect" the proportion of 1s to stabilize around 1/6th. Why? Since the die is balanced, we expect each side to

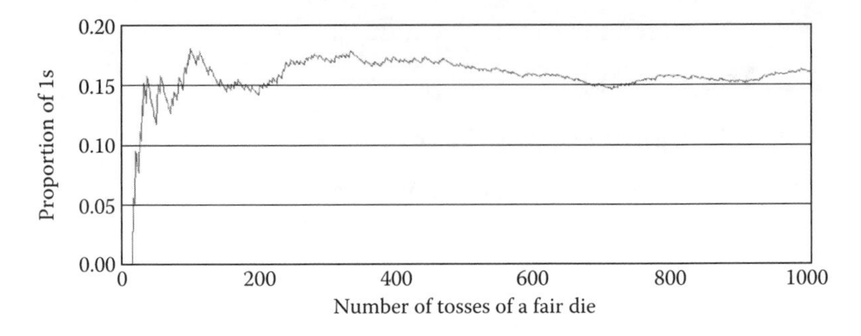

Figure 4.1 The proportion of 1s, in X rolls of a fair die.

come up an equal proportion of the time. And, since there are six sides, this means that each side should come up 1/6th of the time. This suggests another notion of probability based on **a priori** knowledge. With X different equally likely simple events, we conclude, a priori, that the probability of each is $1/X$.

Still another notion is **subjective**. Probabilities are the weights that generate "fair bets." Think of a fair bet as one for which you would be indifferent as to the side of the bet you took. Assume you are risk neutral; all you care about is expected value. In the die example, suppose you had to bet either 1 or not 1. Perhaps you would choose not 1 if the payoff for 1 were less than five times as great, and 1 if the payoff for 1 were more than five times as great. And perhaps you would be indifferent if the payoff for 1 were exactly five times as great. If so, you believe that not 1 is five times as likely. Your subjective probabilities are 1/6th for 1, and 5/6th for not 1.

Different people will have different subjective probabilities for the same situation because they differ in the facts they know and in their interpretations of these facts. This makes the subjective notion especially apt in many social science and business applications. One person may think that there is a better than 0.50 probability that a new product will be popular with consumers; another person may think otherwise. They disagree because they differ in what they know about the product and consumers, and because they interpret the facts differently.

In practice, these various probability notions are complementary. In the real world, we rarely look at repeated rolls of a die to decide whether the probability of a particular outcome is 1/6th. We look at the die and make a subjective judgment as to whether it looks balanced. If it does, we apply our a priori notion that—with six equally-likely outcomes—the probability of a particular outcome is 1/6th. In effect we assume that—had we actually rolled it many, many times—the results would have agreed with our frequency notion of probability. We assume that each outcome would have occurred 1/6th of the time. Now, though, suppose we begin rolling this die over and over and observe that some outcomes are occurring far more often than others. This result violates our a priori notion, and we begin to revise our subjective judgment that the die is balanced.

Or consider the example of deciding the probability that a new product will be popular with consumers. In this case, there is no obvious a priori probability. Moreover, it is not possible to repeat the same experiment over and over. Each new product is different. Consumers' tastes change. Still, even subjective probabilities have to come from somewhere. And often, they will represent attempts to extrapolate from admittedly different experiments. That is, individuals who have seen a higher frequency of success among products they judge to be similar will have a higher subjective probability of success for this one.

In the material to come, either you will be given information on the population from which you can generate an a priori probability or you will be given a subjective probability. As an example of the first, you may be told that a fair die has been rolled, or that one item has been chosen at random from a bin of 12 items. You will know that each simple event is equally likely and, with X simple events, the probability of each is $1/X$. As an example of the second, you may simply be told that there is thought to be a 0.05 chance that a part will be defective. And you will assume, for the purpose of the problem, that this probability is correct.

4.2.4 Complementary Events

In the die example above, we looked at the probability of 1 versus the probability of not 1. Not 1 was the **complement** of 1. More generally, if A is an event, A' (not A) is its complement, and includes all events within the sample space not included in A. And because, together, A and A' exactly cover the sample space, $P(A) + P(A') = 1$.

This result implies that if you know either $P(A)$ or $P(A')$ you actually know them both. You will have many opportunities to take advantage of this relationship. Sometimes, one of the two is easier than the other to calculate. Suppose, for example, you wanted to find the probability of finding at least 1 defective item in a sample of 10. There are all sorts of ways in which you could end up with at least one defective item; accounting for them all would be a real chore. But consider its complement. The complement of at least one defective item is zero defective items. And there is only one way of getting zero defective items; you have to get a nondefective item every time. So you would calculate the probability of getting zero defective items and then subtract that from 1 for the probability you actually want.

Figure 4.2 shows A and its complement in a **Venn diagram**. The rectangle represents the sample space. Circles within the sample space represent various events. Venn diagrams can be useful in clarifying probability relationships.

4.2.5 Conditional Probabilities

Suppose we roll a fair die once. Let A be the compound event that we get an even number; let B be the compound event that we get a number greater than three. A contains three—{2, 4, 6}—of the six simple events, so $P(A) = 3/6 = 0.50$. B also contains three—{4, 5, 6}—of the six simple

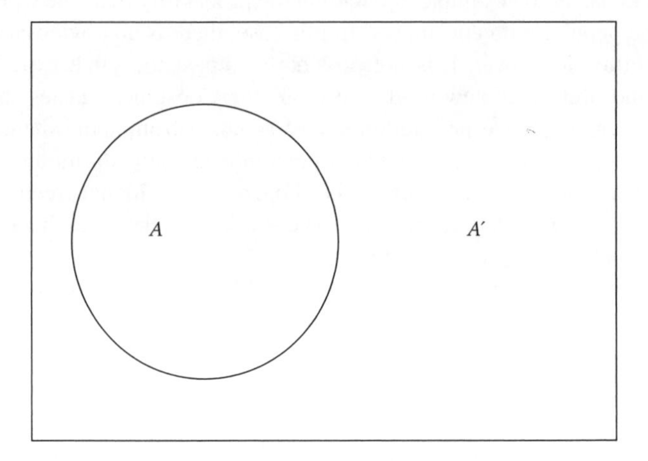

Figure 4.2 A Venn diagram showing A and its complement.

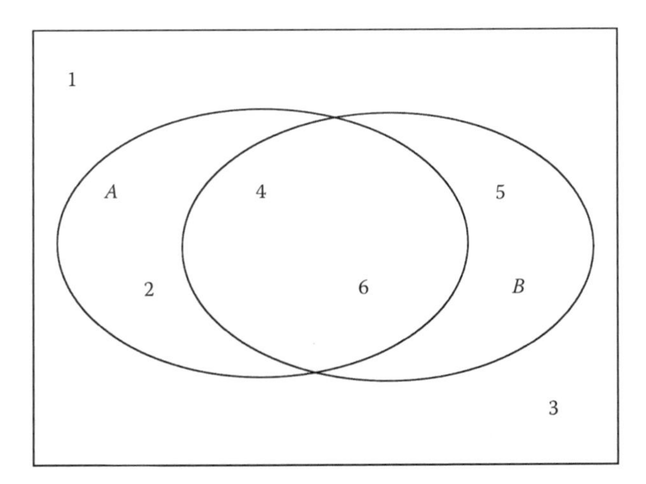

Figure 4.3 A Venn diagram showing *A* and *B*.

events, so $P(B) = 3/6 = 0.50$. These probabilities are sometimes called "marginal probabilities," for their position in two-way tables (Section 4.4.1).

But suppose you know that *B* is true. Now, what is the probability of *A*? In a simple example like this, we can figure it directly. The sample space has been reduced to {4, 5, 6} and, since two of these are even, $P(A|B) = 2/3 = 0.67$. The probability of "*A* given *B*" is 2/3.

This is an example of a **conditional probability**. The vertical line means "given." The *A* ahead of the line is the event for which you want the probability; the *B* after the line is the condition you are given as true.

Figure 4.3 shows the example above in a Venn diagram. Oval *A* contains the even numbers. Oval *B* contains the numbers greater than three. Since *A* contains three of the six numbers, $P(A) = 3/6 = 0.50$. For $P(A|B)$, though, *B* becomes the sample space; the only events to consider are {4, 5, 6}. And since *A* contains two of these three values, $P(A|B) = 2/3 = 0.67$.

4.2.6 Independent Events

In the example above, when we were told that *B* was true, the probability of *A* changed from 0.50 to 0.67. The probability of *A* was not independent of *B*. But that need not be the case. In many cases, knowing that one event has happened does nothing to change the probability of another event. Such events are known as **independent events**. Continuing the example above, let *C* be the compound event that we get a number greater than four. Figure 4.4 shows the Venn diagram. $P(A) = 3/6 = 0.50$ as before. Now, suppose we know that *C* is true. $P(A|C) = 1/2 = 0.50$. Knowing that *C* is true does not change the probability that *A* is true. $P(A|C) = P(A)$.

In the preceding examples, it was really not obvious ahead of time that events *A* and *C* were independent while *A* and *B* were not. We needed to calculate the relevant probabilities and find that $P(A|C) = P(A)$ while $P(A|B) \neq P(A)$. The more important cases are those we can recognize ahead of time, since recognizing them simplifies certain calculations a lot. In particular, for many interesting experiments the results of repeated trials are independent of each other. Thus, in repeated rolls of a

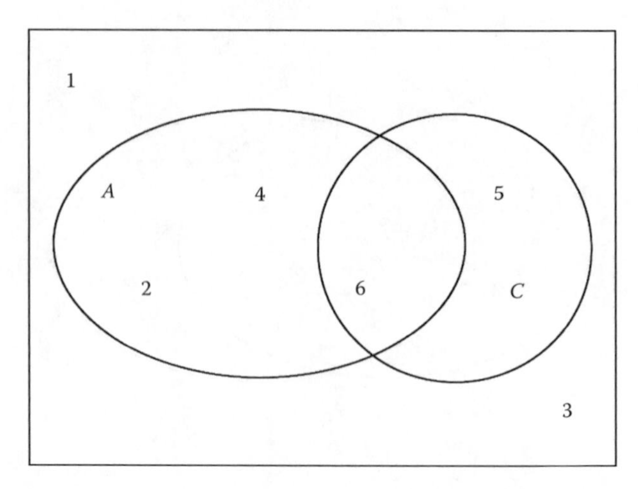

Figure 4.4 A Venn diagram showing *A* and *C*.

fair die, the probability of a 1 remains 1/6th each time. If E_1 and E_2 are the events that we get 1 on the 1st and 2nd rolls of a die, respectively, then the probability of E_2 does not depend on whether or not E_1 happened. $P(E_2|E_1) = P(E_2|E_1') = P(E_2) = 1/6$. While you may not be that interested in dice, many more important real-world experiments are at least roughly the same.

One must be careful though. Suppose, instead of the die above, we have a bin with six balls numbered 1 through 6. We are going to draw two and are interested in the probability of getting 1. If E_1 and E_2 are now the events that we get 1 on the 1st and 2nd draws, respectively, then the probability of E_2 does depend on whether or not E_1 happened. There are only five balls left on the second draw and if E_1 happened, the 1 is no longer among them, while if E_1 did not happen, it is $P(E_2|E_1) = 0/5$, $P(E_2|E_1') = 1/5$, and $P(E_2) = 1/6$. $P(E_2|E_1) \neq P(E_2|E_1') \neq P(E_2)$. A key in many probability problems is to distinguish those cases for which it is reasonable to assume independence from those cases for which it is not.

4.2.7 Mutually Exclusive Events

Recall complementary events from Section 4.2.4. Complementary events are a special case of **mutually exclusive events**. Clearly, A and A' cannot both be true; if one is true, the other is not. In the terms of Section 4.2.5, $P(A|A') = P(A'|A) = 0$. A and A' are mutually exclusive. Now, consider Figure 4.5, which continues the die example of the last two sections. As before, A contains {2, 4, 6}, so $P(A) = 3/6 = 0.50$. D contains just {3}, so $P(D) = 1/6 = 0.17$. However, A and D do not overlap; if one is true, the other is not. Again, in the terms of Section 4.2.5, $P(A|D) = P(D|A) = 0$. Again, A and D are mutually exclusive. The only difference is that, in this case, it is possible for neither to be true.

Mutually exclusive events are important special cases. Recognizing them simplifies certain calculations a lot. Suppose you are adding probabilities, for example. You cannot just add $P(A)$ and $P(C)$, in Figure 4.4,

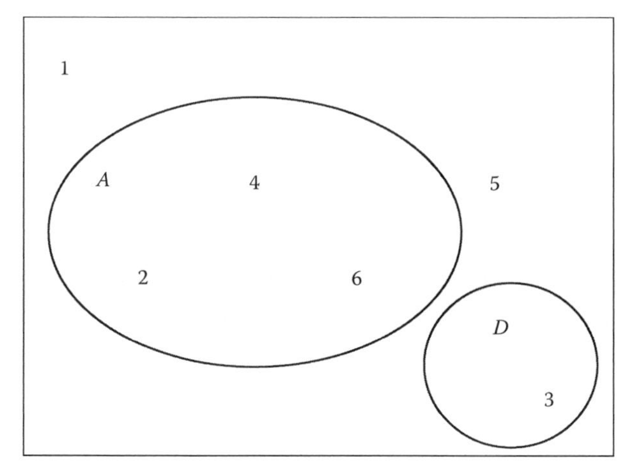

Figure 4.5 A Venn diagram showing *A* and *D*.

since you would be double counting the overlap. But you can just add P(A) and P(D) in Figure 4.5 because there is no overlap.

A common problem for beginners is to distinguish between independent and mutually exclusive events. While they may seem somewhat similar, they are two entirely different conditions. Think of independent events as unrelated events—your grade on your next exam and the price of tea in China, for example. One has no bearing on the other. Mutually exclusive events are related, albeit in a negative way. One precludes the other.

4.3 Computing Probabilities

It is fairly common to want to compute the probability of some event from the probabilities of related events. We have already dealt with one simple example—the probability of a complement. $P(A') = 1 - P(A)$. Another example would be wanting to know the probability of two things both being true. This would be the probability of an **intersection**—the overlap of the circles in a Venn diagram. Words like "both" and "and" should make you think intersection—and multiplication. Still another example would be wanting to know the probability of at least one of two things being true. This would be the probability of a **union**—the combined area of the circles in a Venn diagram. Words like "and/or" or just "or" (the "and" implied) should make you think union—and addition. Each has a general and a special case, as follows.

4.3.1 The Probability of an Intersection

4.3.1.1 The General Case

The probability of an intersection of two things is the probability of them both being true. Figure 4.6 first gives the formula for the general case in terms of *A* and *B*. To understand what this formula actually means and the intuition behind it, we will continue with the die example begun in

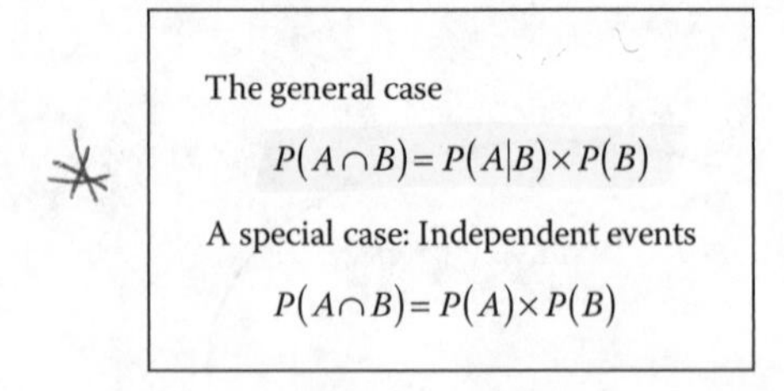

Figure 4.6 The probability of the intersection of *A* and *B*.

Section 4.2.5. Figure 4.7 just repeats Figure 4.3. The intersection of *A* and *B* is the overlap and, because this example is so simple, we can find its probability just by inspection. The intersection clearly contains two—{4, 6}—of the six simple events, so $P(A \cap B) = 2/6 = 0.33$. The correct answer is 0.33; this is the answer we should get using the formula as well.

Now, using the formula, *A* contains 2 of the 3 events in *B*, so $P(A|B) = 2/3 = 0.67$. And *B* contains three of the six simple events, so $P(B) = 3/6 = 0.50$. Thus, $P(A \cap B) = P(A|B) \times P(B) = 0.67 \times 0.50 = 0.33$. It works.

But what is the logic? What is the intuition? $P(A|B) = 0.67$ is the probability of choosing *A* given that we already have *B*. But, of course, we may not have *B*; indeed, there is only a 0.50 chance that we have *B*. Thus, we need to weight the 0.67 by 0.50. Thought of in this way, the formula makes sense.

Two other facts may be worth spelling out explicitly. First, *A* and *B* can be reversed. That is, $P(A|B) \times P(B)$ must equal $P(B|A) \times P(A)$. That is pretty easy to see in this particular example, since areas *A* and *B* are of equal size. What may not be as obvious, though, is that it is true in general. This fact is important since you will often know only one of the conditional probabilities.

Second, the equation can be rearranged to solve for the conditional probability. If you know both $P(A)$ and $P(A \cap B)$, you can calculate $P(A|B)$. Indeed, given any two of the three numbers, you can calculate the third. This fact is important since there is no telling what probabilities you will know and what ones you will want to know in a particular situation.

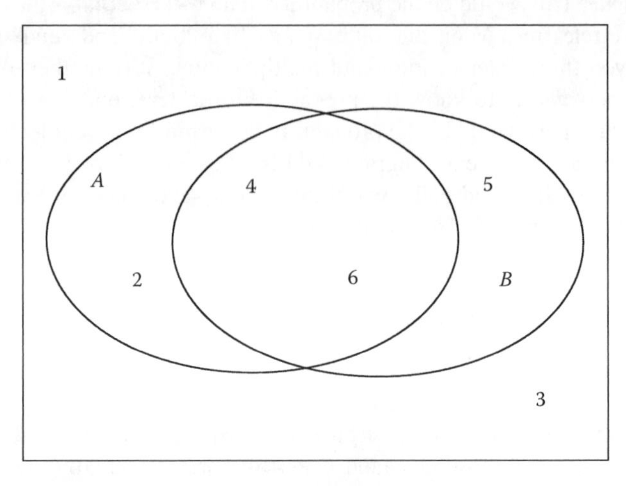

Figure 4.7 A Venn diagram showing *A* and *B*.

4.3.1.2 A Special Case: Independent Events

Figure 4.6 also gives a second formula for the special case in which we have independent events. Do not think of it as an additional formula that you need to learn. Recall that independent events are those for which $P(A|B) = P(A)$. So all this second formula really says is the obvious—you need not worry about whether B occurred when calculating the probability of A in those cases in which it does not affect the probability of A anyway.

As noted in Section 4.2.6, for many interesting experiments the results of repeated trials are independent of each other. For these, finding the probability of their intersection is a simple matter of multiplying their probabilities:

- In repeated rolls of a fair die, the probability of a 1 remains 1/6th each time. Therefore, the probability of getting two 1s in two tries is just $1/6 \times 1/6 = (1/6)^2 = 1/36$; the probability of getting five 1s in five tries is just $(1/6)^5 = 1/7776$.
- Or suppose you believe that a random 2% of the parts turned out by a particular machine are defective. The probability of drawing two parts and finding both defective is just $0.02 \times 0.02 = (0.02)^2 = 0.0004$.
- Or perhaps you have both a favorite pro football team and a favorite college football team. You have subjective probabilities of 0.60 and 0.80, respectively, that each will win its next game. The two games are unrelated, so it seems reasonable to assume independence. Thus, the probability that they will both win their next game is $0.60 \times 0.80 = 0.48$.

One must be careful though, considering the following:

- Suppose you have a bin of five parts, two of which are defective. If you draw two parts, what is the probability of them both being defective? If you assumed the two draws were independent, your answer would be $2/5 \times 2/5 = (2/5)^2 = 0.16$. But the two draws are not independent. The probability of getting a defective on the second draw depends on what you got on the first. The 2/5 is the probability on the first draw. But drawing that first defective one leaves only four parts, just one of which is defective, so 1/4 is the probability on the second draw. The probability of them both being defective is $2/5 \times 1/4 = 0.10$.
- Or suppose you are trying to sell your house and have two prospects, A and B. You estimate subjective probabilities of 0.50 that each will buy it. That is, $P(A) = P(B) = 0.50$. Does this mean that the probability of them both buying it is $(0.50)^2 = 0.25$? Of course not. They cannot both buy your house; these are mutually exclusive and not independent events, $P(A|B) = P(B|A) = 0$.

To summarize, when you are looking for the probability that two or more things are *all* true, you are looking for the probability of an *intersection*; and you are looking to *multiply* probabilities. If these things are *independent*—and in many interesting cases they are—you can simply multiply their probabilities. But independence is not always a reasonable assumption; you need to ask yourself whether it is appropriate in a

The general case

$$P(A \cup B) = P(A) + P(B) - P(A \cap B)$$

A special case: Mutually exclusive events

$$P(A \cup B) = P(A) + P(B)$$

Figure 4.8 The probability of the union of *A* and *B*.

particular situation. And if knowing that *B* is true changes the probability of *A*, you need to take that into account.

4.3.2 The Probability of a Union

4.3.2.1 The General Case

The probability of a union of two things is the probability of either or both being true. Figure 4.8 first gives the general formula in terms of *A* and *B*. To understand what this formula actually means and the intuition behind it, we will continue with Figure 4.7. The union is the area in either *A* or *B* and, because the example is so simple, we can find its probability just by inspection. The union contains four—{2, 4, 5, 6}—of the six simple events, so $P(A \cup B) = 4/6 = 0.67$. The correct answer is 0.67; this is the answer we should get using the formula as well.

Now, using the formula, *A* and *B* both contain three of the six events, so $P(A) = P(B) = 3/6 = 0.50$. And, as we calculated in the previous section, $P(A \cap B) = P(A|B) \times P(B) = 0.67 \times 0.50 = 0.33$. Thus, $P(A \cup B) = P(A) + P(B) - P(A \cup B) = 0.50 + 0.50 - 0.33 = 0.67$. It works.

But what is the logic? What is the intuition? We want the probability of being in either *A* or *B*, so adding their probabilities makes sense. But notice that in doing so we have added in the probability of their intersection twice. We want to include that probability, but not twice; hence, we subtract it out once. Thought of in this way, the formula makes sense.

4.3.2.2 A Special Case: Mutually Exclusive Events

Figure 4.8 also gives a second formula for the special case in which we have mutually exclusive events. Again, do not think of it as an additional formula that you need to learn. Recall that mutually exclusive events are those for which $P(A \cap B) = 0$. So, again, all this second formula really says is the obvious—you need not worry about subtracting out the probability of the intersection in those cases in which it is zero.

As noted in Section 4.2.7, for many interesting experiments, the possible outcomes are mutually exclusive. For these, finding the probability of their union is a simple matter of adding their probabilities:

- Suppose you have an unbalanced die, with probabilities for {1, 2, 3, 4, 5, 6} of 1/4, 1/6, 1/6, 1/6, 1/6, 1/12, and you want to know the probability of an even number. You can no longer just count and say that the probability is 3/6 since the outcomes are not equally likely. But since the outcomes are mutually exclusive and you know the probability of each you can simply add their probabilities. The probability of an even number is just $P(2) + P(4) + P(6) = 1/6 + 1/6 + 1/12 = 5/12 = 0.4167$.

- Or suppose you believe that a random 2% of the parts turned out by a particular machine are defective, and you want to know the probability that a sample of 10 will contain fewer than two defectives. "Fewer than 2" means zero or one. And zero and one are mutually exclusive; you cannot have both zero and one defectives in a single sample. So the probability of fewer than two defectives is just $P(0) + P(1)$.
- Or suppose you want to know the probability that the sample of 10 will contain 2 or more defectives. "2 or more" means 2, 3,…, 10 and, again, these are all mutually exclusive. So the probability of two or more defectives is just $P(2) + P(3) + \ldots + P(10)$. However, better yet, "2 or more" and "fewer than 2" are mutually exclusive; they are complements. So the probability of two or more defectives is just $1 - (P(0) + P(1))$.

Again, though, one must be careful, consider the following:

- Suppose that 50% of local households get the morning newspaper and 30% get the evening one. Can you conclude that 80% get one or the other? Probably not. These would not seem to be mutually exclusive events. There are probably some households that get both and you will have counted them twice.

To summarize, when you are looking for the probability that at least one out of two or more things is true, you are looking for the probability of a union; and you are looking to add probabilities. If these things are mutually exclusive—and in many interesting cases they are—you can simply add their probabilities. But mutual exclusive is not always a reasonable assumption; you need to ask yourself whether it is appropriate in a particular situation. And if both A and B can be true, you need to take that into account.

4.4 Some Tools That May Help

In principle, the equations of Figures 4.6 and 4.8 are about all you need to know in order to work most discrete probability problems. In practice, probability problems can get confusing, and aids have been developed that may help in organizing the information. The Venn diagram that we have already used is one. Others are **two-way tables** and **tree diagrams**.

4.4.1 Two-Way Tables

Figure 4.9 shows the layout of a two-way table and may suggest why it is useful. The columns represent A and A'; the rows represent B and B'. The cell where, for example, A and B cross holds the probability of their intersection. The probabilities of the intersections sum in both directions to the marginal probabilities. The marginal probabilities sum in both directions to one. Conditional probabilities do not show explicitly, but each can be found as just the probability of an intersection over the marginal probability. $P(A|B) = P(A \cap B)/P(B)$, for example.

	A		A'		
B	$P(A \cap B)$	+	$P(A' \cap B)$	=	$P(B)$
	+		+		+
B'	$P(A \cap B')$	+	$P(A' \cap B')$	=	$P(B')$
	‖		‖		‖
	$P(A)$	+	$P(A')$	=	1

Figure 4.9 A two-way probability table.

Consider the following example:

Suppose, among tourists visiting Washington, DC for the first time, 60% visit the Jefferson Memorial, 75% visit the Lincoln Memorial, and 80% of those visiting the Jefferson Memorial also visit the Lincoln Memorial. What is the probability that a randomly chosen, first-time tourist will visit both memorials? Neither memorial? Exactly one memorial?

It often helps to start by writing out what we know and what we want:

We know:	We want:
$P(J) = 0.60$	$P(L \cap J) = ??$
$P(L) = 0.75$	$P(L' \cap J') = ??$
$P(L\|J) = 0.80$	$P(L \cap J') + P(L' \cap J) = ??$

Beginners often have difficulty recognizing the difference between a conditional probability and the probability of an intersection. Note that the 0.80 is not the probability of a tourist visiting both memorials; it is the probability of the tourist visiting the Lincoln Memorial, given that he or she visits the Jefferson Memorial. Note also that there are two ways of visiting exactly one memorial.

Figure 4.10 steps through the process of filling in the two-way table. Since we know $P(J)$ and $P(L)$, we also know their complements. Hence, we can immediately fill in all the marginal probabilities.

Now we need to know the probability of one of the intersections. If visits to the two memorials were independent events, we could simply multiply marginal probabilities. You should be suspicious that they are not, though, and indeed they are not. $P(L|J) \neq P(L)$. Since we know $P(L|J)$ and $P(J)$, though, we can calculate $P(L \cap J) = P(L|J) \times P(J) = 0.80 \times 0.60 = 0.48$.

Panel A

	J	J'	
L			**0.75**
L'			**0.25**
	0.60	**0.40**	**1.00**

Panel B

	J	J'	
L	**0.48**		0.75
L'			0.25
	0.60	0.40	1.00

Panel C

	J	J'	
L	0.48	**0.27**	0.75
L'	**0.12**	**0.13**	0.25
	0.60	0.40	1.00

Figure 4.10 A two-way probability table: An example.

Finally, since the rows and columns all need to add to their marginal probabilities, we can calculate the probabilities of the three other intersections. The probability that the tourist will visit both memorials is 0.48; the probability that he or she will visit neither is 0.13; and the probability that he or she will visit exactly one is $0.27 + 0.12 = 0.39$.

A two-way table need not be 2×2. Consider the following example:

If a pair of fair dice is rolled once, what is the probability of getting a 7? 9? 11?

Figure 4.11 shows the two-way table. We know that dice have six sides and, if they are fair, each side has the same probability. Hence, the marginal probability of each outcome for each die must be 1/6. Moreover, the outcomes for the two dice should be independent—the outcome for one should not affect the outcome for the other—so the probabilities of each intersection should be just $1/6 \times 1/6 = 1/36$.

Note here that there are actually 36 ways that the pair can land, each with a probability of 1/36. However, some totals are more likely than others because they can occur in more ways. There are six different outcomes that total 7, so $P(7) = 6 \times 1/36 = 1/6$. There are just four different outcomes that total 9, so $P(9) = 4 \times 1/36 = 1/9$. And there are just two different outcomes that total 11, so $P(11) = 2 \times 1/36 = 1/18$.

Hopefully, two-way probability tables have reminded you of the two-way relative frequency tables back in Section 2.2.2. They are very similar. The difference is that, in two-way relative frequency tables, we were describing the tendencies of an observed sample—presumably in order to learn something about the population from which it came. In two-way probability tables, we are describing the tendencies of a theoretical population—presumably in order to learn about the probability of drawing particular samples.

4.4.2 Tree Diagrams

An alternative to a two-way table is a tree diagram. Figure 4.12 shows the general layout. Tree diagrams are especially useful in cases for which

Panel A

Die 2	Die 1						
	1	2	3	4	5	6	
1	**1/36**						**1/6**
2							**1/6**
3							**1/6**
4							**1/6**
5							**1/6**
6							**1/6**
	1/6	**1/6**	**1/6**	**1/6**	**1/6**	**1/6**	**1**

Panel B

Die 2	Die 1						
	1	2	3	4	5	6	
1	1/36	1/36	1/36	1/36	1/36	**1/36**	**1/6**
2	1/36	1/36	1/36	1/36	**1/36**	1/36	**1/6**
3	1/36	1/36	1/36	**1/36**	1/36	**1/36**	**1/6**
4	1/36	1/36	**1/36**	1/36	**1/36**	1/36	**1/6**
5	1/36	**1/36**	1/36	**1/36**	1/36	**1/36**	**1/6**
6	**1/36**	1/36	**1/36**	1/36	**1/36**	1/36	**1/6**
	1/6	**1/6**	**1/6**	**1/6**	**1/6**	**1/6**	**1**

Figure 4.11 A two-way table: The roll of two dice.

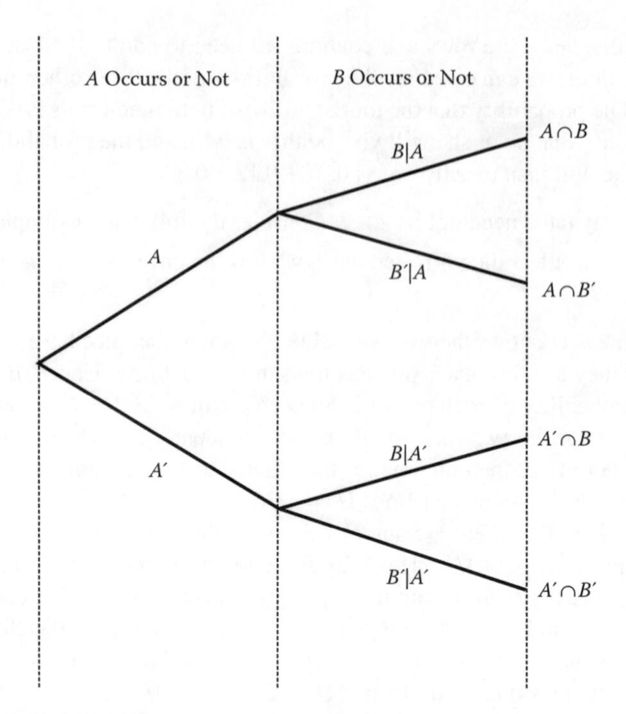

Figure 4.12 A tree diagram.

experiments are, or can be thought of as being, sequential. First A occurs or it does not. Then B occurs or it does not.

In a tree diagram, each event along a particular branch after the first is conditional on the previous ones. Each complete branch is the intersection of the events along it. Thus, in the top branch in Figure 4.12, $P(A)$ times $P(B|A)$ gives $P(A \cap B)$. Moreover, the branches are mutually exclusive. Hence, if there is more than one way to arrive at the outcome of interest, the probabilities of the separate branches can be added. The probabilities of all the branches sum to 1.

Consider the following example:

> Suppose your firm has two suppliers of a particular part, used in the assembly of your product. You get 60% of the parts from supplier A and the rest from supplier B. 2% of the parts from A and 5% of the parts from B are defective. If you select a part at random, what is the probability that it will be defective?

Again, we start by writing out what we know and what we want:

We know:	We want:
$P(A) = 0.60$	$P(D) = ??$
$P(B) = 0.40$	
$P(D|A) = 0.02$	
$P(D|B) = 0.05$	

Figure 4.13 shows the tree. In the first step the part is supplied by either A or B; we know the probabilities for these are 0.60 and 0.40, respectively, and in the second step the part is either defective or not; the probabilities

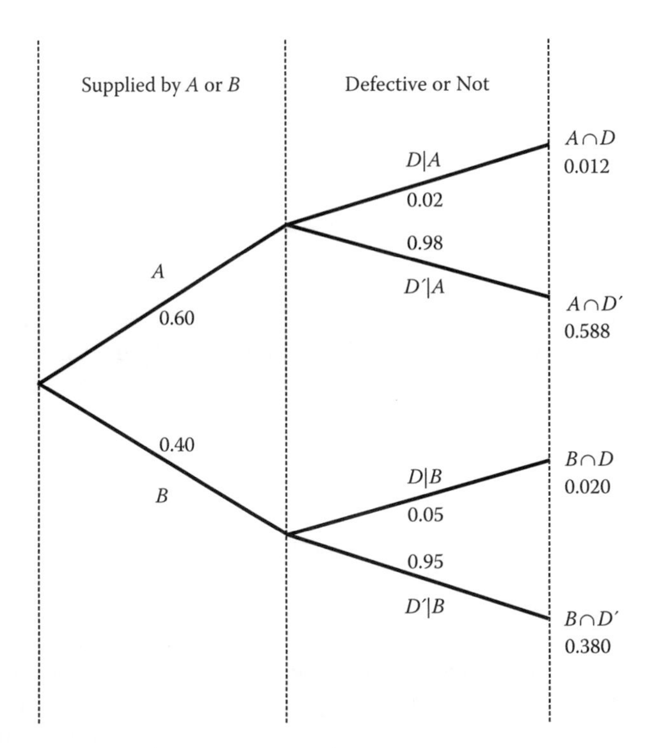

Figure 4.13 A tree diagram: An example.

of these are conditional on whether A or B happened. If A, they are 0.02 and 0.98; if B, they are 0.05 and 0.95.

Once you have assigned the probabilities to the branches, the rest is pretty mechanical. $P(A \cap D) = 0.60 \times 0.02 = 0.012$ and $P(B \cap D) = 0.40 \times 0.05 = 0.020$. These are the only branches that involve getting a defective and, since they are mutually exclusive—the part can not have come from both A and B—we can simply add $P(D) = 0.012 + 0.020 = 0.032$.

How did we know that the first step was A and B rather than D and D'? Because the probabilities of A and B are the ones we knew unconditionally, whereas we knew the probabilities of D and D' only conditionally on whether the supplier was A or B. Thus, A and B had to come first.

As with two-way tables, tree diagrams can show experiments with more than two possible outcomes, though this can make them unwieldy. A tree diagram of the two dice example in the last section would involve six branches for the first die, each with six branches for the second, or 36 branches in all. For just two experiments, each with multiple possible events, the two-way table is probably the more useful.

On the other hand, a tree diagram can easily handle additional experiments—C, D, and so on. This is an advantage of tree diagrams over two-way tables, for which three or more experiments—the roll of three or more dice instead of two, for example—would require three or more dimensions.

Finally, a tree diagram can have branches of varying lengths. Consider the following example:

> Suppose a sales representative calls on a customer to make a sale; if unsuccessful, she calls on him again a second time; if still

unsuccessful, she calls on him a third time before giving up. If her subjective probabilities of success on her first, second, and third calls are 0.30, 0.20, and 0.10, what is the probability that she will (at some point) make the sale.

Figure 4.14 shows the tree diagram. In the first step, she either succeeds or fails to make a sale. The second step occurs only if she fails on the first one; the third only if she fails on the second. I have simplified the notation. Remember, though, the probabilities for the second visit—0.20 and 0.80—must be conditional on her failing in her first visit. Thus, the probability of succeeding on the second visit must be weighted by the probability of actually making a second visit $P(F_1 \cap S_2) = P(F_1) \times P(S_2|F_1) = 0.70 \times 0.20 = 0.14$. And the probabilities for the third visit—10 and 0.90—must be conditional on her failing in her first two. Thus, the probability of succeeding on the third visit must be weighted by the probability of actually making a third visit. $P(F_1 \cap F_2 \cap S_3 = P(F_1) \times P(F_2|F_1) \times P(S_3|F_2) = 0.70 \times 0.80 \times 0.10 = 0.056$. As always with a tree diagram, the branches are mutually exclusive. Thus, the $P(S) = 0.30 + 0.14 + 0.056 = 0.496$.

Hopefully, you noted that there is actually an easier way of working this problem. The tree has only four branches, three leading to success and one not doing so. Thus, finding the complement—the probability of not succeeding—would have involved finding the probability along just one branch. $P(S') = P(F_1 \cap F_2 \cap F_3) = P(F_1) \times P(F_2|F_1) \times P(F_3|F_2) = 0.70 \times 0.80 \times 0.90 = 0.504$. And so, $P(S) = 1 - 0.504 = 0.496$.

Perhaps it is worth noting explicitly that the formulas above have gotten rather messy. The tree diagram, on the other hand, is still pretty intuitive. There are three, mutually exclusive branches that include a success. Just multiply the probabilities along each of those branches and add them up. Better yet, notice that there is just one branch that does not include a success. Just multiply the probabilities along this branch, and subtract from one. This is the virtue of the tree diagram.

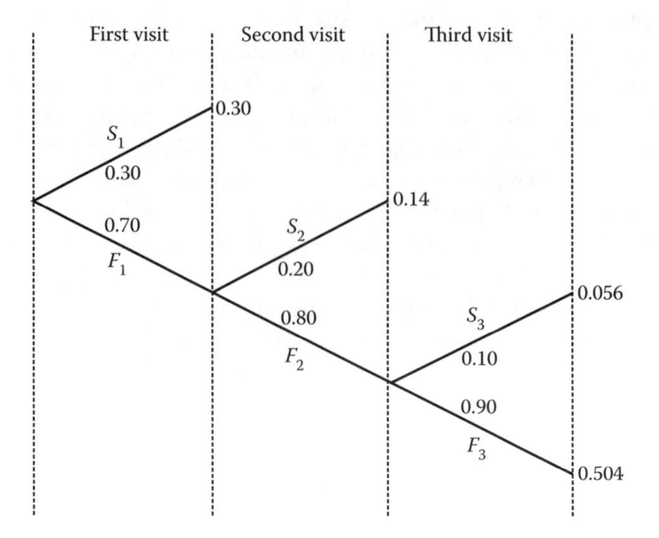

Figure 4.14 A tree diagram: Another example (a).

Finally, consider the following example:

> A fair die is rolled three times. What is the probability of getting 1 three times? twice? once? not at all?

Figure 4.15 shows the tree diagram. A tree diagram works fairly well in this case, because we are not interested in all the possible outcomes—just the 1s and not 1s. Since the die is fair, we know that $P(1) = 1/6$ and $P(1') = 5/6$ on each roll. Note that, unlike the last two examples, we have independent events; the probabilities on the second and third rolls are the same as on the first and the same no matter what has happened previously.

The all-or-nothing cases are the easiest to dispose of. The top branch is the only one that yields three 1s; its probability is just $(1/6)^3 = 1/216$. The bottom branch is the only one that yields no 1s at all; its probability is just $(5/6)^3 = 125/216$. The mixed cases are the more interesting, because there are multiple possible orders. There are three orders in which you can get two 1s; moreover, notice that all three of these orders have the same probability. This is because all three have 1/6 twice and 5/6 once. What this means is that, instead of calculating each one individually and adding to get the probability of two 1s, you can calculate just one—$(1/6)^2(5/6)^1$—and then multiply by three. The probability of two 1s is $3 \times (1/6)^2(5/6)^1 = 15/216$. Likewise, there are three orders in which you can get just a single 1, and they all have the same probability. So again, you can calculate just one—$(1/6)^1(5/6)^2$—and then multiply by three. The probability of just a single 1 is $3 \times (1/6)^1(5/6)^2 = 75/216$.

This is not a general result; it works here because we have independent events and, thus, a constant probability of success on each try. Still, as

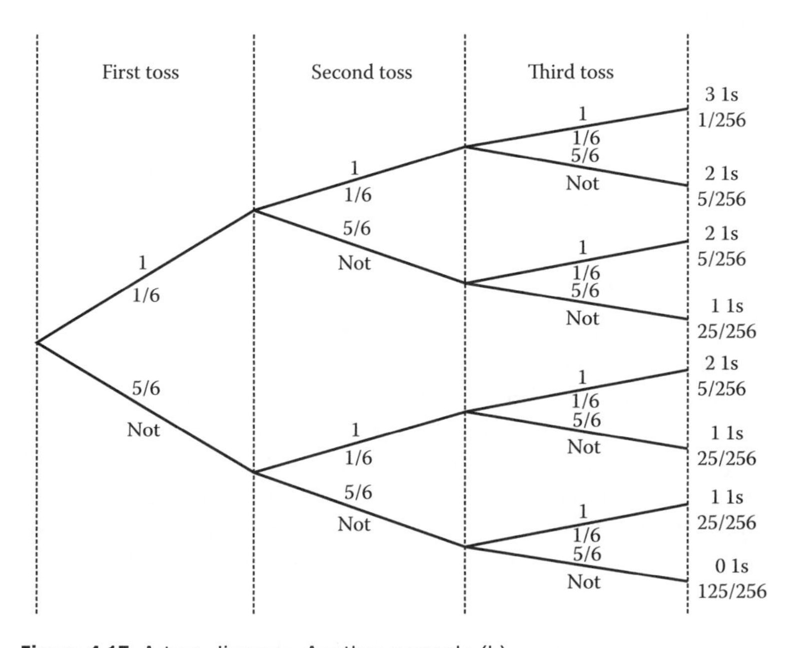

Figure 4.15 A tree diagram: Another example (b).

I have indicated, there are many situations for which independence is a good assumption. And for these, all orders that yield a certain number of successes have the same probability. If you know the probability of one order, and the number of orders, you can find the probability of that many successes by just multiplying. This result, which may not look very important now, will become so as we start to look at problems too big for you to want to draw a tree diagram. We will deal with such problems in Chapter 5.

4.5 Revising Probabilities with Bayes' Theorem

Consider the following example:

> Suppose your firm has two suppliers of a particular part used in the assembly of your product. You get 60% of the parts from supplier *A* and the rest from supplier *B*. 2% of the parts from *A* and 5% of the parts from *B* are defective. If you select a part at random and it is defective, what is the probability that it came from *A*? *B*?

You should recognize the basic scenario; we used it in the last section on tree diagrams. Back then, though, the problem was to find $P(D)$; this time it is to find $P(A|D)$ and $P(B|D)$.

Think of the problem in these terms. You had initial, or prior probabilities of a part coming from *A* or *B*. But now you are being given new information—information that the part in question is defective—that should change those probabilities. Just intuitively, since *B* supplies a larger percentage of defective parts, the probability that this part is from *B* should increase. But how much? **Bayes' Theorem** provides the answer.

Again, we start by writing out what we know and what we want:

We know:	We want:
$P(A) = 0.60$	$P(A\|D) = ??$
$P(B) = 0.40$	$P(B\|D) = ??$
$P(D\|A) = 0.02$	
$P(D\|B) = 0.05$	

Figure 4.16, which just repeats Figure 4.13, can help with the intuition. Previously, we found that $P(D) = P(A \cap D) + P(B \cap D)$, since these are the only ways of getting a defective. Hence, $P(D) = 0.012 + 0.020 = 0.032$.

Now, though, you are given that a part was drawn and was indeed defective, so you know that either $A \cap D$ or $B \cap D$ did in fact happen. And, since one or the other did in fact happen, you know that their probabilities need to add up to one. How do you make them add up to one? Take $P(D) = P(A \cap D) + P(B \cap D)$ and divide through by $P(D)$:

$$\frac{P(D)}{P(D)} = \frac{P(A \cap D)}{P(D)} + \frac{P(B \cap D)}{P(D)}.$$

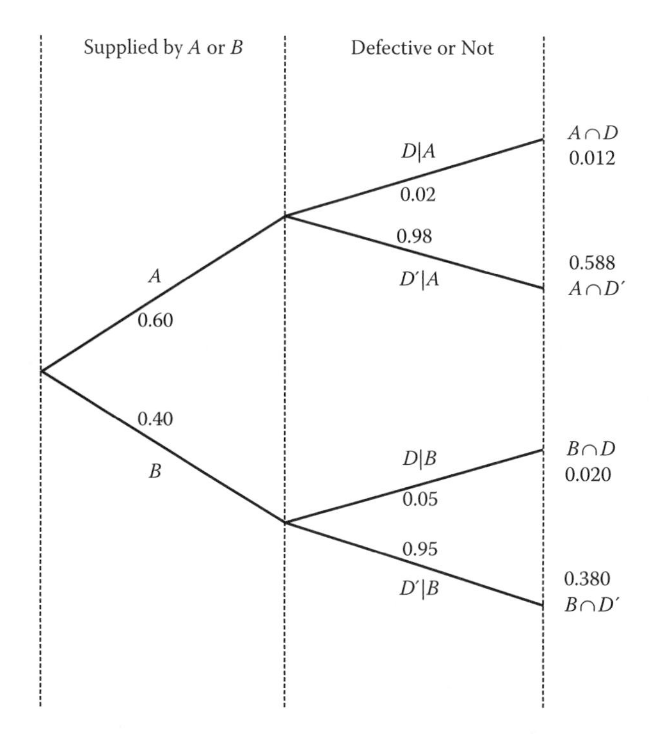

Figure 4.16 A tree diagram: An example.

Now, recognize that

$$P(A|D) = \frac{P(A \cap D)}{P(D)} = \frac{0.012}{0.032} = 0.375 \quad \text{and}$$

$$P(B|D) = \frac{P(B \cap D)}{P(D)} = \frac{0.020}{0.032} = 0.625,$$

by the formula for an intersection.

Note that $0.375 + 0.625 = 1$, as it must, since the part must still have come from *A* or *B*. Note, as well, that our intuition was right. *B* is now more probable. $P(B) = 0.40$, but $P(B|D) = 0.625$.

Figure 4.17 sums up more formally what we have done, with increasing levels of detail. I have replaced *B* with *A′* just to reflect the fact that there could be more than one other supplier. The level of detail in (2) probably connects best with one's intuition; this is how I used the tree diagram in the example. The level of detail in (3) accords best with the raw information in the problem.

In fact, though, I have found that people have a hard time making direct use of the formula in Figure 4.17. Thus, Figure 4.18 offers an alternative tabular version.

Columns (1) and (2) list all possible suppliers and their prior probabilities. Note that these add up to one, as they must. Column (3), then, gives the probability of the defective outcome, given that we are in that row. Note that each of these numbers relates only to the row it is in and not to the other.

$$P(A|D) = \overset{(1)}{\frac{P(A \cap D)}{P(D)}} = \overset{(2)}{\frac{P(A \cap D)}{P(A \cap D) + P(A' \cap D)}} = \overset{(3)}{\frac{P(A) \times P(D|A)}{P(A) \times P(D|A) + P(A') \times P(D|A')}}$$

Figure 4.17 A formal statement of Bayes' Theorem.

(1)	(2)		(3)		(4)	(5)
Supplier	$P(S)$	×	$P(D\|S)$	=	$P(S \cap D)$	$P(S\|D)$
A	0.60		0.02		0.012	0.375
A'	0.40		0.05		0.020	0.625
	1.00		$P(D)$	=	0.032	1.000

Figure 4.18 A tabular version of Bayes' Theorem.

Once we have these first three columns, the rest is mechanical. Column (4) multiplies (2) and (3) to find the probabilities of the intersections. Their sum, then, is the probability that a defect occurs, one way or the other. But we know that a defect did occur; thus, we want to reweight these numbers so that they sum to one. Column (5) accomplishes this by dividing each of the numbers in (4) by their sum.

Since the last two columns are mechanical and are the same for every problem, the key is to get the first three columns set up right. How did we know to make the rows A and A', instead of D and D'? Look back at the question. We are asked for the probability of A and B (that is, A') given the new information in the problem (that the part was defective). Thus, the rows must be A and B (that is, A').

As for column (3), these represent the probability of the event that did, in fact, occur—in this case one part was drawn and it was defective. These probabilities may not simply be given as they were in this example. Calculating these may be sort of a problem within a problem. A simple example would be if the problem had specified that two parts were drawn and both were defective, in this case, the numbers in column (3) would have been $(0.02)^2 = 0.0004$ and $(0.05)^2 = 0.0025$.

Consider the following example:

Suppose three bins of parts look identical, but bins A and B each contain seven good and three defective parts, while bin C contains four of each. You select a bin and draw two parts at random. If both are good, what is the probability that you have bin C?

We know:	We want:
$P(A) = 1/3$	$P(C\|2G) = ??$
$P(B) = 1/3$	
$P(C) = 1/3$	
$P(G\|A) = 7/10$	
$P(G\|B) = 7/10$	
$P(G\|C) = 4/8$	

(1)	(2)		(3)		(4)	(5)
Bin	P(Bin)	×	P(2G|Bin)	=	P(Bin ∩ 2G)	P(Bin|2G)
C'	2/3		7/10 × 6/9		0.3111	0.8133
C	1/3		4/8 × 3/7		0.0714	**0.1867**
	1		P(2G)	=	0.3825	1.0000

Figure 4.19 Bayes' Theorem: Another example.

Figure 4.19 shows the Bayes' Theorem table. Since it is the probability of having a particular bin that we want to revise based on new information, the rows have to represent the bins. Bins A and B could be listed separately and if their contents were different we would need to do so. In this case, though, since their contents are identical, we can just treat them as a single bin with twice the probability of being selected.

Note that the event that occurred—on the basis of which we wish to revise probabilities—was a draw of two parts, both good. Thus, column (3) needs to contain the probability of drawing two parts, both good, given a particular bin. These probabilities need to be calculated. We need a good one on the first draw *and* on the second draw. *And* means intersection and multiplication. Moreover, the events are not independent. What we draw first changes what is left that can be drawn second. So $P(G_1 \cap G_2) = P(G_1) \times P(G_2|G_1)$. For bins A and B, this is 7/10 × 6/9; for C it is 4/8 × 3/7.

As always, column (4) just multiplies (2) and (3) to get the intersections. Column (5) just reweights those intersections, by dividing each by their total, to make them sum to one. The probability that we have bin C, given the draw of two parts, both good, is only 0.1867.

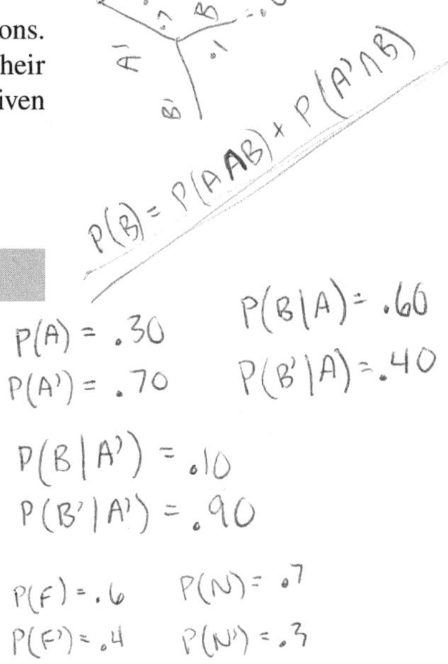

4.6 Exercises

4.1 If you make a sale to Able, the probability is 0.60 that you will also make a sale to Baker. If you fail to make a sale to Able, the probability is 0.10 that you will make a sale to Baker anyway. The probability is 0.30 that you will make a sale to Able. What is the probability that you will make a sale to Baker?

4.2 On a Festival Cruise Line cruise to the Bahamas, guests on the ship have the option of getting off for the day at two ports, Freeport and Nassau. Of all the guests on the ship, 60% get off in Freeport and 70% get off in Nassau. Moreover, of those who get off in Freeport, 90% also get off in Nassau. What proportion of the ship's guests:

a. Get off at both ports?
b. Get off at most at one of the two?
c. Get off at neither?

4.3 A state claims that 20% of the cards in its instant scratch-and-win game are winners (mostly of break-even prizes—just another card). Assuming independence, if three tickets are selected at random, what is the probability that

a. All three are winners?
b. Exactly one is a winner?
c. At least one is a winner?

4.4 The Trace household has triplets who have just turned 16, and all of them want to drive the family car to school. To determine who drives, they each toss a coin and the odd person gets to drive. If all coins show heads or tails, they toss again. What is the probability that a decision will be reached in four or fewer tosses?

4.5 A machine runs properly 80% of the time but, to ensure product quality, samples of its output are checked at regular intervals. When the machine is running properly, the probability is 0.90 that the sample will be good. When it is not running properly, the probability is only 0.10 that the sample will be good. If a sample is taken and is good, what is the probability that the machine is running properly?

4.6 A department store reports that 30% of its customer transactions are in cash, 20% are by check, and the rest are by credit card. 20% of the cash transactions, 90% of the check transactions, and 60% of the credit card transactions are for more than $75. A customer has just made a $125 purchase. What is the probability that she paid cash?

4.7 A company puts its product through three independent tests to check the quality of its product before it leaves the factory. Historically, the chances of failing to catch a defect are just 8% for the first test, 12% for the second test, and 15% for the third. Assume all three tests are run on every unit. If there is a defect, what is the probability that

a. All three tests find it.
b. All three tests fail to find it.
c. Only one test finds it.
d. At least one test finds it.

4.8 A clothing store, Threads R Us, has 900 regular customers. Of these, 60% are college students, and half of these college students do not receive a catalog in the mail. On the other hand, 30% of the customers who are not college students do receive a catalog in the mail.

a. How many customers receive catalogs in the mail?
b. What is the probability of a customer being a college student, given that he or she receives a catalog in the mail?
c. What is the probability of a customer receiving a catalog in the mail, given that he or she is a college student?

4.9 A firm has three assembly lines, each producing one-third of its product. At the end of each line, the output is packed, two items to a carton, and shipped to storage. Lines A and B produce output with just 5% defective, while C produces output with 20% defective.

a. If a carton is chosen at random from those in storage, what is the probability that it will contain two defective items?
b. If a carton chosen at random from those in storage contains two defective items, what is the probability that it came from line C?

4.10 Four dice look alike, but one is "loaded" so that the probability of getting a 1 is 1/3.

a. If you select a die at random and roll it once, what is the probability of getting a 1?
b. If you select a die at random and roll it twice, getting a 1 both times, what is the probability that you have selected the loaded die?

4.11 Three bins look alike and each contains two parts. However, one of the bins has only good parts, another has only defective parts, and the third has one of each. Suppose you select a bin at random and draw a part. If that part is good, what is the probability that the other part in the bin is good too?

4.12 A hospital reports that 50% of its patients have private insurance, 40% have Medicare or Medicaid, and the rest are uninsured. 80% of those who have private insurance, 30% of those with Medicare or Medicaid and 60% of the uninsured are released from the hospital in less than two days. A patient has just been released from the hospital in one day. What is the probability that she has private insurance?

5

Probability Distributions

Chapter 4 introduced some of the basic ideas of probability. This chapter builds on it in two main respects. First, if you think of the examples of Chapter 4, things started to get fairly messy, even with small numbers. Recall the example of rolling a die three times, and finding the probabilities of getting 0, 1, 2, or 3 1s. There were eight branches to our tree diagram; one branch each led to 0 and 3 successes; three branches each led to 1 and 2. Hopefully it occurred to you that the number of branches was doubling with each roll, and that such trees were going to get unwieldy really fast. Indeed that is the case. For example, suppose there were eight rolls, and we wanted to find the probabilities of 0, 1, 2, 3, 4, 5, 6, 7, or 8 1s. There would be 256 branches to our tree. You certainly would not want to draw such a tree, let alone count how many branches lead to 5 1s. We need a more formal, systematic approach. You will learn it in this chapter.

Second, Chapter 4 dealt just with discrete probabilities—situations in which there are a countable number of possible outcomes. We explicitly put off problems in which possible outcomes are continuous. Such problems require thinking of probabilities a little differently. We will address them in this chapter.

5.1 Discrete Random Variables

5.1.1 Discrete Random Variables and Probability Distributions

We begin with a continuation of discrete probability. A little formalism will help. We define a **discrete random variable** as a variable that takes on a countable number of *numerical* values, each with a certain probability. In the example of three rolls of a die, the number of 1s would be a random variable. It can take on four different numerical values—{0, 1, 2, or 3}.

What is new here? Not much, really. We do restrict ourselves to numerical outcomes. Thus, technically, the result of flipping a single coin—{H or T}—does not qualify. And while the result of rolling a single die—{1, 2, 3, 4, 5, 6}—qualifies, it no longer does if the outcomes of interest are simply 1 or not 1, since not 1 is not a number. In fact though, such yes–no, success–failure problems are important and are dealt with by recoding yes or success as 1, and no or failure as 0.

A **discrete probability distribution**, then, relates each numerical outcome to its probability. This can be accomplished in tables or graphs. Figures 5.1 and 5.2 both show the probability distribution for *X*, three rolls of a die. And just as with frequency distributions, probability distributions can also be displayed in ordinary or cumulative form. Figures 5.1 and 5.2 show the information both ways. Of course, these are no real advance on what we did in Chapter 4. The advance will come because we will also be able to describe many of these probability distributions much more efficiently by formula.

5.1.2 The Mean and Standard Deviation of a Probability Distribution

Figures 5.1 and 5.2 show the entire frequency distribution. As problems get larger though, doing this will become impractical. It will be useful to summarize frequency distributions more compactly. Since we have restricted

	x	0	1	2	3
Ordinary	$P_X(x)$	0.579	0.347	0.069	0.005
Cumulative	$\sum P_X(x)$	0.579	0.926	0.995	1.000

Figure 5.1 The probability distribution for *X*, the number of 1s in three rolls of a die.

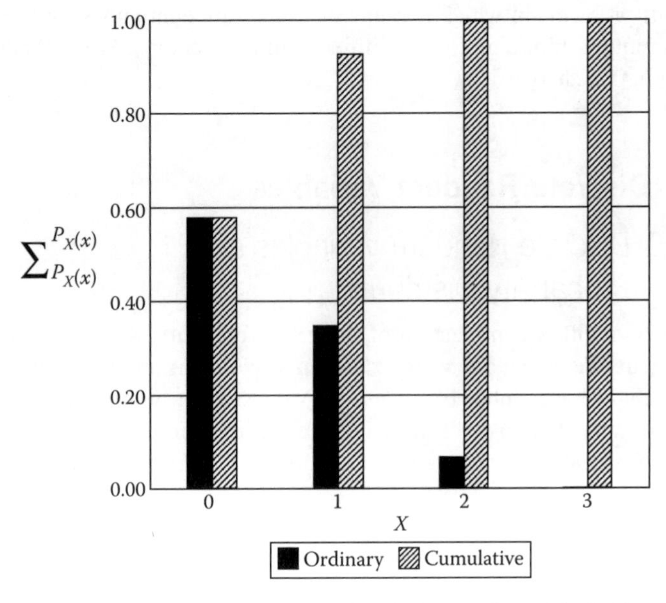

Figure 5.2 The probability distribution for *X*, the number of 1s in three rolls of a die.

$$\mu_X = \sum x P_X(x)$$
$$\sigma_X^2 = \sum (x-\mu_X)^2 P_X(x)$$
$$\sigma_X = \sqrt{\sigma_X^2}$$

Figure 5.3 The mean and standard deviation of a random variable, *X*.

ourselves to numerical outcomes, the mean and standard deviation—the measures we used to summarize samples back in Chapter 3—should come to mind. Of course, we are not working with a sample here. Our probability distribution describes an infinite population. The experiment of rolling the die three times can be repeated and repeated forever. Still, there are similarities.

Figure 5.3 gives the formulas. Think in particular of the case in which we needed to estimate the mean from a relative frequency distribution when we did not know n. We did that in Figure 3.9. One of our approaches was to weight each midpoint by its relative frequency as a proportion and then sum. We will do essentially the same thing here, weighting each possible value by its probability and then summing.

The same approach applies to the variance and standard deviation. Recall that the variance is just a mean of squared deviations from the mean. In this case, we will weight each possible squared deviation by its probability. The standard deviation, as always, is just the square root of the variance. For the example in Figures 5.1 and 5.2,

$$\mu_X = (0\times 0.579)+(1\times 0.347)+(2\times 0.069)+(3\times 0.005)=0.500$$

$$\sigma_X^2 = (0-0.500)^2\times 0.579+(1-0.500)^2\times 0.347+(2-0.500)^2$$
$$\times 0.069+(3-0.500)^2\times 0.005=0.417$$

$$\sigma_X = \sqrt{\sigma_X^2} = \sqrt{0.417} = 0.645.$$

The mean is often called the **expected value**, and written $E(X)$. This does not mean that we necessarily expect it to occur in any individual trial. In our example of three rolls of a die, the expected value of 0.500 is not even a possible outcome. If we were to repeat the experiment over and over, though, we would expect our results to average out to 0.500.

5.1.3 Special Cases

So far we have dealt with discrete random variables in general. The formula for the mean, for example, is true in general. There are a number of special cases, though, which are very common, and for which we can develop special formulas that organize and simplify things considerably.

The most important special case is the one in which a random variable follows the **binomial probability distribution**. In binomial problems, we are looking for a number of successes (x) in a number of trials (n), when the probability of success (π_X) is the same on each trial. Our example of three rolls of a die is an example of a random variable that follows a binomial probability distribution. Successes were 1s. We wanted to know the probability of 0, 1, 2, or 3 successes in $n = 3$ trials, when the probability of a success was a constant $\pi_X = 1/6$. We will develop, in the next section, the formula for calculating probabilities in these situations. This formula will generate the probabilities in Figures 5.1 and 5.2 directly, without the need for a tree diagram. It will make tractable the larger problems suggested in the introduction to this chapter, for which a tree diagram is out of the question.

A slightly different special case is the one in which we are looking for a number of successes (x) in a number of trials (n), when drawing from a finite population. We had an example in Chapter 4 in which two parts were drawn from a bin with seven good and three defective parts. The probability of getting a good part on the first draw was 7/10, but the probability of getting a good part on the second draw—6/9 or 7/9—depended on what was left after the first draw. This is not a binomial problem because the probability of success varies from draw to draw. Still, it varies in a very systematic way, and the number of successes in a problem like this follows the **hypergeometric probability distribution**.

We will not cover the hypergeometric formula in this course. However, you should recognize the difference between the hypergeometric and binomial cases, and know that there is such a formula if you need it. You should also recognize that, as the finite population gets large, the probabilities change very little from draw to draw, and so the binomial becomes a very good approximation for the hypergeometric. That is, suppose the bin above had 700 good and 300 defective parts. The probability of getting a good part on the first draw is 700/1000 = 0.7000. The probability of getting a good part on the second draw is either 699/999 = 0.6997 or 700/999 = 0.7007. The probability from draw to draw is practically constant. So if you were to assume a constant probability of 0.7, your answer would be close enough for almost any practical situation.

There are many other special cases, in which still other discrete probability distributions apply. However, we will not deal with them in this course.

5.2 The Binomial Probability Distribution

5.2.1 The Binomial Formula

5.2.1.1 The Probability of a Single Branch of a Tree

Back in Chapter 4, we used a tree diagram to find the probability of 0, 1, 2, or 3 1s in three rolls of a die. Figure 5.4 reproduces that tree. We noted in Chapter 4 that all branches leading to a certain number of successes had the same probability. Indeed, we can formalize that result. Let π_X be the probability of a success—1/6 in this example. Then $(1 - \pi_X)$ must be the probability of a failure—5/6 in this example. Moreover, any branch with x successes must also have $(n - x)$ failures—$(3 - x)$ in this

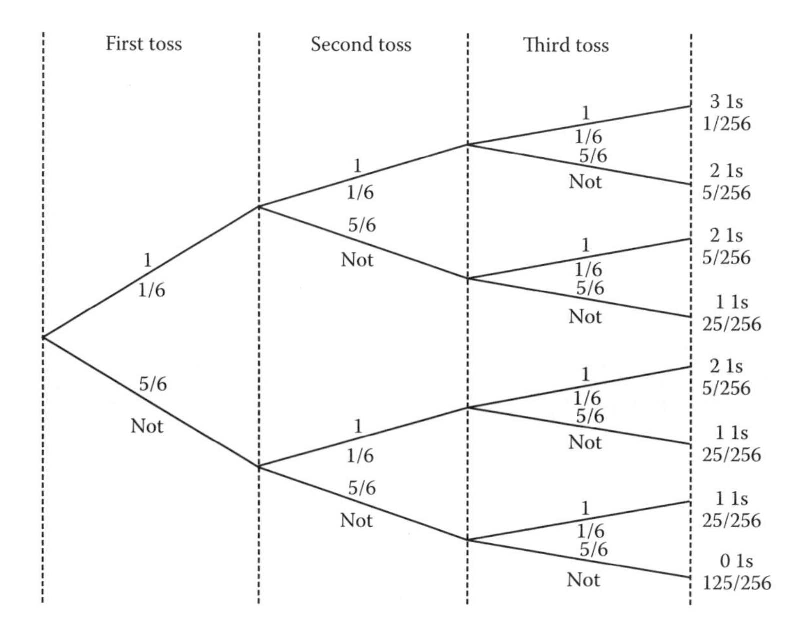

Figure 5.4 A tree diagram: An example of a binomial.

example. Thus, in this example, the probability along any branch with x successes must be $(1/6)^x(5/6)^{(3-x)}$. As we found in Chapter 4, the branches with two successes all have probabilities of $(1/6)^2(5/6)^1$ and the branches with one success all have probabilities of $(1/6)^1(5/6)^2$. Moreover, since anything to the zero power is 1, the formula even works for the all or nothing cases. The branch with three successes has a probability of $(1/6)^3(5/6)^0 = (1/6)^3$, and the branch with no successes has a probability of $(1/6)^0(5/6)^3 = (5/6)^3$.

More generally, for any *binomial* tree diagram we can imagine drawing, with n tries and a probability of success each time of π_X, the probability of a branch with x successes is

$$(\pi_X)^x(1-\pi_X)^{(n-x)}.$$

We need no longer actually draw the tree to find the probability along a branch with x successes. What we need now is a way, without actually drawing the tree, of knowing the number of branches with x successes.

5.2.1.2 The Number of Branches: Combinations

C_x^n represents the number of **combinations** of n things taken x at a time. It is the number of different orders in which we can get x successes and $(n-x)$ failures. This is exactly what we need. C_x^n is calculated as follows:

$$C_x^n = \frac{n!}{x!(n-x)!}.$$

The "!" here stands for **factorial**, and indicates that the number in question is multiplied by 1 less, then by 2 less, and so on down to 1. That is, $x!$

("x factorial") $= x \times (x-1) \times (x-2) \times \ldots \times 1$, and $n!$ is calculated likewise. $0!$ is a special case and equals 1.

We can try out the formula for three rolls of a die. We know from the tree diagram that there is only one order in which we can get 3 or 0 1s, while there are three orders in which we can get 2 or 1. These are the answers that we should get using the formula and we do:

$$C_3^3 = \frac{3!}{3! \times 0!} = \frac{(\cancel{3} \times \cancel{2} \times \cancel{1})}{(\cancel{3} \times \cancel{2} \times \cancel{1}) \times (1)} = 1 \quad C_1^3 = \frac{3!}{1! \times 2!} = \frac{(3 \times \cancel{2} \times \cancel{1})}{(1) \times (\cancel{2} \times \cancel{1})} = 3$$

$$C_2^3 = \frac{3!}{2! \times 1!} = \frac{(3 \times \cancel{2} \times \cancel{1})}{(\cancel{2} \times \cancel{1}) \times (1)} = 3 \qquad C_0^3 = \frac{3!}{0! \times 3!} = \frac{(\cancel{3} \times \cancel{2} \times \cancel{1})}{(1) \times (\cancel{3} \times \cancel{2} \times \cancel{1})} = 1.$$

Note that the numbers in the denominator—3 and 0, 2 and 1, etc.—always sum to 3, the number in the numerator. This makes sense. The number of successes plus the number of failures must add up to the number of trials. Note too the symmetry. The number of ways of getting two successes and one failure, for example, equals the number of ways of getting one success and two failures.

Finally, note that the factorials in the denominator inevitably cancel part or all of the numerator. Hence there is no need to calculate out the full $n!$. Indeed, you will find it unnecessary to write it all out. Use the larger factorial in the denominator to cancel out as much of the numerator as possible. In the formula for C_2^3, for example, you know that 2! cancels out all of 3! except the 3, so you should be able to write simply

$$C_2^3 = \frac{{}^{3}\cancel{3}!}{{}_{1}\cancel{2}! \times \cancel{1}!_{1}} = 3.$$

Suppose we consider a larger problem; suppose we roll the die eight times. I pointed out at the beginning of the chapter that a tree for this experiment would include 256 branches. How many of these branches would include five successes and three failures?

$$C_5^8 = \frac{8!}{5! \times 3!} = \frac{{}^{8\times7\times\cancel{6}}\cancel{8}!}{{}_{1}\cancel{5}! \times \cancel{3}!_{1}} = 8 \times 7 = 56.$$

Note that I used 5! to cancel all of 8! except $8 \times 7 \times 6$. Then, I used 3! to cancel the 6, leaving just $8 \times 7 = 56$. Of the 256 branches, 56 include five successes and three failures.

5.2.1.3 Putting It All Together

We have a formula for the probability of a branch having x successes in n tries with a constant probability π_X of success each time. And we have a formula for the number of branches having x successes in n tries. Putting them together gives us the binomial formula. Figure 5.5 does so.

Consider the following example:

> A fair die is rolled eight times. What is the probability of getting 1 eight times? not at all? once? more than once?

$n = 8$

$\pi_x = 1/6$

$$P_X(x|n,\pi_X)=C_x^n(\pi_X)^x(1-\pi_X)^{(n-x)}$$

Figure 5.5 The binomial formula.

We know:	We want:
$n = 8$	$P_X(x = 8 \mid n = 8, \pi_X = 1/6) = ??$
$\pi_X = 1/6$	$P_X(x = 0 \mid n = 8, \pi_X = 1/6) = ??$
	$P_X(x = 1 \mid n = 8, \pi_X = 1/6) = ??$
	$P_X(x > 1 \mid n = 8, \pi_X = 1/6) = ??$

First, eight successes:

$$P_X(x=8|n=8,\pi_X=1/6)=C_8^8(1/6)^8(5/6)^0$$

$$=\frac{{}^{1}\not{8}!}{{}_{1}\not{8}!\times\not{0}!_{1}}(1/6)^8(1)=(1)(0.0000006)=0.0000006.$$

There is just one order in which we can get eight successes out of eight tries—just one branch of the tree. And clearly, this branch has a very low probability of occurring.

Next, zero successes:

$$P_X(x=0|n=8,\pi_X=1/6)=C_0^8(1/6)^0(5/6)^8$$

$$=\frac{{}^{1}\not{8}!}{{}_{1}\not{0}!\times\not{8}!_{1}}(1)(5/6)^8$$

$$=(1)\times(0.2326)=0.2326.$$

Again, there is just one order in which we can get zero successes—just one branch of the tree. However, this branch has a much greater probability of occurring.

Next, one success:

$$P_X(x=1|n=8,\pi_X=1/6)=C_1^8(1/6)^1(5/6)^7$$

$$=\frac{{}^{8}\not{8}!}{{}_{1}\not{1}!\times\not{7}!_{1}}(1/6)^1(5/6)^7$$

$$=(8)\times(0.0465)=0.3721.$$

Compared to the branch with zero successes, the branches with one success are only 1/5th as likely but, since there are eight different branches that yield one success, the probability of one success is actually greater.

Finally, more than one success. Note that this question calls for the probability of a range of outcomes. In this problem, "more than one" means 2, 3, 4, …, or 8. We could work the binomial formula for each of these and—since the outcomes are mutually exclusive—simply add. But hopefully you see an easier way. More than one also means everything but 0 and 1. We can find the probabilities of 0 or 1, add them, and subtract from 1. In this case, we already know the probabilities of 0 and 1. Thus,

$$P_X\left(x>1 \mid n=8,\pi_X=1/6\right)=1-\left[P_X(0)+P_X(1)\right]=1-0.6047=0.3953.$$

Or consider the following example:

> A factory machine produces a large number of parts. A quality assurance employee checks the machine each hour by taking a sample of 10 parts, and stops production if more than 1 is defective. If the machine is actually producing 5% defectives, what is the probability that he will stop the machine after taking his next sample?

We know:	We want:
$n = 10$	$P_X(x > 1 \mid n = 10, \pi_X = 0.05) = ??$
$\pi_X = .05$	

In this problem, more than one means 2, 3,…, or 10; again it is easier to find the complement—the probability of 0 or 1—and subtract from 1.

$$P_X\left(x=0 \mid n=10,\pi_X=0.05\right)=C_0^{10}(0.05)^0(0.95)^{10}=\frac{{}^{1}1\cancel{0}!}{{}_{1}\cancel{0}!\times 1\cancel{0}!_{1}}(1)(0.95)^{10}$$

$$=(1)\times(0.5987)=0.5987.$$

$$=\frac{{}^{10}1\cancel{0}!}{{}_{1}\cancel{1}!\times\cancel{9}!_{1}}(0.05)^1(0.95)^9$$

$$P_X\left(x=1 \mid n=10,\pi_X=0.05\right)=C_1^{10}(0.05)^1(0.95)^9$$

$$=(10)\times(0.03151)=0.3151.$$

$$P_X(x>1 \mid n=10,\ \pi_X=.05)=1-[P_X(0)+P_X(1)]=1-.9138=.0862.$$

5.2.2 Aids in Finding Binomial Probabilities

Once you understand its underlying logic, the binomial formula is not difficult to remember or work. You should work a number of problems by formula to assure that you really understand it. Once you really understand it, though, using the formula becomes tedious, especially if you have a large n and want a wide range of possible values of x. For example, suppose we changed the die problem above to ask for the probability of more than 10 1s in 20 rolls of the die. We would need to compute separately $P_X(11)$, $P_X(12)$,…, $P_X(20)$, each of which would involve some messy numbers and then sum. Aids would be most welcome.

There are three aids that do away with much of the tedium: the binomial table, the binomial special function, and the normal approximation. We will deal with the first two in this section. The third will need to await our introduction to the normal distribution later in the chapter.

5.2.2.1 The Binomial Table

Suppose someone were to calculate the binomial for all possible combinations of x, n, and π_X and arranged them in a table. From then on, we would not need to do such calculations again; we could just look up the answer. That is the essence of the binomial table. Table 1, in Appendix C, shows such a table as it is usually laid out. Find it now.

Since our probabilities depend on three different things—x, n, and π_X—we are working in three dimensions and, of course, the page is two dimensional. The usual way of overcoming this problem is to create separate sub-tables for each n. So first, we find the n we want; then, we read across to find π_X; finally, we read down to find the value or values of x.

We can use Table 1 to check our last answer above. We needed to find $P_X(x = 0|n = 10, \pi_X = 0.05)$ and $P_X(x = 1|n = 10, \pi_X = 0.05)$. First, we find the 10 sub-table; then, we read across to find the 0.05 column; finally, there in rows 0 and 1 are our answers, 0.5987 and 0.3151.

Of course, Table 1 does not actually give all possible combinations of x, n, and π_X. That would be impossible. First, there is no limit to how large n can be. Our table stops at 20. Some might go higher, but they all have to stop at some point.

Second, there are an infinite number of π_X values in which we might be interested. Table 1 gives just 10. Actually, indirectly, it gives another nine. If you are interested in $\pi_X = 0.75$, for instance, you can still use the table. Just look for failures instead of successes. That is, use $(1 - \pi_X) = 0.25$ instead of $\pi_X = 0.75$, and look up $(n - x)$ instead of x. Still, there are an infinite number of π_X values between each pair of columns. Consider our ongoing die problems, for example. For those problems, π_X was 1/6, or 0.1667, which is not in the table. We could use 0.15 if we did not care too much about accuracy. But there are better alternatives.

Finally, recall that probability distributions can be presented in ordinary or cumulative form.

Table 1 presents the binomial distribution in ordinary form. When you look up $x = 3$, for example, you find the probability of "just 3." If you want the probability of "3 or less," you need to find the probabilities of 0, 1, 2, and 3 separately and add.

In a cumulative table, on the other hand, when you look up $x = 3$, you get the probability of "3 or less," directly, In a cumulative table, if you want to find the probability of "just 3," you need to find the probabilities of "3 or less" and "2 or less" separately and find the difference.

While some problems are answered more directly with one form of the table than with the other, any problem that can be answered with one form can be answered with the other as well. For this reason and because of the increasing availability of computer spreadsheets, I have not bothered to include the cumulative form in this book.

5.2.2.2 The Binomial Spreadsheet Functions

As with the binomial formula, you should become familiar with at least one of the binomial tables. If you have a statistics book handy, and no computer, it is likely to be the easiest way to find the answer to a binomial problem. On the other hand, if you have a computer with a spreadsheet

	Formula	Example
Ordinary:	=BINOMDIST(x, n, π_x, 0)	=BINOMDIST(1, 8, 1/6, 0) = 0.3721
Cumulative:	=BINOMDIST(x, n, π_x, 1)	=BINOMDIST(1, 8, 1/6, 1) = 0.6047
(For Lotus/Quattro Pro, replace "=" with "@")		

Figure 5.6 The binomial special functions in common spreadsheet programs.

handy, you can find the answer quite easily and precisely using the spreadsheet's binomial special functions. Any relatively up-to-date spreadsheet has such a function, though different spreadsheets may implement it differently.

Figure 5.6 shows how it works for several of the most common spreadsheets. The first three arguments are the values for x, n, and π_X. The last is a 0 or 1, indicating an ordinary or a cumulative value. Recall the example in which we figured the probabilities of various outcomes from eight rolls of a die. $P_X(x = 1|n = 8, \pi_X = 1/6)$ = BINOMDIST(1, 8, 1/6, 0) = 0.3721. And $P_X(x > 1|n = 8, \pi_X = 1/6) = 1 -$ BINOMDIST(1, 8, 1/6, 1) = 1 – 0.6047 = 3953.

These binomial special functions are now the easiest means of calculating binomial probabilities. Indeed, I created the binomial table in the Appendix using a spreadsheet.

5.2.3 The Mean and Standard Deviation of the Binomial Distribution

In the last section we covered two aids in finding binomial probabilities—tables and special functions—and alluded to a third—the normal approximation. The normal distribution—the famous bell-shaped curve—becomes a good approximation for the binomial distribution as the sample size gets big. In order to use it, though, we need to know the mean and standard deviation of the binomial distribution we are approximating. Thus, before leaving our coverage of the binomial, we need to be sure we can find the mean and standard deviation of a binomial distribution.

Earlier, we found the general formulas for the mean and standard deviation of a random variable, X. These formulas, set out in Figure 5.3, work for the binomial probability distribution. Indeed, the example used to demonstrate their use was a binomial. In fact, though, these formulas can be simplified tremendously in the case of the binomial probability distribution. Figure 5.7 gives the simplified forms.

$$\mu_X = n\pi_X$$

$$\sigma_X^2 = n\pi_X(1 - \pi_X)$$

$$\sigma_X = \sqrt{\sigma_X^2} = \sqrt{n\pi_X(1 - \pi_X)}$$

Figure 5.7 The mean and standard deviation of the binomial distribution.

Recall we found the mean for the number of 1s in three rolls of a die by weighting each possible outcome by its probability. And we found the variance by weighting each possible squared deviation from the mean by its probability. Those calculations are repeated below for easy comparison:

$$\mu_X = (0 \times 0.579) + (1 \times 0.347) + (2 \times 0.069) + (3 \times 0.005) = 0.500$$

$$\sigma_X^2 = (0 - 0.500)^2 \times 0.579 + (1 - 0.500)^2 \times 0.347$$

$$+(2 - 0.500)^2 \times 0.069 + (3 - 0.500)^2 \times 0.005 = 0.417$$

$$\sigma_X = \sqrt{\sigma_X^2} = \sqrt{0.417} = 0.645.$$

This approach was not terribly burdensome in that example because there were only four possible outcomes, 0, 1, 2, and 3. Still, consider the simplified binomial formulas:

$$\mu_X = n \times \pi_X = 3 \times 1/6 = 0.500$$

$$\sigma_X^2 = n \times \pi_X \times (1 - \pi_X) = 3 \times 1/6 \times 5/6 = 0.417$$

$$\sigma_X = \sqrt{\sigma_X^2} = \sqrt{n \times \pi_X \times (1 - \pi_X)} = \sqrt{3 \times 1/6 \times 5/6} = \sqrt{0.417} = 0.645.$$

We no longer need even list all possible outcomes, let alone find each of their probabilities. This is a tremendous simplification as problems get bigger. And, recall, it is primarily for big samples that we will want means and standard deviations, so that we can use the normal approximation.

Why do these simplified formulas work? We can develop a little intuition, at least for the mean. Since π_X is the probability of success on any individual toss, it is actually the expected value, or mean, for an individual toss. In three tosses, then, π_X is the expected value each time. The expected value of their sum, then, is $\pi_X + \pi_X + \pi_X = 3\pi_X$. It does make sense.

5.3 Continuous Random Variables

So far, we have worked only with discrete probabilities. There have always been a countable number of possible outcomes, each with a given probability of occurring. But consider selecting someone at random and measuring his or her height in inches. Since we are selecting the person at random, the height, *H*, is a random variable. But because height is continuous—a person might be 67 inches, or 67.005, or 66.99995, etc.—there are an infinite number of possible outcomes. The sum of the probabilities of all possible outcomes must still be one, though. So dividing up one into an infinite number of possible outcomes suggests that the probability of any one outcome must be zero. Clearly that will not do. We need to approach continuous probabilities a bit differently. We need to think of the probability of being *between* two values.

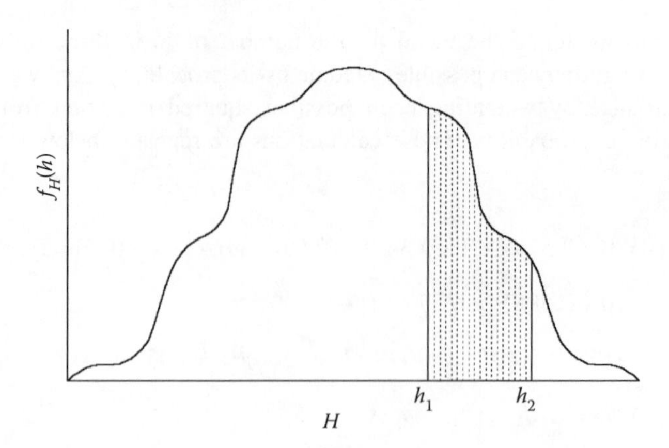

Figure 5.8 A continuous probability distribution.

Figure 5.8 shows a hypothetical continuous probability distribution. I have drawn it wavy to suggest that it might have any number of shapes. Such a continuous probability distribution is often called a **probability density function**. The probability is the area under the curve. The area under the whole curve must equal one. The probability of being between h_1 and h_2, then, is the area under the probability distribution between those two values.

If you have had some calculus, you will recognize this as a calculus problem. The area between h_1 and h_2 is the integral of the probability density function between these points. Fortunately, these areas—these probabilities—have been calculated and tabulated for all the important special cases. In this chapter, we deal with just one special case, the normal distribution.

5.4 The Normal Distribution: The Bell-Shaped Curve

Everyone has heard of the **normal distribution**—the famous bell-shaped curve. It is extremely important. Many real-life random variables follow this distribution, at least approximately. Moreover, as we will see in the chapters to come, interesting sample statistics—like the possible means one can get in sampling from a population—converge to a normal distribution for large samples, even if the random variable itself does not follow this distribution.

All normal distributions have the same basic shape. They are symmetric around their means. They differ only in the values of their means and standard deviations. However, since there are an infinite number of possible means and standard deviations, there are an infinite number of different normal distributions. Tabulating all is an impossible task.

We get around this apparent problem as follows. First, we rescale the normal distribution of interest to have the same mean and standard deviation as some "standard." Then we use these rescaled values and the table for that standard normal distribution. We take up these steps in reverse order, below. First, we look at how to find probabilities for various outcomes for the standard normal. Then, we look at how to turn any other normal distribution into a standard one.

5.4.1 The Standard Normal Distribution

We start with the special case of a normal random variable, Z, *that has a mean of zero and a standard deviation of one.* This is the **standard** normal distribution. Its formula is:

$$f_Z(Z) = \frac{1}{\sqrt{2\pi}} e^{-z^2/2}.$$

However, the standard normal table and spreadsheet function make it unnecessary to work with this formula for our purposes.

5.4.1.1 The Standard Normal Table

Again, if we want the probability of being between z_1 and z_2, we want the area under this curve between these two points. Let us arbitrarily set one of these two points equal to zero, the mean. We can then tabulate, for many possible values of Z, the area under the curve between those values and the mean. Table 2, in Appendix C, shows such a table as it is usually laid out. Find it now.

Look first at the diagram at the top. It indicates that when you look up a value z_0 in this table, the value you will find is the shaded area—the area under the curve from z_0 to the mean. If you pick up another statistics book, you might find a different area shaded. For example, the area shaded might be from z_0 all the way down to negative infinity. Or it might be from z_0 up to positive infinity. The reference point—negative infinity, zero, or positive infinity—is arbitrary, but you need to make sure you know what it is.

The table has Z values to two decimal places. Read down the left-hand column to find your value to one decimal place; then read across the top to find the second decimal. For example, suppose you wanted to know the probability of Z being between 1.75 and the mean. You would read down the left-hand column to 1.70; then you would read across to the 0.05 column. The probability of Z being between 1.75 and the mean is 0.4599.

Since the normal distribution is symmetrical, we actually know some other probabilities as well. The probability of Z being between -1.75 and the mean is also 0.4599; the probability of Z being in the range mean ± 1.75 is $0.4599 \times 2 = 0.9198$. Since the total area under the curve—the total probability—is 1, the area on either side of the mean is 0.5000. So the probability of Z being greater than 1.75 or less than -1.75 is $0.5000 - 0.4599 = 0.0401$. And the probability of Z being less than 1.75 or greater than -1.75 is $0.4599 + 0.5000 = 0.9599$. Figure 5.9 shows some of these probabilities.

The table can be used to find the area under the curve between any two values for Z. If the values are on opposite sides of the mean—the

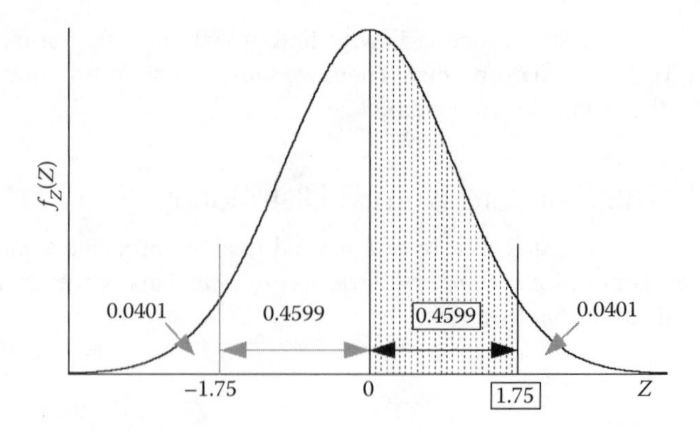

Figure 5.9 Areas under the standard normal distribution: An example.

reference point for this table—find the area from each to the mean and then sum. If they are on the same side of the mean—the reference point for this table—find the area from each to the mean and then find the difference. Figures 5.10 and 5.11 illustrate. In Figure 5.10, we find the probability of Z being between −1.00 and + 1.75. As the figure makes clear, because they are on opposite sides of the mean, we want the sum of the areas. Due to symmetry, the probability of Z being between −1.00 and zero is the same as the probability of Z being between +1.00 and the zero, 0.3413. And the probability of Z being between + 1.75 and zero is 0.4599. Thus, the probability of Z being between −1.00 and +1.75 is 0.3413 + 0.4599 = 0.8012. In Figure 5.11, we find the probability of Z being between + 1.00 and +1.75. As the figure makes clear, this time we want the difference in areas. The probability of Z being between +1.00 and +1.75 is 0.4599 − 0.3413 = .1186.

I indicated earlier that some tables will use a reference point other than the mean; I mentioned negative infinity as a possibility. Notice that this would put all points of interest on the same side of the reference point. Working with such a table, you would always be finding a difference, never a sum. In Figure 5.10, the probability of Z being between negative infinity and −1.00 would be given as 0.1587; the probability between negative infinity and +1.75 would be given as 0.9599. The probability of Z being between −1.00 and +1.75 would be calculated 0.9599 − 0.1587 = 0.8012, the same answer we got by summing.

So far we have looked for the probability of Z being in the range z_1 to z_2. But often we will want to know the values of z_1 and z_2 such that the probability between them is some given amount. That is, instead of working from Z values to a probability, we will want to work from a probability to Z values. Suppose, for example, we want the values, $\pm z_1$, such that the probability of being between them is 0.80. Figure 5.12 illustrates. We want the area, between $+ z_1$ and $-z_1$ to be 0.80. That means we want the area between $+z_1$ and the mean to be 0.40. This is a probability, not a Z value; we need to look for it in the body of the table, not the margin. Looking through the body of the table, 0.3997 is the closest. And the corresponding Z value is 1.28. Thus, ±1.28 are the values, $\pm z_1$ such that the probability of being between them is 0.80.

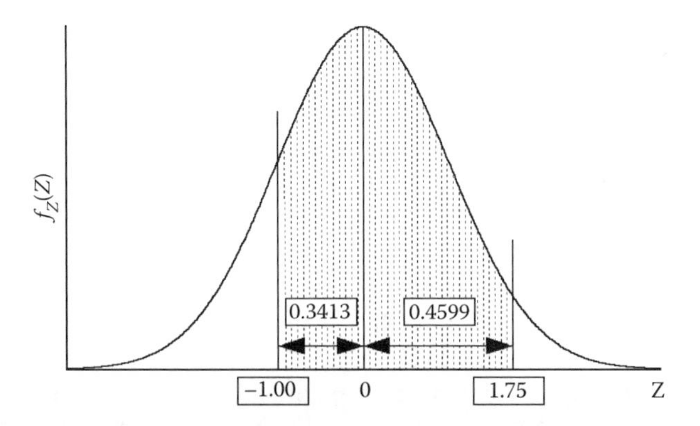

Figure 5.10 Areas under the standard normal distribution: Another example (a).

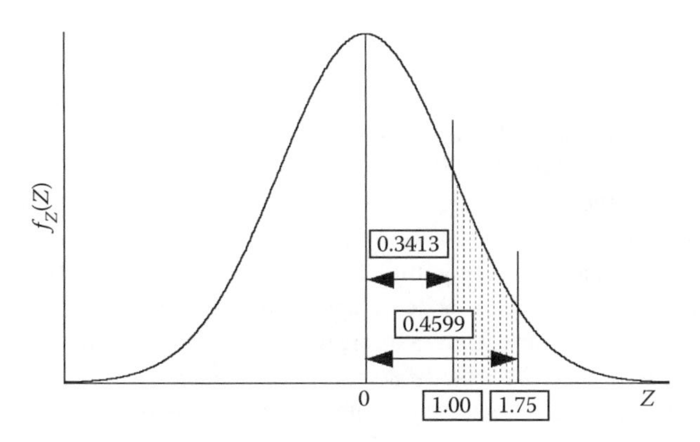

Figure 5.11 Areas under the standard normal distribution: Another example (b).

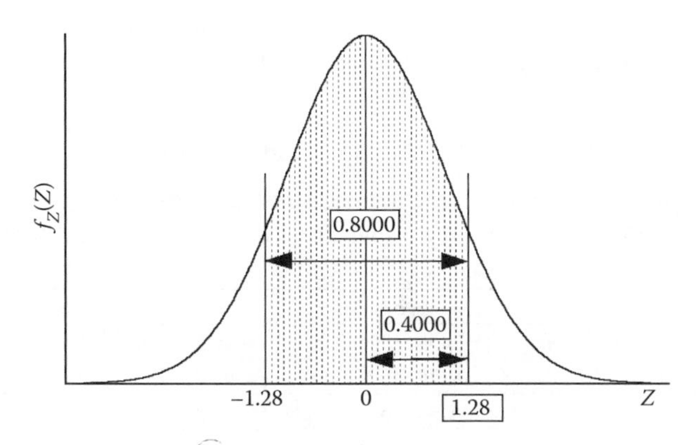

Figure 5.12 Areas under the standard normal distribution: Working backward.

5.4.1.2 *The Standard Normal Spreadsheet Functions*

Unlike the binomial table, the standard normal table is compact and quite complete. Still, there are times when it would be nice to find a standard normal probability directly in a spreadsheet. Figure 5.13 shows the special functions for several of the most common spreadsheets. Typically, they use negative infinity, not the mean, as their reference points. =NORMSDIST(z_0), then, gives the probability from negative infinity up to z_0. And =NORMSINV(p) gives the z_0 for which the probability from negative infinity to z_0 is p. Thus, to find the probability in Figure 5.10, you would find =NORMSDIST(+1.75) − NORMSDIST(−1.00) = 0.9599 − 0.1587 = 0.8012. And to find $\pm z_1$ in Figure 5.12 such that the probability of being between them is 0.80, we would note that 0.10 has to be in each tail. =NORMSINV(0.10) = −1.28, and =NORMSINV(0.90) = +1.28.

5.4.1.3 *Two Final Points*

First, a point of interpretation. Since the standard normal distribution has a standard deviation of 1, *Z* values represent the distance from the mean in standard deviations units. Thus, the probabilities of *Z* being in the ranges ±1, ±2, and ±3 are the probabilities of being within 1, 2, and 3 standard deviations of the mean of a normal distribution. These probabilities are $0.3413 \times 2 = 0.6826$, $0.4772 \times 2 = 0.9544$, and $0.4987 \times 2 = 0.9974$; or, rounding off, 68, 95, and 100%.

Actually, these percentages should seem familiar. Back in Chapter 3, I introduced the empirical rule, which gave these as the rough percentages of the data to expect within 1, 2, and 3 standard deviations of the mean, for data that were at least approximately bell-shaped. Now you know where they came from. Sample data, as opposed to theoretical populations, will never be exactly normally distributed. But if the data are approximately bell-shaped, the standard normal distribution can help in their interpretation.

Finally, a bit of practical advice. Never proceed with a normal distribution problem without first drawing the curve, and marking the mean, *Z* values, and areas of interest. Without a graph, it is easy to become confused between *Z* values and their probabilities, between cases in which you want the probability of *Z* being more than or less than a certain amount, and between cases in which you want a sum or a difference. Drawing and labeling a graph organizes your thinking.

5.4.2 Standardizing a Normal Distribution

Now that we have covered the standard normal distribution, we need to deal with the fact that most normal distributions are not standard. That

	Formula	Example
From z_0 to a probability:	=NORMSDIST(z_0)	=NORMSDIST(−1.00)=0.1587
From a probability to z_0:	=NORMSINV(p)	=NORMSINV(.10)=−1.28
(For Lotus/Quattro Pro, replace "=" with "@")		

Figure 5.13 The standard normal special functions in common spreadsheet programs.

$$z_0 = \frac{x_0 - \mu_X}{\sigma_X}$$

Figure 5.14 Standardizing a normal distribution.

is, they do not have a mean of zero and a standard deviation of one. It turns out to be fairly simple to standardize any other normal distribution. Figure 5.14 illustrates. Suppose X is normally distributed with a mean of μ_X and a standard deviation of σ_X. By subtracting μ_X, we shift the distribution so that the mean is zero; by dividing by σ_X, we change the spread of the distribution so that the standard deviation is one.

Consider the following example:

> Suppose the heights of young adult males are normally distributed with a mean of 70 inches and a standard deviation of 3 inches. If an individual is chosen at random from this population, (a) what is the probability that his height will be within 4 inches of the mean? (b) how tall would he need to be in the tallest 5%?

We know:	We want:
$\mu_H = 70$	$P(66 < H < 74)$
$\sigma_H = 3$	The h_0 for which $P(H > h_0) = 0.05$

Figure 5.15 shows the normal distribution for (a). We want the area between 66 and 74. We need first to rescale the key values of H to their Z equivalents:

For $H = 66$	For $H = 70$	For $H = 74$
$Z = \frac{66-70}{3} = -1.33$	$Z = \frac{70-70}{3} = 0$	$Z = \frac{74-70}{3} = 1.33$

Of course, we do not really need to do it for the mean; by design, the mean always rescales to zero. In this case, 66 and 74 rescale to ±1.33. Using the

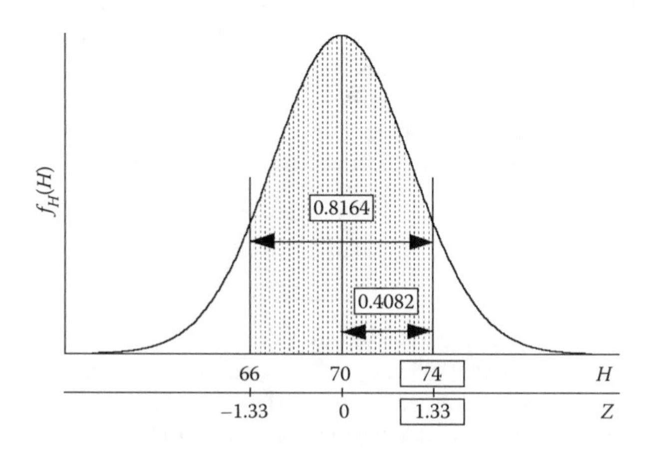

Figure 5.15 The normal distribution: An example.

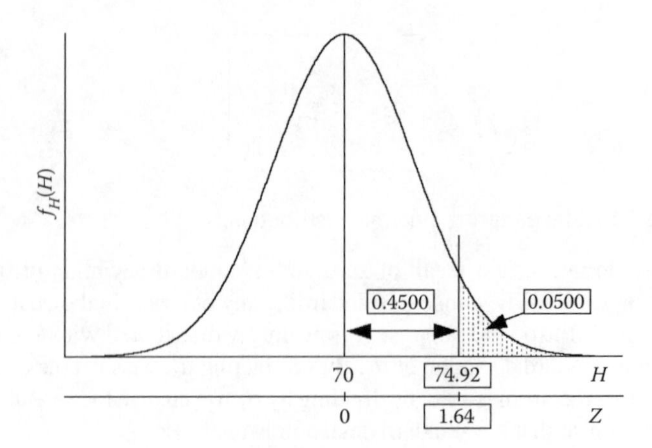

Figure 5.16 The normal distribution: An example.

standard normal table, the probability of being between 1.33 and the mean is 0.4082. Doubling that, to reflect the negative side as well, the probability that the individual's height will be between 66 and 74 inches is 0.8164.

In (a), we went from an H-value to its Z equivalent and then used the standard normal table to find our probability. In (b), we need to reverse that. Figure 5.16 illustrates. We know the probability we want—0.05. If only 0.05 is to be above our number, it must be on the upper end of the normal distribution; and the area between it and the mean must be 0.45. Using the standard normal table backward, we find that 1.64 (or 1.65) comes closest to giving us 0.45. Thus, Z must equal 1.64 (or 1.65). We can now rescale in the opposite direction—from Z to H:

$$Z = \frac{H - 70}{3} = 1.64, \text{ so } H - 70 = 4.92, \text{ and } H = 70 + 4.92 = 74.92.$$

He would need to be 74.92 inches or taller to be in the tallest 5%.

Or consider the following example:

> Suppose you take over a store selling TVs and, to build a reputation for quality, decide to offer your own warranty on the TVs you sell. You figure that you will be able to afford the warranty costs as long as only 5% of the sets require service within the warranty period. If the trouble-free life of the TVs, X, is normally distributed, with a mean of 48 months, and a standard deviation of 10 months, for how many months can you let the warranty run?

We know:	We want:
$\mu_X = 48$	The x_0 for which $P(X < x_0) = 0.05$
$\sigma_X = 10$	

Figure 5.17 illustrates. As in the last example, we know the probability we want—0.05. If only 0.05 is to be below our number, it must be on the lower end of the normal distribution; and the area between it and the mean must be 0.45. Using the standard normal table backward, we find that 1.64 (or 1.65) comes closest to giving us 0.45. Thus, Z must equal –1.64.

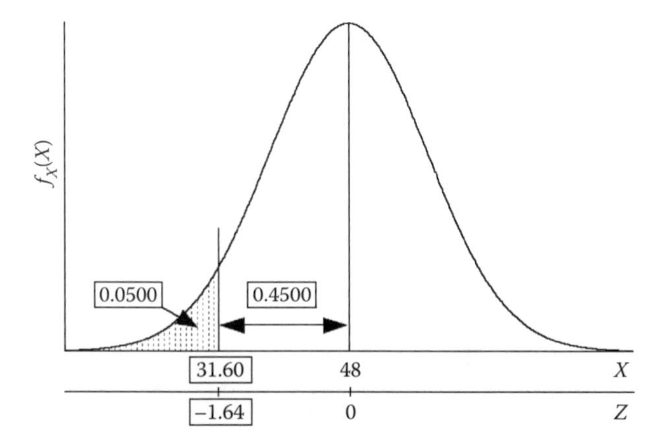

Figure 5.17 The normal distribution: Another example (a).

Do not forget to make it negative, or you will end up covering 95% of the TVs! We can now rescale from Z to X:

$$Z = \frac{X - 48}{10} = -1.64, \text{ so } X - 48 = -16.40, \text{ and } X = 48 - 16.40 = 31.60$$

You would not want your warranty to run more than 31.6 months.

Forgetting to make Z negative is an easy mistake to make, especially if you have not drawn the diagram. So always draw it. Also always ask yourself whether your answer makes sense. Without the negative, your answer would have been 64.40, which is obviously much too high.

Finally, consider the following example:

A bottle-filling machine can be set to fill "32-ounce" bottles to any level, on average, but the levels in individual bottles, X, will be normally distributed around that average, with a standard deviation of 0.2 ounces. To what level must the average be set to ensure that 98% of the bottles are filled to at least 32 ounces?

We know:	We want:
$x_0 = 32$	The μ_0 for which $P(X > 32) = 0.98$
$\sigma_X = 0.2$	

Figure 5.18 illustrates. Again, we know the probability we want—0.98. And this time we know the X value—32. What we do not know is the mean. If 0.98 is to be above 32, it must be on the lower end of the normal distribution; and the area between it and the mean must be 0.48. Using the standard normal table backward, we find that 2.05 comes closest to giving us 0.48. Thus, Z must equal –2.05. Again, do not forget to make it negative; we are in the lower end of the distribution. We can now rescale from Z to X:

$$Z = \frac{32 - \mu_X}{0.2} = -2.05, \text{ so } 32 - \mu_X = -.41, \text{ and } \mu_X = 32 + .41 = 32.41.$$

You would want to set the average to 32.41 to ensure that 98% of the bottles are filled to at least 32 ounces.

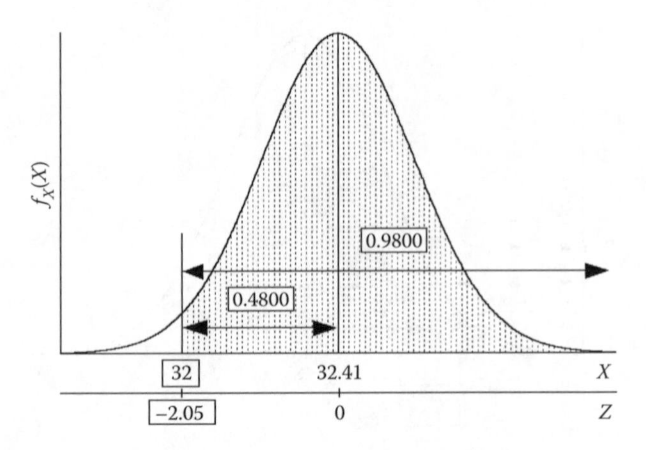

Figure 5.18 The normal distribution: Another example (b).

5.5 The Normal Approximation to the Binomial

Earlier in the chapter, I alluded to the normal approximation to the binomial distribution, but postponed coverage until we had covered the normal distribution. We are now in a position to deal with it. Recall that the binomial formula gets messy for large numbers. Moreover, the binomial table is cumbersome and incomplete as compared to the normal table. The binomial table in Appendix C stretches to six pages, and covers only a small number of π_X values, and sample sizes only as large as 20. Fortunately, since the binomial approaches the normal as n gets large, we will often be able to use the normal table instead.

How large is large enough? As always that depends on the accuracy we require. In this case, it also depends on π_X. Figures 5.19 and 5.20 show binomial distributions for n of 10 and 50, for π_X of 0.10 and 0.50. The distribution for $\pi_X = 0.50$ is always symmetrical around the mean. And, as Figure 5.19 shows, even with an n of just 10, it is starting to look quite a bit like the normal. With an n of just 10, though, the distribution for $\pi_X = 0.10$ is quite skewed. As Figure 5.20 shows, with an n of 50, the distribution for $\pi_X = 0.50$ looks very much like the normal, and even the distribution for $\pi_X = 0.10$ is looking fairly symmetrical and bell-shaped.

A common rule of thumb is that the normal is a good approximation for the binomial distribution when both $n \times \pi_X$ and $n \times (1 - \pi_X)$ are greater than 5. Thus, if $\pi_X = 0.50$, an n of just 10 is probably large enough for the normal to be a good approximation; for $\pi_X = 0.10$ or 0.90, an n of 50 would be required; for $\pi_X = 0.05$ or 0.95, an n of 100 would be necessary.

The normal distribution we use for our approximation is the one with the same mean and standard deviation as our binomial. Recall that these can be calculated for any binomial as $\mu_X = n \times \pi_X$ and $\sigma_X = \sqrt{n \times \pi_X \times (1 - \pi_X)}$.

One issue that we need to face is that the binomial is discrete, with probabilities only at integers, while the normal is continuous, with probabilities measured between points. How do we convert from one to the other? The answer is quite simple. We count as the probability of 6, for example, the area under the normal curve from 5.5 to 6.5. Figure 5.21 illustrates. Note, this means that to approximate the binomial for “6 or

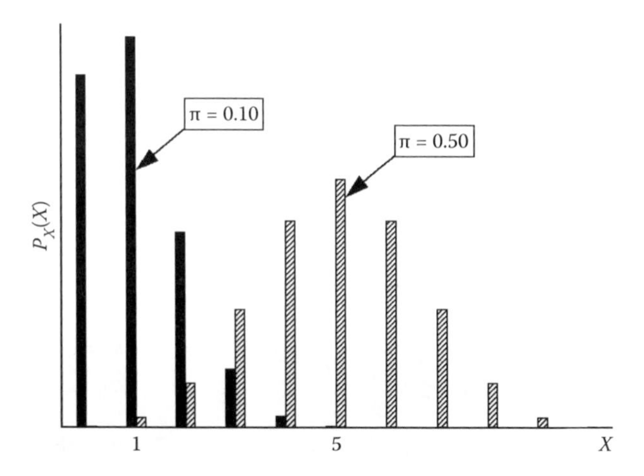

Figure 5.19 The binomial distribution with $n = 10$.

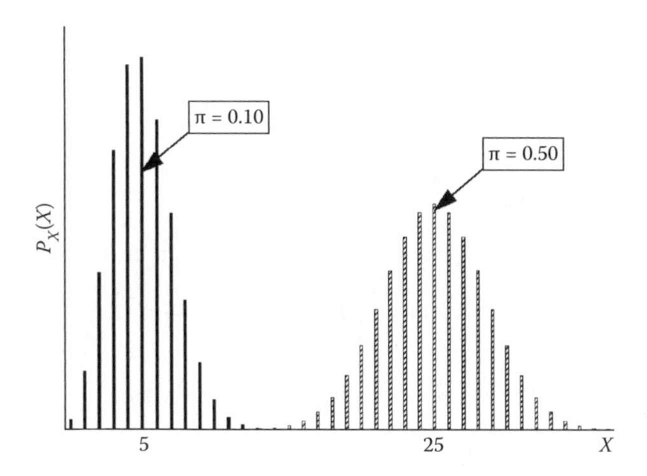

Figure 5.20 The binomial distribution with $n = 50$.

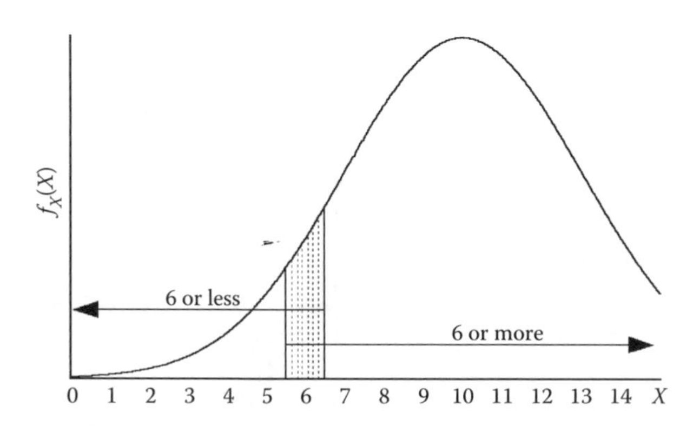

Figure 5.21 Approximating the binomial distribution with the normal.

less" requires that we start at 6.5, while to approximate the binomial for "6 or more" requires that we start at 5.5. Otherwise, we are not including all of the area associated with 6.

Consider the following example:

> Suppose a production process produces 7.5% defective items. What is the probability of getting more than 10 defectives in a sample of 100 from this process? Use the normal approximation.

We know:	We want:
$n = 100$	$P_D(D > 10 \mid n = 100, \pi_D = 0.075)$
$\pi_D = 0.075$	

First, recognize that this problem is a binomial. We want the probability of a certain number of successes—more than ten—in a certain number of tries—100—where the probability of a success is a constant 0.075. In this problem, "more than ten" means 11, 12,…, 100; if we were doing this as a binomial it would pay to find the probability of the complement (0, 1,…, 10) and subtract from 1. With the normal, though, this is no easier. Figure 5.22 illustrates.

We need to calculate the mean and standard deviation of this binomial:

$$\mu_D = n \times \pi_D = 100 \times 0.075 = 7.50 \text{ and}$$

$$\sigma_D = \sqrt{n \times \pi_D \times (1 - \pi_D)}$$

$$= \sqrt{100 \times 0.075 \times 0.925} = 2.634.$$

Our D here is 10.5, to include all of 11 or more, so $Z = (10.5\text{–}7.50)/2.634 = 1.4$.

Looking up a Z value of 1.14 in the standard normal table gives us a probability of 0.3729 between 1.14 and the mean. We want the probability above 1.14 instead: $0.5000 - 0.3729 = 0.1271$.

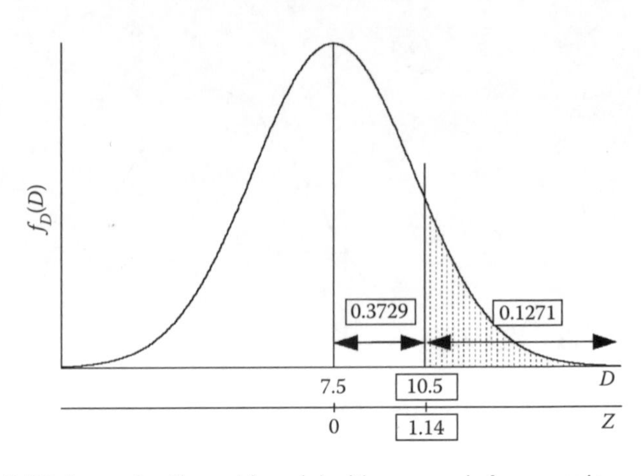

Figure 5.22 Approximating a binomial with a normal: An example.

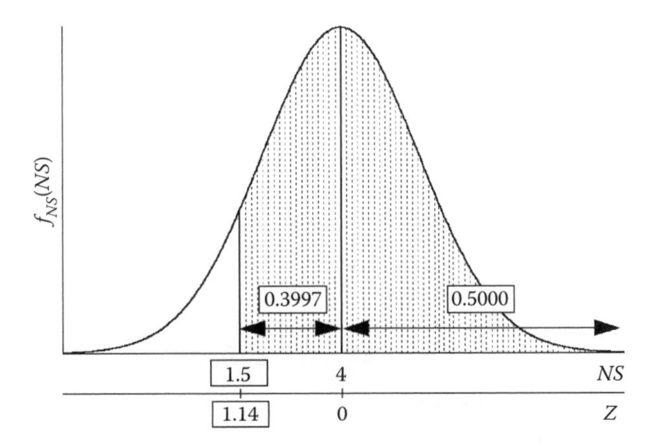

Figure 5.23 Approximating a binomial with a normal: Another example.

The probability of getting more than 10 defectives in such a sample is approximately 0.1271. (The approximation is pretty good; the actual binomial probability would have been 0.1293.)

Finally, consider the following example:

> Suppose, on average, 5% of the people with airline tickets fail to show up for their flights. If 80 tickets are sold for a flight with only 78 seats, what is the probability that there will be enough seats for all the people who actually show up? Use the normal approximation:

We know:	We want:
$n = 80$	$P_{NS}(NS \geq 2 \mid n = 80,\ \pi_N = 0.05)$
$\pi_{NS} = 0.05$	

First, note that this problem is again a binomial. Be careful though; it specifies the number of "shows," but gives the probability of a no-show. Do not be tricked. Either we want "78 or fewer shows," with $\pi_S = 0.95$, or we want "2 or more no-shows," with $\pi_{NS} = 0.05$. I have chosen the no-shows, but either works, as long as you are consistent. Figure 5.23 illustrates.

We need to calculate the mean and standard deviation of this binomial:

$$\mu_{NS} = n \times \pi_{NS} = 80 \times 0.05 = 4.00 \text{ and}$$

$$\sigma_{NS} = \sqrt{n \times \pi_{NS} \times (1 - \pi_{NS})} = \sqrt{80 \times 0.05 \times 0.95} = 1.95.$$

The NS here is 1.50, to include all of 2 or more, so $Z = (1.50\text{–}4.00)/1.95 = -1.28$.

Looking up a Z value of 1.28 in the standard normal table gives us a probability of 0.3997 between −1.28 and the mean. We want all the probability above −1.28, so 0.3997 + 0.5000 = 0.8997.

The probability that there will be enough seats for all the people who actually show up is approximately 0.8997. (The actual binomial probability would have been 0.9139.)

This final example might be used to make a couple of points. First, the approximation was not as good as in the previous example. This is because $n \times \pi_{NS} = 4$, not 5, as we offered for our rule of thumb. Still, we were off by only 0.0142, and this may be close enough in many practical cases. The rule of thumb is only that—a general guide.

On the other hand, approximating the binomial in cases like this is of less practical importance than it was in the days before special spreadsheet functions. I easily computed the exact binomial probability of 0.9139 by opening a spreadsheet and typing in a cell:

```
= 1-BINOMDIST(1,80,0.05,1)
```

The real and continuing importance of the relationship between the binomial and the normal will become apparent when we get to sampling distributions in the next chapter.

5.6 Exercises

5.1 A pair of dice is tossed five times.
  a. What is the probability of getting 9s on the first and last, but not the second, third, or fourth tosses?
  b. What is the probability of getting 9s on two of the five tosses?

5.2 If an insurance sales representative has a 0.20 probability of making a sale on every call, what is the probability that he will make at least two sales in six calls?

5.3 A pop quiz consists of five multiple-choice questions, each with four possible answers. If you are totally unprepared and must guess randomly on all five, what is the probability that you will get the majority right?

5.4 A process produces 15% defective items.
  a. If we take a sample of 10, what is the probability of getting more than 2 defectives.
  b. If we take a sample of 100, what is the probability of getting more than 20 defectives. Use the normal approximation.

5.5 The life of a particular type of light bulb is normally distributed with a mean of 1000 and a standard deviation of 150 hours. The manufacturer is considering a warranty.
  a. What proportion of the bulbs have a life of more than 1150 hours?
  b. What proportion of the bulbs have a life of between 925 and 1150 hours?
  c. If the manufacturer wants to replace no more than 5% of the bulbs under warranty, how many hours should the warranty cover?

5.6 A candy-making machine can be set to produce candy bars of any weight, on average, but the weights on the individual candy bars will be normally distributed around that average, with a standard deviation of 0.2 ounces. If the candy bars are supposed to weigh 8 ounces, and you want to ensure that 99% of them weigh at least that much, to what should you set the average?

5.7 The sales manager for a national shoe manufacturer has found that 33% of the company's orders come from specialty shoe stores and the rest come from general retailers. Assuming that the next five orders are independent of each other, what is the probability that at least two will be from specialty shoe stores?

5.8 The quality-control officer for an electronics firm claims that 96% of the parts shipped out are in good working order. Assuming that this claim is correct, what is the probability that a shipment of 100 parts will contain more than two defectives? Use the normal approximation.

5.9 A beer-bottling machine can be set to fill bottles to any level, on average, but the levels in individual bottles will be normally distributed with a standard deviation of 0.5 ounces. The nominally 12-ounce bottles can actually hold a maximum of 13 ounces. What is the maximum level to which the average can be set, if no more than 2% of the bottles are to overflow?

5.10 A firm has three assembly lines, each producing one-third of its product. At the end of each line, the output is packed, four items to a carton, and shipped to storage. Lines A and B produce output with just 5% defective, while C produces output with 20% defective. If a carton chosen at random from those in storage contains two defective items, what is the probability that it came from C?

5.11 A motel will have 38 rooms available on a certain night. Since, on average, 10% of the people making reservations do not show up, the motel makes 40 reservations for this night. What is the probability that there will be enough rooms for all who do show up? Use the normal approximation.

6

Sampling and Sampling Distributions

This chapter is pivotal. It connects the probability of the last two chapters with the statistical inference of Chapters 7 through 13. In this chapter we will be asking about the probability of getting a sample statistic—a sample proportion or sample mean—within given ranges of the comparable population parameter. For example, suppose we know that the population proportion preferring political candidate Smith is 0.55. That is, π_S is 0.55. We will be able to say something like "given this population, if we take a random sample of 1000, the probability is 0.95 that the sample proportion, p_S, will be within the range 0.55 ± 0.03." Said differently, in repeated samples of 1000, the sample proportion preferring Smith, p_S, will be in this range 95% of the time and will be outside this range 5% of the time.

Starting in Chapter 7, it will be p_S that we know, not π_S. But now consider the 95% of the cases for which p_S is within the range 0.55 ± 0.03. In these cases, the range $p_S \pm 0.03$ will include 0.55. In these cases, the range $p_S \pm 0.03$ will be a correct estimate of π_S. More generally, since π_X $\pm$ some interval includes p_X A% of the time, then $p_X \pm$ that same interval will be correct A% of the time. Using this reasoning, we will be able to make estimates of unknown population parameters like π_S with probabilities attached that they are correct. This is where we are headed. But we have some work to do to get there. First we look, in very general terms, at sampling technique. Then we examine the probability distributions of the sample proportion and sample mean.

6.1 Sampling*

6.1.1 Random and Judgment Samples

Suppose you are an elected official and want to know your constituents' views on a particular issue. You know your constituents well. You might simply talk to a number of people whom you think constitute a "representative sample."

What you have done is take a **judgment sample**. And, if your judgment is good, your sample may be quite representative. Indeed, it would probably seem strange at first to take a sample that made no use of your judgement as to what is true of your constituents. But there are two flaws to this approach. First, your sample is likely to build in any biases and misconceptions you have concerning your constituents. Hence, it will seem to confirm those biases and misconceptions rather than correcting them. And second, since your sample was not generated in a probabilistic manner, there is no way of making the sort of probabilistic statements suggested in paragraph two. For these, we need to take a **random sample**.

We defined a random sample back in Chapter 1. It is a sample taken in a manner such that all members of the population have an equal probability of being selected for the sample. Note that there is no guarantee that a random sample will be representative. Indeed, it is pretty much guaranteed that occasionally it will not be. But first, the probabilities are on our side. And second, the probabilities can be calculated and reported along with our sample results.

6.1.2 Techniques for Random Sampling

Actually taking a random sample is harder than it might seem at first. If you just stop every 10th person who passes at the mall, your sample is unlikely to be truly random, since some people are more likely than others to be at the mall. If you select every 50th telephone number from the telephone book, your sample is unlikely to be truly random, since some people have unlisted numbers.

It helps a lot if the population is numbered, or at least listed so that numbers can be applied. A sample of employees of a large firm would be facilitated if all employees have an employee ID number. A sample of registered voters can be taken from a list of all such voters by first numbering them from 1 to *N*.

Once the population is numbered, the trick is to choose randomly from those numbers. The two most common tools for doing so are random number tables and computer random number generators.

6.1.2.1 Random Number Tables

Figure 6.1 shows a portion of a **random number table**, as it is usually laid out. Think of these numbers as 500 single digits, 0 through 9, arranged in 5 by 5 blocks just for ease in reading. And they are arranged so that each possible digit is equally likely in each position.

93933	98098	55364	06128	57175	70872	50972	59506	42073	62224
66448	29775	89545	09214	34206	24892	71595	54092	30986	38387
47589	42120	04115	66890	26021	85250	80815	71242	31207	85656
21572	52873	12198	12955	35338	40592	62036	22692	89019	94798
16343	77453	02769	77338	28137	75931	45638	99154	08408	08154
82557	07665	81700	64827	61340	34989	64822	18634	36243	38563
39679	32044	91960	20318	42206	50150	61652	74810	05699	27818
76885	61763	04672	66197	99931	87736	78800	20410	07719	21502
34003	66610	58576	90462	68248	97580	24893	60737	40372	71906
91520	98931	55263	30250	69905	28333	65322	85853	79922	49561

Figure 6.1 A small section of a random number table.

To use the table, we first determine how many digits we need to cover the entire population. Suppose, for example, the employee ID numbers we are going to use are four digits. Then we need to take strings of four digits from the random number table. Starting anywhere in the table, we read off the first four digits. If the number corresponds to an actual employee's ID number, that employee is in our sample. We then move on to the next four digits. Again, if the number corresponds to an actual employee's ID number, that employee is in our sample. If we get a number that does not correspond to an actual employee's ID number, we ignore it and move on. If we get a repeat, we ignore it and move on. We can move through the table in any systematic way—up, down, left, right—until we reach the sample size we want.

Until recently, statistics books included tables of random numbers that went on for pages, because these provided the surest tool for generating a sample that was truly random. They are no longer the tool of choice, though, and I have not bothered to include them in this book.

6.1.2.2 Random Number Generators

The tables of random numbers referred to in Section 6.1.2.1 were actually generated by computers back at a time when computers were relatively new. Back then, computers were very expensive and very unfriendly. Thus, it made sense for someone to generate a large set of random numbers once and then publish them for others to use.

Today, any common spreadsheet or statistics program can generate random numbers for us. In the most common spreadsheets, = RAND() or @RAND will generate a random number between 0 and 1. Suppose we want a list of 100 random four digit integers. Typing =INT(RAND() * 9999) or @INT(@RAND * 9999) once gives us the first; copying it to 99 more cells gives us the rest. Indeed, that is exactly how I generated the random numbers in Figure 6.1, except that I multiplied by 99999 to make them five digits instead of four.

How do we make use of these random number capabilities? It is easiest if we have the entire population in a spreadsheet or similar database, with the ability to generate random numbers. First, we would use the random number function to assign a random number to each case. We would not need to worry about making them integers. Then, we would sort by these random numbers. We now have the cases in random order so we can just take the first n cases.

6.1.2.3 Systematic Random Sampling

Of course, life is not always so easy; suppose we do not have the population computerized. Suppose what we have is an alphabetical list of 100,000 registered voters, and we want to survey a random sample of 1000 about their political views. That means that we want 1 out of 100. We might take what is called a **systematic random sample**. In such a sample, we would use either a random number table or a random number generator to chose an individual from the first 100; we would then simply take every 100th individual thereafter. Since the starting point is random, the sample itself should be random.

A systematic random sample involves a lot less work than choosing each individual randomly. One does need to be careful, though. In the example, we are implicitly assuming that there is no systematic relationship between the voters' political views and their place in the alphabet. This is probably a pretty safe assumption. However, it is possible that different ethnic groups, whose names tend to fall at different points in the alphabet, might also have different political views. More problematic would be taking a systematic random sample based on employee ID numbers. These ID numbers have probably been assigned chronologically. Thus, those employees with low numbers will have been around longer than those with high numbers, and there are likely to be systematic differences between them.

6.1.3 More Advanced Techniques

We will sometimes refer to samples generated in the preceding ways as "simple" random samples to distinguish them from samples generated in more sophisticated ways. And we will assume, from here on, that all the samples we deal with are simple random samples. But you should be aware at least of a couple of more advanced techniques that can be more efficient in certain cases. We introduce them here very briefly.

6.1.3.1 Stratified Sampling

In the example of sampling registered voters, we might be interested in knowing about differences in political views among groups, as well as political views overall. And, if we take a simple random sample, we may end up with so few members of certain groups that we cannot infer anything meaningful. For example, suppose the groups we want to compare are whites and African Americans. A simple random sample will tend to have far more whites than African Americans. We may well end up with more whites than we really need to make inferences about their political views, but not nearly enough African Americans to make inferences about theirs. To avoid this problem, we could use **stratified sampling**. We could identify the groups of interest ahead of time and take separate random samples within each group. And we would sample a larger proportion of the smaller group.

We might also have reason to believe that certain groups are more homogeneous in their political views than others. For such groups, even a small sample is likely to be enough. For groups that are more heterogeneous in their views, a larger sample would be necessary to be equally confident of our results. Again, we could use stratified sampling. We could identify these groups ahead of time and take separate random samples within each group. And we would sample a larger proportion of the more heterogeneous group.

In general, we would want to "over-sample" groups that are (1) small within the population or (2) heterogeneous, and to "under-sample" the opposites.

Can we then say anything about what is true overall? Are our results not biased, since not all individuals had an equal chance of being chosen? We do need to be careful. But, since we did the over- and under-sampling knowingly, we need not be fooled. As long as we weight our results properly when we combine them, we can make overall inferences as well. The formulas for doing so can be found in more comprehensive texts. They are not difficult, but they are beyond the scope of this text.

6.1.3.2 Cluster Sampling

So far, we have assumed a list of population members. What if there is no list? Sometimes **cluster sampling** is the answer. Again, consider the survey of political views, but now suppose we have no list of all registered voters. Do we at least have a map? If so, we could divide the map into small areas—perhaps blocks—and number each area. Each of these areas will contain a cluster of registered voters. Using a random number table or random number generator, we can then select a random sample of clusters. Then, we can go to these areas and interview everyone in those clusters.

We will have a random sample of the political views of the clusters, but can we reason from these back to the views of individuals? Yes we can. Again, this involves weighting our results properly when we combine them. Again, the formulas for doing so can be found in more comprehensive texts. They are not difficult, but they are beyond the scope of this text.

6.2 What Are Sampling Distributions and Why Are They Interesting?

Imagine taking repeated random samples of a particular size, n, from the same population and calculating some sample statistic. Perhaps we are interested in p_{LH}, the proportion who are left-handed. Perhaps we are interested in $\overline{X}_{\text{Age}}$, their average age. As we calculated p_{LH} or $\overline{X}_{\text{Age}}$ in sample after sample, we would get a range of answers, just because each sample would include different individuals. The **sampling distribution** is just the probability distribution of the p_{LH} or $\overline{X}_{\text{Age}}$ values we could get. See Figure 6.2.

It should be clear that these are theoretical distributions, not ones that we would ordinarily generate ourselves. We are not going to take sample after sample after sample. We are going to take just one sample and hope for the best. However, for many common statistics, like the proportion and the mean, we do know something about how these distributions would look. We know, for example, that the mean of the probability distribution of p_{LH} would be π_{LH}, the true proportion of left-handers in the population. The mean of the probability distribution of $\overline{X}_{\text{Age}}$ would be μ_{Age}, the true average age of individuals in the population. Indeed, as we will see, below, we can describe such distributions in some detail.

> A **sampling distribution** is a probability distribution for a sample statistic, like p_X or $\overline{X}$, for all samples of a given size that can be selected from a population.

Figure 6.2 The definition of a sampling distribution.

Why would we want to do so? Recall the introduction to this chapter on where we are headed. If, in this chapter, we can find the interval around π_X or μ_X that includes $A\%$ of all possible p_X or $\overline{X}$ values, we will be able, in the next chapter, to use that same interval around our individual p_X or $\overline{X}$ value, to create an estimate of π_X or μ_X that has an $A\%$ chance of being correct.

6.3 The Sampling Distribution of a Proportion

We start with proportions because they link most directly to the last chapter. Recall the binomial distribution—the distribution for the number of successes, X, in n trials, when the probability of a success is a constant π_X. One of our examples found the probability of 0 through 8 1s in eight rolls of a die. The first two columns of Figure 6.3 give the probability distribution, computed using the binomial spreadsheet function $P(X = 2|n = 8, \pi_X = 1/6) = 0.2605$, for example.

Note, since proportions are just counts divided by n, and n is fixed, it would change nothing to express this in proportions instead of counts. If the probability of getting $X = 2$ in eight rolls is 0.2605, the probability of getting $p_X = 2/8 = 0.250$ in eight rolls must also be 0.2605. They are the exact same thing. The exact sampling distribution for a proportion is a binomial.

Remember where we are headed, though. We are going to want to know the probability of being within the range $\pi_X \pm$ some interval. And we are going to want high probabilities of being within narrow intervals. Why? Since that way, when we know p_X not π_X, we will at least know that there is a high probability that our p_X is within that narrow interval of π_X. With a sample of only eight observations, though, there are only nine *possible* values for p_X, and only one of them is even close to $\pi_X = 1/6 = 0.167$. The interval 0.167 ± 0.05, for example, stretching from 0.117 to 0.217, includes only $p_X = 0.125$. And $p_X = 0.125$ has a probability of only 0.3721. If we take a sample of eight, and calculate p_X, there is only a 0.3721 chance that our p_X will be within ± 0.05 of π_X. This is not nearly precise enough to be useful. Small samples will be of little use for inferences about π_X. So, we are going to need large samples to make meaningful inferences

(1) X	(2) $P(X)$	(3) $X/n = p_X$	(4) $P(P_X)$
0	0.2326	0/8 = 0.000	0.2326
1	0.3721	1/8 = 0.125	0.3721
2	0.2605	2/8 = 0.250	0.2605
3	0.1042	3/8 =0.375	0.1042
4	0.0260	4/8 = 0.500	0.0260
5	0.0042	5/8 = 0.625	0.0042
6	0.0004	6/8 = 0.750	0.0004
7	0.0000	7/8 = 0.875	0.0000
8	0.0000	8/8 = 1.000	0.0000

Figure 6.3 The binomial distribution for both X and p_X.

$$p_X = \frac{X}{n}$$

$$\mu_p = \frac{n\pi_X}{n} = \pi_X$$

$$\sigma_p = \frac{\sqrt{n\pi_X(1-\pi_X)}}{n} = \sqrt{\frac{\pi_X(1-\pi_X)}{n}}$$

Figure 6.4 The mean and standard error of the sampling distribution of p_X.

about π_X. And, recall, for large samples, the normal distribution is a very good approximation for the binomial. Thus, we will use the normal rather than the binomial as our sampling distribution of a proportion.

Recall that to convert a binomial to a normal, we need to know the mean and standard deviation of the underlying binomial. Figure 6.4 illustrates that, just as p_X is simply X, rescaled by dividing through by the fixed n, so are the mean and standard deviation of p_X simply the mean and standard deviation of X, rescaled in the same way. So the sampling distribution of a proportion, p_X, is a normal distribution with mean, $\mu_p = \pi_X$, and standard deviation, $\sigma_p = \sqrt{\pi_X \times (1-\pi_X)/n}$.

Note that σ_p is our measure of the spread in the sample p_X values we could get from a population with a mean of π_X. If σ_p is small, there is little spread in possible p_X values we could get. This means there is a high probability that our p_X is close to π_X. This will be good when we get to inference, since then, of course, we will not know π_X and will be using p_X as our estimate. Indeed, because of its role as a measure of how wrong p_X could be as an estimate of π_X, σ_p is generally referred to, not as the standard deviation of p_X, but as the **standard error of p_X.**

We will want the standard error, σ_p, to be small. Examining the formula means we will want the numerator, $\pi_X \times (1 - \pi_X)$, to be small, or the denominator, n, to be large.

We do not control π_X, but it does matter for the size of the standard error, σ_p. If π_X is 0.05, for example, the numerator is $0.05 \times 0.95 = 0.0475$. If π_X is 0.50, on the other hand, the numerator is $0.50 \times 0.50 = 0.25$, more than five times as large. In fact, $\pi_X = 0.50$ is the worst case, causing the largest numerator possible. As π_X rises above 0.50, $(1 - \pi_X)$ falls below it. If π_X is 0.95, for example, we have just the mirror image of when π_X is 0.05; the numerator is back down to $0.95 \times 0.05 = 0.0475$.

The thing we can actually control is n, the sample size. Since it is in the denominator, an increase in sample size decreases the standard error, σ_p. Of course, n is within the square root, so it takes a fourfold increase in sample size to cut σ_p in half. Precision is not cheap. But if it is required, it can be had. Indeed, we will look shortly at the problem of determining how large a sample needs to be to achieve any desired level of precision.

Consider the following example:

> Your supplier claims that only 5% of the parts it supplies are defective. Suppose this is true. What is the probability that you will get

between 2 and 8% defectives—that is, a p_X within the range $\pi_X \pm 0.03$—if you take a sample of 100? if you take a sample of 400?

We know:	We want:
$\mu_p = \pi_X = 0.05$	$P(0.02 < p_X < 0.08) = ??$

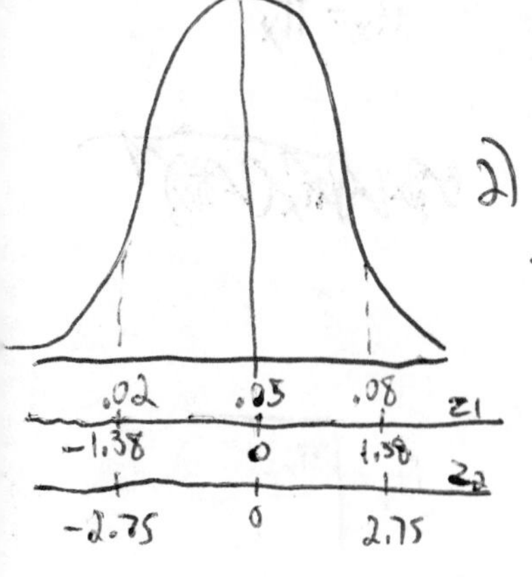

First we need to calculate the standard error, σ_p, for each sample:

For $n_X = 100$	For $n_X = 400$:
$\sigma_p = \sqrt{0.05 \times 0.95/100} = 0.0218$	$\sigma_p = \sqrt{0.05 \times 0.05/400} = 0.0109$

Since the second sample is four times as large, it has only half as large a σ_p. Thus, the sampling distribution is narrower and numbers far from π_X are less likely. Figure 6.5 illustrates.

Now, we need to find the Z value equivalents of 0.02 and 0.08. Of course, since they are symmetric about the mean of 0.05, we need calculate only one:

$$Z_1 = \frac{p_X - \mu_p}{\sigma_p} = \frac{0.08 - 0.05}{0.0218} = 1.377 \qquad Z_2 = \frac{p_X - \mu_p}{\sigma_p} = \frac{0.08 - 0.05}{0.0109} = 2.752$$

Finally, we need to look up Z in the standard normal table and double the probability we find to include both sides:

$$P(0.02 < p_X < 0.08) = 0.4162 \times 2 = 0.8324 \qquad P(0.02 < p_X < 0.08) = 0.4970 \times 2 = 0.9940$$

Again, because of the smaller σ_p with the larger sample, 0.02 and 0.08 rescale to Z values further from the mean. Therefore, the probability of p_X being within this range ($\pi_X \pm 0.03$) is higher.

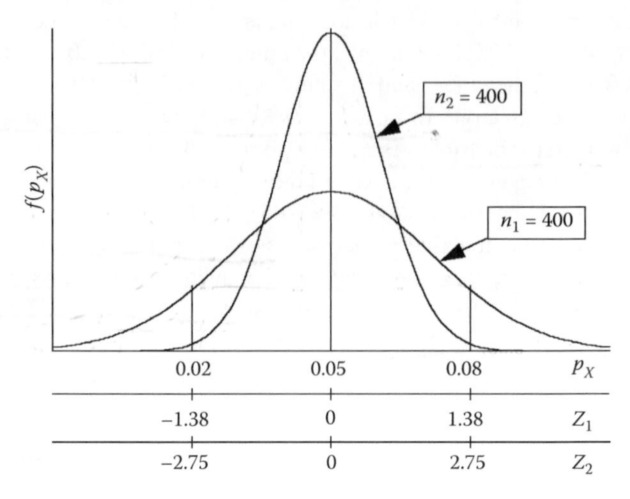

Figure 6.5 The sampling distribution of a proportion: An example.

Consider the following example:

> Suppose, in the previous example, you had wanted a 0.95 probability that p_X would be within the range $\pi_X \pm$ some interval. How wide an interval would you have needed with the sample of 100? with the sample of 400?

We know:	We want:
$\mu_p = \pi_X = 0.05$	$p_1 = ??$
$P(p_1 < p_X < p_2) = 0.95$	$p_2 = ??$

In the last problem we started with a desired interval and samples sizes and found probabilities; in this one we start with a desired probability and sample sizes and need to find intervals. Figure 6.6 illustrates.

We already know the standard errors, σ_p:

For $n_X = 100$	For $n_X = 400$:
$\sigma_p = \sqrt{0.05 \times 0.95/100} = 0.0218$	$\sigma_p = \sqrt{0.05 \times 0.05/400} = 0.0109$

Since we want the probability within the interval to be 0.95, the probability from either p_1 or p_2 to the mean should be half of that, 0.4750. Using the standard normal table backward, the Z value that gives us this probability is 1.96. We want Z to equal ± 1.96.

Since $Z = (p_i - \mu_p)/\sigma_p = \pm 1.96$, we can solve for p_i.$p_i = \mu_p \pm 1.96 \times \sigma_p$:

$p_i = 0.05 \pm 1.96 \times 0.0218$	$p_i = 0.05 \pm 1.96 \times 0.0109$
$= 0.05 \pm 0.0427$	$= 0.05 \pm 0.0214$

This time, because of the smaller σ_p in the larger sample, the 0.95 probability interval is narrower.

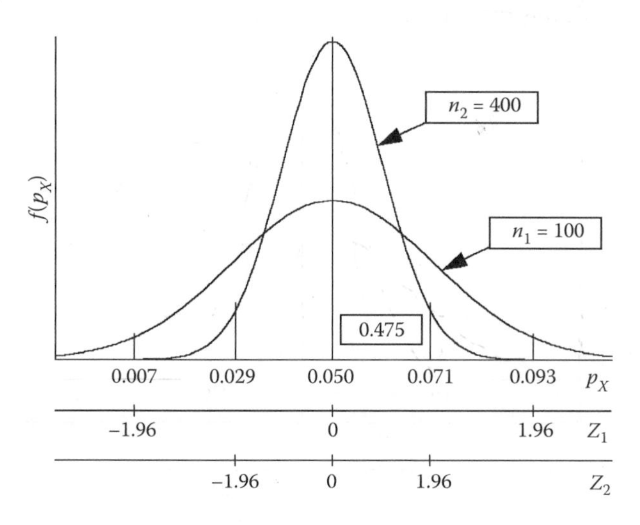

Figure 6.6 The sampling distribution of a proportion: Another example (a).

Finally, consider the following example:

> Suppose, in the previous examples, you had wanted a 0.98 probability of between 3 and 7% defectives—that is, a p_X within the range $\pi_X \pm 0.02$. How large a sample would you have needed?

We know:	We want:
$\mu_p = \pi_X = 0.05$	$n = ??$
$P(0.03 < p_X < 0.07) = 0.98$	

In the last two problems we started with given sample sizes and either a desired interval or a desired probability and found the other; in this one, we start with a desired interval *and* probability and need to find the sample size that will give both. Figure 6.7 illustrates.

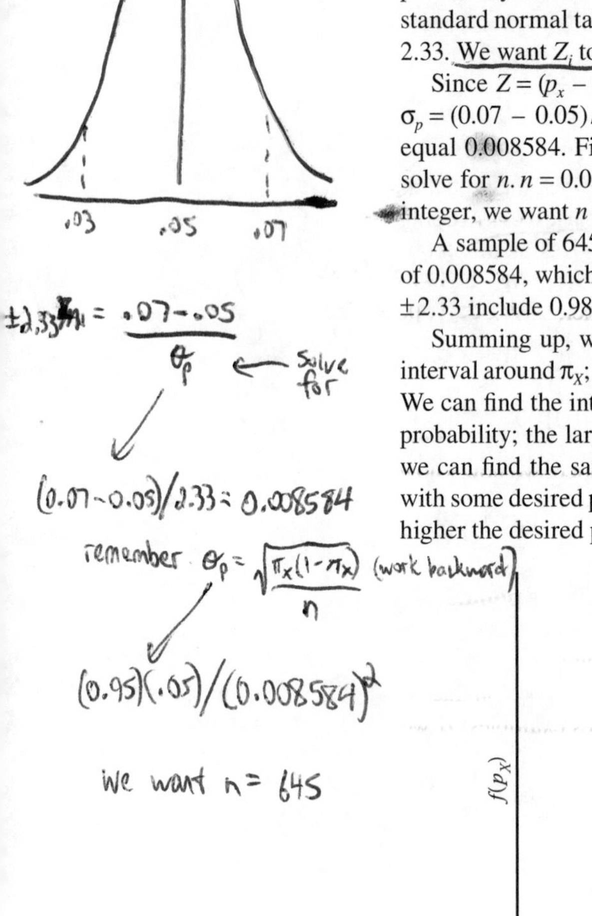

Since we want the probability between 0.03 and 0.07 to be 0.98, the probability from each to the mean should be half of that, 0.4900. Using the standard normal table backward, the Z value that comes closest (0.4901) is 2.33. We want Z_i to equal ± 2.33.

Since $Z = (p_x - \mu_p)/\sigma_p = (0.07 - 0.05)/\sigma_p = 2.33$, we can solve for σ_p. $\sigma_p = (0.07 - 0.05)/2.33 = 0.008584$. We want the standard error, σ_p, to equal 0.008584. Finally, since $\sigma_p = \sqrt{(0.05 \times 0.95)/n} = 0.008584$, we can solve for n. $n = 0.05 \times 0.95/(0.008584)^2 = 644.7$. Rounding up to the next integer, we want n to equal 645.

A sample of 645 would be large enough to give us a standard error, σ_p, of 0.008584, which is small enough to give us a Z of 2.33. And Z values of ± 2.33 include 0.98 of the area under the standard normal.

Summing up, we can find the probability of p_X being in some desired interval around π_X; the larger the sample, the higher this probability will be. We can find the interval around π_X that will contain p_X with some desired probability; the larger the sample, the narrower this interval will be. And we can find the sample size required for a desired interval to contain p_X with some desired probability. The narrower the desired interval and/or the higher the desired probability, the larger this sample will need to be.

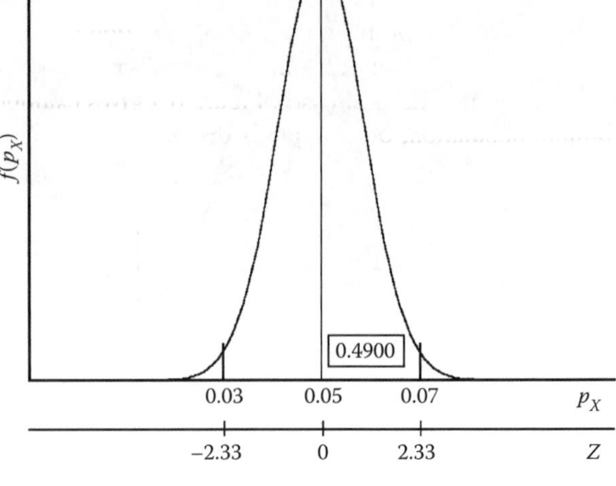

Figure 6.7 The sampling distribution of a proportion: Another example (b).

6.3.1 Three Complications

Before leaving the sampling distribution of a proportion, there are three complications that need to be addressed. First, recall that in Chapter 5 (Section 5.5), when we used the normal to approximate a binomial, we divided halfway between the numbers of successes. The probability of 6 successes was the area between 5.5 and 6.5; "6 or less" started at 6.5, "6 or more" started at 5.5. Since the true sampling distribution of a proportion is this same binomial, just rescaled by dividing through by the constant n, and we are approximating it with the normal, logic suggests that we should still be dividing between numbers of successes.

In the example on Section 6.3, we could have noted that $p_X = 0.08$ is actually $X_1 = 8$ successes when $n_1 = 100$, and $X_2 = 32$ successes when $n_2 = 400$. Then, instead of using $p_X = 0.08$ in calculating our Z values, we could have used $p_X = 8.5/100 = 0.085$, and $p_X = 32.5/400 = 0.08125$. In the case of the smaller sample, this would indeed have made for a somewhat better approximation; in the case of the larger sample, the difference is trivial and may not even be an improvement. In practice, when dealing with the sampling distribution of a proportion, this continuity correction is seldom applied.

Second, the formula we have been using for the standard error, σ_p, assumes that the population from which we are drawing is infinite. In the example of rolling a die, this is certainly the case; the number of possible rolls of a die is infinite. In the example of looking for defective parts, this may not literally be true; your supplier is not going to supply you with an infinite number of parts. But if it is an ongoing supplier, we still do not know what the population size is.

On the other hand, suppose we know the population size. Suppose we have just received a shipment of 500 parts and we are going to evaluate this shipment by inspecting a randomly selected sample of 50. Our sample size, n, is 50; our population size, N, is 500. Notice, we now know 50/500 = 10% of our population for sure. This should reduce our chances of getting a sample very different from the population; this should reduce the standard error, σ_p. And indeed it does.

Figure 6.8 updates the formula for σ_p to include the **finite population correction factor**. Ignoring the –1 in the denominator, the ratio within the square root is just the proportion of the population *not* sampled.

How much difference does this finite population correction factor make? Surprisingly little in most cases. Figure 6.9 gives examples. If we have an infinite population, our sample is 0% of the population and the

$$p_X = \frac{X}{n}$$

$$\mu_p = \pi_X$$

$$\sigma_p = \sqrt{\frac{\pi_X(1-\pi_X)}{n}} = \sqrt{\frac{N-n}{N-1}}$$

Figure 6.8 The mean and standard error of the sampling distribution of p_X when the population is finite.

Percentage of population sampled	0%	2%	5%	10%	25%	50%	100%
Correction factor	1.00	0.99	0.97	0.95	0.87	0.71	0.00
Percentage reduction in standard error	0%	1%	3%	5%	13%	29%	100%

Figure 6.9 The effect of the finite population correction factor.

finite population correction factor is 1. And, of course, multiplying by 1 does not change anything; this is why leaving it out is equivalent to assuming that the population is infinite.

Now note how slowly the finite population correction factor takes effect. Sampling 10% of the population reduces the correction factor, and thus the standard error, by just 5%. Sampling even 25% of the population reduces the correction factor, and thus the standard error, by just 13%.

Moreover, sampling this much of a population is probably pretty rare. Generally we resort to statistical sampling because the population is fairly large. So as we sample a larger and larger proportion of the population, our sample itself gets large. Recall the effect of sample size that we saw in the previous examples. As the sample size gets larger, the standard error gets smaller regardless of how small it may still be compared to the population. A sample of 1000, which is often likely to be just a small percentage of the population, is still likely to be a large enough number in the denominator of the standard error to give us all the precision we need.

Perhaps you have heard someone say—or said yourself—"*I've* never been contacted by a political pollster. How can these polls be reliable when they sample only a tiny fraction of U.S. voters?" Now you know. A sample of 1000 is, for many purposes, a large sample. The fact that it is still "a tiny fraction of U.S. voters" makes very little difference.

In most cases the population size makes little difference and we can just treat it as infinite. Still, it is never incorrect to use the finite population correction factor when we know the population size. And, if the population is *not* large, it will make a difference.

Finally, I included the 100% case in Figure 6.9 just to make a point. If we have sampled 100% of the population, the finite population correction factor is zero. The standard error, σ_p, is zero. This makes sense. In this case our sample is the population; p_X is π_X. There is no sampling error.

The third complication that needs to be addressed looks again at where we are headed. In the next chapter, we will know the sample p_X and want to use it as an estimate of the unknown π_X. And we will want to argue that, since there is an A% chance of p_X being within $\pi_X \pm z \times \sigma_p$, then $p_X \pm z \times \sigma_p$

$$s_p = \sqrt{\frac{p_X(1-p_X)}{n}}\sqrt{\frac{N-n}{N}}$$

Figure 6.10 Estimating the standard error from a sample when π_X is not known.

has an $A\%$ chance of including π_X. But if we do not know π_X, we cannot calculate σ_p. We will need to estimate it. Figure 6.10 shows how.

The statistic s_p is our estimate of the true standard error, σ_p. The main change is the one you should have expected. Since p_X is our best estimate of π_X, it makes sense that we would use it wherever we need π_X and do not know it. A minor secondary change is that the -1 disappears from the finite population correction factor.

Can we just replace the true π_X with an estimate without changing anything else? If we were dealing with small samples, the answer would be no. But since we are dealing only with large samples here, the answer is yes.

6.4 The Sampling Distribution of a Mean: σ_X Known

For the same reasons that we will want to know the probability that p_X is within some interval of π_X, we will also want to know the probability that $\bar{X}$ is within some interval of μ_X. Again, then, we will need to know its sampling distribution. If we were to take repeated samples of size n, and calculate the mean, $\bar{X}$, for each, what would the resulting distribution of $\bar{X}$s look like?

A special case is the one in which X itself has a normal distribution, with mean μ_X and standard deviation σ_X. This is a special case, though a useful one, since many real-world variables have approximately normal distributions. In this case, $\bar{X}$ also has a normal distribution, with mean $\mu_{\bar{X}} = \mu_X$ and standard deviation $\sigma_{\bar{X}} = \sigma_X/\sqrt{n}$, no matter how large or small the sample size.

Note that $\sigma_{\bar{X}} = \sigma_X/\sqrt{n}$ is our measure of the spread in the sample $\bar{X}$ values we could get from a population with a mean of μ_X. If $\sigma_{\bar{X}}$ is small, there is little spread in possible $\bar{X}$ values we could get. This means there is a high probability that our $\bar{X}$ is close to μ_X. This will be good when we get to inference, since then, of course, we will not know μ_X and will be using $\bar{X}$ as our estimate. Indeed, because of its role as a measure of how wrong $\bar{X}$ could be as an estimate of μ_X, $\sigma_{\bar{X}}$ is generally referred to, not as the standard deviation of $\bar{X}$, but as the **standard error of $\bar{X}$.**

We will want the standard error, $\sigma_{\bar{X}}$, to be small. Examining the formula, this means we will want the numerator, σ_X, to be small, or the denominator, n, to be large.

We do not control σ_X, but it does matter for the size of the standard error, $\sigma_{\bar{X}}$. The more spread out the original distribution, the more spread out is the distribution of sample means. Certainly this makes intuitive sense.

The thing we can actually control is n, the sample size. Since it is in the denominator, an increase in sample size decreases the standard error, $\sigma_{\bar{X}}$. Since it is within the square root, it takes a fourfold increase in sample size to cut $\sigma_{\bar{X}}$ in half.

Suppose, for example, the distribution for heights of young men is normal, with $\mu_H = 70$ inches and $\sigma_H = 3$ inches. Figure 6.11 illustrates this population distribution ($n = 1$), along with the distributions of averages for samples of size 4 and 16. Notice that, for the population, the range 70 ± 3

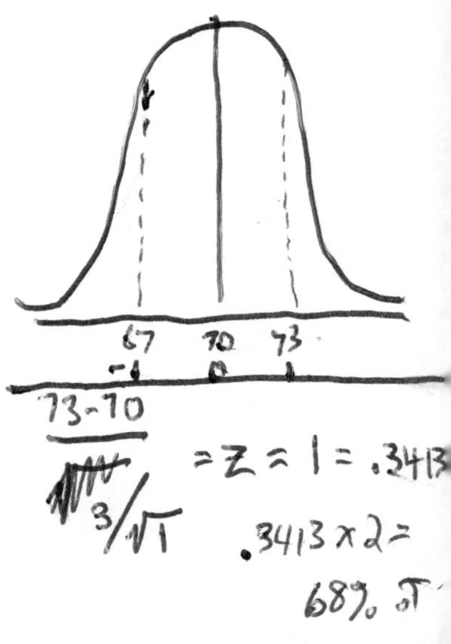

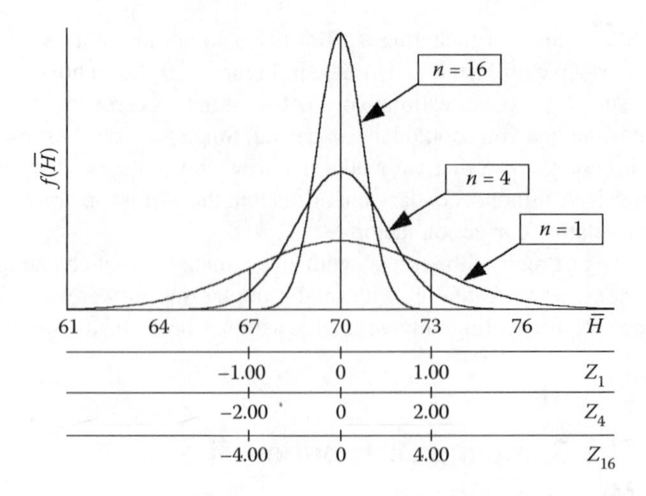

Figure 6.11 Normal distribution for heights of young men: The effect of sample size.

For any variable, X, with mean μ_X and standard deviation σ_X as the sample size, n, increases, the sampling distribution of the sample mean, $\bar{X}$, becomes normally distributed with mean μ_X and standard deviation $\sigma_X / \sqrt{n}$. That is, for large enough samples, $\bar{X}$ is normally distributed with:

$$\mu_{\bar{X}} = \mu_X$$

$$\sigma_{\bar{X}} = \sigma_X / \sqrt{n},$$

even if the original variable, X, is not normally distributed.

Figure 6.12 The central limit theorem.

inches is $70 \pm 1\ \sigma_H$ and includes only 68% (0.3413×2) of the population. For possible averages of 4, though, $\sigma_{\bar{H}} = 3/\sqrt{4} = 1.5$, so 70 ± 3 inches is $70 \pm 2\sigma_{\bar{H}}$ and includes about 95% (0.4772×2) of the possibilities. And for possible averages of 16, $\sigma_{\bar{H}} = 3/\sqrt{16} = .75$, so 70 ± 3 inches is $70 \pm 4\sigma_{\bar{H}}$ and includes essentially all of the possibilities. For a random sample of 16 from this population, the sample average is almost certain to be within 3 inches of the population mean.

So far, we have assumed that the population distribution is normal. For large enough samples, though, the **Central Limit Theorem** (Figure 6.12) allows us to assume that the sampling distribution of the sample mean is normal, even when the population distribution is not.

How large must samples be to invoke the central limit theorem? That really depends on how far the population distribution departs from normal. Generally, 30 is considered large enough; in many practical cases, half of that may be enough. Figures 6.13 through 6.15 illustrate.

Figure 6.13 shows a uniform population distribution ($n = 1$), along with the distributions of averages for samples of size 4 and 16. The distribution for samples of 4 appears a little too broad in the "shoulders" to be normal; the distribution for samples of 16 appears quite normal.

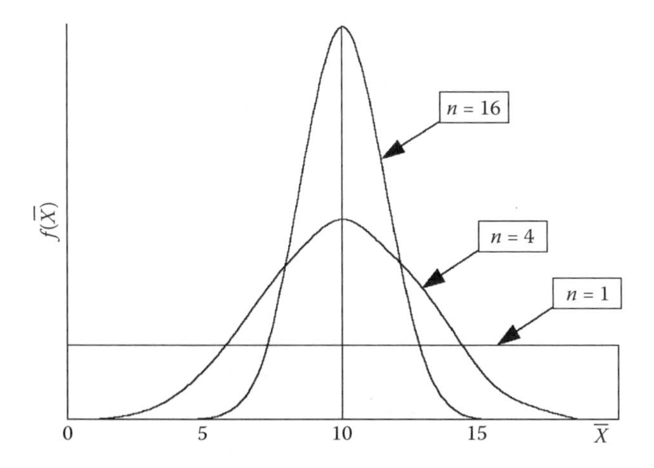

Figure 6.13 Uniform population distribution: The effect of sample size.

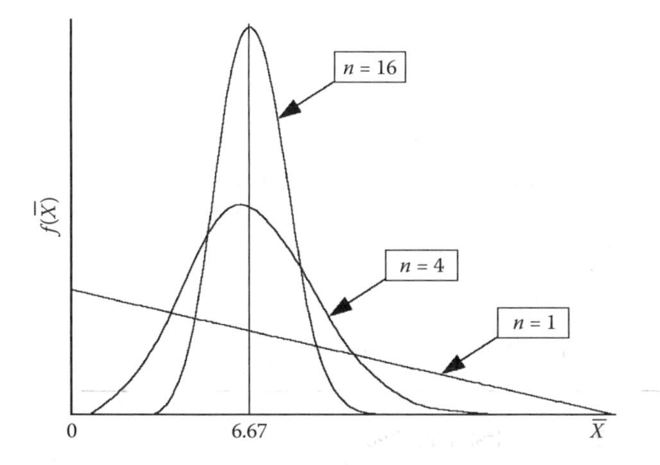

Figure 6.14 Skewed population distribution: The effect of sample size.

Figure 6.14 shows a skewed population distribution ($n = 1$), along with the distributions of averages for samples of size 4 and 16. The distribution for samples of 4 is clearly still a little skewed; indeed, the distribution for samples of 16 remains a bit skewed, though not enough to show in this diagram; a somewhat larger sample would be better. But clearly we are close.

Figure 6.15 shows a population distribution in which the probabilities are at the two extremes ($n = 1$), along with the distributions of averages for samples of size 4 and 16. The distribution for samples of 4 is clearly not normal; indeed, the distribution for samples of 16 remains a bit broad, though not enough to show in this diagram; a somewhat larger sample would be better. Again, though, we are close.

This last distribution would be a strange one actually to observe. Still, we could probably assume that samples of 30 are large enough.

Consider the following example:

A supplier claims that a gasket it supplies to you in large numbers averages 4 inches in inside diameter, with a standard deviation of 0.2 inches. Suppose this is true. What is the probability that you

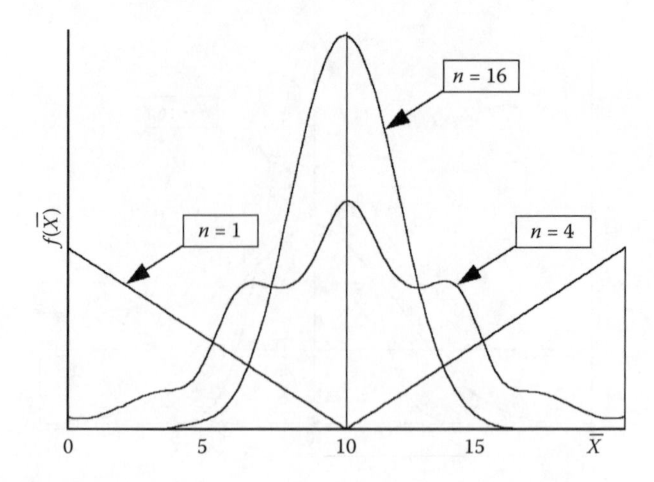

Figure 6.15 Extreme population distribution: Effect of sample size.

will get an average diameter of between 3.95 and 4.05–that is, an $\bar{X}$ within the range $\mu_X \pm 0.05$, if you take a sample of 25? if you take a sample of 100?

We know:

$$\mu_{\bar{X}} = \mu_X = 4.00$$

$$\sigma_X = 0.2$$

We want:

$$P(3.95 < \bar{X} < 4.05) = ??$$

First we need to calculate the standard error, $\sigma_{\bar{X}}$, for each sample:

$$\text{For } n_1 = 25: \sigma_{\bar{X}} = 0.2/\sqrt{25} = 0.04 \quad \Big| \quad \text{For } n_2 = 100: \sigma_{\bar{X}} = 0.2/\sqrt{100} = 0.02$$

Since the second sample is four times as large, it has only half as large a $\sigma_{\bar{X}}$. Thus, the sampling distribution is narrower, and numbers far from μ_X are less likely. Figure 6.16 illustrates.

Now, we need to find the Z value equivalents of 3.95 and 4.05. Of course, since they are symmetric about the mean of 4.00, we need calculate only one:

$$Z_1 = \frac{\bar{X} - \mu_{\bar{X}}}{\sigma_{\bar{X}}} = \frac{4.05 - 4.00}{0.04} = 1.25 \quad \Big| \quad Z_2 = \frac{\bar{X} - \mu_{\bar{X}}}{\sigma_{\bar{X}}} = \frac{4.05 - 4.00}{0.02} = 2.50.$$

Finally, we need to look up Z in the standard normal table and double the probability we find to include both sides:

$$P(3.95 < \bar{X} < 4.05) = 0.3944 \times 2 = 0.7888 \quad \Big| \quad P(3.95 < \bar{X} < 4.05) = 0.493\,8 \times 2 = 0.9876.$$

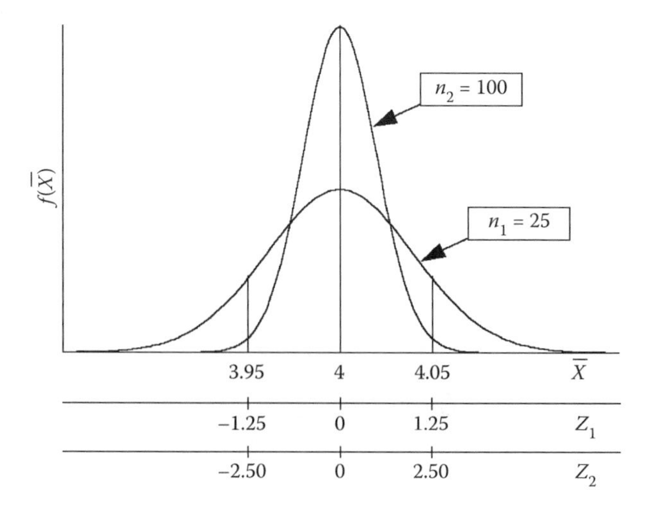

Figure 6.16 The sampling distribution of a mean: An example.

Again, because of the smaller $\sigma_{\overline{X}}$ with the larger sample, 3.95 and 4.05 rescale to Z values further from the mean. Therefore, the probability of $\overline{X}$ being within this range—$\mu_X \pm 0.05$—is higher.

Consider the following example:

> Suppose, in the previous example, you had wanted a 0.95 probability that $\overline{X}$ would be within the range $\mu \pm$ some interval. How wide an interval would you have needed with the sample of 25? with the sample of 100?

We know:	We want:
$\mu_{\overline{X}} = \mu_X = 4.00$	$\overline{X}_1 = ??$
$\sigma_X = 0.2$	$\overline{X}_2 = ??$
$P(\overline{X}_1 < X < \overline{X}_2) = .95$	

In the last problem, we started with a desired interval and samples sizes and found probabilities; in this one, we start with a desired probability and sample sizes and need to find intervals. Figure 6.17 illustrates.

We already know the standard errors, $\sigma_{\overline{X}}$:

$$\text{For } n_1 = 25{:}\ \sigma_{\overline{X}} = 0.2/\sqrt{25} = 0.04 \quad \Big| \quad \text{For } n_2 = 100{:}\ \sigma_{\overline{X}} = 0.2/\sqrt{100} = 0.02.$$

Since we want the probability within the interval to be 0.95, the probability from either $\overline{X}_1$ or $\overline{X}_2$ to the mean should be half of that, 0.4750. Using the standard normal table backward, the Z value that gives us this probability is 1.96. We want Z_i to equal ± 1.96. Since $Z_i = (\overline{X}_i - \mu_{\overline{X}})/\sigma_{\overline{X}} = \pm 1.96$, we can solve for $\overline{X}_i$. $\overline{X}_i = \mu_{\overline{X}} \pm 1.96 \times \sigma_{\overline{X}}$:

$\overline{X}_i = 4.00 \pm 1.96 \times .04$	$\overline{X}_i = 4.00 \pm 1.96 \times .02$
$= 4.00 \pm 0.0784$	$= 4.00 \pm 0.0392$

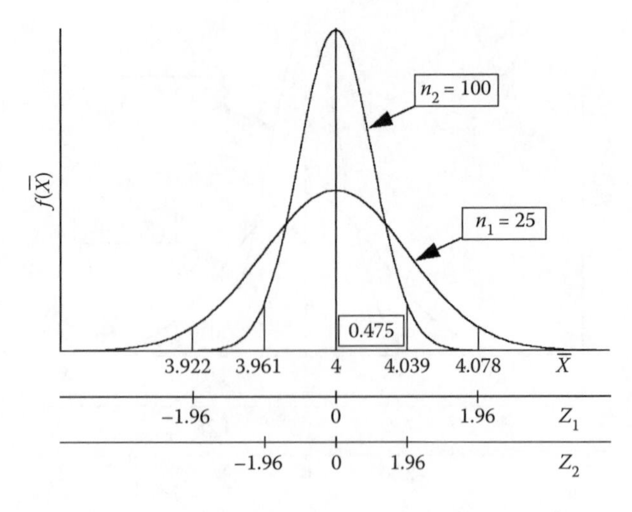

Figure 6.17 The sampling distribution of a mean: Another example (a).

This time, because of the smaller $\sigma_{\bar{X}}$ in the larger sample, the 0.95 probability interval is narrower.

Finally, consider the following example:

> Suppose, in the previous examples, you had wanted a 0.98 probability of an $\bar{X}$ between 3.95 and 4.05 inches—that is, an $\bar{X}$ within the range $\mu_X \pm .05$ inches. How large a sample would you have needed?

We know:	We want:
$\mu_{\bar{X}} = \mu_X = 4.00$	$n = ??$
$\sigma_X = 0.2$	
$P(3.95 < \bar{X} < 4.05) = 0.98$	

In the last two problems, we started with given sample sizes and either a desired interval or a desired probability and found the other; in this one, we start with a desired interval *and* probability and need to find the sample size that will give both. Figure 6.18 illustrates.

Since we want the probability between 3.95 and 4.05 inches to be 0.98, the probability from each to the mean should be half of that, 0.4900. We want Z_i to equal ±2.33. Since $Z_i = (\bar{X} - \mu_{\bar{X}})/\sigma_{\bar{X}} = (4.05 - 4.00)/\sigma_{\bar{X}} = 2.33$, we can solve for $\sigma_{\bar{X}}$. $\sigma_{\bar{X}} = (4.05 - 4.00)/2.33 = 0.02146$. We want the standard error, $\sigma_{\bar{X}}$, to equal 0.02146.

Finally, since $\sigma_{\bar{X}} = \sigma_X/\sqrt{n} = 0.2/\sqrt{n} = 0.02146$, we can solve for n. $n = 0.2^2/0.02146^2 = 86.86$. Rounding up to the next integer, we want n to equal 87. A sample of 87 would be large enough to give us a standard error, $\sigma_{\bar{X}}$, of 0.02146, which is small enough to give us a Z of 2.33. And Z values of ±2.33 are large enough to include 0.98 of the area under the standard normal.

Summing up, we can find the probability of $\bar{X}$ being in some desired interval around μ_X; the larger the sample, the higher this probability will be. We can find the interval around μ_X that will contain $\bar{X}$ with some desired probability; the larger the sample, the narrower this interval will

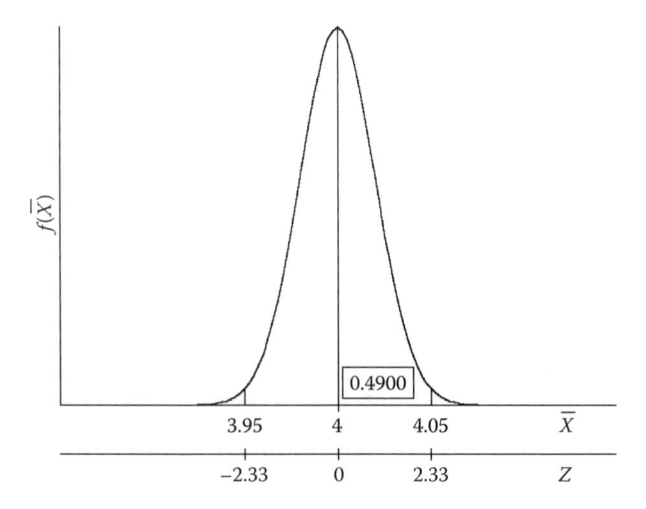

Figure 6.18 The sampling distribution for a mean: Another example (b).

be. And we can find the sample size required for a desired interval to contain $\overline{X}$ with some desired probability. The narrower the desired interval and/or the higher the desired probability, the larger this sample will need to be.

6.4.1 Two Complications

Before leaving the sampling distribution of a mean, there are two complications that need to be addressed. First, the formula we have been using for the standard error, $\sigma_{\overline{X}}$, assumes that the population from which we are drawing is infinite. Ordinarily, this is at least approximately right.

However, suppose we know the population size. Suppose we have just received a shipment of 500 gaskets, and we are going to evaluate this shipment by inspecting a randomly selected sample of 50. Our sample size, *n*, is 50; our population size, *N*, is 500. Notice, we now know $50/500 = 10\%$ of our population for sure. This should reduce our chances of getting a sample very different from the population; this should reduce the standard error, $\sigma_{\overline{X}}$. And indeed it does.

Figure 6.19 updates the formula for $\sigma_{\overline{X}}$ to include the **finite population correction factor**. Ignoring the −1 in the denominator, the ratio within the square root is just the proportion of the population *not* sampled.

Notice that the finite population correction factor here is exactly the same as the one on Section 6.3.1, for the standard error of a proportion. Thus, everything we said there about the effect of population size on the standard error applies here, as well. Ordinarily, the population is very large compared to the sample; hence, as we sample a larger and larger percentage of the population, our sample size itself gets large. And as the sample size gets larger, the standard error gets smaller regardless of how small it may still be compared to the population. A sample of 1000, which is often likely to be just a small percentage of the population, is still likely to be a large enough number in the denominator of the standard error to give all the precision we need.

$$\mu_{\bar{X}} = \mu_X$$

$$\sigma_{\bar{X}} = \frac{\sigma_X}{\sqrt{n}}\sqrt{\frac{N-n}{N-1}}$$

Figure 6.19 The mean and standard error of the sampling distribution of $\bar{X}$ when the population is finite.

In most cases, then, the population size makes little difference, and we can just treat it as infinite. Still, it is never incorrect to use the finite population correction factor when we know the population size. And, if the population is *not* large, it will make a difference.

The second complication that needs to be addressed looks again at where we are headed. In the next chapter, we will know the sample $\bar{X}$ and want to use it as an estimate of the unknown μ_X. And we will want to argue that, since there is an $A\%$ chance of $\bar{X}$ being within $\mu_X \pm z \times \sigma_{\bar{X}}$, then $\bar{X} \pm z \times \sigma_{\bar{X}}$ has an $A\%$ chance of including μ_X. But if we do not know μ_X, we are unlikely to know σ_X either. And without σ_X, we cannot calculate $\sigma_{\bar{X}}$. We will need to estimate it.

We ran into this same problem with proportions. There, when we did not know π_X, the population proportion, we simply substituted p_X, the sample proportion. Almost nothing else changed; however that was because we were working just with large samples. With means, we have not limited ourselves to large samples. And with small samples, things are a bit more complicated.

6.5 The Sampling Distribution of a Mean: σ_X Unknown

Figure 6.20 summarizes the changes when σ_X is unknown. The statistic $s_{\bar{X}}$ is our estimate of the true standard error, $\sigma_{\bar{X}}$. The first changes are the ones you should have expected. Since s_X is our best estimate of σ_X, it makes sense that we would use it where ever we need σ_X and do not know it. A minor secondary change is that the −1 disappears from the finite population correction factor, just as it did with proportions.

But there is more. With small samples, we cannot ignore the fact that $s_{\bar{X}}$ is just an estimate of $\sigma_{\bar{X}}$. If X is normally distributed, so is $\bar{X}$, no matter how small the sample; hence, so is $\bar{X} - \mu_{\bar{X}}$. When we divide through by $s_{\bar{X}}$, though, we are dividing through by another random variable, with its own probability distribution and the result is no longer a normal distribution.

Fortunately, the resulting distribution is known. It is the "*t* distribution," sometimes known as "Student's *t*," because it was first published by W. S. Gosset under the pseudonym "Student." The distribution looks a fair amount like the normal distribution. It is, however, broader than the normal. This makes sense because the use of an estimate, $s_{\bar{X}}$, instead of the actual standard error, $\sigma_{\bar{X}}$, introduces an additional source of variability.

$$s_{\bar{X}} = \frac{s_X}{\sqrt{n}}\sqrt{\frac{N-n}{N}} \quad \text{where } s_X = \sqrt{\frac{\sum(X-\bar{X})^2}{n-1}}$$

$$t = \frac{\bar{X} - \mu_{\bar{X}}}{s_{\bar{X}}}$$

Figure 6.20 Summary of changes when σ_X is unknown.

How much broader depends on the size of the sample. Again, this makes sense because $s_{\bar{X}}$ should become a better and better estimate of $\sigma_{\bar{X}}$, as the size of the sample from which it is calculated increases. The *t* distribution becomes normal as the sample size goes to infinity; indeed, it becomes essentially normal fairly quickly.

Actually, it is not the sample size, *per se*, but the related **degrees of freedom** that matters for *t*. Figure 6.20 repeats the formula for s_X, to remind you that we divide by $n - 1$, not n, in calculating the sample standard deviation. Recall from Chapter 3 (Section 3.2.2.2) that we do this because we are using $\bar{X}$, not μ in the numerator; hence, there are only $n - 1$ independent deviations from the mean. This $n - 1$ is our degrees of freedom (df). Figure 6.21 shows the *t* distribution with 5 df along with the standard normal (*z*). The *t* distribution is shorter, with bigger tails.

6.5.1 The *t* Table

Table 3, in Appendix C, shows the *t* table as it is usually laid out. Find it now. Notice that the *t* table is laid out quite differently from the standard normal (*z*) table; this is partly of necessity and partly just convention.

First, while the standard normal table relates just two things, *z* and *P*(*z*), the *t* table must relate three, df, *t*, and *P*(*t*, df). Thus, each row represents a df, and each column represents a probability. You can think of each row as

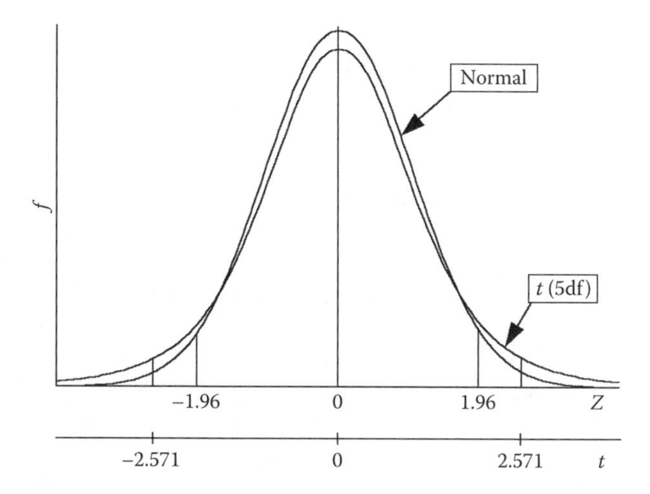

Figure 6.21 The standard normal (*z*) and *t* distribution with five degrees of freedom.

giving points along a particular t distribution, combinations of t and $P(t)$, for that particular df.

Second, look at the diagram at the top. It indicates that the probabilities in the table are not those between t_0 and the mean, but rather those between t_0 and positive infinity. While this change is not necessary, it is conventional. Still, if you pick up another statistics book, you might find another area shaded. Just as with the normal, it really does not matter what area under the curve the table gives, just so long as you know and take it into account.

We can compare the t with the standard normal (z) using Figure 6.21. Suppose we want the $\pm z_0$ values that include 95% of the standard normal. We have done this many times now. To include 95% the area from $+z_0$ to the mean must be half of that −0.475; looking up 0.475 in the standard normal table gives 1.96; the range ±1.96 includes 95% of the standard normal.

Now suppose we want the $\pm t_0$ values that include 95% of the t distribution with 5 df. To include 95%, it must *exclude* 5%. To exclude 5%, the area from t_0 to positive infinity must be half of that, 0.025; looking at the 0.025 column in the t table and reading down to 5 df, gives 2.571; the range ±2.571 includes 95% of the t with 5 df.

Recall now that the t distribution becomes normal as df goes to infinity. We can see this in the t table. Read down the 0.025 column; notice how the numbers get smaller and smaller; notice that the bottom number is 1.960, the same number that we got from the standard normal.

6.5.2 The *t* Spreadsheet Functions

Like the standard normal table, the t table is simple and widely available. Still, there are times when it would be nice to find the t value for a probability or df that is not tabulated. Figure 6.22 shows the special functions for several of the most common spreadsheets.

Typically, =TDIST requires that we specify values for t_0 and df, as well as whether we want one or two tails, and returns either the probability beyond t_0 (one tail) or beyond $\pm t_0$ (two tails). In the last example of t with 5 df, =TDIST(2.571,5,1) = 0.025, since 0.025 is the probability either above 2.571 or below −2.571. Then =TDIST(2.571,5,2) = 0.05 since it includes both tails.

For some reason, =TINV typically works a bit differently. It requires that we specify values for a probability and df, and simply assumes that the probability should be divided between the two tails. In the last example of t with 5 df, we wanted the $\pm t_0$ values that included 95% of the distribution; that meant we needed to *exclude* 5%, half in each tail. Using Table 3,

	Formula	Example
From t_0 to a probability:	=TDIST(t_0, df, tails)	=TDIST(2.571, 5, 2) = 0.05
From a probability to t_0:	=TINV(p, df)	=TINV(.05, 5) = 2.571
(For Lotus/Quattro Pro, replace " = " with "@")		

Figure 6.22 The t distribution special functions in common spreadsheet programs.

we divided 0.05 in half and used the 0.025 column. This was because the probabilities in Table 3 refer to a single tail. Using =TINV, we would not have divided the 0.05 in half. This is because =TINV assumes the probability refers to both tails, =TINV(0.05,5) = 2.571.

Whenever you switch books or computer programs, it is important to make sure you know what assumptions the tables or programs make. With tables, look for a diagram that shows what area it is associating with the t_0 values. With tables or programs, try out some cases for which you already know the answer.

Consider the previous examples, but now assume we do not know the standard deviation, σ_X:

> A supplier claims that a gasket it supplies to you in large numbers averages 4 inches in inside diameter. Suppose this is true. What is the probability that you will get an average diameter of between 3.95 and 4.05 if you take a sample of 25? if you take a sample of 100? Assume that, in both cases, when you find the sample standard deviation, s_X, it equals 0.2 inches.

We know:	We want:
$\mu_{\bar{X}} = \mu_X = 4.00$	$P(3.95 < \bar{X} < 4.05) = ??$
$s_X = 0.2$	

First we need to calculate the estimated standard error, $s_{\bar{X}}$, for each sample:

For $n_1 = 25: s_{\bar{X}} = 0.2/\sqrt{25} = 0.04$ | For $n_2 = 100: s_{\bar{X}} = 0.2/\sqrt{100} = 0.02$.

Since the second sample is four times as large, it has only half as large a $s_{\bar{X}}$. Thus, the sampling distribution is narrower, and numbers far from μ_X are less likely. Figure 6.23 illustrates.

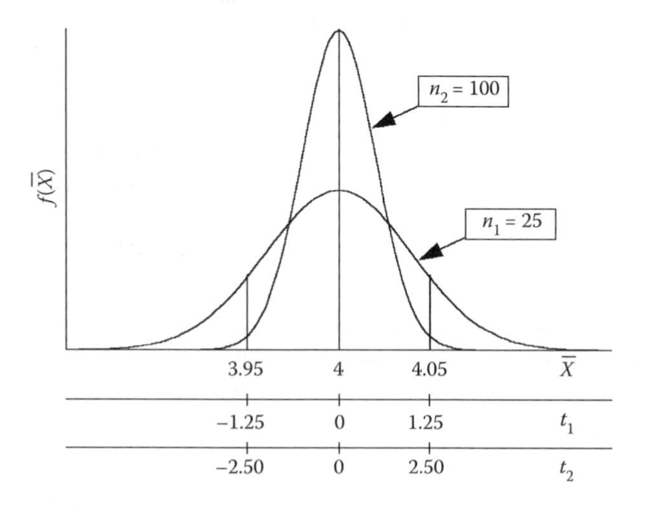

Figure 6.23 The sampling distribution of a mean: An example.

Now, we need to find the t value equivalents of 3.95 and 4.05. Of course, since they are symmetric about the mean of 4.00, we need calculate only one:

$$t = \frac{\bar{X} - \mu_{\bar{X}}}{s_{\bar{X}}} = \frac{4.05 - 4.00}{0.04} = 1.25 \quad \Bigg| \quad t = \frac{\bar{X} - \mu_{\bar{X}}}{s_{\bar{X}}} = \frac{4.05 - 4.00}{0.02} = 2.50.$$

Finally, we need to find t in the t table or with the spreadsheet function, for the proper degrees of freedom. Relying on the table, we can get only approximate results.

The smaller sample has 24 df. Looking across that row, the closest we can get to 1.25 is 1.318, which has a probability of 0.1000 above it. Doubling that, to account for both tails, and subtracting from 1, we get a probability between 3.95 and 4.05 of approximately 0.8000. Using the spreadsheet function, we get = 1 – TDIST(1.25,24,2) = 0.7766.

The larger sample has 99 df, but 99 is not in the table, so we use 100. Looking across that row, the closest we can get to 2.50 is 2.626, which has a probability of 0.0050 above it. Doubling that, to account for both tails, and subtracting from 1, we get a probability between 3.95 and 4.05 of approximately 0.99. Using the spreadsheet function, we get = 1 – TDIST(2.50,99,2) = 0.9859.

If it strikes you that the t table is not especially well laid out for dealing with this problem, you are right. Indeed, you would have found more accurate estimates for these probabilities by treating these t-values as Z-values and looking them up in the standard normal table. However, this will be less of a problem than it may appear. Ordinarily, we will start with a desired probability and look for the corresponding t. The probabilities we will usually want are the nice round ones that are tabulated in the t table.

Consider the following example:

Suppose, in the previous example, you had wanted a 0.95 probability that $\bar{X}$ would be within the range $\mu \pm$ some interval. How wide an interval would you have needed with the sample of 25? with the sample of 100?

We know:	We want:
$\mu_{\bar{X}} = \mu_X = 4.00$	$\bar{X}_1 = ??$
$s_X = 0.2$	$\bar{X}_2 = ??$
$P(\bar{X}_1 < \bar{X} < \bar{X}_2) = 0.95$	

In the last problem we started with a desired interval and samples sizes and found probabilities; in this one we start with a desired probability and sample sizes and need to find intervals. Figure 6.24 illustrates.

We already know the estimated standard errors, $s_{\bar{X}}$

$$\text{For } n_1 = 25 : s_{\bar{X}} = 0.2 / \sqrt{25} = 0.04 \quad \Big| \quad \text{For } n_2 = 100 : s_{\bar{X}} = 0.2 / \sqrt{100} = 0.02.$$

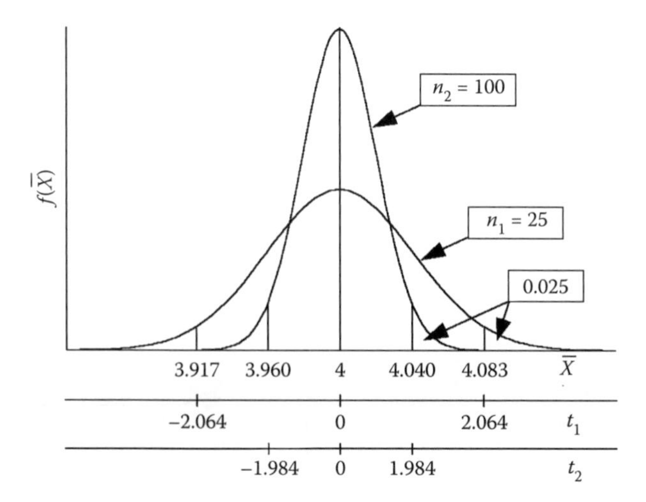

Figure 6.24 The Sampling Distribution of a Mean: Another example (a).

Since we want the probability within the interval to be 0.95, the probability above $\bar{X}_2$ should be 0.05/2 = 0.025. We use the 0.0250 column of the t table.

The smaller sample has 24 df. The t value in the 0.0250 column for 24 df is 2.064. We want t to equal 2.064. Since $t = (\bar{X}_i - \mu_{\bar{X}}) / s_{\bar{X}} = \pm 2.064$, we can solve for $\bar{X}_i$. $\bar{X}_i = \mu_{\bar{X}} \pm 2.064 \times s_{\bar{X}}$. $\bar{X}_i = 4.00 \pm 2.064 \times 0.04 = 4.00 \pm 0.08256$.

The larger sample has 99 df, but 99 is not in the table, so we use 100. The t value in the 0.0250 column for 100 df is 1.984. We want t to equal 1.984. Since $t = (\bar{X}_i - \mu_{\bar{X}}) / s_{\bar{X}} = \pm 1.984$, we can solve for $\bar{X}_i$. $\bar{X}_i = \mu_{\bar{X}} \pm 1.984 \times s_{\bar{X}}$. $\bar{X}_i = 4.00 \pm 1.984 \times 0.04 = 4.00 \pm 0.03968$.

This time, the larger sample gives the narrower interval because t as well as $s_{\bar{X}}$ is smaller. But both are at least a little wider than the one we found using Z because t is greater than Z.

Finally, consider the following example:

> Suppose, in the previous examples, you had wanted a 0.98 probability that $\bar{X}$ would be between 3.95 and 4.05 inches—that is, within the range $\mu_X \pm 0.05$ inches. How large a sample would you have needed?

We know:	We want:
$\mu_{\bar{X}} = \mu_X = 4.00$	$n = ??$
$P(3.95 < \bar{X} < 4.05) = 0.98$	

In the last two problems we started with given sample sizes and either a desired interval or a desired probability and found the other; in this one we start with a desired interval *and* probability and need to find the sample size that will give both. Figure 6.25 illustrates.

Since we want the probability between 3.95 and 4.05 inches to be 0.98, the probability above 4.05 should be 0.02/2 = 0.01. We use the 0.0100 column of the t table. Here we hit a snag, though. We need

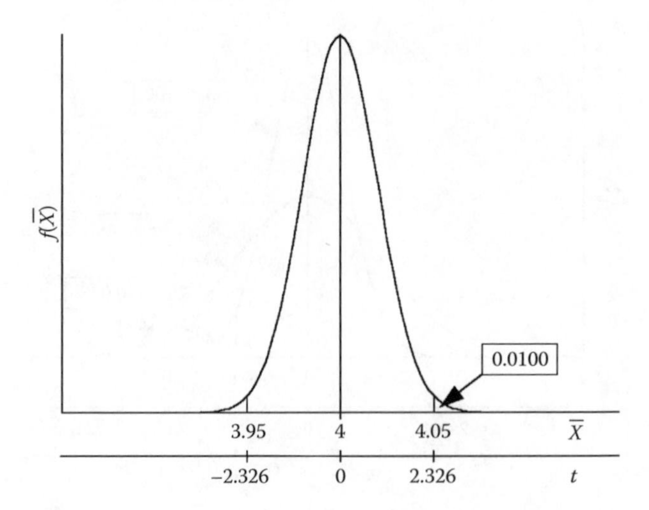

Figure 6.25 The sampling distribution for a mean: Another example (b).

degrees of freedom and that depends on n, which is what we are trying to find. We need to guess at approximately what our answer is going to be, to come up with a t value. Suppose we think there is reason to believe that the sample will end up needing to be large; we might go ahead and use the Z value equivalent, which we can get from the bottom of the t table, 2.326. We will assume (for now) that we want t to equal ±2.326.

Since $Z = \bar{X}_i - \mu_{\bar{X}} / s_{\bar{X}} = 4.05 - 4.00 / s_{\bar{X}} = 2.326$, we can solve for $s_{\bar{X}}$. $s_{\bar{X}} = 4.05 - 4.00 / 2.326 = 0.02150$. We want the standard error, $s_{\bar{X}}$ to equal 0.02150.

Since $s_{\bar{X}} = s_X / \sqrt{n} = 0.02150$, we can solve for n. $n = s_X^2 / 0.02150^2$. However, here we hit a second snag. s_X is the standard deviation of the sample that we have not yet taken. Recall, we are still trying to decide how big it should be. We again need to guess. Perhaps there is evidence from past studies as to roughly how big s_X will turn out to be.

Suppose we guess that it will turn out to be about 0.2. Then, $n = 0.2^2 / .02150^2 = 86.53$. Rounding up to the next integer, we want n to equal 87.

Of course, one of our initial guesses has already proven wrong. An n of 87 implies 86 df; we should have chosen a slightly larger t value. We could go back through the calculations above or just round our n up a little more. Then, assuming our guess for s_X turns out to have been large enough, $s_{\bar{X}}$ will be small enough, and our interval of $\mu_X \pm 0.05$ will be wide enough to include 0.98 of the area under the t distribution.

Summing up, we can find the probability of $\bar{X}$ being in some desired interval around μ_X; the larger the sample, the higher this probability will be. We can find the interval around μ_X that will contain $\bar{X}$ with some desired probability; the larger the sample, the narrower this interval will be. And we can find the sample size required for a desired interval to contain $\bar{X}$ with some desired probability. The narrower the desired interval and/or the higher the desired probability, the larger this sample will need to be. Not knowing the population standard deviation, σ_X, complicates some of these, but they can still be done.

6.6 Other Sampling Distributions

We have paid a great deal of attention to the sampling distributions for p_X and $\bar{X}$, because these are the ones we will be making use of first. We will be exploring, in the next two chapters, what we can say is probably true about π_X and μ_X, based on just sample values p_X and $\bar{X}$. We will certainly want to go beyond just π_X and μ_X, though. We may want to test whether there is a *difference* in π_1 and π_2, the proportion of defective parts turned out by two machines. We will be able to address this question because we will know the sampling distribution of $p_1 - p_2$, the difference in sample proportions we could get from a population in which the true proportions are the same. We may want to know if there is a *difference* in μ_m and μ_w, the average salaries of men and women. We will be able to address this question because we will know the sampling distribution of $\bar{X}_m - \bar{X}_w$, the difference in sample means we could get from a population in which the true means are the same. Or we may want to know if salaries go up with additional years of education. We will be able to address this question because we will know the sampling distribution of the sample slopes we could get from a population in which the true slope is zero.

In each case, we will have a sample statistic that is an estimate of the unknown population parameter. And since we will know the sampling distribution of that statistic, we will be able to say what we think is true of the population parameter, with a related probability that we are correct.

6.7 Exercises

6.1 Your supplier claims that only 5% of the parts it supplies are defective. Suppose this is true.

a. If you take a sample of 10 parts, what is the probability that it will contain at least 8% defective parts?

b. If you take a sample of 100 parts, what is the probability that it will contain at least 8% defective parts?

c. If you take a sample of 400 parts, what is the probability that it will contain at least 8% defective parts?

6.2 Suppose, in the situation above, you are not really sure that your supplier's claim is true.

a. If your sample of 10 has 8% defectives, would you be concerned?

b. If your sample of 100 has 8% defectives, would you be concerned?

c. If your sample of 400 has 8% defectives, would you be concerned?

6.3 The Energetic Corporation produces electric light bulbs. When everything is going properly, the lengths of these light bulbs are normally distributed, with a mean of 3.00 inches, and a standard deviation of 0.10 inches. Suppose everything is going properly.

a. If you take a random sample of just 1 bulb, what is the probability that its length will be within the range of 2.95–3.05 inches?
b. If you take a random sample of 10 bulbs, what is the probability that its mean length will be within the range of 2.95–3.05 inches?
c. If you take a random sample of 40 bulbs, what is the probability that its mean length will be within the range of 2.95–3.05 inches?

6.4 Suppose, in the situation above, you do not really know that everything is going properly.
a. If your sample of 1 has a length outside the range of 2.95–3.05, would you be concerned?
b. If your sample of 10 has a mean length outside the range of 2.95–3.05, would you be concerned?
c. If your sample of 40 has a mean length outside the range of 2.95–3.05, would you be concerned?

6.5 Suppose a random sample of 40 students is drawn from a population of 80 students whose IQs average 118, with a (population) standard deviation of 10.
a. What is the probability that the mean of this sample exceeds 120?
b. If the population contained 800 students, how would your answer change?
c. If the population contained 8000 students, how would your answer change?
d. If the population were infinite, how would your answer change?

6.6 Suppose you are an auditor, wishing to estimate the true mean dollar amount of a company's accounts receivable. You know that the standard deviation of these dollar amounts is $1000.
a. If you take a random sample of 100 accounts, what is the probability that your sample mean will be within $100 of the true mean?
b. If you take a random sample of 500 accounts, what is the probability that your sample mean will be within $100 of the true mean?

6.7 Continue with the situation above.
a. If you want the probability of being within $100 to be 0.95, how big a random sample do you need?
b. If you take a sample of this size, how many times out of 100 will your sample mean be within $100 of the true mean?
c. If you take a sample of this size, how many times out of 100 will the true mean be within $100 of your sample mean?

6.8 A company samples delinquent accounts, in order to estimate the mean amount owed. They want a 0.95 probability that their estimate is within ±$A of the true mean.
a. If 25 are sampled and the sample standard deviation is $20.00, how large is $A?

b. If 100 are sampled and the sample standard deviation is \$20.00, how large is \$A?

6.9 Continue with the situation above; suppose they find the results above and are dissatisfied.

a. If they want \$A to be only \$2.00, how big a random sample do they need?

b. If they take a sample of this size, how many times out of 100 will their sample mean be within \$2.00 of the true mean?

c. If they take a sample of this size, how many times out of 100 will the true mean be within \$2.00 of their sample mean?

7

Estimation and Confidence Intervals

With this chapter, we begin formal inference. From here on, we will want to say something about an unknown population parameter based solely on a sample.

There are two sorts of questions we can address. First, we can ask, "based on this sample, what is our best estimate for some population parameter; and how good is this estimate?" Second, we can ask, "based on this sample, can we conclude that something is true of a population parameter; and how sure are we?"

We will start by building on what we know from Chapter 6 about the sampling distributions of the proportion and mean. In this chapter, we will answer the first question. We will arrive at estimates of π_X and μ_X that have attached probabilities of being right. In Chapter 8, we will answer the second question. We will arrive at conclusions concerning the size of π_X and μ_X that have attached probabilities of being wrong. Chapters 9 through 13, then, extend the same basic reasoning to additional, more interesting cases.

7.1 Point and Interval Estimators of Unknown Population Parameters

Point estimators are simply sample statistics that we can use to estimate unknown population parameters. It should come as no surprise that our point estimator of π_X, the population proportion, is going to be p_X, our sample proportion, and our point estimator of μ_X, the population mean, is going to be $\bar{X}$, our sample mean. Indeed, these may seem so intuitively obvious that they need no justification. However, intuition can sometimes be misleading; it is best to have criteria for choosing our estimators.

7.1.1 Qualities of a Good Point Estimator

There are a number of criteria for evaluating estimators. We will consider just two: unbiasedness and efficiency.

7.1.1.1 Unbiasedness

A statistic is an unbiased estimator of a parameter if its expected value equals that parameter. There is, then, no systematic tendency for the statistic to be too high or low.

Recall from Chapter 6 that the sampling distribution of p_X centered on π_X, and the sampling distribution of $\bar{X}$ centered on μ_X. The average, or expected value, of p_X is π_X; the average, or expected value, of $\bar{X}$ is μ_X. Both p_X and $\bar{X}$ are unbiased estimators of their respective parameters.

In some cases, though, we need to calculate a sample statistic in a non-intuitive way in order for it to be unbiased. Recall that when we calculate the sample standard deviation, s_X, we divide by $n - 1$, instead of the more intuitively-appealing n, to avoid it being too small on average. Dividing by n would have made s_X a biased estimator of σ_X; dividing by $n - 1$ makes it unbiased.

7.1.1.2 Efficiency

There is often more than one unbiased estimator of a parameter. When there is, the best is the one that is most efficient. Consider perhaps a silly example. We could estimate μ_X by taking just the first case in a sample, X_1, and ignoring all the rest. Since there is no tendency for the first case to be too high or low, X_1 would be, like $\bar{X}$, an unbiased estimator of μ_X. But, clearly, we would not be using the sample information available to us efficiently. The X_1 would have a larger standard error than $\bar{X}$. Its sampling distribution would be wider. The chance would be greater of it giving us a misleading estimate of μ_X. We want the statistic with the smallest standard error, hence the narrowest sampling distribution. This reduces, as much as possible, the chance of getting a sample value that is a misleading estimate of the parameter.

Both p_X and $\bar{X}$ are efficient, as well as unbiased, estimators of their respective parameters. Indeed, they are sometimes referred to as "BLUE," the Best (most efficient) Linear Unbiased Estimators of π_X and μ_X.

7.1.2 Point versus Interval Estimators

The problem with point estimators is that they are unlikely ever to be *exactly* right. Hence, we want some way of attaching to an individual estimate some measure of how good it might be. We do this by establishing an interval around the point estimate for which we can make some probability statement. For example, we saw in Chapter 6 that for large samples the interval $\pi_X \pm 1.96\ \sigma_p$ contains 95% of all possible p_X values. For those 95% of all p_X values, the interval $p_X \pm 1.96\ \sigma_p$ is correct in the sense that it contains the parameter π_X within it. And of course, for the other 5% of values, the interval $p_X \pm 1.96\ \sigma_p$ is incorrect in the sense that it does not contain the parameter π_X within it. The $p_X \pm 1.96\ \sigma_p$ is called a "95% confidence interval" estimator of π_X. It yields correct estimates 95% of the time.

In any particular case, we will not know whether we have been lucky or unlucky, but the narrower the interval, or the higher the probability associated with it, the better. Consider two 95% confidence intervals, $p_X \pm 0.03$ and $p_X \pm 0.01$. With the first, there is a 5% chance that our estimate is off by as much as 0.03 whereas with the second there is a 5% chance that our estimate is off by as much as 0.01. Clearly, the second is the more precise. Likewise, consider two intervals of $p_X \pm 0.02$, but suppose the first is a 90% confidence interval and the second is a 99% confidence interval. With the first, there is a 10% chance that our estimate is off by as much as 0.02 whereas with the second, there is just a 1% chance that our estimate is off by that much. Clearly, the second is the better estimate.

Confusion over confidence intervals is common, so we should take some time to clarify what they are and are not.

First, a 95% confidence interval *does not* "contain 95% of the data." This is a confusion of confidence intervals with the empirical rule in Chapter 3. The empirical rule said that an interval of $\pm 2s_X$—that is, ± 2 sample standard deviations—contains approximately 95% of the data. It was a way of thinking about what the sample data looked like. In contrast, confidence intervals are ways of thinking about what some unknown population parameter might be. They are based on the theoretical sampling distributions that we studied in Chapter 6.

Second, a 95% confidence interval *is not* "95% correct." A particular interval is either correct or incorrect.

Third, a 95% confidence interval *does not* imply that we have done something wrong 5% of the time. Assuming that we do everything right, we will still be unlucky 5% of the time.

Rather, a 95% confidence interval is simply an estimate of an unknown population parameter generated in a way that will be correct 95% of the time and incorrect 5% of the time.

As a philosophical aside, some authors (and instructors) object to saying that *a particular* 95% confidence interval has a 0.95 probability of being right on the grounds that, once it is taken, it is either right or it is not. However, I will not avoid such language.

An analogy may help explain the issue. Suppose we were gamblers, and I offered you a bet on the outcome from the flip of a fair coin; to ensure fairness a third person will flip the coin. A fair coin lands heads 50% of the time. Hence, 0.50 is the probability we would each use in deciding what bets we were willing to make. Now, suppose the third person flips the coin, but hides the result. What is the probability that the result is heads now? In one sense, either one or zero; the result is either heads or it is not. Still, since we have not seen the coin, we do not know what came up. In this sense, nothing has changed. And 0.50 is still the probability we would have to use in deciding what bets we were willing to make.

Now, suppose that, instead of flipping a coin, the third person is going to estimate a confidence interval in a way that will be right 95% of the time. Then 0.95 is the probability we would each use in deciding what bets we were willing to make. Now, suppose the third person actually estimates the interval. What is the probability that it is right now? In one sense, either one or zero; the result is either right or it is not. Still, we do not know what came up. In this sense, nothing has changed. And 0.95 is still the probability we would have to use in deciding what bets we were willing to make. In this sense, it seems reasonable to think that the probability is still 0.95.

7.2 Estimates of the Population Proportion

We began already in the last section using estimates of proportions as examples of the reasoning we would be using in general. Figure 7.1 illustrates it graphically.

Assume we want a 95% confidence interval for π_X. We start by taking a sample, and calculating our best point estimate, p_X. This will be the midpoint of our confidence interval.

To make our interval a 95% confidence interval, we find the range, $\pm Z$, which includes 95% of the standard normal distribution. By now, you know that this range is ± 1.96. Let i stand for the upper and lower bounds of this range. Since we know that $Z_i = (p_i - \pi_X)/\sigma_p = \pm 1.96$, we can rearrange terms for the $p_i = \pi_X \pm 1.96\sigma_p$ that bracket 95% of all possible p_X values. In Chapter 6, when we knew π_X, we were able to solve for these p_i. Now, though, we cannot because we do not know π_X—that is what we are trying to estimate. Neither, then, do we know whether our particular sample p_X is within the interval. But we do know the approximate width of the interval. And we do know that 95% of the sample p_X values we *could* have are, like p_1 within this interval, and that 5% of the sample p_X values we *could* have are, like p_2 outside this interval.

Finally, notice that, for the 95% of possible sample p_X values that are, like p_1, in the interval $\pi_X \pm 1.96\ \sigma_p$, the interval $p_X \pm 1.96\ \sigma_p$ is right; it includes π_X. For the 5% of possible sample p_X values that are, like p_2, not in the interval $\pi_X \pm 1.96\ \sigma_p$, the interval $p_X \pm 1.96\ \sigma_p$ is not right; it does not include π_X. Thus, the interval $p_X \pm 1.96\ \sigma_p$ is right 95% of the time. There is a 95% probability that it contains the true population value of π_X. It is a 95% confidence interval.

I said above that we know the *approximate* width of the interval that includes 95% of all possible p_X values. The width is $\pm 1.96\ \sigma_p$; however, calculating σ_p requires knowing π_X, and if we knew π_X we would not be estimating it. We will need to use p_X to calculate s_p, and our 95% confidence interval will be $p_X \pm 1.96\ s_p$, instead. Since we are working with

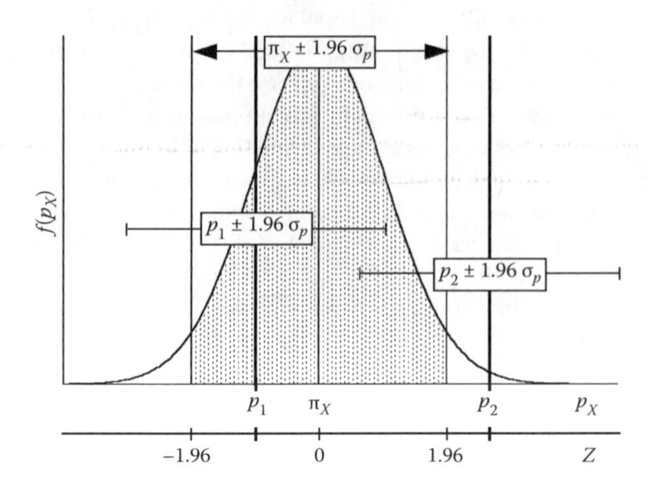

Figure 7.1 Confidence intervals for a proportion.

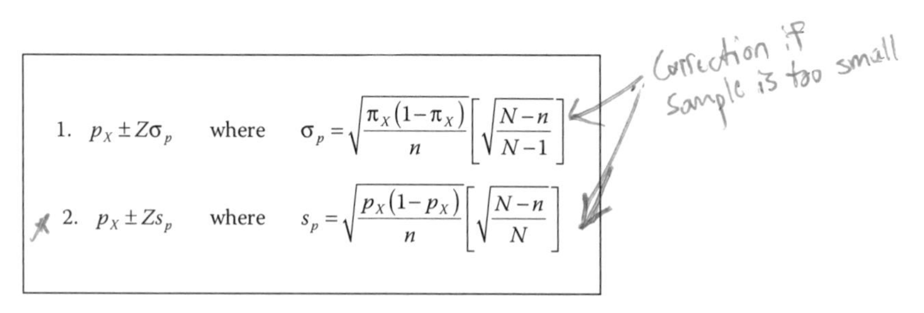

1. $p_X \pm Z\sigma_p$ where $\sigma_p = \sqrt{\frac{\pi_X(1-\pi_X)}{n}}\left[\sqrt{\frac{N-n}{N-1}}\right]$

2. $p_X \pm Zs_p$ where $s_p = \sqrt{\frac{p_X(1-p_X)}{n}}\left[\sqrt{\frac{N-n}{N}}\right]$

Figure 7.2 Confidence intervals for a proportion: The formulas.

large samples, it is a good approximation. Figure 7.2 summarizes; we will be using version 2.

The discussion so far has been in terms, specifically, of a 95% confidence interval; hence, the ± 1.96. And 95% is a commonly used level of confidence. However, any level of confidence can be had (except 100%). If you want a 80%, 90%, or 99% confidence interval, you just look up the Z values that correspond ($\pm 1.28, \pm 1.65, \pm 2.58$). Of course, given a particular sample, you have only a certain amount of information. Higher levels of confidence are obtained by making the interval wider—by being less precise.

Suppose you want a confidence interval to have both a certain probability and a certain precision. Perhaps you want a 99% confidence interval that is no wider than ± 0.03. The Z value for 99% confidence is 2.58, so you want $\pm 2.58\ s_p \leq 0.03$. This means you want $s_p \leq 0.03/2.58 = 0.01163$. Looking at the formula for s_p (and ignoring the finite population correction factor) you want $s_p = \sqrt{p_X(1-p_X)/n} = 0.01163$ Solving for n, $n = p_X(1 - p_X)/(0.01163)^2$. For any p_X there is a sample size n that will give you both the probability and the precision you want.

One final difficulty in this sort of problem is that p_X is the proportion of the sample whose size you are currently determining. You have not yet taken the sample, so you do not yet know p_X. You will need to guess. Perhaps, you have information on a previous sample p_X that you can use. If not, the worst case—the case that requires the largest n—is 0.50. If 0.50 is—a plausible p_X to get in your sample, use it in calculating n. This way you are sure to get the confidence and precision you want, no matter what p_X turns out to be. Otherwise, ask yourself what the worst plausible case is—the plausible p_X closest to 0.50. Perhaps you are estimating the proportion of defectives produced by a particular machine, and think surely it will be no worse than 0.10. Then the plausible range is 0.00–0.10. The worst plausible case is 0.10; use it in calculating n. In this case, using 0.50 would lead you to take an unnecessarily large (and expensive) sample.

Consider the following example:

> Senator Smith appears to be in a very close race for reelection; you take a random sample of 400 likely voters. Suppose you find that 220 of the 400 likely voters prefer Smith. Construct a 98% confidence interval for the proportion of all likely voters who prefer Smith.

We know:	We want:
$n = 400$	98% CI for π_{Smith}
$X = 220$ prefer Smith	

We know that the confidence interval will be of the form $p_{Smith} \pm Zs_p$; we need values for p_{Smith}, Z, and s_p. Then $p_{Smith} = 220/400 = 0.55$; $s_p = \sqrt{0.55 \times 0.45 / 400} = \sqrt{0.0006188} = 0.0249$. And the $\pm Z$ values that include 98% of the standard normal distribution are ± 2.33. Piecing these together

$$98\% \text{ CI: } 0.55 \pm 2.33 \times 0.0249$$

$$\text{or } 98\% \text{ CI: } 0.55 \pm 0.058$$

$$\text{or } 98\% \text{ CI: } 0.492 < \pi_{Smith} < 0.608.$$

Consider the following example:

In the previous example, the 98% confidence interval stretched from 0.492 to 0.608. That is not very precise; you cannot even say with 98% confidence whether Smith has more or less than half the vote. If you want a 98% confidence interval no wider than ± 0.03, by how much do you need to increase your sample?

We know:	We want:
98% CI for π_{Smith}	$n = ??$
$Zs_p \le 0.03$	

Since the confidence level desired is unchanged from before, the $\pm Z$ values are still ± 2.33. Since $2.33\ s_p \le 0.03$, then, $s_p \le 0.03/2.33 = 0.01288$. Thus, $s_p = \sqrt{p_{Smith}(1 - p_{Smith})/n} = 0.01288$ or, rearranging terms, $n = p_{Smith}(1 - p_{Smith})/0.01288^2$.

The proportion p_{Smith} is the proportion we will find in our new larger sample; thus, we do not know it yet. We need to guess. We might use the value $p_{Smith} = 0.55$ from the previous sample. Or we might reason that, with the race this close, 0.50 is plausible and use that.

Using $p_{Smith} = 0.55$ as our guess, $n = 0.55 \times 0.45/0.01288^2 = 1493$.
Using $p_{Smith} = 0.50$ as our guess, $n = 0.50 \times 0.50/0.01288^2 = 1509$.

Since 0.55 is so close to 0.50, the answers are not very different. Taking the larger one, we need to increase our sample by $1509 - 400 = 1109$. Once we have done so, we will find the actual p_{Smith} for this sample; we will use this actual p_{Smith} in calculating the actual s_p for this sample. And that s_p will be 0.01288 or less, meaning that $2.33\ s_p$ will be 0.03 or less.

Consider the following example:

Your supplier claims that only 0.05 of the parts it supplies are defective. However, you are not sure you believe this claim; you want an independent estimate of your own. You want it to be a 99% confidence interval, no wider than ± 0.02. You think the worst plausible case would be that 0.10 of the parts might be defective. How large a sample do you need?

We know:	We want:
99% CI for π_{def}	$n = ??$
$Zs_p \le 0.02$	
$p_{def} = 0.10$ is the worst plausible value for p_{def}	

First, we need the $\pm Z$ values that include 0.99 of the standard normal distribution; these are ± 2.58. So, $2.58\, s_p \le 0.02$, and $s_p \le 0.02/2.58 = 0.007752$. Thus, $s_p = \sqrt{p_{def}(1-p_{def})/n} = 0.007752$ or rearranging terms, $n = p_{def}(1-p_{def})/0.007752^2$.

Finally, you think that p_{def} is going to turn out to be somewhere in the range 0.00–0.10. This means that 0.10 is the worst plausible value, not only in the sense that it is the highest plausible proportion of defectives, but in the sense that it is the plausible proportion that is closest to 0.50. Using 0.10 as our guess for p_{def} in the formula above, $n = 0.10 \times 0.90/0.007752^2 = 1498$.

For this much confidence and precision, we need a sample of about 1500. Note, though, if we had used 0.50 instead of 0.10, we would have come up with an n of well over 4000. Ruling out such a value as a plausible p_{def} value has saved us from taking a very much larger (and more expensive) sample than necessary.

Consider the following example:

You take the sample of 1500 parts, above, and find 105 defectives. Construct the 99% confidence interval estimate.

We know:	We want:
$n = 1500$	99% CI for π_{def}
def = 105	

We know that the confidence interval will be of the form $p_{def} \pm Zs_p$; we need values for p_{def}, Z, and s_p. The $p_{def} = 105/1500 = 0.07$; $s_p = \sqrt{0.07 \times 0.93/1500} = 0.00659$. And the $\pm Z$-values that include 99% of the standard normal distribution are ± 2.58. Piecing these together,

$$99\%\text{ CI: } 0.07 \pm 2.58 \times 0.00659$$

$$\text{or } 99\%\text{ CI: } 0.07 \pm 0.017$$

$$\text{or } 99\%\text{ CI: } 0.053 < \pi_{def} < 0.087.$$

Notice that, because p_{def} turned out to be less than our worst plausible case, our interval is actually a little bit narrower than the ± 0.02 we originally sought.

7.3 Estimates of the Population Mean

The logic of confidence intervals we have developed for proportions applies for means as well. Figure 7.3 illustrates.

Assume we want a 95% confidence interval for μ_X. We start by taking a sample and calculating our best point estimate, $\bar{X}$. This will be the midpoint of our confidence interval.

To make our interval a 95% confidence interval, we find the range $\pm Z$ that includes 95% of the standard normal distribution. By now, you know that this range is ± 1.96. Let i stand for the upper and lower bounds. Since we know that $Z_i = (\bar{X}_i - \mu_X)/\sigma_{\bar{X}} = \pm 1.96$, we can rearrange terms for the $\bar{X}_i = \mu_X \pm 1.96\sigma_{\bar{X}}$ that bracket 95% of all possible $\bar{X}$-values. In Chapter

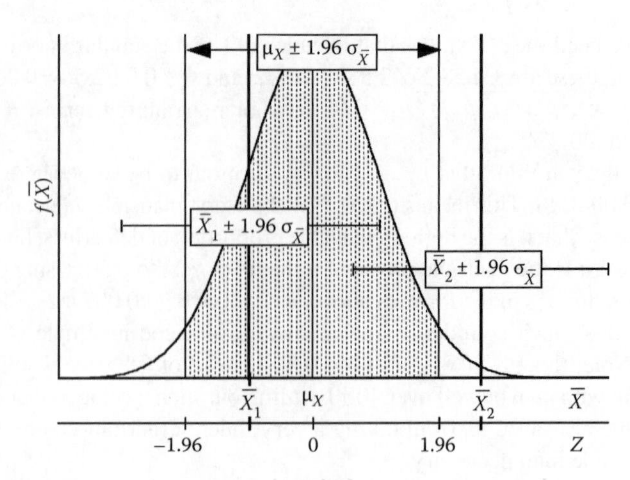

Figure 7.3 Confidence intervals for a mean.

6, when we knew μ_X, we were able to solve for these $\bar{X}_i$. Now, though, we cannot because we do not know μ_X, which is what we are trying to estimate. Neither, then, do we know whether our particular sample $\bar{X}$ is within the interval. But we do know at least the approximate width of the interval. And we do know that 95% of the sample $\bar{X}$-values we could have are, like $\bar{X}_1$, within this interval, and that 5% of the sample $\bar{X}$-values we could have are, like $\bar{X}_2$, outside this interval.

Finally, notice that, for the 95% of possible sample $\bar{X}$-values that are like $\bar{X}_1$, in the interval $\mu_X \pm 1.96\sigma_{\bar{X}}$, the interval $\bar{X} \pm 1.96\sigma_{\bar{X}}$ is right; it includes μ_X. For the 5% of possible sample $\bar{X}$-values that are like $\bar{X}_2$, not in the interval $\mu_X \pm 1.96\sigma_{\bar{X}}$, the interval $\bar{X} \pm 1.96\sigma_{\bar{X}}$ is not right; it does not include μ_X. Thus, the interval $\bar{X} \pm 1.96\sigma_{\bar{X}}$ is right 95% of the time. There is a 95% probability that it contains the true population value of μ_X. It is a 95% confidence interval.

I previously said that we know *at least the approximate* width of the interval that includes 95% of all possible $\bar{X}$-values. The width is $\pm 1.96\,\sigma_{\bar{X}}$. Recall that for a proportion, the true standard error required knowing π_X, the very thing we were trying to estimate. Hence, we could calculate only an estimated standard error based on p_X. In terms of Figure 7.2, we needed to use version 2. For an average, the true standard error requires knowing σ_X, not μ_X, so it is at least possible that we will be able to calculate it. If we know σ_X, we can use version 1 in Figure 7.4.

Of course, if we do not know μ_X, we are unlikely to know σ_X either. And if we do not know σ_X, we will be able to calculate only an estimated standard error based on s_X. Moreover, recall from Chapter 6 that when we use $s_{\bar{X}}$ instead of $\sigma_{\bar{X}}$, the sampling distribution is no longer the standard normal; it is t with $n - 1$ degrees of freedom. Hence, we need to replace Z with t. Figure 7.4 summarizes.

The discussion so far has been in terms, specifically, of a 95% confidence interval; hence, the ± 1.96. And 95% is a commonly used level of confidence. However, any level of confidence can be had (except 100%). If you want a 80, 90, or 99% confidence interval, you just look up the Z-or t-values that correspond. Of course, given a particular sample, you

1. $\bar{X} \pm Z\sigma_{\bar{X}}$ where $\sigma_{\bar{X}} = \frac{\sigma_X}{\sqrt{n}}\left[\sqrt{\frac{N-n}{N-1}}\right]$

2. $\bar{X} \pm t_{(n-1)}s_{\bar{X}}$ where $s_{\bar{X}} = \frac{s_X}{\sqrt{n}}\left[\sqrt{\frac{N-n}{N}}\right]$

Figure 7.4 Confidence intervals for a mean: The formulas.

have only a certain amount of information. Higher levels of confidence are obtained by making the interval wider—by being less precise.

Suppose you want a confidence interval to have both a certain probability and a certain precision. Perhaps, you want a 99% confidence interval for the average salary of some population that is no wider than \$50. If you know the σ_X, you would use version 1 of Figure 7.4. The Z-value for 99% confidence is 2.58, so you want $\pm 2.58\,\sigma_{\bar{X}} \leq \$50$. This means you want $\sigma_{\bar{X}} \leq \$50/2.58 = \$19.38$. Looking at the formula for $\sigma_{\bar{X}}$ (and ignoring the finite population correction factor) you want $\sigma_{\bar{X}} = \sigma_X / \sqrt{n} = \$19.38$. Solving for n, $n = \sigma_X^2/(\$19.38)^2$. And we assumed you know σ_X.

If you do not know σ_X you will need to use version 2 from Figure 7.4. This causes two problems. First, the exact t depends on degrees of freedom, $n - 1$, and you do not yet know n. You will need to make a ballpark guess. Second, s_X is the standard deviation of the sample whose size you are currently determining. You have not yet taken the sample, so you do not yet know s_X. Again you will need to guess. Perhaps you have information on a previous sample s_X that you can use. Or think of salaries plausibly close to the maximum and minimum salaries for this population. According to the empirical rule, this range should be about $\pm 2\ s_X$, so divide it by 4 to get s_X.

Consider the following example:

> Suppose that the process by which certain shafts are created produces shafts whose lengths have a standard deviation of 2.5 millimeters. If you want a 99% confidence interval for their average length that is no wider than ± 0.5 mm, how large a sample do you need?

We know:	We want:
99% CI for μ_X	$n = ??$
$Z\sigma_{\bar{X}} \leq 0.5$	
$\sigma_X = 2.5$ mm	

Note first that the standard deviation is for all shafts created with this process. It is the population standard deviation, σ_X; hence, the confidence interval will be of the form $\bar{X} \pm Z\sigma_{\bar{X}}$. The $\pm$ Z-values that include 99% of the standard normal distribution are $\pm$ 2.58. Since $2.58\ \sigma_{\bar{X}} \leq 0.5$, $\sigma_{\bar{X}} \leq 0.5/2.58 = 0.1938$. Thus, $\sigma_{\bar{X}} = 2.5 / \sqrt{n} = 0.1938$ or, rearranging, $n = 2.5^2/.1938^2 = 166.4$.

Rounding up to the next higher integer, a sample size of 167 combined with a σ_X of 2.5, will give us a small enough $\sigma_{\bar{X}}$ that $\pm 2.58\ \sigma_{\bar{X}}$ will be only ± 0.5.

Consider the following example:

> Suppose you take the sample of 167 shafts. You find their lengths to average 100.2 mm, with a standard deviation of 2.4 mm. Construct the 99% confidence interval estimate.

We know:	We want:
$\bar{X} = 100.2$ mm	99% CI for μ_X
$s_X = 2.4$ mm	
$n = 167$	
$\sigma_X = 2.5$ mm	

Note first that, while we can certainly calculate the sample standard deviation, we do not need it; we already knew the population standard deviation, and σ_X trumps s_X. We would not use an estimate when we are lucky enough to know the real thing. Since we have σ_X, the confidence interval will be of the form $\bar{X} \pm Z\sigma_{\bar{X}}$. We need values for $\bar{X}$, Z, and $\sigma_{\bar{X}}$. $\bar{X} = 100.2$ mm, $\sigma_{\bar{X}} = 2.5/\sqrt{167} = 0.1935$. And the $\pm Z$-values that include 99% of the standard normal distribution are ± 2.58. Piecing these together

$$\text{99\% CI: } 100.2 \pm 2.58 \times 0.1935$$
$$\text{or 99\% CI: } 100.2 \pm 0.5$$
$$\text{or 99\% CI: } 99.7 < \mu < 100.7$$

Since we were not required to guess any of the relevant values in the previous problem, our interval here is (almost) exactly the ± 0.5 we sought. The only source of error arose from having to round off n to an integer value. That caused $\sigma_{\bar{X}}$ to equal 0.1935, when we had found in the previous problem that 0.1938 would have been small enough.

Consider the following example:

> As the personnel manager of a large firm wanting an estimate for the average age of your firm's employees, you take a random sample of 25 employees and record the age of each. Suppose you find a mean of 44.2 years and a standard deviation of 12.4 years. Construct a 95% confidence interval for the average age of all your firm's employees.

We know:	We want:
$\bar{X} = 44.2$	95% CI for μ_X
$s_X = 12.4$	
$n = 25$	

Note first that this time we do not know the population standard deviation; we know only the sample standard deviation, s_X. Hence, the confidence interval will be of the form $\bar{X} \pm ts_{\bar{X}}$. We need values for $\bar{X}$, t, and $s_{\bar{X}}$. $\bar{X} = 44.2$ years, $s_{\bar{X}} = 12.4/\sqrt{25} = 2.48$.

We want the $\pm$ t-values that include 95% of the t distribution. Recall that the probabilities in the t table are the probabilities in one of the tails. So, if we want 95% between $\pm t$, we want 5% outside of this range, with 2.5% in each tail. The probability column we need to use is 0.025. And, with $n = 25$, we have 24 degrees of freedom. Our t-value is 2.064. Piecing all this together

$$95\%\ \text{CI: } 44.2 \pm 2.064 \times 2.48$$
$$\text{or } 95\%\ \text{CI: } 44.2 \pm 5.12$$
$$\text{or } 95\%\ \text{CI: } 39.08 < \mu < 49.32$$

Consider the following example:

Suppose that you want the 95% confidence interval above to be no wider than ±3.0 years. By how much must you increase the sample?

We know:	We want:
95% CI for μ_X	$n = ??$
$ts_{\overline{X}} \leq 3.0$	

The confidence interval will be of the form $\overline{X} \pm ts_{\overline{X}}$. Since $ts_{\overline{X}} \leq 3.0$, we want $s_{\overline{X}} \leq 3.0/t$. Not knowing n, though, we do not know the appropriate degrees of freedom. We know only that it is greater than 24. We might guess that the sample will need to be big and go to the bottom of the t table, 1.960. Or we might be more conservative and continue to use 24. If we do the latter, we want $s_{\overline{X}} \leq 3.0/2.064 = 1.453$. Thus, $s_{\overline{X}} = s_X/\sqrt{n} = 1.453$ or, rearranging, $n = s_{\overline{X}}^2/1.453^2$. But s_X is the sample standard deviation of the sample whose size we are currently determining. We do not yet know s_X; we will need to guess. In this case, a sensible guess would be the 12.4 from the previous sample. If we use that, $n = 12.4^2/1.453^2 = 72.8$. We will need to increase the sample size by $73 - 25 = 48$. If we take a sample of 73, and s_X turns out to be 12.4 or less, our interval will be no wider than ±3.0.

7.4 A Final Word on Confidence Intervals

In this chapter, we have looked at confidence interval estimates for the proportion and for the mean. We have computed confidence intervals with desired probabilities of being correct. We have computed, at least approximately, the sample size necessary for a confidence interval with a desired probability to also have a desired precision. Figure 7.5 summarizes.

In working problems, the first question to address, always, is whether we are trying to estimate a proportion or a mean. If it is a proportion that we are trying to estimate, there is only one workable approach, version 2 in Figure 7.5, since version 1 requires knowing π_X, the very thing we are trying to estimate. If the sample has already been taken, we simply calculate p_X and s_p from the sample, and find the Z-value that corresponds to the probability we want. If the sample has not yet been taken and the question is how large it should be, we find the Z-value that corresponds to the probability we want; that and the desired precision determine how small s_p needs to be; and how small s_p needs to be determines how large n needs to be.

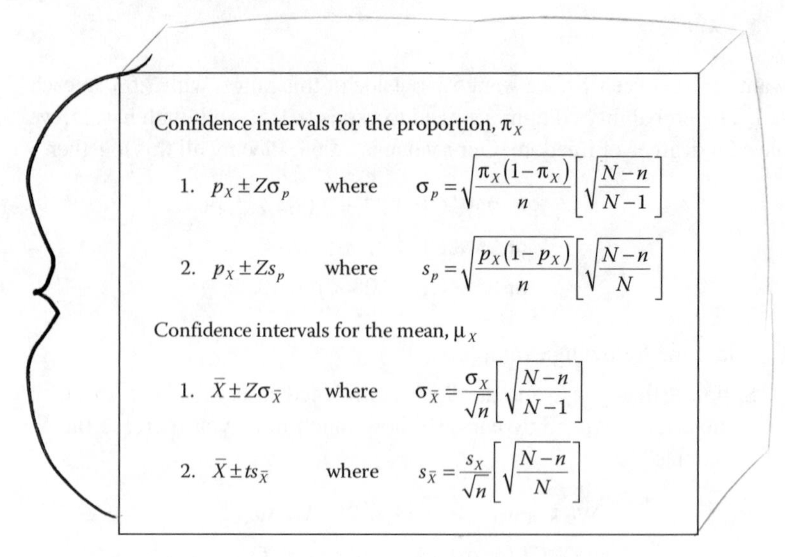

Confidence intervals for the proportion, π_X

1. $p_X \pm Z\sigma_p$ where $\sigma_p = \sqrt{\frac{\pi_X(1-\pi_X)}{n}}\left[\sqrt{\frac{N-n}{N-1}}\right]$

2. $p_X \pm Zs_p$ where $s_p = \sqrt{\frac{p_X(1-p_X)}{n}}\left[\sqrt{\frac{N-n}{N}}\right]$

Confidence intervals for the mean, μ_X

1. $\bar{X} \pm Z\sigma_{\bar{X}}$ where $\sigma_{\bar{X}} = \frac{\sigma_X}{\sqrt{n}}\left[\sqrt{\frac{N-n}{N-1}}\right]$

2. $\bar{X} \pm ts_{\bar{X}}$ where $s_{\bar{X}} = \frac{s_X}{\sqrt{n}}\left[\sqrt{\frac{N-n}{N}}\right]$

Figure 7.5 Confidence intervals for a proportion and a mean: Summary.

If it is a mean that we are trying to estimate, there are two approaches, depending on whether or not we know σ_X. It is possible though not likely that we will. If we know σ_X, we will use Z-values, as in version 1 in Figure 7.5; if we do not know σ_X, we will use t-values, as in version 2. Again, if the sample has already been taken, we simply calculate $\bar{X}$ and $\sigma_{\bar{X}}$ or $s_{\bar{X}}$ from the sample, and find the Z- or t-value that corresponds to the probability we want. If the sample has not yet been taken and the question is how large it should be, we find the Z- or t-value that corresponds to the probability we want; that and the desired precision determine how small $\sigma_{\bar{X}}$ or $s_{\bar{X}}$ needs to be; and how small $\sigma_{\bar{X}}$ or $s_{\bar{X}}$ needs to be determines how large n needs to be.

Finally, if we know the population size, it is always correct to include the finite population correction factor. However, we seldom know the population size and, as long as the population is large compared to the sample, the correction makes very little difference anyway.

We have limited ourselves to confidence interval estimates for the proportion and mean. These are the population parameters for which we are most likely to want estimates. Still, the logic that we have applied here is quite general. For estimates of any population parameter, we need a corresponding sample statistic and we need to know the sampling distribution for that sample statistic. Then, if we can identify an interval around the population parameter that would include 95% of the sample values, we can create an interval around our sample value that has a 95% probability of including that population parameter.

7.5 Exercises

7.1 If the inside of a pump seal is greater than 3.05 inches, the seal is considered defective. Suppose a random sample of 100 seals from a very large lot of pump seals turns up 5 defective seals.

a. Construct a 75% confidence interval estimate for the proportion of defective seals in the entire lot.

b. Change the interval above to a 95% confidence interval estimate for the proportion of defective seals in the entire lot.
c. If you want this 95% confidence interval estimate to be no wider than ± 0.02, by how much must you increase the sample size?

7.2 A mill packs flour in "25 lb" sacks. Its quality control inspectors check the process by randomly selecting 20 sacks each hour and recording their weights. The weights of the latest sample have a mean of 25.45 lbs and a standard deviation of 0.80 lbs.
a. Construct a 95% confidence interval estimate for the average weight of all such 25 lb sacks.
b. Change the interval above to a 99% confidence interval estimate for the average weight of all such sacks.
c. If they want this 99% confidence interval to be no wider than ± ¼ lb, by how much must they increase their sample?

7.3 Congressman Able is in a very tight race for reelection. If he wants a 95% confidence interval estimate for the proportion of voters who support him that is no wider than ±0.02, how large a sample does he need?

7.4 In Exercise 3.1 you calculated summary statistics for the sample of 50 student GPAs in the file Students1.xls. Return to those data.
a. Construct a 95% confidence interval estimate for the mean GPA.
b. Change the interval above to a 99% confidence interval estimate for the mean GPA.
c. If you want this 99% confidence interval estimate to be no wider than ± 0.1, by how much must you increase the sample size?

7.5 Continue with these same data.
a. Construct an 85% confidence interval for the proportion of students who are female.
b. Change the interval above to a 95% confidence interval estimate for the proportion of students who are female.
c. If you want this 95% confidence interval estimate to be no wider than ± 0.05, by how much must you increase the sample size?

7.6 In Exercises 3.4 and 3.5, using NLSY1.xls, you calculated the mean and standard deviation of Height for the entire sample and for men and women separately. Return to those data and construct a 95% confidence interval around each of those means.

7.7 Continue with these data. In Exercise 3.4, you also calculated a range that should include approximately 95% of all values according to the empirical rule. In Exercise 7.6, you estimated a 95% confidence interval. How do these intervals compare? Explain what each one represents.

7.8 In Exercises 3.6 and 3.7, using Employees1.xls, you calculated the mean and standard deviation of Salary for the entire sample and for men and women separately. Return to those data and construct a 95% confidence interval around each of those means.

7.9 Continue with these data. In Exercise 3.6, you also calculated a range that should include approximately 95% of all values according to the empirical rule. In Exercise 7.8, you estimated a 95% confidence interval. How do these intervals compare? Explain what each one represents.

7.10 Continue with these data.

a. Construct a 95% confidence interval for mean education.
b. Construct a 95% confidence interval for mean experience.

7.11 Continue with these data.

a. Compute a 95% confidence interval estimate for the proportion female.
b. If you want this 95% confidence interval estimate to be no wider than ±0.05, by how much must you increase the sample size?

7.12 In Exercises 3.9 and 3.10 using Students2.xls, you found means (among other statistics) for student heights, weights, entertainment spending, study time, and college GPAs. Construct a 95% confidence interval estimate around each mean.

7.13 Continue with these data.

a. Compute a 95% confidence interval estimate for the proportion female.
b. If you want this 95% confidence interval estimate to be no wider than ±0.05, by how much must you increase the sample size?

7.14 Continue with these same data.

a. Repeat Exercise 7.13 for the proportion majoring in economics.
b. Repeat Exercise 7.13 for the proportion holding a job.
c. Repeat Exercise 7.13 for the proportion participating in a varsity sport.
d. Repeat Exercise 7.13 for the proportion participating in a music ensemble.
e. Repeat Exercise 7.13 for the proportion belonging to a fraternity or sorority.

7.15 Suppose you have been hired by Nickels, a local department store, to do an analysis of its market. You undertake a survey, collecting the information below for a random sample of 50 consumers. The file Nickels1.xls contains their responses.

Customer: Nickels customer (1 = yes; 0 = no);
Female: Female (1 = yes; 0 = no);

Age: Age (years);
Income: Income ($ thousands);
Source: Primary source of market information (1 = newspaper; 2 = radio; 3 = other).

a. Construct a 95% confidence interval estimate for the proportion of Nickels customers who are female.
b. Construct a 95% confidence interval estimate for the mean age of Nickels customers.
c. Construct a 95% confidence interval estimate for the mean income of Nickels customers.
d. Construct a 95% confidence interval estimate for the proportion of Nickels customers whose primary source of market information is the newspaper.

8

Tests of Hypotheses: One-Sample Tests

Chapter 7 introduced formal inference and suggested two types of questions we could address. The first was "based on this sample, what is our best estimate for some population parameter; and how good is this estimate?" The second was "based on this sample, can we conclude that something is true of a population parameter; and how sure are we?"

Chapter 7 addressed the first question. We were able to construct confidence interval estimates for π_X and μ_X, the population proportion and mean—estimates with known probabilities of being correct. In this chapter, we will address the second question. We will arrive at conclusions concerning π_X and μ_X, the population proportion and mean—conclusions with known probabilities of being wrong. Chapters 9 through 13, then, extend the same basic reasoning to additional, more interesting cases.

8.1 Testing a Claim: Type I and Type II Errors

Suppose you are called to serve on a jury. The defendant is either innocent or guilty; the jury will find him innocent or guilty. As Figure 8.1 illustrates, there are two ways in which the jury can be right and two ways in which the jury can be wrong.

We generally regard convicting an innocent person to be the worse error to make. Thus, as a juror, you are told to start with the assumption that the defendant is innocent; this is your **null hypothesis (H_0)**. Then, only if you decide there is evidence beyond a reasonable doubt do you reject the null hypothesis, innocent, in favor of the **alternative hypothesis (H_a)**, guilty. And the rules of evidence are intended to assure a low probability that you will reject the null hypothesis when it is true. This sort of error—rejecting the null hypothesis when it is true—is called a **type I error**, and its probability is $\boldsymbol{\alpha}$.

		The Truth	
		Innocent	Guilty
Your	H_0: Innocent	Correct $(1 - \alpha)$	Type II error (β)
Decision	H_a: Guilty	Type I error (α)	Correct $(1 - \beta)$

Figure 8.1 Decision making with incomplete information.

We want α to be small. Still, we need to recognize that the harder we make it to convict an innocent person, the harder we make it to convict a guilty person too. This sort of error—failing to reject the null hypothesis when it is false—is called a **type II error**, and its probability is $\boldsymbol{\beta}$. And, for a given amount of information, the smaller we make α, the larger we make β.

Notice that we either reject H_o or we fail to reject H_o. If we fail to reject H_o, it could be that this is because H_o is true. It could also be because there was simply not enough evidence. You may not believe that the defendant is actually innocent. But if there is not enough evidence to establish his guilt beyond a reasonable doubt, you fail to convict.

8.2 A Two-Tailed Test for the Population Proportion

8.2.1 The Null and Alternative Hypotheses

Suppose someone claims that 12% of all college students are left-handed. This is a claim about a population parameter, π_{LH}, which may or may not be true. We can test it in much the same way we tested the claim of innocence in the trial. A claim that we are "testing" must be the null hypothesis. The null hypothesis is H_o: $\pi_{LH} = 0.12$.

The alternative hypothesis is what we will believe if we succeed in rejecting the null hypothesis. A general, "two-tailed" alternative would be simply that the null hypothesis is wrong. That is, H_a: $\pi_{LH} \neq 0.12$.

The first step in testing hypotheses is to always write down your null and alternative hypotheses. There is no point in doing the work of testing a hypothesis if, at the end, you do not know what your answer means.

1. H_o: $\pi_{LH} = 0.12$
   H_a: $\pi_{LH} \neq 0.12$.

8.2.2 The Decision Criterion: Setting the Probability of a Type I Error

We need to decide on a probability, α, which we are willing to accept of making a type I error—of declaring that H_o is false when it is true. The probability we are willing to accept will vary with the seriousness of making such a mistake. Often, though, a probability of 0.05 will be acceptable. As you will see shortly, choosing $\alpha = 0.05$ is very much akin to choosing a 95% level for a confidence interval in Chapter 7.

We are going to test H_o by taking a sample and finding p_{LH}, the proportion of left-handers in the sample. If we were to get $p_{LH} = 0.12$, then that would certainly offer support for the null hypothesis. But, of course, we are unlikely to get exactly $p_{LH} = 0.12$, even if the null hypothesis is true. As we know, there is a whole sampling distribution of p_{LH}-values we could get from a population for which $\pi_{LH} = 0.12$.

Still, we know what that distribution looks like. We know that, for large samples, it is a normal distribution, with mean, $\mu_p = \pi_{LH} = 0.12$,

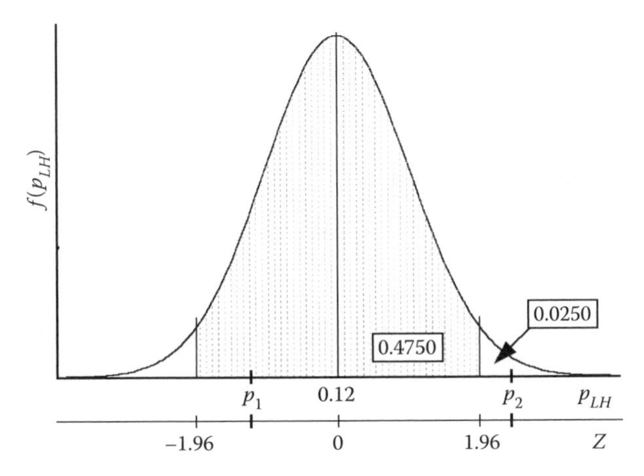

Figure 8.2 The sampling distribution for a proportion, assuming $\pi_{LH} = 0.12$.

and standard error, $\sigma_p = \sqrt{\pi_{LH}(1-\pi_{LH})/n}$. And we know that 0.12 $\pm$ 1.96σ_p includes all but 5% of the possible p_{LH} values. Figure 8.2 illustrates.

So, suppose we adopt the following procedure. We take our sample and find our sample p_{LH} value. We convert this p_{LH} value to its Z value equivalent, assuming that the null hypothesis, $\pi_{LH} = 0.12$, is true. Then, if this Z value is outside the ± 1.96 range, we conclude that our sample p_{LH} value is too far from the hypothesized π_{LH}. We should not have come up with a sample p_{LH} value this far from the true value. Therefore, the null hypothesis must not be true.

Figure 8.2 assumes the null hypothesis is true that $\pi_{LH} = 0.12$. If it is, we know that 95% of the time we will get a p_{LH} value like p_1, close enough to 0.12 that it will standardize to a Z value in the ± 1.96 range. We will fail to reject H_o, which is the correct decision since H_o is true. The other 5% of the time, though, we will be unlucky and get an unusual p_{LH} value like p_2, far enough from 0.12 that it will standardize to a Z value outside the ± 1.96 range. In these cases, we will reject H_o, which is a type I error. This criterion, then, gives us an α of 0.05.

This criterion can be expressed as follows: Reject H_o if $|Z_c| > 1.96$ ($\alpha = 0.05$). The c subscript indicates that this Z value is the one we calculate, based on our sample p_{LH} value, rather than the criterion (1.96) that we find in the Z table.

We can, of course, choose different values for α. If we are not willing to reject wrongly 5% of the time, we can choose a smaller probability. If we choose $\alpha = 0.01$, we would not reject unless $|Z_c| > 2.58$, and we would reject wrongly only 1% of the time. Of course, this would also increase the probability of making a type II error—failing to reject when H_o is actually false.

Adding our criterion to the hypotheses, we have now defined the problem. We have specified what we are going to assume, for the purpose of the test (H_o), what we will believe if we reject that assumption (H_a), what we are going to calculate in order to decide (Z_c), what our criterion is going to be ($|Z_c| > 1.96$), and what probability we have accepted of rejecting wrongly (α).

1. H_o: $\pi_{LH} = 0.12$
   H_a: $\pi_{LH} \neq 0.12$,
2. Reject H_o if $|Z_c| > 1.96$ ($\alpha = 0.05$).

8.2.3 The Calculations

So far, we have not consulted the data; that is step 3. Suppose we choose a random sample of 200 college students and find that 21 are left-handed. Then, $p_{LH} = 21/200 = 0.105$. Our sample proportion is not 0.12, but is it so far from 0.12 that we can conclude the population proportion must not be 0.12 either? To decide, we need to calculate Z_c.

$$3.\ \ Z_c = \frac{(p_{LH} - \pi_{LH})}{\sigma_p} \quad \text{where} \quad \sigma_p = \sqrt{\frac{\pi_{LH}(1-\pi_{LH})}{n}}\left[\sqrt{\frac{N-n}{N-1}}\right].$$

Notice a change from Chapter 7 here. In Chapter 7, when we were estimating π_X, we had no value of π_X to plug into the formula for σ_p. We had to substitute s_p, an estimate based on p_X. In this case, we do have a hypothesized true value of π_X, so we can calculate σ_p as follows (ignoring the finite population factor since, as usual, we do not know the population size).

$$3.\ \ \sigma_p = \sqrt{\frac{0.12 \times 0.88}{200}} = 0.02298, \quad \text{and}$$

$$Z_c = \frac{(0.105 - 0.12)}{0.02298} = -0.653.$$

8.2.4 The Conclusion

The $|Z_c|$, then, is not greater than 1.96, our criterion for rejecting H_o with an α of 0.05. Apparently, our sample p_{LH} could be just randomly different from 0.12. Our conclusion, then, is:

4. $\therefore$ Fail to Reject H_o.

We do not generally say that we accept H_o. We certainly have not shown it to be true. It may well be that a larger sample, with its smaller σ_p would have allowed us to reject H_o. All we can say for sure is that this particular sample does not provide enough evidence to reject H_o with a probability of rejecting wrongly that we were willing to accept.

Consider the following example:

> A company, considering where to place its advertising, commissions a poll of a random sample of 500 households in a city. In the past, 22% of all households in the city received the morning newspaper. If, in this poll, 130 of the 500 households received the morning paper, can the company conclude that the percentage has changed? Use $\alpha = 0.05$.

In the previous example, we "tested" the claim that $\pi_{LH} = 0.12$. A claim we are testing is the null hypothesis. But, since we assume it is true, there is no way to show or conclude it is true. Recall, we did not even accept it as true; we just "failed to reject" it. Something we want to show or conclude is true must be the alternative hypothesis. In this example, then, if we want to show or conclude that the percentage has changed, $\pi_{\text{Paper}} \neq 0.22$ must be the alternative; and we show or conclude it is true by rejecting the null hypothesis that $\pi_{\text{Paper}} = 0.22$.

1. H_o: $\pi_{\text{Paper}} = 0.22$
   H_a: $\pi_{\text{Paper}} \neq 0.22$.

Since we are willing to reject H_o wrongly 5% of the time, we choose our $\pm Z$ criterion to include 95% of the Z_c-values we could get if H_o is true. That $\pm Z$, as you know, is ± 1.96.

2. Reject H_o if $|Z_c| > 1.96$ ($\alpha = 0.05$).

Doing the calculations:

3. $$p_{\text{Paper}} = 130/500 = 0.26; \quad \sigma_p = \sqrt{\frac{0.22 \times 0.78}{500}} = 0.01853;$$

$$Z_c = \frac{0.26 - 0.22}{0.01853} = 2.16.$$

Finally, $2.16 > 1.96$, so:

4. ∴ Reject H_o. Conclude that the proportion has changed.

Consider the following example:

Redo the previous example, but use $\alpha = 0.02$.

The hypotheses do not change; but the rejection criterion changes to

2. Reject H_o if $|Z_c| > 2.33$ ($\alpha = 0.02$).

The calculations do not change either, but the conclusion changes since $2.16 < 2.33$.

4. ∴ Fail to Reject H_o. Do not conclude that the proportion has changed.

We reach opposite conclusions from the same data depending on the probability, α, we are willing to accept as being wrong. Our Z_c is extreme enough to be outside the range of values we would get 95% of the time when the null hypothesis is true. But it is not extreme enough to be outside the range of values we would get 98% of the time when the null hypothesis is true. Figure 8.3 illustrates.

Clearly, one answer is right and one answer is wrong. The value of π_{Paper} is either 0.22 or it is not. If π_{Paper} is really 0.22, our first answer was wrong—we made a type I error, rejecting the null hypothesis when it was true. If π_{Paper} is not really 0.22, our second answer was wrong—we made a type II error, failing to reject the null hypothesis when it was false. Which was it this time? We do not know; we only know the probabilities.

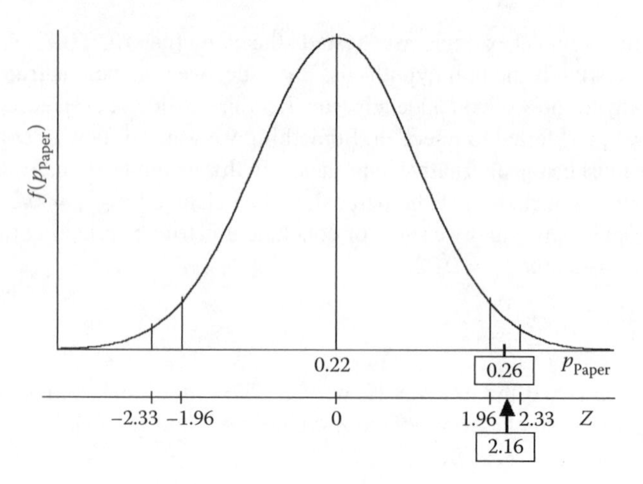

Figure 8.3 Two different criteria, two different decisions.

8.2.5 The *P*-Value of the Test

Our approach so far is sometimes referred to as the "classical" approach. We (i) decide the probability we are willing to accept; (ii) run the test; and (iii) either reject or fail to reject. And this approach has a lot of merit, especially when a real-world decision must be made based on the outcome of this particular test.

But suppose we are reading academic research on some topic say, the use of illicit drugs by high school students. Different studies will have different data and use different techniques. Probably no one study is definitive; rather, it is by replication and the testing of previous work that we gradually begin to understand. In this context, suppose an older study claims 22% of high school students have used illicit drugs; a newer study claims to reject that number.

First, of course, we would want to know by what criterion the claim was rejected. That is, what was α? And we would tend to be more impressed if the α were small, since that means a smaller chance that the rejection was just a type I error. Any such study should report the α used, often called the **level of significance**. The language can become a little convoluted here since an α of 0.01 will often be called a higher level of significance than an α of 0.05, even though it is a smaller number.

Sometimes you will see calculated Z_c values presented, along with *, **, or ***, representing significance at the 0.10, 0.05, and 0.01 levels, respectively. If we were reporting our own results above using this system we would report our Z value as 2.16**, since we were able to reject at the 0.05 level but not at the more demanding 0.01 (or even 0.02) level.

A second approach is to present the ***p*-value** of the test. Also referred to as the **prob-value** or the **sig-value,** this is the α value we would need to be willing to accept in order to reject the null hypothesis. In our examples above, we found Z_c equaled 2.16. We can look up this Z_c value in the standard normal table and find the probability associated with it. We get 0.4846. So, the area between ± 2.16 in the standard normal distribution is $2 \times 0.4846 = 0.9692$. The area not in this interval is $1 - 0.9692 = 0.0308$.

So, if we are willing to accept a probability of rejecting wrongly, α, of 0.0308 or greater, we are able to reject. If we are not willing to accept this great a probability of rejecting wrongly, we are not able to reject.

The advantage of presenting things this way is that it allows you to present a more complete account of your results. In most serious research, you will test a range of hypotheses and you will get a range of results. Some results may be so strong that you can reject the null hypothesis even with a very small α. There is very little chance that these are type I errors. Others may be somewhat weaker; you can reject at higher αs, but not at very small ones. There is a greater chance that these are type I errors. By presenting a p-value along with each result, you help the reader understand how strong a result it is.

The way the standard normal table is set up makes it easy to calculate p-values; other tables tend to have fewer probabilities, making it less possible to find exact p-values. However, spreadsheet functions will work in such cases. And statistical packages give p-values routinely.

8.2.6 The Probability of a Type II Error*

Return to the examples testing H_o: $\pi_{Paper} = 0.22$ in Chapter 8, Section 2.4, where we came to opposite conclusions depending on whether we chose an α of 0.05 or 0.02. I said we did not know which conclusion was right; we only knew the probabilities. So far, though, we have dealt only with α, the probability of a type I error. What about β, the probability of a type II error?

The probability of a type II error, β, depends on several things. First, we know that it varies inversely with α. Other things equal, the lower we make α, the probability of rejecting wrongly, the higher we make β, the probability of failing to reject when we should.

Second, β depends on the sample size. The β we must accept, for any given α, goes down as sample size increases. More information does help.

And third, β depends on the true value of π_{Paper}. Suppose the true value of π_{Paper} were not 0.22 but very close—say 0.23. The distribution of p_{Paper} values we could get for π_{Paper} values of 0.22 and 0.23 are very similar. The sample p_{Paper} value we would get is likely to be consistent with our null hypothesis, even though our null hypothesis is wrong. Hence, we would be likely to make a type II error. On the other hand, suppose the true value of π_{Paper} were much further from 0.22—say 0.29.

The distribution of p_{Paper} values we could get for π_{Paper} values of 0.22 and 0.29 are not very similar. The sample p_{Paper}-value we would get is not likely to be consistent with our null hypothesis. Hence, we would not be likely to make a type II error. Indeed, there is a whole distribution of β-values, depending the true value of π_{Paper}.

Figure 8.4 illustrates the situation if H_o:$\pi_{Paper} = 0.22$ and the true π_{Paper} is 0.23. Our criterion for $\alpha = 0.05$ was to reject H_o if $|Z_c| > 1.96$. The ± 1.96 range, figured around our hypothesized value of 0.22, is shown on the $Z_{0.22}$ scale. Anything within this range was deemed plausible, given H_o; hence, we would fail to reject. Notice, though, that there is still a strong chance of getting a value in this range, even though H_o is not correct, and the true π_{Paper} is 0.23. To find this probability, we need to (i) take our ± 1.96 range, figured around our hypothesized value of 0.22, and translate it into actual p_{Paper} value

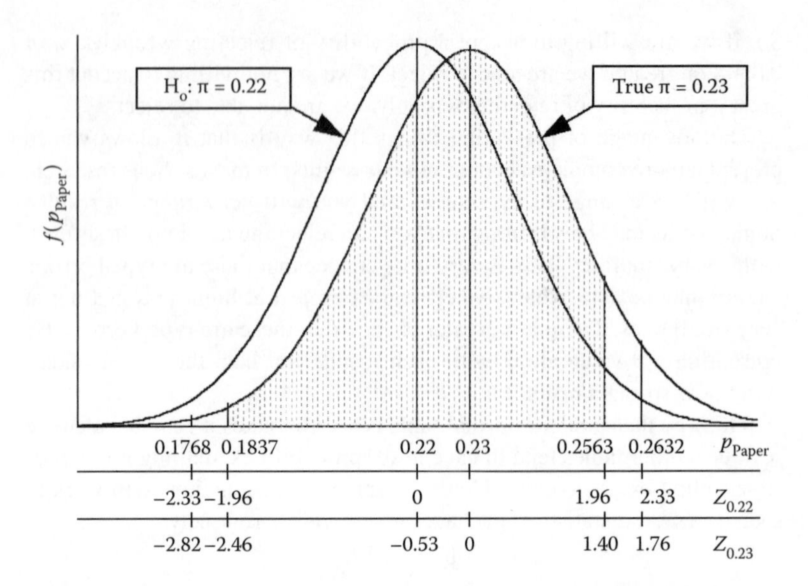

Figure 8.4 Type II error for H$_o$: $\pi_{\text{Paper}} = 0.22$ when the true $\pi_{\text{Paper}} = 0.23$.

equivalents; (ii) re-standardize these p_{Paper} values using the true π_{Paper} of 0.23; and (iii) find the probability between these corrected Z values.

First, to find the p_{Paper} value equivalents of ±1.96, as we would have calculated them using our hypothesized π_{Paper} of 0.22:

$$Z_{0.22} = \frac{p_{\text{Lower}} - 0.22}{\sqrt{\dfrac{0.22 \times 0.78}{500}}} = \frac{p_{\text{Lower}} - 0.22}{0.01853} = -1.96 \quad \text{and}$$

$$Z_{0.22} = \frac{p_{\text{Upper}} - 0.22}{\sqrt{\dfrac{0.22 \times 0.78}{500}}} = \frac{p_{\text{Upper}} - 0.22}{0.01853} = 1.96.$$

$$\text{So,} \quad p_{\text{Lower}} = 0.22 - 1.96 \times 0.01853 = 0.1837 \quad \text{and}$$

$$p_{\text{Upper}} = 0.22 + 1.96 \times 0.01853 = 0.2563.$$

Second, to calculate the Z value equivalents of these p_{Paper} values, using the true π_{Paper} of 0.23:

$$Z_{0.23} = \frac{0.1837 - 0.23}{\sqrt{\dfrac{0.23 \times 0.77}{500}}} = \frac{0.1837 - 0.23}{0.01882} = -2.46 \quad \text{and}$$

$$Z_{0.23} = \frac{0.2563 - 0.23}{\sqrt{\dfrac{0.23 \times 0.77}{500}}} = \frac{0.2563 - 0.23}{0.01882} = 1.40.$$

Since 0.23 is between p_{Lower} and p_{Upper}, the Z values are on opposite sides of the mean. Looking them up in the standard normal table, the probability between them is 0.4931 + 0.4192 = 0.9123. That is, if we test H_o: $\pi_{\text{Paper}} = 0.22$, using $\alpha = 0.05$ and a sample of 500, and the true π_{Paper} is 0.23, the probability is 0.9123 that we will make a type II error, failing to reject the null hypothesis, even though it is false. The qualifications in the previous sentence matter. If we were to use $\alpha = 0.02$, the overlap of the two distributions would be even greater. Substituting ±2.33 for ±1.96 in the previous work, the probability rises to 0.9584 that we will make a type II error. On the other hand, if we were to double the sample to 1000, using $\alpha = 0.05$, the two distributions would get narrower and thus overlap less. Substituting $n = 1000$ for $n = 500$ in the previous work, the probability falls to 0.8773 that we will make a type II error. You should verify these probabilities.

The main reason our probability of a type II error is so high, of course, is that the true π_{Paper}, while different from that in our null hypothesis, is so close that it is likely to give evidence consistent with our null hypothesis. As the true π_{Paper} differs more and more from our null hypothesis, it is less and less likely to give evidence consistent with our null hypothesis.

Figure 8.5 illustrates the situation if H_o: $\pi_{\text{Paper}} = 0.22$ and the true π_{Paper} is 0.29. Our criterion for $\alpha = 0.05$ was to reject H_o if $|Z_c| > 1.96$. The ±1.96 range, figured around our hypothesized value of 0.22, is again shown on the $Z_{0.22}$ scale. This time, though, notice that only a little of the lower tail of the true distribution spills over into this range.

The p_{Lower} and p_{Upper} based on our null hypothesis and α are unchanged at 0.1837 and 0.2563.

Then,

$$Z_{0.29} = \frac{0.1837 - 0.29}{\sqrt{\dfrac{0.29 \times 0.71}{500}}} = \frac{0.1837 - 0.29}{0.0203} = -5.24$$

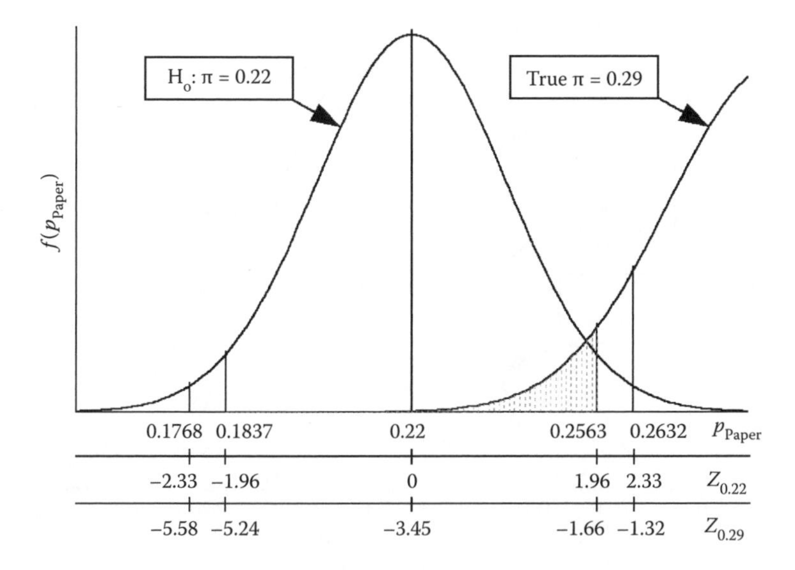

Figure 8.5 Type II error for H_o: $\pi_{\text{Paper}} = 0.22$ when the true $\pi_{\text{Paper}} = 0.29$.

and

$$Z_{0.29} \frac{0.2563 - 0.29}{\sqrt{\frac{0.29 \times 0.71}{500}}} = \frac{0.2563 - 0.29}{0.0203} = -1.66.$$

Notice that, since 0.29 is above $p_{\text{Upper}} = 0.2563$, both Z values are now negative. We must now subtract probabilities to find the probability between them. The probability from the lower value, –5.24, up to the mean is so large that it rounds off to 0.5000; the probability from the upper value, –1.66, up to the mean is 0.4515. The probability we want, then, is 0.5000 – 0.4515 = 0.0485.

Again, if we use $\alpha = 0.02$, instead of $\alpha = 0.05$, the overlap would be greater; substituting +2.33 for +1.96 in the previous work, the probability rises to 0.0934 that we will make a type II error. And again, if we were to double the sample to 1000, using $\alpha = 0.05$, the two distributions would get narrower and thus overlap less. Substituting $n = 1000$ for $n = 500$ in the previous work, the probability falls to 0.0010 that we will make a type II error. You should verify these probabilities.

Imagine repeating this procedure over and over, for H_o: $\pi_{\text{Paper}} = 0.22$, using $\alpha = 0.05$ and a sample of 500, varying the true π_{Paper} value from well below 0.22 to well above it. We could then plot the probability of a type II error as a function of the true π_{Paper}. The result would be an **operating-characteristic (OC) curve.** The middle curve in Figure 8.6 shows what it would look like.

For any possible true π_{Paper}, the distance up to the OC curve represents the probability of not rejecting; the distance from the OC curve up is the probability of rejecting. They add up to one because we will always do one or the other. For all possible true π_{Paper} values except 0.22, not rejecting constitutes a type II error, so we would want the curve to be low. For 0.22, of course, not rejecting is correct; the error would be in rejecting.

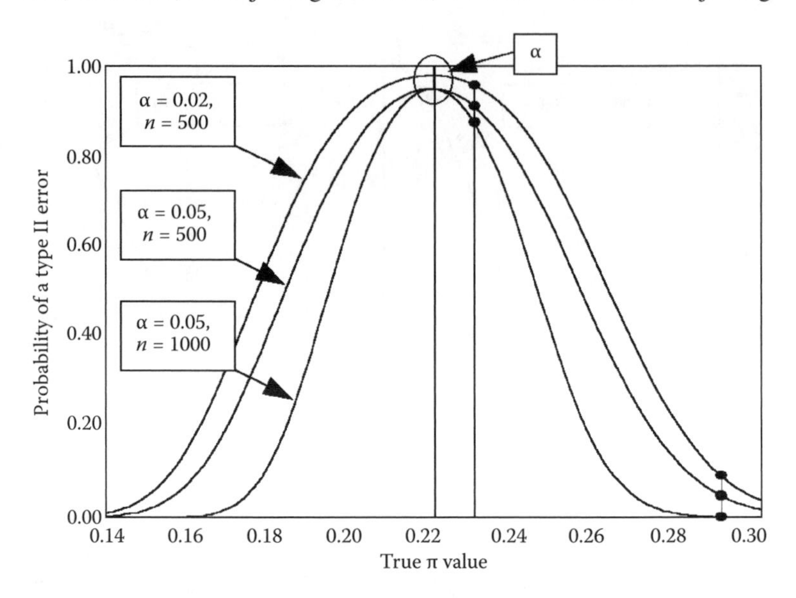

Figure 8.6 Three OC curves for H_o: $\pi_{\text{Paper}} = 0.22$.

So, at 0.22, the distance from the curve up is the probability of rejecting wrongly, α. At 0.22, we would want the curve to be high.

Figure 8.6 shows three curves, to show the effect of changes in α and n. Reducing α to 0.02 shifts the curve up. At 0.22, of course, this is good; we want the curve to be high there. We want to fail to reject there. Everywhere else, though, this is bad; we have increased the probability of a type II error.

Increasing the sample size reduces the probability of a type II error, for any α. Notice that the lower curve agrees with the middle curve at 0.22; they are both calculated for $\alpha = 0.05$. The lower curve, though, is based on a sample twice as large. It falls off much faster, meaning that the probability of a type II error is less. Having more information increases our ability to discriminate between cases in which the null hypothesis is true and cases in which it is false but fairly close.

Consider the following example:

> In a major market study for Walton's, a local department store, you select 400 customers at random and interview them about their attitudes toward the store. You also collect demographic data on these customers. Suppose 222 of these customers are female. Can you reject the hypothesis that 50% of all Walton's customers are female? Use $\alpha = 0.05$. Use $\alpha = 0.01$. In which case might you be making an type I error? In which case might you be making a type II error? What is the p-value of this test?

To reject a hypothesis, it must be the null hypothesis, so the hypotheses are:

1. H_o: $\pi_F = 0.50$
   H_a: $\pi_F \neq 0.50$.

The rejection criterion depends on α:

2. Reject H_o if $|Z_c| > 1.96$ $(\alpha = 0.05)$ Reject H_o if $|Z_c| > 2.58$ $(\alpha = 0.01)$.

Our calculations are as follows; recall that we use our hypothesized population π_F, not our sample p_F, in calculating σ_p.

3.
$$p_F = \frac{222}{400} = 0.555, \quad \sigma_p = \sqrt{\frac{0.50 \times 0.50}{400}} = 0.025,$$
$$Z_c = \frac{0.555 - 0.50}{0.025} = 2.20.$$

Our conclusion again depends on our rejection criterion:

4. $|Z_c| > 1.96$ ∴ Reject H_o $|Z_c| < 2.58$ ∴ Fail to Reject H_o.

Using $\alpha = 0.05$, we reject the null hypothesis; either $\pi_F \neq 0.50$ or we are making a type I error—rejecting a true hypothesis. Using $\alpha = 0.01$, we fail to reject the null hypothesis; either $\pi_F = 0.50$ or we are making a type II error—failing to reject a false hypothesis.

The p-value of the test can be found by looking up our $|Z_c|$ of 2.20 in the standard normal table. It gives us a probability of 0.4861 between 2.20 and the mean. So, the range ±2.20 includes $2 \times 0.4861 = 0.9722$ of the p_F values we could have gotten if the null hypothesis is true; it does not include the other $1 - 0.9722 = 0.0278$ we could have gotten if the null hypothesis is true. So, the *p*-value is 0.0278. If we are willing to accept $\alpha \geq 0.0278$, we should reject H_o. If we are not, we should fail to reject H_o.

Consider the following example:

In the previous example, suppose you used $\alpha = 0.05$; what is the probability of making a type II error if the true $\pi_F = 0.48$ or 0.52? 0.46 or 0.54? 0.44 or 0.56?

First, we need to translate our $|Z_c|$ rejection criterion into the equivalent values for p_F:

$$Z_{0.50} = \frac{p_{\text{Lower}} - 0.50}{\sqrt{\dfrac{0.50 \times 0.50}{400}}} = \frac{p_{\text{Lower}} - 0.50}{0.0250} = -1.96 \quad \text{and}$$

$$Z_{0.50} = \frac{p_{\text{Upper}} - 0.50}{\sqrt{\dfrac{0.50 \times 0.50}{400}}} = \frac{p_{\text{Upper}} - 0.50}{0.0250} = 1.96.$$

So, $p_{\text{Lower}} = 0.50 - 1.96 \times 0.0250 = 0.4510$ and $p_{\text{Upper}} = 0.50 + 1.96 \times 0.0250 = 0.5490$.

Now, we need to re-standardize these p_F values using the various π_F values listed. Since the OC curve is symmetrical around 0.50, each pair listed will carry the same probability of a type II error. We can calculate just one side or the other; suppose we arbitrarily choose the lower values.

For 0.48 (or 0.52):

$$Z_{0.48} = \frac{0.4510 - 0.48}{\sqrt{\dfrac{0.48 \times 0.52}{400}}} = \frac{0.4510 - 0.48}{0.0250} = -1.16 \quad \text{and}$$

$$Z_{0.48} = \frac{0.5490 - 0.48}{\sqrt{\dfrac{0.48 \times 0.52}{400}}} = \frac{0.5490 - 0.48}{0.0250} = 2.76.$$

Since 0.48 is between p_{Lower} and p_{Upper}, the *Z* values are on opposite sides of the mean. Looking them up in the standard normal table, the probability between them is $0.3770 + 0.4971 = 0.8741$.

For 0.46 (or 0.54):

$$Z_{0.46} = \frac{0.4510 - 0.46}{\sqrt{\dfrac{0.46 \times 0.54}{400}}} = \frac{0.4510 - 0.46}{0.0249} = -0.36 \quad \text{and}$$

$$Z_{0.46} = \frac{0.5490 - 0.46}{\sqrt{\dfrac{0.46 \times 0.54}{400}}} = \frac{0.5490 - 0.46}{0.0249} = 3.57.$$

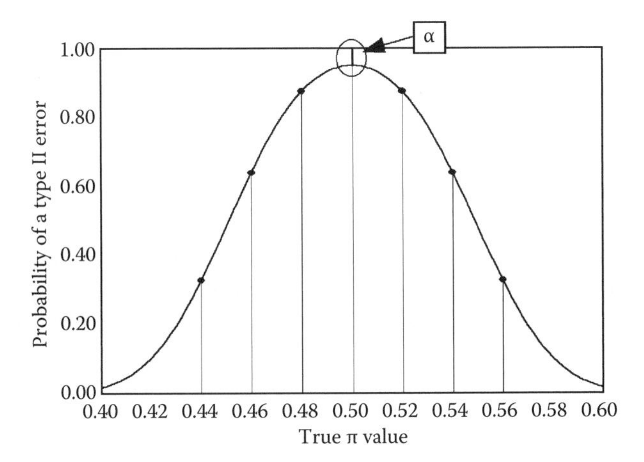

Figure 8.7 An OC curve for H_o: $\pi_F = 0.50$.

Since 0.46 is between p_{Lower} and p_{Upper}, the Z values are on opposite sides of the mean. Looking them up in the standard normal table, the probability between them is 0.1406 + 0.4998 = 0.6404.

And for 0.44 (or 0.56):

$$Z_{0.44} = \frac{0.4510 - 0.44}{\sqrt{\frac{0.44 \times 0.56}{400}}} = \frac{0.4510 - 0.44}{0.0248} = 0.44 \quad \text{and}$$

$$Z_{0.44} = \frac{0.5490 - 0.44}{\sqrt{\frac{0.44 \times 0.56}{400}}} = \frac{0.5490 - 0.44}{0.0248} = 4.39.$$

Since 0.44 is below $p_{\text{Lower}} = 0.4510$, both Z values are now positive. We must now subtract probabilities to find the probability between them. The probability from the upper value, 4.39, down to the mean is so large that it rounds off to 0.5000; the probability from the lower value, 0.44, down to the mean is 0.1700. Thus, the probability we want is 0.5000 – 0.1700 = 0.3300.

Figure 8.7 shows our calculated points along the OC curve for this test.

8.3 A One-Tailed Alternative for the Population Proportion

So far, we have been looking at two-tailed tests. We have been willing to reject H_o in favor of H_a if our sample p_X value turned out to be either too high or too low. Often, though, we are looking for evidence in just one direction or the other. There are two main reasons why.

First, we may be looking for support for some theory that suggests a particular value or range. Consider the previous example concerning the proportion of female customers at Walton's department store. Perhaps some theory from marketing or psychology allows us to predict—before we ever see any data on Walton's—that this sort of store attracts more women than men. We are, then, looking explicitly for a p_F greater than 0.50. A p_F greater than 0.50 supports this theory; a p_F less than or equal to 0.50 does not.

In fact, most theories relate more to relationships among variables than to levels of a single variable. Thus, this theoretical motivation for a one-tailed test will matter more when we begin looking at relationships, starting in the next chapter.

Second, only one of the two tails may be of any concern to us. Suppose our suppliers guarantee that no more than 5% of the parts they supply to us are defective. We may want to check to make sure. However, we will not be concerned if less than 5% turn out to be defective; only if more than 5% do. A lot of "quality control" tests are one-tailed.

8.3.1 The Null and Alternative Hypotheses

Until now " = " had to be the null hypothesis; " ≠ " had to be the alternative. It has been pretty straightforward. With one-tailed tests, though, we need to think about which tail goes with the null hypothesis, and which tail goes with the alternative.

Think in terms of "proof by contradiction." If we want to show, support, or conclude that something is true, it needs to bear the burden of proof. It needs to be the alternative. We need to assume it is false and then reject that assumption.

In the Walton's example, if we are looking for evidence to support the theory, we would set up the hypotheses as follows:

$$H_o: \pi_F \leq 0.50 \qquad H_o: \pi_F = 0.50$$
$$H_a: \pi_F > 0.50 \qquad H_a: \pi_F > 0.50.$$

I have written H_o two different ways to make a point. The version on the left clearly and logically lays out the choice. Theory says $\pi_F > 0.50$. Evidence for that would be rejecting the opposite, $\pi_F \leq 0.50$. Stating things this way makes it clear that one tail goes with the alternative hypothesis and the other tail goes with the null. No matter how far *below* 0.50 our sample p_F turns out to be, we will not reject the null hypothesis in favor of the alternative. Indeed, it will need to be *enough above*.

Of course, to actually conduct the test we need to compare our sample p_F to a specific number and the specific number we will use is 0.50; the version on the right makes that clear. This is also the way H_o is usually presented. However, I will generally follow the first convention.

Like the Walton's example, in which we are looking for evidence to support a theory, the theory should always be the alternative. That is, the theory suggested that $\pi_F > 0.50$, so $\pi_F > 0.50$ needs to be the alternative. In other cases, such as the supplier whose parts should include no more than 5% defective, it is not so clear. It depends on who bears the burden of proof.

Clearly, it would be to our advantage for the burden of proof to be on the supplier:

H_o: $\pi_{Def} \geq 0.50$ $\quad$ H_o: $\pi_{Def} = 0.50$
H_a: $\pi_{Def} < 0.50$ $\quad$ H_a: $\pi_{Def} < 0.50$.

The sample p_{Def} value would need to be *enough below* 0.05 that we can reject H_o with just an α probability that we were rejecting wrongly.

Clearly, it would be to our suppliers advantage for the burden of proof to be on us:

H_o: $\pi_{Def} \leq 0.50$ $\quad$ H_o: $\pi_{Def} = 0.50$
H_a: $\pi_{Def} > 0.50$ $\quad$ H_a: $\pi_{Def} > 0.50$.

The sample p_{Def} value would need to be *enough above* 0.05 that we can reject H_o with just an α probability that we were rejecting wrongly. The burden of proof and the α to be used might well be the subject of negotiations.

8.3.2 The Decision Criterion

Return to the Walton's example, where we were looking for evidence to support the theory that predicted π_F to be greater than 0.50. Our hypotheses were:

1. H_o: $\pi_F \leq 0.50$
   H_a: $\pi_F > 0.50$.

Figure 8.8 shows the sampling distribution for p_F on the assumption that the null hypothesis is true. Suppose we are willing to accept a probability of rejecting wrongly, α, of 0.05.

With a two-tailed test, we needed to split α between the two tails. There was a 0.025 probability of getting an unusually low sample p_F, like p_1 that translated into a Z value below −1.96, even though the null hypothesis was true. And there was a 0.025 probability of getting an unusually high sample p_F that translated into a Z value above +1.96, even though the null hypothesis was true. Since we would reject the null hypothesis in both of these cases, we would make type I errors in both. The two probabilities added up to our α of 0.05.

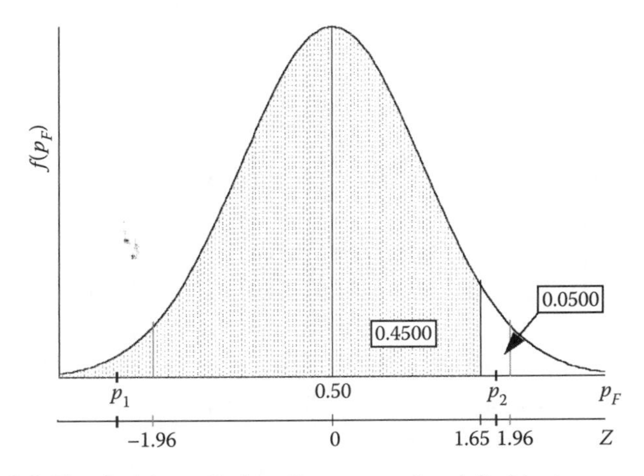

Figure 8.8 The decision criterion: One- versus two-tailed tests.

With the one-tailed test, above, there is still a 0.025 probability of getting an unusually low sample p_F, like p_1 that translates into a Z value below −1.96, even though the null hypothesis is true. But, now, getting it should not lead us to reject H_o, because it is not evidence in favor of H_a. And, if we do not reject H_o in favor of H_a when p_F is unusually low, we will never do so wrongly; we will never make a type I error. Only unusually high sample p_F values are evidence in favor of H_a. We should reject H_o only if our p_F is unusually high; hence, we will make type I errors only if p_F is unusually high. Thus, we should choose a rejection criterion such that the full α of 0.05 is in the upper tail. That is:

2. Reject H_o if $Z_c > 1.65$ ($\alpha = 0.05$).

Consider the following example:

> Suppose your supplier guarantees that the parts it supplies in very large lots include no more than 5% defectives. You can reject a shipment if you can show, based on a Sample of 200 and an α of 0.05 that it contains more. Suppose a sample from the next lot contains 14 defective parts. Can you reject the shipment? What is the p-value of this test? How many out of 200 does it take to reject the null hypothesis?

The burden of proof is on you to show that too many are defective; hence, too many must be the alternative hypothesis. Figure 8.9 illustrates.

1. H_o: $\pi_{Def} \leq 0.05$
   H_a: $\pi_{Def} > 0.05$,
2. Reject H_o if $Z_c > 1.65$ ($\alpha = 0.05$),
3.
$$p_{Def} = \frac{14}{200} = 0.07, \quad \sigma_p = \sqrt{\frac{0.05 \times 0.95}{200}} = 0.0154,$$
$$Z_c = \frac{0.07 - 0.05}{0.0154} = 1.30,$$
4. ∴ Fail to Reject H_o.

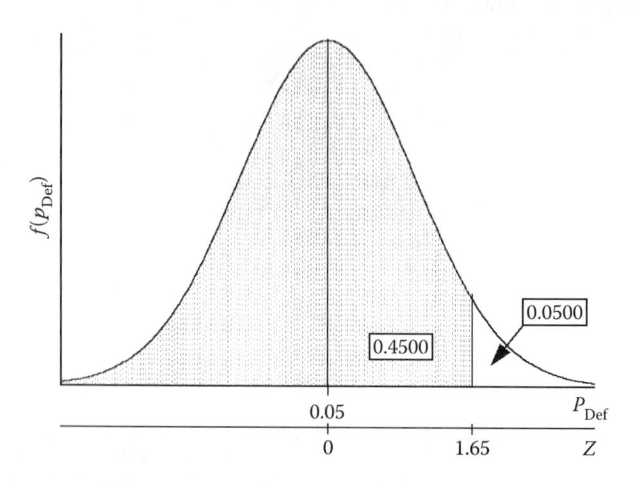

Figure 8.9 A one-tailed test for a proportion: The upper tail.

This sample p_{Def} is not large enough to reject the null hypothesis that the true π_{Def} is 0.05.

To find the p-value, look up Z_c in the standard normal table. The probability from 1.30 to the mean is 0.4032. Since this is a one-tailed test, we do not need to worry about the other side of the distribution. The p-value is just $0.5000 - 0.4032 = 0.0968$. If we had been using an α of 0.10, we could have rejected the shipment.

Finally, we can solve for the number of defectives required to reject H_o. We know that, to reject H_o, Z_c must be greater than 1.65. Thus, $Z_c = ((X/200)-0.05)/0.0154 > 1.65$. Solving for X,

$$\frac{X}{200} - 0.05 > 0.0154 \times 1.65 = 0.0254, \quad \frac{X}{200} > 0.05 + 0.0254 = 0.0754,$$

$$X > 0.0754 \times 200 = 15.09.$$

It takes more than 15 defectives to establish (with an α of 0.05) that there really are too many defectives in the shipment.

Consider the following example:

> Suppose you are able to shift the burden of proof onto your supplier. You do not need to show that there are too many defectives; your supplier needs to show that there are not. If a sample from the next lot contains six defectives, have they met their burden of proof?

To show that too many are not defective, not too many has to be the alternative hypothesis. Figure 8.10 illustrates.

1. H_o: $\pi_{\text{Def}} \geq 0.05$
   H_a: $\pi_{\text{Def}} < 0.05$,
2. Reject H_o if $Z_c < -1.65$ ($\alpha = 0.05$),
3. $$p_{\text{Def}} = \frac{6}{200} = 0.03, \quad \sigma_p = \sqrt{\frac{0.05 \times 0.95}{200}} = 0.0154,$$
   $$Z_c = \frac{0.03 - 0.05}{0.0154} = -1.30,$$
4. ∴ Fail to Reject H_o.

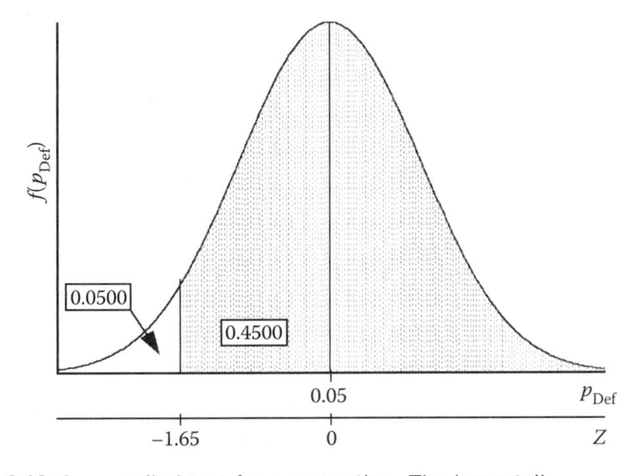

Figure 8.10 A one-tailed test for a proportion: The lower tail.

This sample p_{Def} is not small enough to reject the null hypothesis that the true $\pi_{\text{Def}} \geq 0.05$.

8.3.3 The Probability of a Type II Error*

In the previous examples, it took more than 15 defectives to show that π_{Def} was greater than 0.05; it took fewer than five defectives to show that π_{Def} was less than 0.05. This disparity illustrates fairly starkly the asymmetry between the null and alternative hypotheses. Either null hypothesis ($\pi_{\text{Def}} \leq 0.05$ or $\pi_{\text{Def}} \geq 0.05$) was consistent with a range of possible sample values. And since it was simply assumed, there was no need to show it to be true. The burden of proof was on the alternative. The outcome needed to be so extreme in the direction of the alternative that the null was no longer tenable.

This asymmetry between the null and alternative hypotheses is not unique to one-tailed tests. In all tests of hypothesis, the burden of proof is on the alternative. Indeed, provided that the alternative is in the correct direction, a one-tailed test is stronger than a two-tailed test. Provided that the alternative is in the correct direction, it is easier to reject the null hypothesis. After all, the criterion for rejecting with a two-tail test would have been $|Z_c| > 1.96$, not 1.65.

Figure 8.11 compares the OC curve for H_o: $\pi_{\text{Def}} \leq 0.05$ with that for the two-tailed test with the same α. Recall that an OC curve shows β, the probability of making a type II error—of not rejecting H_o when it is false—for every possible true value of π_{Def}. The two-tailed curve looks similar to those we saw earlier. The probability of failing to reject wrongly is fairly low for true π_{Def} values far above or below 0.05. The probability of failing to reject wrongly rises as the true π_{Def} value gets closer and closer to 0.05, and the true π_{Def} becomes harder to distinguish from 0.05.

The OC curve for H_o: π_{Def}, ≤ 0.05 differs in two ways. First, it exists just for true π_{Def} values above 0.05. Since H_o includes true π_{Def} values below 0.05, failing to reject is correct for these. Hence, we cannot make a type II error. Second, for true π_{Def} values above 0.05, the one-tailed curve is below the two-tailed curve. For any such true π_{Def}, we are less likely to fail

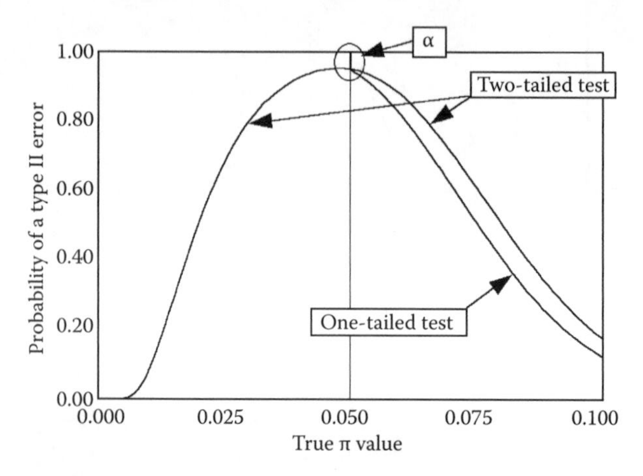

Figure 8.11 The OC curve for H_o: $\pi_{\text{Def}} \leq 0.05$ ($\alpha = 0.05$).

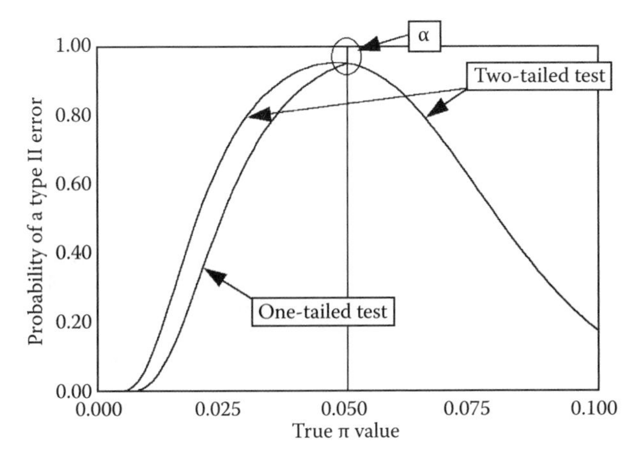

Figure 8.12 The OC curve for H_o: $\pi_{Def} \geq 0.05$ ($\alpha = 0.05$).

to reject when we should have. Again, this makes sense since the rejection criterion associated with any particular α is less extreme.

Of course, when the burden of proof changes from one side to the other, so do the opportunities for type II errors. Figure 8.12 illustrates. The two-tailed curve is exactly the same as in Figure 8.11; the one-tail curve swaps sides. Now, it exists just for true π_{Def} values below 0.05. Since H_o includes true π_{Def} values above 0.05, failing to reject is correct for these. Hence, we cannot make a type II error. And, again, for true π_{Def} values below 0.05, the one-tailed curve is below the two-tailed curve. For any such true π_{Def}, we are less likely to fail to reject when we should have. Again, this makes sense since the rejection criterion associated with any particular α is less extreme.

8.3.4 When a One-Tailed Test is Legitimate

As we have just seen, the rejection criterion associated with a one-tail test is less extreme than the comparable tail in a two-tailed test. We are less likely to make type II errors. Hence, we should use one-tailed tests whenever they are legitimate. However, it is just as important not to use them when they are not. Always ask yourself whether you are truly interested in just one of the tails; and always ask yourself *before you have looked at your data.*

In the problems dealing with π_{Def}, the proportion of defective parts, there is a fundamental difference between there being too many and too few defectives. Indeed, it is not clear that there is any such thing as too few defectives. Certainly we would not reject a shipment because it had too few defectives. The choice of tail had to do with who bore the burden of proof. Did we need to show something was wrong? Or did our supplier need to show that nothing was wrong? There was no dispute over what it meant for something to be wrong. It meant there were too many defectives.

By contrast, return to the Walton's example, where we were looking for evidence to support the theory that predicted π_F, the proportion of customers who were female would be greater than 0.50. In cases like this,

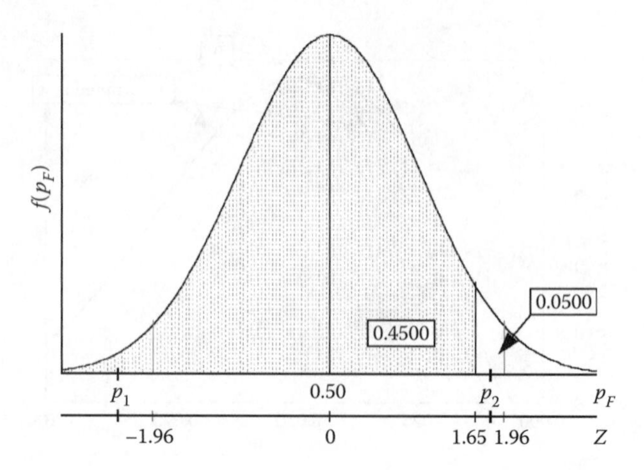

Figure 8.13 The decision criterion: One- versus two-tailed tests.

we generally hope to reject H_o. We want support for our theory. And, as we have seen, the one-tailed criterion allows us to reject H_o in some cases in which the two-tailed criterion does not. In Figure 8.13, which simply repeats Figure 8.8 for convenience, a sample p_F-value of p_2 is great enough to allow us to reject H_o using a one-tailed criterion, but not a two-tailed criterion. Used correctly, though, the one-tailed test is just as rigorous. After all, there are other sample p_F-values, like p_1, which allow us to reject H_o using a two-tailed criterion, but not using a one-tailed criterion, because we ended up with the wrong tail.

The key to correct use of the one-tail test is that you genuinely decide the tail without regard to the data. If you let the data determine the tail you are interested in, you have really already used both tails. In the Walton's example, did you really decide that you would be uninterested in p_F values like p_1, *before you looked at the data?* If you would have found p_1 interesting as well—perhaps as evidence for some competing theory—then you really need to use a two-tailed test. If you get a sample p_F value like p_2, but would have found p_1 interesting as well, you cannot honestly claim to reject the null hypothesis with a one-tailed α of 0.05. You can, though, still honestly claim to reject it with a two-tailed α of 0.10.

8.4 Tests for the Population Mean

We have covered quite a few concepts concerning hypothesis testing—null and alternative hypotheses, one- and two-tail tests, decision criteria, type I and type II errors with probabilities of α and β, and the *p*-value of the test—all using hypotheses concerning π_X, the population proportion. What about hypotheses concerning μ_X, the population mean? It is tempting to say "ditto" and be done with it. Instead, we will go through most of it again. There are a few differences. As you already know from Chapters 6 and 7, we usually do not know σ_X, the population standard deviation, necessary to calculate the Z_c. As a result, we usually need to estimate it using s_X, the sample standard deviation and what we calculate is t_c instead. So there will be some minor differences. Still, you should be struck by how similar the reasoning is. Indeed, this same reasoning and these same ideas—null and alternative hypotheses, one- and two-tail tests, decision criteria, type I and type II errors with probabilities of α and β, and the p-value of the test—should carry you through pretty much the rest of the book.

8.5 A Two-Tailed Test for the Population Mean

8.5.1 The Null and Alternative Hypotheses

Suppose someone claims that the average height of male college students is 70 inches. This is a claim about a population parameter, μ_H, which may or may not be true. And we can test it in much the same way we tested claims about π_X earlier. The null hypothesis is H_o: $\mu_H = 70$ inches. The alternative hypothesis is what we will believe if we succeed in rejecting the null hypothesis. A general, two-tailed hypothesis would be simply that the null hypothesis is wrong. That is, H_a: $\mu_H \neq 70$ inches.

As always, the first step is to write down these hypotheses:

1. H_o: $\mu_H = 70$ inches
   H_a: $\mu_H \neq 70$ inches.

8.5.2 The Decision Criterion

We need to decide on a probability, α, which we are willing to accept of making a type I error—of declaring that H_o is false when it is true. As with tests of proportions, the probability we are willing to accept will vary with the seriousness of making such a mistake. Suppose we decide on $\alpha = 0.05$.

We also need to check whether or not we know σ_H, the population standard deviation of these heights. Recall from the last two chapters that if we know σ_H we will be able to calculate the standard error, $\sigma_{\bar{H}}$; and $(\bar{H} - \mu_H)/\sigma_{\bar{H}}$ follows the standard normal distribution. But if, as would be typical, we do not know σ_H, we will need to estimate it using s_H and use s_H to find an estimate of the standard error, $s_{\bar{H}}$. And $(\bar{H} - \mu_H)/s_{\bar{H}}$ follows the t distribution with $n - 1$ degrees of freedom.

Suppose (a) that we know σ_H, the population standard deviation. We are going to test H_o by taking a sample and finding $\bar{H}$, the average height in the sample. Of course, we are unlikely to get exactly $\bar{H} = 70$ inches, even if the null hypothesis is true. As we know, there is a whole sampling distribution of $\bar{H}$ values we could get from a population for which $\mu_H = 70$ inches.

Still, we know what that distribution looks like. We know that, if the population distribution is normal or the sample is large, the sampling distribution is normal with mean $\mu_{\bar{H}} = \mu_H = 70$ and standard error $\sigma_{\bar{H}} = \sigma_H / \sqrt{n}$. We also know that $70 \pm 1.96\ \sigma_{\bar{H}}$ includes all but 5% of the possible $\bar{H}$ values. Figure 8.14 illustrates.

We follow the same procedure as we did for proportions. We take our sample and find our sample $\bar{H}$. We convert this $\bar{H}$ value to its Z_c equivalent,

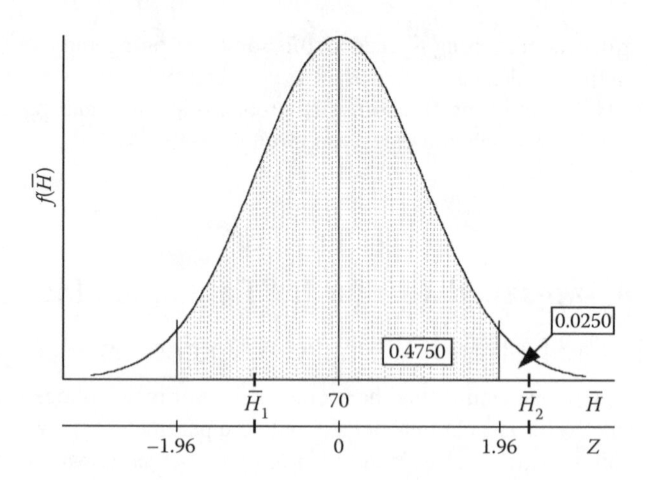

Figure 8.14 The sampling distribution for a mean, $\mu_H = 70$, σ_H known.

assuming that the null hypothesis, $\mu_H = 70$ inches is true. Then, if this Z_c is outside the ±1.96 range, we conclude that our sample $\bar{H}$ value is too far from the hypothesized μ_H. We should not have come up with a sample $\bar{H}$ value this far from the true value. Therefore, the null hypothesis must not be true.

Figure 8.14 assumes the null hypothesis is true and that $\mu_H = 70$ inches. If it is, we know that 95% of the time we will get an $\bar{H}$ value like $\bar{H}_1$, close enough to 70 that it will standardize to a Z_c in the ±1.96 range. We will fail to reject H_o, which is the correct decision since H_o is true. The other 5% of the time, though, we will be unlucky and get an unusual $\bar{H}$ value like $\bar{H}_2$, far enough from 70 that it will standardize to a Z_c outside the ±1.96 range. In these cases, we will reject H_o, which is a type I error. This criterion then (rejecting H_o if $|Z_c| > 1.96$) gives us an α of 0.05.

As before, we can choose different values for α. If we are not willing to reject wrongly 5% of the time, we can choose a smaller probability. If we choose $\alpha = 0.01$, we would not reject unless $|Z_c| > 2.58$ and we would reject wrongly only 1% of the time. Of course, this would also increase the probability of making a type II error—of failing to reject H_o when it is actually false. Adding this criterion to our hypotheses, we have now formally defined the test:

1. H_o: $\mu_H = 70$ inches
   H_a: $\mu_H \neq 70$ inches,

2a. Reject H_o if $|Z_c| > 1.96$ ($\alpha = 0.05$).

Instead, suppose (b) that we do not know σ_H, the population standard deviation.

We are still going to test H_o by taking a sample and finding $\bar{H}$, the average height in the sample. And there is still a whole sampling distribution of $\bar{H}$ values that we could get from a population for which $\mu_H =$ 70 inches. Now, though, we cannot calculate the actual standard error, $\sigma_{\bar{H}} = \sigma_H / \sqrt{n}$, because we do not know σ_H, the population standard deviation. We can only estimate it based on s_H, the sample standard deviation, $s_{\bar{H}} = s_H / \sqrt{n}$. And calculating $(\bar{H} - \mu_H)/s_{\bar{H}}$ gives us not Z_c but t_c.

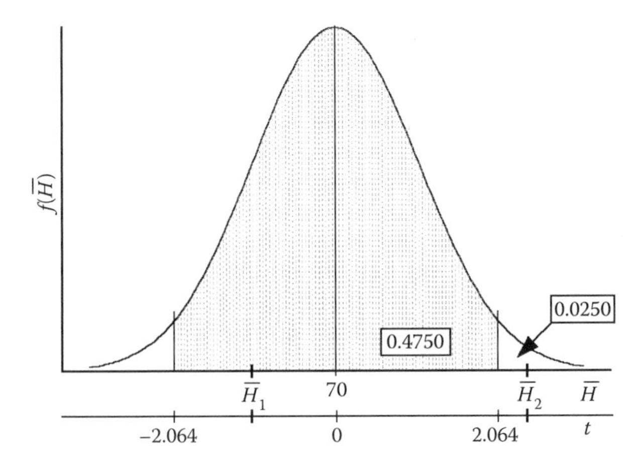

Figure 8.15 The sampling distribution for a mean, $\mu_H = 70$, σ_H unknown, $n = 25$.

Our criterion, then, is different. Suppose our sample size is 25. Then, we have 24 degrees of freedom. Looking in the 0.025 column of the *t* table, and 24 df (degrees of freedom), our criterion is not ±1.96 but ±2.064. Otherwise, our reasoning is exactly the same.

Figure 8.15 illustrates. The only difference from Figure 8.14 is that the *t* distribution is slightly wider than the normal, meaning that we need to go further out—to ±2.064 instead of just ±1.96—to include 95% of the $\bar{H}$-values we might get if the null hypothesis is true.

Again, adding our criterion to the hypotheses, we have now defined the problem:

1. H_o: $\mu_H = 70$ inches
   H_a: $\mu_H \neq 70$ inches,

2b. Reject H_o if $|t_c| > 2.064$ (df = 24; $\alpha = 0.05$).

8.5.3 The Calculations

So far, we have not consulted the data; that is step 3. Suppose our random sample of 25 male college students has a sample mean height, $\bar{H}$, of 71.5 inches, and a sample standard deviation in heights, s_H, of 2.95 inches. Our sample $\bar{H}$ is not 70, but is it so far from 70 that we can conclude the population mean must not be 70 either? To decide we need to calculate Z_c or t_c.

Suppose (a) that we know that $\sigma_H = 3$ inches.

3a. $$Z_c = \frac{(\bar{H} - \mu_H)}{\sigma_{\bar{H}}} \quad \text{where} \quad \sigma_{\bar{H}} = \frac{\sigma_H}{\sqrt{n}}\left[\frac{N-n}{N-1}\right].$$

Doing the calculations (again ignoring the finite population factor since we do not know the population size),

3a. $$\sigma_{\bar{H}} = \frac{3}{\sqrt{25}} = 0.6, \quad \text{and} \quad Z_c = \frac{(71.5-70)}{0.6} = 2.500.$$

Instead, suppose (b) that we do not know σ_H, the population standard deviation.

$$3b.\ t_c = \frac{(\bar{H} - \mu_H)}{s_{\bar{H}}} \quad \text{where} \quad s_{\bar{H}} = \frac{s_H}{\sqrt{n}}\left[\sqrt{\frac{N-n}{N}}\right].$$

Doing the calculations,

$$3b.\ s_{\bar{H}} = \frac{2.95}{\sqrt{25}} = 0.59, \quad \text{and} \quad t_c = \frac{(71.5 - 70)}{0.59} = 2.542.$$

8.5.4 The Conclusion

In this example we get the same answer whether we are calculating Z_c or t_c. $|Z_c| = 2.500 > 1.96$ and $|t_c| = 2.542 > 2.064$.

4. $\therefore$ Reject H_o.

The average height of male college students is not 70 inches.

Suppose as in the previous example we had decided on an α of 0.01 instead. Our hypotheses would have been unchanged. However our criteria would have been different.

Suppose (a) that we know that $\sigma_H = 3$ inches.

2a. Reject H_o if: $|Z_c| > 2.58$ ($\alpha = 0.01$).

Instead, suppose (b) that we do not know σ_H, the population standard deviation.

2b. Reject H_o if: $|t_c| > 2.797$ (df = 24; $\alpha = 0.01$).

The calculations do not change either, but the conclusion does in this case whether we are calculating Z_c or t_c, $|Z_c| = 2.500 < 2.58$ and $|t_c| = 2.542 < 2.797$.

4. $\therefore$ Fail to Reject H_o.

Clearly, one answer is right and one answer is wrong. The value of μ_H is either 70 inches or it is not. If μ_H is really 70 inches, our first answer was wrong—we made a type I error rejecting the null hypothesis when it was true. If μ_H is not really 70 inches, our second answer was wrong—we made a type II error failing to reject the null hypothesis when it was false. Which was it this time? We do not know; we only know the probabilities.

8.5.5 The *P*-Value of the Test

Again, because we can reject or not reject, depending on our choice of α, we may wish to calculate the *p*-value of our test. Again, this is especially true when we are presenting results to others. If we tell them the α they would have to accept in order to reject H_o, they can decide for themselves whether they are willing to accept this risk of rejecting wrongly.

Suppose (a) that we knew σ_H, and so calculated Z_c.
In our examples above, we found Z_c equaled 2.500. We can look up this Z_c-value in the standard normal table and find the probability associated with it. We get 0.4938. The area between ±2.500 in the standard normal distribution is $2 \times 0.4938 = 0.9876$. The area not in this interval is $1 - 0.9876 = 0.0124$. So, if we are willing to accept a probability of rejecting wrongly, α, of 0.0124 or greater, we are able to reject. If we are not willing to accept this great a probability of rejecting wrongly, we are not able to reject.

Suppose (b) that we did not know σ_H, and so calculated t_c.
In our examples above, we found t_c equaled 2.542. We can look up this t_c-value in the *t* table and find the (approximate) probability associated with it. Reading across the line for 24 degrees of freedom, 2.542 is between 2.492 and 2.797 in the 0.0100 and .0050 columns, respectively; hence the probability in each tail is somewhere between 0.01 and 0.005. The area outside the range ± 2.542 in the *t* distribution is somewhere between $2 \times 0.01 = 0.02$ and $2 \times 0.005 = 0.01$.

As we noted in Chapter 6, with most tables other than the standard normal, finding *p*-values can be imprecise because relatively few probabilities are listed. The previous result may be precise enough. It tells us that we can reject with an α of 0.02, but not with an α of 0.01. Still, if we want to be more precise, we can use the spreadsheet special functions from Chapter 6. In this case, with a t_c of 2.542, 24 degrees of freedom, and a two-tailed test, the actual *p*-value is =TDIST(2.542,24,2) = 0.0179.

So, if we are willing to accept a probability of rejecting wrongly, α, of 0.0179 or greater, we are able to reject. If we are not willing to accept this great a probability of rejecting wrongly, we are not able to reject.

Consider the following example:

> A mill packs flour in 25 lb sacks. Its quality control inspectors check the process by randomly sampling 10 sacks each hour and recording the weight of each. If, in the next sample, they find a mean of 24.3 and a standard deviation of 0.75, can they conclude that something has gone wrong and the true average is no longer 25 lb? Use $\alpha = 0.05$. Use $\alpha = 0.01$. In which case might you be making an type I error;? In which case might you be making a type II error? What is the *p*-value of this test?

To conclude something, it must be the alternative hypothesis, so the hypotheses are

1. H_o: $\mu_W = 25$
   H_a: $\mu_W \neq 25$.

The rejection criterion depends on whether we are able to calculate Z_c. Do we know σ_W, the true standard deviation in weights? Reading carefully, the only standard deviation given is s_W, the sample standard deviation in weights. We will need to use s_W as our estimate for σ_W, and we will be calculating not Z_c but t_c, with $n - 1 = 10 - 1 = 9$ degrees of freedom.

The rejection criterion also depends on α:

2. Reject H_o if $|t_c| > 2.262$ (df = 9; $\alpha = 0.05$)   Reject H_o if $|t_c| > 3.250$ (df = 9; $\alpha = 0.01$).

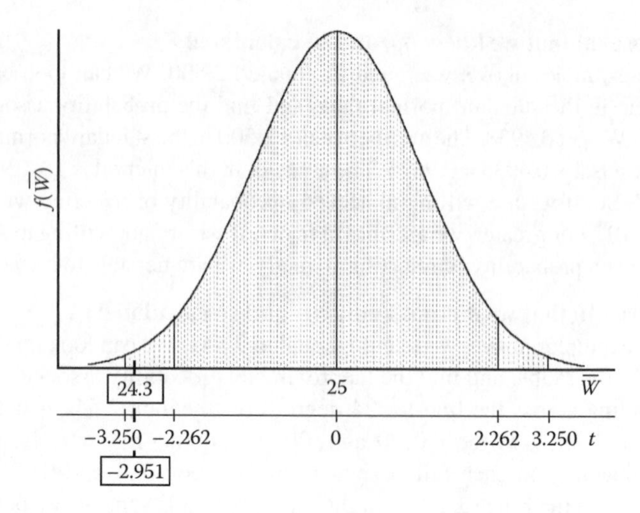

Figure 8.16 Two different criteria, two different decisions.

Figure 8.16 illustrates the sampling distribution and the two rejection criteria on the assumption that H_o is correct.

Our calculations are as follows:

3. $s_{\overline{W}} = \dfrac{0.75}{\sqrt{10}} = 0.2372, \quad t_c = \dfrac{24.3 - 25}{0.2372} = -2.951$.

Our conclusion again depends on our rejection criterion.

4. $|-2.951| > 2.262$ $\qquad$ $|-2.951| < 3.250$

$\therefore$ Reject H_o $\qquad$ $\therefore$ Fail to Reject H_o.

Using $\alpha = 0.05$, we reject the null hypothesis; either $\mu_W \neq 25$ or we are making a type I error—rejecting a true hypothesis. Using $\alpha = 0.01$, we fail to reject the null hypothesis; either $\mu_W = 25$ or we are making a type II error—failing to reject a false hypothesis.

The *p*-value of the test can be found by looking up our $|t_c|$ of 2.951 in the *t* table. Reading across at 9 degrees of freedom, our t_c-value falls between 2.821 and 3.250, in the 0.0100 and 0.0050 columns, respectively. The *p*-value is between $2 \times 0.0100 = 0.02$ and $2 \times 0.0050 = 0.01$. To be more precise, the *p*-value is =TDIST(2.951,9,2) = 0.0162. If we are willing to accept $\alpha \geq 0.0162$, we should reject H_o. If we are not, we should fail to reject H_o.

8.5.6 The Probability of a Type II Error*

What about β, the probability of a type II error? We find the answer in exactly the same manner as we did for proportions. And it depends on the same things. The β varies inversely with α. Other things equal, the lower we make α, the probability of rejecting wrongly, the higher we make β, the probability of failing to reject when we should.

And β depends on the sample size. The β we must accept, for any given α, goes down as sample size increases. More information helps.

And finally, β depends on the true value of μ. Continuing with the preceding example, suppose the true value of μ_W were not 25, but very close—say 25.2. The distribution of $\bar{W}$ values we could get for μ_W-values of 25 and 25.2 are very similar. The sample $\bar{W}$ value we would get is likely to be consistent with our null hypothesis, even though our null hypothesis is wrong. Hence, we would be likely to make a type II error. On the other hand, suppose the true value of μ_W were much further from 25—say 27. The distribution of $\bar{W}$ values we could get for μ_W-values of 25 and 27 are not very similar. The sample $\bar{W}$ value we would get is not likely to be consistent with our null hypothesis. Hence, we would not be likely to make a type II error. Indeed, there is a whole distribution of β values, depending the true value of μ_W.

Figure 8.17 illustrates the situation if H_o: $\mu_W = 25$ and the true μ_W is 25.2. Our criterion for $\alpha = 0.05$ was to reject H_o if $|t_c| > 2.262$. The ±2.262 range, figured around our hypothesized value of 25, is shown on the t_{25} scale. Anything within this range was deemed plausible, given H_o; hence we would fail to reject. Notice, though, there is still a strong chance of getting a value in this range, even though H_o is not correct and the true μ_W is 25.2. To find this probability, we need to (i) take our ±2.262 range figured around our hypothesized value of 25; and translate it into actual $\bar{W}$ value equivalents; (ii) re-standardize these $\bar{W}$ values using the true μ_W of 25.2; and (iii) find the probability between these corrected t-values.

First, to find the $\bar{W}$ value equivalents of ±2.262, as we would have calculated them using our hypothesized μ_W of 25:

$$t_{25} = \frac{\bar{W}_{\text{Lower}} - 25}{0.75/\sqrt{10}} = \frac{\bar{W}_{\text{Lower}} - 25}{0.2372} = -2.262 \quad \text{and}$$

$$t_{25} = \frac{\bar{W}_{\text{Upper}} - 25}{0.75/\sqrt{10}} = \frac{\bar{W}_{\text{Upper}} - 25}{0.2372} = 2.262.$$

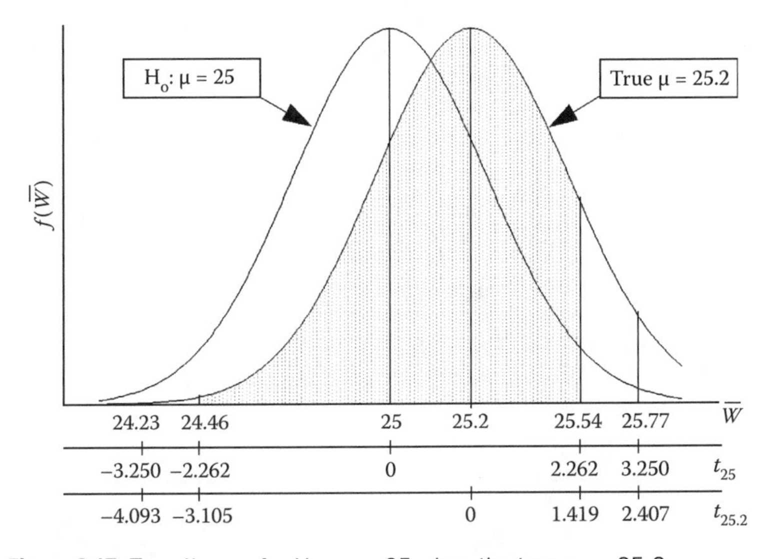

Figure 8.17 Type II error for H_o: $\mu_W = 25$ when the true $\mu_W = 25.2$.

So, $\bar{W}_{\text{Lower}} = 25 - 2.262 \times 0.2372 = 24.46$ and $\bar{W}_{\text{Upper}} = 25 + 2.262 \times 0.2372 = 25.54$.

Second, to calculate the *t*-value equivalents of these $\bar{W}$ values, using the true μ_W of 25.2 – $t_{25.2} = (24.46 - 25.2)/0.2372 = -3.105$ and $t_{25.2} = (25.54 - 25.2)/0.2372 = 1.419$.

Since 25.2 is between $\bar{W}_{\text{Lower}}$ and $\bar{W}_{\text{Upper}}$, the *t*-values are on opposite sides of the mean. Looking them up in the *t* table does not give us much precision. Using the line for 9 degrees of freedom, 3.105 falls somewhere between 2.821 and 3.250 in the 0.0100 and 0.0050 columns; 1.419 falls somewhere between 1.383 and 1.833 in the 0.1000 and 0.0500 columns. Using the spreadsheet special function instead, =TDIST(3.105,9,1) = 0.0063 and =TDIST(1.419,9,1) = 0.0948. Adding these, 0.0063 + 0.0948 = 0.1011. Recall, these are the probabilities in the tails; we want the shaded area—the overlap. Thus, 1.0000 – .1011= 0.8989 is β, the probability of getting a $\bar{W}$ between 24.46 and 25.54, and accepting H_o: $\mu_W = 25$ when the true μ_W is 25.2.

The probability of failing to reject when the true mean is this close then is quite high. As the true mean gets further and further from our hypothesized true mean, though, the distribution around the true mean overlaps less and less with our fail-to-reject region. The probability of a type II error shrinks.

Again, as we did with proportions, we could repeat this procedure over and over, for H_o: $\mu_W = 25$, using $\alpha = 0.05$ and a sample of 10, varying the true μ_W-value from well below 25 to well above it. We could then plot the probability of a type II error as a function of the true μ_W. The middle curve in Figure 8.18 shows the resulting OC curve.

Figure 8.18 also shows two other OC curves to show the effect of changes in α and *n*. Reducing the α to 0.02 shifts the curve up. At 25 this is good since we want the curve to be high there. We want to fail-to-reject there since H_o would be right. Everywhere else, though, this is bad; we have increased the probability of a type II error.

Increasing the sample size reduces the probability of a type II error for any α. The lower curve agrees with the middle curve at 25; they are

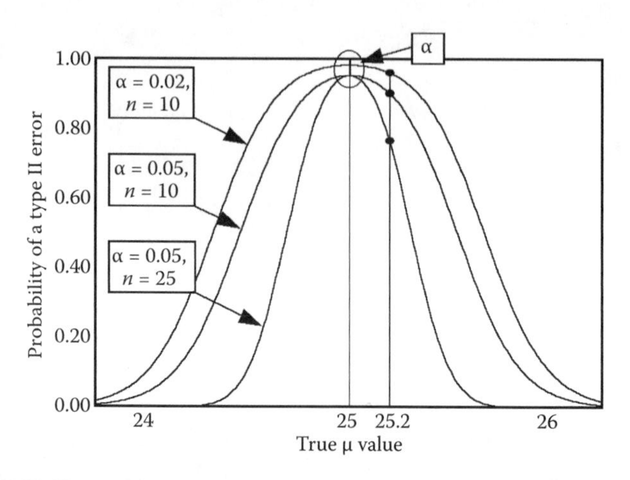

Figure 8.18 Three OC curves for H_o: $\mu_W = 25$.

both calculated for $\alpha = 0.05$. The lower curve, though, is based on a larger sample of 25. It falls off much faster, meaning that the probability of a type II error is less. Having more information increases our ability to discriminate between cases in which the null hypothesis is true, and cases in which it is false but fairly close.

8.6 A One-Tailed Alternative for the Population Mean

So far, we have been looking at two-tailed tests of the mean. We have been willing to reject H_o in favor of H_a if our sample $\bar{W}$ value turned out to be either too high or too low. As with proportions, though, we are often looking for evidence in just one direction or the other. The reasons are the same.

First, we may be looking for support for some theory that suggests a particular value or range for the mean. As already noted, though, most theories relate more to relationships among variables than to levels of a single variable. Thus, this theoretical motivation for a one-tailed test will matter more when we begin looking at relationships, starting in Chapter 9.

Second, only one of the two tails may be of any concern to us. Suppose, in the previous examples on the weights of 25 lb sacks of flour, we were actually government prosecutors investigating a complaint that the mill was "short weighting" its products. We would then be looking only for evidence that $\mu_W < 25$. We would not be concerned (though the mill would) if $\mu_W > 25$.

8.6.1 The Null and Alternative Hypotheses

Continue with the short-weighting example. To decide on the null and alternative hypotheses, we need to decide where the burden of proof should lie. In our justice system, the mill does not need to prove it is innocent; the prosecution needs to prove that it is not. The burden of proof lies with the prosecution. Hence, innocence is the null hypothesis; short weighting is the alternative.

1. H_o: $\mu_W \geq 25$ lbs
   H_a: $\mu_W < 25$ lbs.

8.6.2 The Decision Criterion

As always for the mean, the rejection criterion depends on whether we are able to calculate Z_c. Do we know σ_W, the true standard deviation in weights? If so, we can calculate $\sigma_{\bar{W}}$ and, from that, Z_c. If not, we must use s_W, an estimate of σ_W, to calculate $s_{\bar{W}}$, an estimate of $\sigma_{\bar{W}}$ and, from that, t_c.

In the flour example we did not know σ_W we knew only s_W, the sample standard deviation. Therefore, what we calculated was t_c. Figure 8.19 repeats Figure 8.16, which showed the sampling distribution on the assumption that H_o is correct.

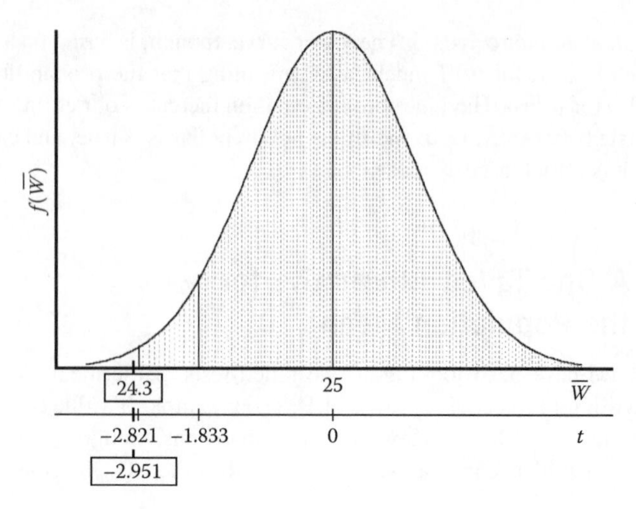

Figure 8.19 A one-tailed test for a mean: The lower tail.

The distribution has not changed; there are still the same probabilities of getting misleadingly high and low sample $\overline{W}$-values, even if the true μ_W is 25. This time, though, because of our one-tailed alternative, we are never going to reject H_o in the upper tail; large values of $\overline{W}$ are not evidence of short weighting. And, since we are never going to reject H_o in the upper tail, we are never going to do so wrongly; we are never going to make a type I error.

The only type I errors we can make are in cases in which our sample $\overline{W}$ is misleadingly small, leading us to reject H_o when it is true. Thus, the whole probability of a type I error is in the lower tail. For the α-values of 0.05 and 0.01, our criteria are

2. Reject H_o if $t_c < -1.833$ (df = 9; $\alpha = 0.05$)   Reject H_o if $t_c < -2.821$ (df = 9; $\alpha = 0.01$).

The calculations do not change:

$$3.\ s_{\overline{W}} = \frac{0.75}{\sqrt{10}} = 0.2372, \quad t_c = \frac{24.3 - 25}{0.2372} = -2.951.$$

Our conclusion for $\alpha = 0.01$ does though:

4. $-2.951 < -1.833$ $-2.951 < -2.821 \therefore$ Reject H_o.

Since we reject only in the lower tail and that is the tail we got, we are able to reject H_o with $\alpha = 0.01$—something we could not do when we used a two-tailed test. What is the p-value of this test? Looking across the t table at 9 degrees of freedom, our value of −2.951 is between −2.821 and −3.250 in the 0.0100 and 0.0050 columns; the p-value is somewhere between 0.01 and 0.005. Using the spreadsheet special function, with a t_c of −2.951, 9 degrees of freedom and a one-tailed test, the actual p-value is =TDIST(−2.951,9,1) = 0.0081.

It is perhaps worth emphasizing that we were able to use the one-tailed criterion in this example because we were government prosecutors, investigating an allegation of short weighting. The mill itself, in setting up its quality control procedures, would probably be concerned with the sacks being too heavy as well as too light. For this purpose, it would want a two-tailed test. So, there is no easy formulaic way of distinguishing one- and two-tailed tests. We need to put ourselves into the position of the decision maker. Between what possibilities is he or she trying to distinguish? For one-tailed tests, then, we need also to determine where the burden of proof lies. This has to be the alternative.

All this can and should be done before consulting the data. Of course, in textbook problems, the data or summary statistics must be given as part of the problem setup; it is tempting to look ahead, see that the t_c statistic is going to come out negative, and decide less than must be the relevant alternative. But there is no reason that the sample $\bar{W}$ in this problem could not have been 25.7 instead of 24.3. It would not have changed what we were looking for, short weighting, so the hypotheses and criterion would be the same. The t_c statistic would have come out positive and just as large. But we would not reject. A positive t_c statistic, no matter how large, would not be evidence of short weighting. The prosecutors have no case.

8.6.3 The Probability of a Type II Error*

As we saw with proportions, the OC curve showing β, the probability of a type II error, also changes with a one-tailed test. Figure 8.20 compares the one-tailed OC curve for H_o: $\mu_W \geq 25$ lbs with the two-tailed one we found earlier (the middle curve in Figure 8.18).

Recall that a type II error is failing to reject a false null hypothesis. And H_o: $\mu_W \geq 25$ lbs is consistent with any true μ_W of 25 or higher; failing to reject is the correct decision not an error. Only if the true μ_W is less than 25 is failing to reject an error; hence, the OC curve for H_o: $\mu_W \geq 25$ lbs exists only for true μ_W values of less than 25.

Moreover, for true μ_W values of less than 25, the one-tailed OC curve is below the two-tailed one with the same α. Again, this makes sense. The rejection criterion for the one-tailed test is less extreme than the comparable criterion for the two-tailed test, –1.833 versus –2.262 (df = 9; α = 0.05).

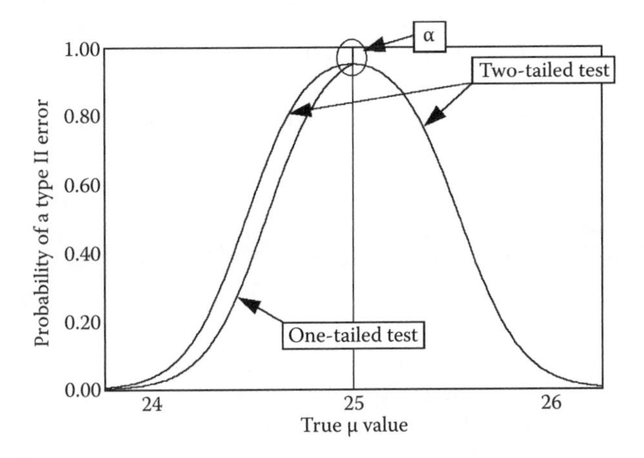

Figure 8.20 The OC curve for H_o: $\mu_W \geq 25$ (df = 9, α = 0.05).

It is easier to reject, assuming you have the correct tail, so it is less likely that you will fail to do so.

8.7 A Final Word on One-Sample Tests

In this chapter, we have looked at one sample tests for the proportion and for the mean. We have tested hypotheses about each, based on a given α—the probability of rejecting a true hypothesis. We have calculated p-values for the tests—the α-values we would need to be willing to accept in order to reject the null hypothesis given our samples. And we have explored the ways in which β—the probability of failing to reject a false hypothesis—varies with α, the sample size, the type of test, and the actual true value of π_X or μ_X.

In working problems, the first question to address always is whether we are testing a claim about a true proportion (π_X) or a true mean (μ_X). We establish null and alternative hypotheses about these true values. These hypotheses could be either two- or one-tailed. Next, we choose α, the probability we are willing to accept of rejecting the null hypothesis wrongly and, based on this α, choose a criterion for rejecting the null hypothesis. For proportions, this criterion is always a Z_c value extreme enough that the probability of getting such a value under the null hypothesis is just α. For means, this criterion is a Z_c value if we know the population standard deviation, σ_X—a possibility, though an unlikely one. It is a t_c value if we do not know σ_X and need to use the sample s_X as an estimate.

Only then do we consult the data—we calculate Z_c or t_c. Finally, we compare this calculated Z_c or t_c-value with our criterion. If this calculated Z_c or t_c is extreme enough, according to our criterion, we reject the null hypothesis in favor of the alternative; if it is not we fail to reject.

Finally, as we have developed the ideas of inference—confidence interval estimates and tests of hypotheses with known probabilities of being right or wrong—many of the examples have been "quality-control" examples. These are the most obvious cases in which we care about the value of a single population parameter like a proportion or mean. No more than a certain proportion of parts should be defective; parts are supposed to be a certain size on average; and so forth. Now that the logic is firmly established, though, the following chapters will develop more interesting examples using this same logic. In particular, we will revisit some of the examples of the early chapters and test whether one variable can explain variation in another.

8.8 Exercises

8.1 A freight railroad company claims that at least 95% of its trains are on time. If a random sample of 200 trains are clocked and 182 are on time, can you reject its claim?

a. Use $\alpha = 0.01$.

b. Use $\alpha = 0.001$.

c. What is the p-value of this test?

8.2 Senator Able is in a very close race with challenger Baker. A poll of 500 likely voters shows 270 for Able and 230 for Baker. Can you conclude that either is actually ahead?
   a. Use $\alpha = 0.10$.
   b. Use $\alpha = 0.05$.
   c. What is the p-value of this test?

8.3 A certain type of thread is manufactured under specifications that its mean tensile strength must be at least 25 lbs. A random sample of 10 specimens yields a mean of 23 lbs and a standard deviation of 3.3 lbs. Can you conclude that something has gone wrong?
   a. Use $\alpha = 0.05$.
   b. Use $\alpha = 0.01$.
   c. What is the p-value of this test?

8.4 Your firm purchases large shipments of a component, subject to the requirement that not more than 4% of a shipment may be defective. If a random sample of 200 parts from a new shipment reveals 12 defectives, can you reject this shipment as not meeting requirements? (Assume that the burden of proof is on you to show that a shipment does not meet this requirement.)
   a. Use $\alpha = 0.10$.
   b. Use $\alpha = 0.05$.
   c. What is the p-value of this test?
   d. How would your analysis change if the burden of proof were on your supplier to show that a shipment does meet this requirement?

8.5 Suppose you're responsible for assuring that a machine produces parts with no more than 5% defective. You instruct an operator to draw a random sample of 50 parts each hour, check them for defects, and shut down the machine if he finds more than "X" defectives. What would "X" be? That is, how many defectives would it take, in a sample of 50, to lead you to conclude that something is wrong? Use $\alpha = 0.05$.

8.6 In Exercise 3.1 you calculated summary statistics for the sample of 50 student GPAs in the file Studentsl.xls. Return to those data.
   a. Can you conclude that more than 50% of all students at this school are female? Use $\alpha = 0.05$. What is the p-value of this test?
   b. Can you conclude that the mean GPA of all students at this school is less than 3.000? Use $\alpha = 0.05$. What is the p-value of this test?

8.7 In Exercise 3.5 using NLSY1.xls, you calculated mean Height for young men and young women separately. Return to those data.
   a. Can you conclude that the mean Height of young men is greater than 67 inches? Use $\alpha = 0.05$. What is the p-value of this test?
   b. Can you conclude that the mean Height of young women is less than 67 inches? Use $\alpha = 0.05$. What is the p-value of this test?

8.8 Continue with these same data.
   a. Can you conclude that the mean Weight of young men is greater than 150 pounds? Use $\alpha = 0.05$. What is the p-value of this test?
   b. Can you conclude that the mean Weight of young women is less than 150 pounds? Use $\alpha = 0.05$. What is the p-value of this test?

8.9 In Exercises 3.9 and 3.10, using Students2.xls, you found means (among other statistics) for student heights, weights, entertainment spending, study time, and college GPAs. Return to those data.
   a. Can you reject the claim that the mean height is 67 inches? Use $\alpha = 0.05$. What is the p-value of this test?
   b. Can you reject the claim that the mean weight is 150 pounds?
   Use $\alpha = 0.05$. What is the p-value of this test?
   c. Can you reject the claim that the mean entertainment spending is $25 per week?
   Use $\alpha = 0.05$. What is the p-value of this test?
   d. Can you reject the claim that the mean study time is 20 hours per week?
   Use $\alpha = 0.05$. What is the p-value of this test?
   e. Can you reject the claim that the mean college GPA is 3.000?
   Use $\alpha = 0.05$. What is the p-value of this test?

8.10 Continue with these same data.
   a. Can you reject the claim that 50% hold jobs?
   Use $\alpha = 0.05$. What is the p-value of this test?
   b. Can you reject the claim that 25% major in economics?
   Use $\alpha = 0.05$. What is the p-value of this test?
   c. Can you reject the claim that 25% participate in varsity sports?
   Use $\alpha = 0.05$. What is the p-value of this test?
   d. Can you reject the claim that 25% participate in music ensembles?
   Use $\alpha = 0.05$. What is the p-value of this test?
   e. Can you reject the claim that 25% belong to fraternities or sororities?
   Use $\alpha = 0.05$. What is the p-value of each test?

8.11 In Exercise 7.15, using Nickelsl.xls, you calculated statistics on several characteristics of consumers, including whether they were customers of Nickels department store. Return to those data. Can you conclude that Nickels has less than half of the market?
Use $\alpha = 0.05$. What is the p-value of this test?

8.12 Continue with these same data.
   a. Can you conclude that more than half of Nickels customers are female?
   Use $\alpha = 0.05$. What is the p-value of this test?

b. Can you conclude that Nickels customers average more than 30 years of age?
Use $\alpha = 0.05$. What is the p-value of this test?
c. Can you conclude that Nickels customers have a mean income of more than $40,000? Use $\alpha = 0.05$. What is the p-value of this test?
d. Can you conclude that more than one-third of Nickels customers rely primarily on newspapers as their source of market information? Use $\alpha = 0.05$. What is the p-value of this test?

8.13 Continue with these data.
a. Can you conclude that less than half of non-Nickels customers are female?
Use $\alpha = 0.05$. What is the p-value of this test?
b. Can you conclude that non-Nickels customers average less than 30 years of age?
Use $\alpha = 0.05$. What is the p-value of this test?
c. Can you conclude that non-Nickels customers have a mean income of less than $40,000? Use $\alpha = 0.05$. What is the p-value of this test?
d. Can you conclude that less than one-third of non-Nickels customers rely primarily on newspapers as their source of market information? Use $\alpha = 0.05$. What is the p-value of this test?

8.14 The file Employees2.xls contains the following information on a random sample of 50 employees of your very large firm. (It is the same as Employees1.xls, but with additional information on job type.)

ID:	Identification number;
Ed:	Education (in years);
Exp:	Job experience (in years);
Type:	type (1 = line; 2 = office; 3 = management);
Female:	Sex (1 = female; 0 = male);
Salary:	Annual salary (in $ thousands).

a. Can you conclude that all the employees of your firm average more than 12 years of education? Use $\alpha = 0.05$. What is the p-value of this test?
b. Can you conclude that all the employees of your firm average more than 18 years of experience? Use $\alpha = 0.05$. What is the p-value of this test?
c. Can you conclude that fewer than 25% of your employees are in management? Use $\alpha = 0.05$. What is the p-value of this test?
d. Can you conclude that fewer than half of your employees are female? Use $\alpha = 0.05$. What is the p-value of this test?
e. Can you conclude that your firm's average salary is less than $50,000? Use $\alpha = 0.05$. What is the p-value of this test?

9

Tests of Hypotheses: Two-Sample Tests

In the last three chapters, we have developed the ideas of inference—confidence intervals and tests of hypotheses with known probabilities of being right or wrong—by exploring what we can say about proportions and means. These intervals and tests involve a single sample proportion or mean, and so are traditionally called "one-sample" intervals and tests. In this chapter, we explore tests about *differences* in proportions and means between two groups. Because we are comparing two groups, these are traditionally called "two-sample" tests.

The careful reader will notice that I have switched from "one-sample intervals and tests" to just "two-sample tests." Are there not two-sample confidence intervals as well? Yes there are. I have omitted them because they seem less important for most actual users of statistics, and because, if you need them, you can probably figure them out without too much extra effort. We will develop estimators and standard errors for the differences in proportions and means needed for testing hypotheses about the true differences. But, of course, these are the same estimators and standard errors we would need for confidence interval estimates of the true differences.

In a technical sense then, two-sample (intervals and) tests are fairly simple extensions of their one-sample counterparts. Indeed, many textbooks combine their treatment of the one- and two-sample cases. I have not, to emphasize the importance of this "simple extension," for the sorts of questions we are able to address. Starting in this chapter, we are looking at more than one variable at a time, and we can start to address rigorously the sorts of questions posed in the first several chapters. We can use one variable to help explain the variation in another. If we believe that Y depends on X, we can collect data on the two variables and see whether the data support our theory with an acceptable probability, α, that we are being misled.

9.1 Looking for Relationships Again

Many of the examples of Chapters 1 though 3 involved looking for relationships in data—of asking whether one variable could help explain some of the variation in another. Does the political party in power help explain the variation in taxation of the rich? Does an employee's salary depend on his or her education, work experience, or sex? Do male and female students differ in their choice of major or their GPA? Can sex, age, or height help explain the weight of a young adult? We have been away from these questions for a while, developing the theoretical ideas of probability and inference that we needed to address them rigorously. You should browse those chapters briefly to refresh your memory. Section 4 of Chapter 3 are especially relevant.

We asked, for example, whether a student's sex could help explain his or her decision to major in Natural Science. We noted that it could not unless the true population proportions of men and women choosing a Natural Science major were different. That is, $\pi_{NS-\text{Female}} - \pi_{NS-\text{Male}}$ could not equal zero. Of course, we did not know the true population proportions. However, we did have sample proportions—$p_{NS-\text{Female}} = 3/30 = 0.10$ and $p_{NS-\text{Male}} = 5/20 = 0.25$. The difference in sample proportions was –0.15.

In terms of the tests of hypotheses developed in the last chapter, then, we wanted to know whether a sample difference this large is so unlikely, given H_o: $\pi_{NS-\text{Female}} - \pi_{NS-\text{Male}} = 0$, that we can reject H_o with an acceptably small probability, α, that we are rejecting wrongly. If we do reject H_o, we conclude that the sample proportions are "significantly different"—that the true population proportions are not the same. And this means that a student's sex *can* help explain his or her decision to major in Natural Science. We will address this problem in the next section.

This example was a *two-sample* test because, when we divided the sample by sex, there were just two categories—male and female. It was a test of *proportions* because we could represent the choice of Natural Science as just a dummy, "yes–no" variable, and calculate the proportion of "yeses" for each sex. When either of these variables includes more than two categories, things change a bit. We asked, for example, whether we could use sex to help explain the whole distribution in majors. And, while sex still had just two categories, major had six. We cannot represent the whole distribution of majors with just a proportion of "yeses." We will address this complication in Chapter 10.

Next, we asked whether a young adult's sex could help explain his or her weight. Again, we noted that it could not unless the true mean weights of men and women were different. That is, $\mu_{\text{Weight–Male}} - \mu_{\text{Weight–Female}}$ could not equal zero. Again, we did not know the population means. However, we did have sample means—$\bar{X}_{\text{Weight}-\text{Male}} = 169.0148$ and $\bar{X}_{\text{Weight}-\text{Female}} = 134.5205$. The difference in sample means was 34.4943.

In terms of the tests of hypotheses developed in the last chapter, then, we wanted to know whether a sample difference this large is so unlikely, given H_o: $\mu_{\text{Weight–Male}} - \mu_{\text{Weight–Female}} = 0$, that we can reject H_o with an acceptably small probability, α, that we are rejecting wrongly. If we do reject H_o, we conclude that the sample means are "significantly different"—that the true means are not the same. And this means that a young adult's sex *can* help explain his or her weight. We will address this problem beginning in Section 3.

This example was a *two-sample* test because, when we divided the sample by sex, there were just two categories—male and female. It was a test of *means* because weight is a numeric variable that can be represented by its mean. If there had been more than two categories, things would change again. Suppose we wanted to know whether a student's GPA depended on his or her major. We could calculate the mean GPA within each major. But the data of Chapter 3 grouped the majors into six categories, not two, so this would not be a two-sample test. We will address this complication in Chapter 11.

Finally, we could have asked, as we did in Chapter 2, whether a young adult's height could help explain his or her weight. We expected that taller people would tend to weigh more. In this case, we cannot divide people into categories like sex or major; the explanatory variable is not categorical but numeric. Chapters 12 and 13 deal with our approach in such cases.

This, then, is a roadmap to the next five chapters. This chapter deals with tests for relationships in which the hypothesized explanatory variable is a dummy, "yes–no" variable, allowing us to divide the data into just two categories, or sub-samples, so that we can calculate and compare sample proportions or means. Chapters 10 and 11 deal primarily with tests for relationships in which the hypothesized explanatory variable is still categorical, but has (potentially) more than two categories. And Chapters 12 and 13 deal with tests for relationships in which the hypothesized explanatory variable (or variables) can be numeric.

9.2 A Difference in Population Proportions

Figure 9.1, which just repeats Figure 2.1 for convenience, shows the data from the file Studentsl.xls on a random sample of 50 students. In Chapters 2 and 3 we sorted by Female to create two subsamples—one of 30 females and one of 20 males—and created relative frequency distributions for each. We found, for instance, that the sample proportion of women majoring in Natural Science was $p_{NS-\text{Female}} = 3/30 = 0.10$ while the sample proportion of men majoring in Natural Science $p_{NS\text{-Male}} = 5/20 = 0.25$. And we wondered whether this sample difference in the choice of a Natural Science major was too great to be due just to chance—whether there was a significant difference between men and women in their choice of a Natural Science major. If so, sex could help explain an individual's choice. We are now ready to answer that question.

9.2.1 The Null and Alternative Hypotheses

Since we are looking for a difference, "difference" needs to be the alternative hypothesis. That is, we need to assume that there is no difference and try to reject that hypothesis:

1a. $H_o: \pi_{NS-\text{Female}} - \pi_{NS-\text{Male}} = 0$ $\quad H_o: \pi_{NS-\text{Female}} = \pi_{NS-\text{Male}}$
$\quad H_a: \pi_{NS-\text{Female}} - \pi_{NS-\text{Male}} \neq 0$ $\quad H_a: \pi_{NS-\text{Female}} \neq \pi_{NS-\text{Male.}}$

I have written the hypotheses in two different forms to make a point. The form on the left emphasizes the similarity with what we have already done. We have already tested the hypothesis that a true proportion, π, equals some amount. We did so by comparing it with a sample p, in terms of σ_p, the standard error of p. We will now test the hypothesis that a true difference in proportions, $\pi_F - \pi_M$, equals some amount. And we will do so by comparing it with a sample $p_F - p_M$. in terms of $\sigma_{p_F - p_M}$. The only

ID	Female	Major	GPA	ID	Female	Major	GPA
1	1	4	2.537	31	1	1	2.717
2	0	2	3.648	32	1	6	1.996
3	1	3	2.981	33	1	3	2.870
4	0	2	2.683	34	0	5	2.986
5	1	2	3.234	35	0	1	3.393
6	1	3	2.467	36	1	2	2.740
7	1	1	3.384	37	1	6	2.499
8	1	3	3.555	38	1	4	3.695
9	0	2	3.263	39	1	4	2.664
10	1	1	3.711	40	1	5	2.306
11	1	5	1.970	41	1	3	3.022
12	1	4	3.406	42	1	6	2.776
13	0	1	2.523	43	1	5	2.175
14	0	3	1.750	44	0	2	3.828
15	1	6	3.191	45	0	6	3.410
16	1	2	2.795	46	0	4	2.330
17	0	1	2.606	47	0	5	3.978
18	0	5	2.397	48	1	3	3.503
19	1	6	3.791	49	1	4	3.253
20	0	4	3.490	50	1	2	2.215
21	0	3	2.421				
22	0	5	3.937				
23	0	5	2.890				
24	1	6	2.246	Codes for Major			
25	1	2	3.371	1	–	Natural Science	
26	1	5	3.114	2	–	Social Science	
27	0	1	3.084	3	–	Humanities	
28	0	5	2.703	4	–	Fine Arts	
29	1	3	3.045	5	–	Business	
30	0	1	1.948	6	–	Nursing	

Figure 9.1 Information on a sample of 50 students (Studentsl.xls).

new things we will need to know are the sampling distribution for $p_F - p_M$ and a formula for $\sigma_{p_F - p_M}$ (or as it will turn out, an estimate, $s_{p_F - p_M}$).

Of course the form on the right means the same thing and is, perhaps, more straightforward and intuitive. We are looking to see if there is a difference in proportions. It is also easier when dealing with one-tailed alternatives to think in terms of one being bigger, rather than the difference being positive or negative.

In the last chapter, I covered two-tailed and one-tailed tests separately because it was important to get the difference straight in your mind. Everything I said about the difference—when it is legitimate to use a one-tailed test, how the criteria change, etc.—applies for two-sample tests as well. There is really nothing new to say. So, to minimize redundancy, I will treat them together from now on.

In this example, then, suppose you had some a priori reason to believe that the proportion of men in Natural Science majors would be greater than the proportion of women in such majors. Perhaps you have read about differences between the sexes in their average verbal and quantitative scores on the SAT and similar tests, and theorize that this will translate into a greater proportion of men than women majoring in Natural Science. That would be a theory that you could test using these data. Only if the proportion of men is higher—and by enough—can you claim support. If the proportion of women is larger you cannot claim support, no matter how big the difference. So your hypotheses would be as follows:

1b. H_o: $\pi_{NS-\text{Female}} - \pi_{NS-\text{Male}} \geq 0$ H_o: $\pi_{NS-\text{Female}} \geq \pi_{NS-\text{Male}}$
H_a: $\pi_{NS-\text{Female}} - \pi_{NS-\text{Male}} < 0$ H_a: $\pi_{NS-\text{Female}} < \pi_{NS-\text{Male}}$.

Again, what we are looking for needs to be the alternative hypothesis. That is, we need to assume that we are wrong and try to reject that hypothesis.

Since the criterion for rejecting is lower, it is important to be honest in deciding to use a one-tailed test. Would you really have found the opposite result just as interesting—perhaps supporting some competing theory? If so, you should use a two-tailed test. And finally, the fact that in this data set the sample proportion is larger for men than for women is irrelevant in setting up your test. You can never decide to use a one-tailed test based on your data.

9.2.2 The Decision Criterion

As Figure 9.2 illustrates, under the null hypothesis, the sampling distribution for $p_F - p_M$ is the familiar normal distribution, with mean equal to $\pi_F - \pi_M = 0$. Thus, we can use everything we already know about setting the decision criterion for a proportion.

Suppose we are conducting the two-tailed test. First, we need to choose α, the probability we are willing to accept of rejecting H_o wrongly. As before, a common one is 0.05. By now the $\pm Z$ value associated with this α is familiar—± 1.96. Recall the reasoning. The ± 1.96 range contains 0.95 of the probability under the normal curve. If our null hypothesis is correct, we will get a Z_c-value in this range 0.95 of the time, and will fail to

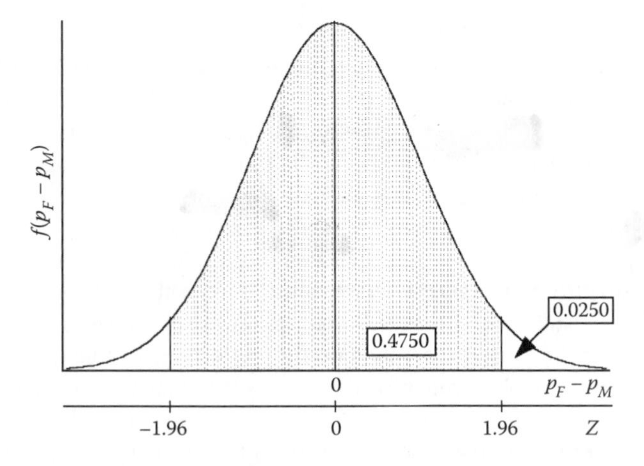

Figure 9.2 The sampling distribution for $p_F - p_M$, assuming $\pi_F - \pi_M = 0$ (two-tailed).

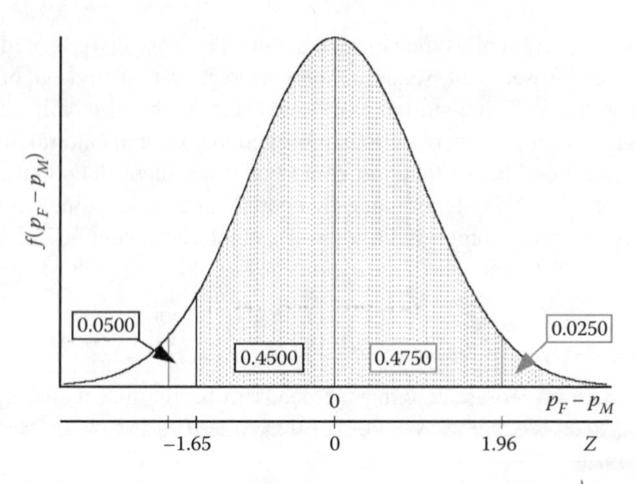

Figure 9.3 The sampling distribution for $p_F - p_M$, assuming $\pi_F - \pi_M = 0$ (one-tailed).

reject—which is the correct decision since the null hypothesis is correct. If our null hypothesis is correct, we will get a Z_c-value outside this range 0.05 of the time, and will reject—which is a type I error since the null hypothesis is correct. So, if we choose $\alpha = 0.05$, our rejection criterion is as follows:

2a. Reject H_o if $|Z_c| > 1.96$ ($\alpha = 0.05$).

Suppose, instead, we are conducting the one-tail test. Figure 9.3 repeats Figure 9.2, with just the change in rejection criterion. The ±1.96 range still contains 0.95 of the probability under the normal curve. There is still a 0.95 probability of getting a Z_c in this range; a 0.05 probability of getting a Z_c outside it. But, in this case, if we get a Z_c above + 1.96 we will not reject; only if we get a Z_c below –1.96 will we reject, since only the negative difference offers support for the alternative. So we need 0.05 in just the lower tail. The Z_c that gives us this is –1.65. Our rejection criterion is as follows:

2b. Reject H_o if $Z_c < -1.65$ ($\alpha = 0.05$).

9.2.3 The Calculations

Whichever our hypotheses and rejection criterion, the calculation of Z_c is the same. And for this we need to calculate the standard error of $p_F - p_M$, $\sigma_{p_F - p_F}$ (or an estimate, $s_{p_F - p_M}$). Figure 9.4 summarizes.

The first version of the formulas would be the complete analog of the formulas for a single proportion—we have $p_F - p_M$ instead of p_X, we have $\pi_F - \pi_M$ instead of π_X, and we have $\sigma_{p_F - p_M}$ instead of σ_{p_X}. Notice that, since our hypothesis is that the true population proportions are the same, the formulas simplify in a couple of ways. Since by hypothesis $\pi_F - \pi_M = 0$, the $\pi_F - \pi_M$ can be dropped from the formula for Z_c. Also, since they are hypothesized to be the same, the formula for $\sigma_{p_F - p_M}$ need not distinguish between them. π_P is the pooled proportion that applies to both men and women. And because it is the same, the formula for $\sigma_{p_F - p_M}$ can be simplified as well.

But there is a problem. Since our hypothesis is about $\pi_F - \pi_M$, not π_P, we have nothing to plug in for π_P. Hence, in the second version we use p_P, the pooled sample proportion, as an estimate for π_P. Why p_P rather than

$$p_F = \frac{x_F}{n_F} \quad p_M \frac{x_M}{n_M} \quad p_p = \frac{x_F + x_M}{n_F + n_M}$$

3a. $$Z_c = \frac{(p_F - p_M) - (\pi_F - \pi_M)}{\sigma_{p_F - p_M}} = \frac{(p_F - p_M) - 0}{\sigma_{p_F - p_M}} = \frac{(p_F - p_M)}{\sigma_{p_F - p_M}}$$

$$\text{where } \sigma_{p_F - p_M} = \sqrt{\frac{\pi_p \times (1 - \pi_p)}{n_F} + \frac{\pi_p \times (1 - \pi_p)}{n_M}} = \sqrt{\pi_p \times (1 - \pi_p) \times \left(\frac{1}{n_F} + \frac{1}{n_M}\right)}$$

3b. $$Z_c = \frac{(p_F - p_M) - (\pi_F - \pi_M)}{s_{p_F - p_M}} = \frac{(p_F - p_M) - 0}{s_{p_F - p_M}} = \frac{(p_F - p_M)}{s_{p_F - p_M}}$$

$$\text{where } s_{p_F - p_M} = \sqrt{\frac{p_p \times (1 - p_p)}{n_F} + \frac{p_p \times (1 - p_p)}{n_M}} = \sqrt{p_p \times (1 - p_p) \times \left(\frac{1}{n_F} + \frac{1}{n_M}\right)}$$

Figure 9.4 Two-sample tests for proportions: The formulas.

p_F and p_M? By hypothesis $\pi_F = \pi_M = \pi_P$, so p_F, p_M, and p_P are all estimates of the same thing, and p_p is the one that makes use of the whole sample. Using p_P, we get $s_{p_F - p_M}$, an estimate of $\sigma_{p_F - p_F}$.

Doing the calculations for our example

3. $$p_{NS\text{-Female}} = \frac{3}{30} = 0.10; \quad p_{NS\text{-Male}} = \frac{5}{20} = 0.25;$$

$$p_{NS\text{-Pooled}} = \frac{3+5}{30+20} = 0.16;$$

$$s_{p_{NS\text{-Female}} - p_{NS\text{-Male}}} = \sqrt{0.16 \times 0.84 \times \left(\frac{1}{30} + \frac{1}{20}\right)}$$

$$= \sqrt{0.16 \times 0.84 \times 0.0833} = \sqrt{0.0112} = 0.1058;$$

$$Z_c = \frac{0.10 - 0.25}{0.1058} = -1.417.$$

9.2.4 The Conclusion

Using our two-tailed criterion, $|Z_c|$ is not greater than 1.96. Using the one-tailed criterion, we have the correct tail, but Z_c is still not large enough. Our conclusion is

4. ∴ Fail to Reject H_o.

Given the small size of these samples, a difference this large could too easily have come from a population in which the true proportions are equal. How easily? What is the p value for this test? Using the standard normal table, the probability between 1.42 and the mean is 0.4222. The probability of being below −1.42, then, is 0.5000 − 0.4222 = 0.0778. This is the p-value for the one-tailed test. This is the probability you would have to accept of being wrong in order to claim support for your theory. The p-value for the two-tailed test is just twice that, 0.0778 × 2 = 0.1556, since it includes the upper tail as well.

Consider the following example:

> Evans is running against Foley in a statewide election. A poll of 1000 Chicago voters shows 520 preferring Evans; a poll of 500 downstate voters shows only 230 preferring Evans. Can you conclude that there is a significant difference in Evans's popularity between Chicago and downstate? Use $\alpha = 0.05$. Use $\alpha = 0.01$. What is the p-value for this test?

We are looking for a "difference." Nothing in the problem suggests a one-tailed alternative (remember, you cannot decide on a one-tailed test by looking at the data!). Hence:

1. H_o: $\pi_{E-C} - \pi_{E-DS} = 0$ H_o: $\pi_{E-C} = \pi_{E-DS}$
   H_a: $\pi_{E-C} - \pi_{E-DS} \neq 0$ H_a: $\pi_{E-C} \neq \pi_{E-DS}$.
2. Reject H_o if $|Z_c| > 1.96$ $(\alpha = 0.05)$ Reject H_o if $|Z_c| > 2.58$ $(\alpha = 0.01)$.
3. $$p_{E-C} = \frac{520}{1000} = 0.52; \quad p_{E-DS} = \frac{230}{500} = 0.46;$$

$$p_{E-Pooled} = \frac{750}{1500} = 0.50;$$

$$s_{p_{E-C}-p_{E-DS}} = \sqrt{0.50 \times 0.50 \times \left(\frac{1}{1000} + \frac{1}{500}\right)}$$

$$= \sqrt{0.50 \times 0.50 \times 0.033} = \sqrt{0.00075} = 0.0274;$$

$$Z_c = \frac{0.52 - 0.46}{0.0274} = -2.191.$$

4. $\therefore$ Reject H_o $(\alpha = 0.05)$ $\therefore$ Fail to Reject H_o $(\alpha = 0.01)$.

$|Z^c|$ is greater than 1.96, but not 2.58; we can reject H_o with an α of 0.05, but not with an α of 0.01. In the first case, either there really is a difference or we are making a type I error—rejecting a true hypothesis. In the second case, either there really is no difference or we are making a type II error—failing to reject a false one.

What is the p-value for the test? Using the standard normal table, the probability between 2.19 and the mean is 0.4857. The probability in a single tail is $0.5000 - 0.4857 = 0.0143$. Doubling that for the two-tailed test, the p-value is $0.0143 \times 2 = 0.0286$. This is the probability we have to accept of being wrong in order to claim that there is a significant difference in Evans's support in the two regions.

Note that I used subscripts "$E-C$" and "$E-DS$" above, to indicate that I was comparing the proportions supporting *Evans* in the two areas. There are actually several ways of making this comparison—Evans' share in each of the two regions, Foley's share in each of the two regions, Chicago's share in each of the candidate's support, and downstate's share in each of the candidate's support. All are correct and give the same answer as long as you are consistent. But it is easy to get confused, so subscripts are a good idea. One, a constant, should indicate what you are counting; the other, should indicate the two groups in which you are counting it.

Consider the following example:

> Suppose a new company claims to be able to supply parts with fewer defectives than your current supplier. Switching suppliers entails costs, though, so you want to be sure that their claim is true. You order a lot of 200 from them and check it against a lot of 500 from your current supplier. If the number of defectives in the two lots are 16 and 50, respectively, can you conclude that the claim is true? Use $\alpha = 0.05$. Use $\alpha = 0.10$. What is the p-value for this test?

The claim is that their lots include *fewer* defects, so this is a one-tailed claim. And to "be sure," or "conclude" that the claim is true, it must be the alternative. Hence,

1. H_o: $\pi_{D\text{-New}} - \pi_{D\text{-Old}} \geq 0$ $\quad$ H_o: $\pi_{D\text{-New}} \geq \pi_{D\text{-Old}}$
   H_a: $\pi_{D\text{-New}} - \pi_{D\text{-Old}} < 0$ $\quad$ H_a: $\pi_{D\text{-New}} < \pi_{D\text{-Old}}$.
2. Reject H_o if $Z_c < -1.65$ (a = 0.05) $\quad$ Reject H_o if $Z_c < -1.28$ (a = 0.10).
3. $$p_{D\text{-New}} = \frac{16}{200} = 0.08; \quad p_{D\text{-Old}} = \frac{50}{500} = 0.10;$$

$$p_{D\text{-Pooled}} = \frac{66}{700} = 0.0943;$$

$$s_{p_{D\text{-New}} - p_{D\text{-Old}}} = \sqrt{0.0943 \times 0.9057 \times \left(\frac{1}{200} + \frac{1}{500}\right)}$$

$$= \sqrt{0.0943 \times 0.9057 \times 0.007}$$

$$= \sqrt{0.000598} = 0.0244;$$

$$Z_c = \frac{0.08 - 0.10}{0.0244} = -0.818.$$

4. ∴ Fail to Reject H_o ($\alpha = 0.05$) $\quad$ ∴ Fail to Reject H_o ($\alpha = 0.10$).

The proportion of defectives from the new supplier is lower—0.08 versus 0.10. But this difference is not significant—even at the fairly lenient α of 0.10.

What is the p-value for the test? Using the standard normal table, the probability between −0.82 and the mean is 0.2939. The probability below −0.82 is 0.5000 − 0.2939 = 0.2061. This is the probability we have to accept of being wrong in order to conclude that the new supplier's claim is true.

This might be a good place to make a couple of related points. First, of course, we have not rejected their claim; we have simply not substantiated it. We may be making a type II error. Moreover, statistical significance is not the same thing as importance. In practice, we would probably look at the sample difference—.08 versus 0.10—and ask ourselves whether it would actually matter even if it were statistically significant. It might be such a small difference that it would not be important, even if it were statistically significant. On the other hand, such a difference—if real—might translate into a huge cost savings. In that case, we might decide to expand our sample to see if we can verify that the difference is real.

9.3 A Difference in Population Means

As was the case with one-sample tests in the last chapter, tests of the mean follow the same logic as tests of proportions. The only new complication arises due to what we may know or assume about σ_1 and σ_2, the two population standard deviations. There are three cases. The first case is the one in which—while we do not know the true values for μ_1 and μ_2 (if we did, we would not be testing hypotheses about them—we do know the true values of σ_1 and σ_2. This case is extremely unlikely and we treat it simply as a jumping off point for considering the other two possibilities. The second case is the one in which we do not know σ_1 and σ_2, but can reasonably assume that they are equal. This assumption is often quite reasonable and tests of means often make it. Sometimes, though, the evidence strongly suggests that the true values of σ_1 and σ_2 are not equal. Hence, the final case is the one in which we do not know σ_1 and σ_2, and cannot reasonably assume that they are equal.

9.4 A Difference in Means: σ_Xs Known

The data set in Figure 9.1 contain data not only on the sex and major of 50 students but also on their GPAs. Suppose we want to know whether GPAs differ by sex. We can sort the data by sex, and calculate $\bar{X}$, the mean GPA for each sex. These sample values are undoubtedly somewhat different. But are they different enough that we can reject the hypothesis that the true mean GPAs are the same?

9.4.1 The Null and Alternative Hypotheses

As always, we begin by writing down the hypotheses:

1. $H_o\text{: } \mu_{GPA-Female} - \mu_{GPA-Male} = 0 \quad H_o\text{: } \mu_{GPA-Female} = \mu_{GPA-Male}$
   $H_a\text{: } \mu_{GPA-Female} - \mu_{GPA-Male} \neq 0 \quad H_a\text{: } \mu_{GPA-Female} \neq \mu_{GPA-Male}.$

Again, I have written the hypotheses in two different but equivalent forms. There are no new issues here. The form on the left emphasizes that we are still estimating a population parameter, $\mu_F - \mu_M$, based on a sample statistic, $\bar{X}_F - \bar{X}_M$, in terms of a standard error, $\sigma_{\bar{X}_F - \bar{X}_M}$. The form on the right is, perhaps, more intuitive.

Again, if there were some a priori reason to expect the average to be higher for one sex than for the other, we could use a one-tailed alternative to see if the data support that expectation. In this case, I know of no a priori reason to predict one way or the other.

9.4.2 The Decision Criterion

As Figure 9.5 illustrates, under the null hypothesis, the sampling distribution for $\bar{X}_F - \bar{X}_M$ is the normal distribution, with mean equal to $\mu_F - \mu_M = 0$. Again, there is essentially nothing new here. We need to choose α, the probability we are willing to accept of rejecting wrongly. If we choose 0.05, our rejection criterion is the familiar ± 1.96. Hence:

2. Reject H_o if $|Z_c| > 1.96$ ($\alpha = 0.05$).

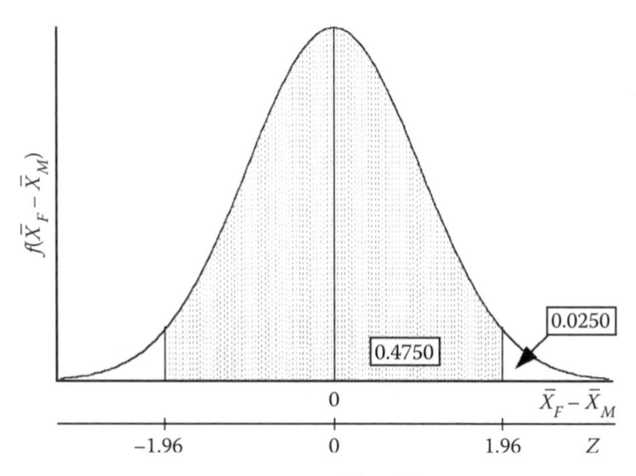

Figure 9.5 The sampling distribution for $\bar{X}_F - \bar{X}_M$, assuming $\mu_F - \mu_M = 0$, σ_Xs known.

9.4.3 The Formulas

Figure 9.6 gives the formulas for calculating Z_c.

Since by hypothesis $\mu_F - \mu_M = 0$, the $\mu_F - \mu_M$ can be dropped from the formula for Z_c. And the formula for $\sigma_{\bar{X}_F - \bar{X}_M}$ is an intuitively plausible extension of the one-sample formula for $\sigma_{\bar{X}}$. But knowing the two true population standard deviations, σ_F and σ_M, is so unlikely that there is no point in pretending it is a real possibility. We again need to substitute estimates based on the sample. And, as was true in the one-sample case, using estimates based on the sample changes the sampling distribution from a normal to a *t*.

The next section describes the *t* test for the important special case in which the unknown population standard deviations can be assumed equal. In this case, the two sample standard deviations can be used to come up with a single, pooled estimate of the population value. The following section describes the test when the unknown population σ_Xs cannot be assumed equal.

9.5 A Difference in Means: σ_Xs Unknown but Equal

9.5.1 The Null and Alternative Hypotheses

The hypotheses have not changed:

1. H_o: $\mu_{\text{GPA-Female}} - \mu_{\text{GPA-Male}} = 0$ $\quad H_o$: $\mu_{\text{GPA-Female}} = \mu_{\text{GPA-Male}}$
   H_a: $\mu_{\text{GPA-Female}} - \mu_{\text{GPA-Male}} \neq 0$ $\quad H_a$: $\mu_{\text{GPA-Female}} \neq \mu_{\text{GPA-Male}}$.

$$Z_c = \frac{(\bar{X}_F - \bar{X}_M) - (\mu_F - \mu_M)}{\sigma_{\bar{X}_F - \bar{X}_M}} = \frac{(\bar{X}_F - \bar{X}_M) - 0}{\sigma_{\bar{X}_F - \bar{X}_M}} = \frac{(\bar{X}_F - \bar{X}_M)}{\sigma_{\bar{X}_F - \bar{X}_M}}$$

$$\text{where } \sigma_{\bar{X}_F - \bar{X}_M} = \sqrt{\frac{\sigma_F^2}{n_F} + \frac{\sigma_M^2}{n_M}}$$

Figure 9.6 Two-sample test for means: The formulas, σ_Xs known.

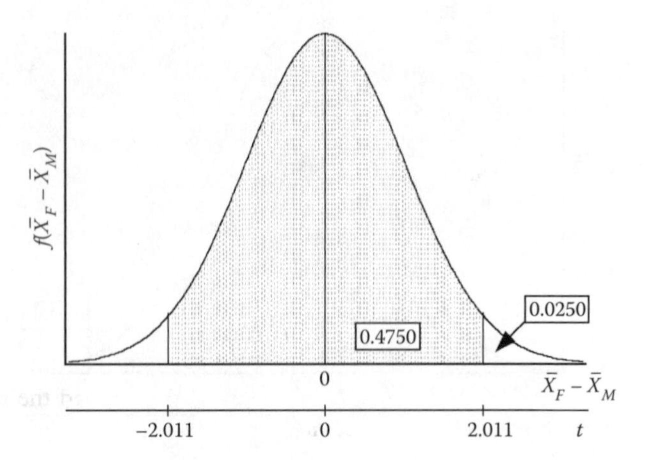

Figure 9.7 The sampling distribution for $\bar{X}_F - \bar{X}_M$, assuming σ_Xs unknown but equal.

9.5.2 The Decision Criterion

Assuming that the true σ_Xs are equal, we can pool our samples to come up with a combined estimate, and use that to calculate an estimate of the standard error, $s_{\bar{X}_F-\bar{X}_M}$. Under the null hypothesis, $(\bar{X}_F - \bar{X}_M)/s_{\bar{X}_F-\bar{X}_M}$ follows the t distribution with $n_F + n_M - 2$ degrees of freedom (df). Figure 9.7 illustrates.

Suppose, as before, that we choose an α of 0.05. For our sample of 30 women and 20 men, we have 48 degrees of freedom. The t table in the back of the book jumps from 45 to 50 degrees of freedom. We could use either 2.014 or 2.009; since the t value changes so slowly, it is unlikely to affect our decision. Or we could use the special spreadsheet function = TINV(.05,48) = 2.011.

2. Reject H_c if $|t_c| > 2.011$ (df = 48; $\alpha = 0.05$).

9.5.3 The Calculations

Figure 9.8 summarizes the calculations. Starting at the bottom, since the $\sigma_X s$ are the same, the sample standard deviations are both estimates of the same thing; hence, we can come up with a better estimate of σ_X by pooling the samples. The first version of the formula makes clear the connection between s_P, based on two, pooled samples, and our ordinary one-sample standard deviation. The numerator contains the sum of squared deviations from the mean for both samples and the denominator contains $n-1$ for

$$t_c = \frac{(\bar{X}_F - \bar{X}_M) - (\mu_F - \mu_M)}{s_{\bar{X}_F-\bar{X}_M}} = \frac{(\bar{X}_F - \bar{X}_M) - 0}{s_{\bar{X}_F-\bar{X}_M}} = \frac{(\bar{X}_F - \bar{X}_M)}{s_{\bar{X}_F-\bar{X}_M}}$$

$$\text{where } s_{\bar{X}_F-\bar{X}_M} = \sqrt{\frac{s_p^2}{n_F} + \frac{s_p^2}{n_M}} = s_p\sqrt{\frac{1}{n_F} + \frac{1}{n_M}}$$

$$\text{and } s_p = \sqrt{\frac{\sum(X_F - \bar{X}_F)^2 + \sum(X_M - \bar{X}_M)^2}{(n_F - 1) + (n_M - 1)}} = \sqrt{\frac{s_F^2(n_F - 1) + s_M^2(n_M - 1)}{n_F + n_M - 2}}$$

Figure 9.8 Two-sample test of means: The formulas, σ_Xs unknown but equal.

both samples. Ordinarily, though, we will either know the one-sample standard deviations or be able to calculate them quickly by computer. Hence, the second version of the formula is usually the easier to calculate. Moreover, it makes clear that we can also think of s_P as a sort of weighted average of the one-sample standard deviations.

Once we have s_P, the rest is fairly straightforward. Since the unknown σ_F and σ_M are the same, s_P is an estimate for both; thus, we can put it into the estimated standard error formula for both. This allows further simplification of the formula. Finally, we are able to calculate t_c.

We are now able to use our sample of 30 women and 20 men in Figure 9.1 to test the hypothesis that the true population mean GPAs of women and men are equal. Sorting the data by sex, we find the sample means, standard deviations, and counts.

3. $\bar{X}_F = 2.9076 \qquad \bar{X}_M = 2.9634$

   $s_F = 0.5277 \qquad s_M = 0.6470$

   $n_F = 30 \qquad n_M = 20.$

Now, in order, we calculate the pooled standard deviation, the estimated standard error, and t_c.

$$s_P = \sqrt{\frac{0.5277^2(30-1)+0.6470^2(20-1)}{30+20-2}} = \sqrt{\frac{16.0282}{48}}$$

$$= \sqrt{0.3339} = 0.5779;$$

$$s_{\bar{X}_F-\bar{X}_M} = 0.5779\sqrt{\frac{1}{30}+\frac{1}{20}} = 0.5779\sqrt{0.0833} = 0.5779(0.2887)$$

$$= 0.1668;$$

$$t_c = \frac{2.9076-2.9634}{0.1668} = -0.3343.$$

9.5.4 The Conclusion

$|t_c|$ is not greater than 2.011. A difference in GPAs this large could very easily have arisen just randomly. Our conclusion is

4. ∴ Fail to Reject H_0.

Consider the following example:

> One way of evaluating stocks is to calculate the mean and standard deviation of their past rates of return. Lacking evidence to the contrary, the mean is taken to be an estimate of a stock's "expected return," and the standard deviation is taken to be an estimate of its "risk." The table below gives summary statistics for two stocks, based on the last 10 and 14 quarters, respectively. Based on these statistics, is there a significant difference in the "expected return" of these two stocks? Assume the true risks are equal. Use $\alpha = 0.05$. Use $\alpha = 0.01$. What is the p-value of this test?

	ABC Inc.	XYZ Corp.
Expected return (mean)	9.50	6.40
Risk (standard deviation)	3.90	2.60
Number of periods	10	14

We are looking for a "difference." Nothing in the problem suggests a one-tailed alternative. Hence,

1. H_o: $\mu_{ABC} - \mu_{XYZ} = 0$ $\quad$ H_o: $\mu_{ABC} - \mu_{XYZ}$
   H_a: $\mu_{ABC} - \mu_{XYZ} \neq 0$ $\quad$ H_a: $\mu_{ABC} \neq \mu_{XYZ}$.
2. Reject H_o if $|t_c| > 2.074$ (df = 22; $\alpha = 0.05$). $\quad$ Reject H_o if $|t_c| > 2.819$ (df = 22; $\alpha = 0.01$).

$$s_P = \sqrt{\frac{3.90^2(10-1)+2.60^2(14-1)}{10+14-2}} = \sqrt{\frac{224.77}{22}}$$

$$= \sqrt{10.217} = 3.196;$$

3. $$s_{\bar{X}_{ABC}-\bar{X}_{XYZ}} = 3.196\sqrt{\frac{1}{10}+\frac{1}{14}} = 3.196\sqrt{.171} = 3.196(.414)$$

$$= 1.323;$$

$$t_c = \frac{9.50-6.40}{1.323} = 2.342.$$

4. $\therefore$ Reject H_o $\quad$ $\therefore$ Fail to Reject H_o.

We can reject H_o with an α of 0.05 but not with an α of 0.01.

What is the *p*-value of this test? Reading across the *t* table at 22 degrees of freedom, our t_c falls between 2.074 and 2.508, in the 0.0250 and 0.0100 columns, respectively. The *p*-value is between $2 \times 0.025 = 0.05$ and $2 \times 0.01 = 0.02$. To be more precise, the *p*-value of this test is =TDIST(2.342,22,2) = 0.0286. We can reject the null hypothesis and conclude that there is a difference if we are willing to accept a probability of 0.0286 or more that we are being misled.

Consider the following example:

> A company's efficiency officer suspects that night-shift workers take longer, on average, to complete common tasks then their more heavily supervised day-shift counterparts. She observes one such task being performed at randomly selected times during each shift. The table below gives summary statistics for the two shifts. Do these statistics support her suspicion? Assume the true standard deviations are equal. Use $\alpha = 0.05$. Use $\alpha = 0.01$. What is the *p*-value of this test?

	Day Shift	Night Shift
Mean (seconds)	26.400	29.900
Standard deviation (seconds)	4.552	5.046
Number of observations	30	20

Her suspicion is not just that there is a difference; it is that the night shift takes *longer*. For the statistics to "support" her suspicion, this needs to be the alternative.

1. H_o: $\mu_{Day} - \mu_{Night} \geq 0$ $\quad$ H_o: $\mu_{Day} \geq \mu_{Night}$
   H_a: $\mu_{Day} - \mu_{Night} < 0$ $\quad$ H_a: $\mu_{Day} < \mu_{Night}$

Note that there would be nothing wrong with reversing the order of the two shifts and looking for a positive sign. The order is entirely arbitrary; the key is to be consistent. Since I have written the alternative hypothesis $\mu_{Day} - \mu_{Night} < 0$, I need to be sure to do my calculation of $\bar{X}_{Day} - \bar{X}_{Night}$ in that same order, and use the negative tail of the t distribution.

Since, again, 48 degrees of freedom is not in the table in the back we would use 45 or 50. Or we could use the special spreadsheet function =TINV(.10,48) = 1.677 and = TINV(.02,48) = 2.407. Recall that this special function assumes a two-tailed test; since we want the full α in just one tail, we need to enter twice the α we really want.

2. Reject H_o if $t_c < -1.677$ (df = 48; $\alpha = 0.05$) $\quad$ Reject H_o if $t_c < -2.407$ (df = 48; $\alpha = 0.01$).

$$S_P = \sqrt{\frac{4.552^2(30-1)+5.046^2(20-1)}{30+20-2}} = \sqrt{\frac{1084.681}{48}}$$

$$= \sqrt{22.598} = 4.754;$$

3. $$s_{\bar{X}_{Day}-\bar{X}_{Night}} = 4.754\sqrt{\frac{1}{30}+\frac{1}{20}} = 4.754\sqrt{0.0833} = 4.754(0.289)$$

$$= 1.372;$$

$$t_c = \frac{26.4 - 29.9}{1.372} = -2.551.$$

4. $\therefore$ Reject H_o $\quad$ $\therefore$ Reject H_o.

Reading across the t table at either 45 of 50 degrees of freedom, t_c is between the values in the 0.0100 and 0.0050 columns. The p-value of the test is between 0.01 and 0.005. To be more precise, the p-value of the test is =TDIST(–2.551,48,1) = 0.0070.

9.6 A Difference in Means: σ_Xs Unknown and Unequal*

As indicated above, it is often reasonable to assume that the population standard deviations are equal. Still, it will not always be. We postpone a formal test of whether they are until Chapter 11 a formal text of whether they are, since it requires the introduction of a new sampling distribution. But certainly, if the sample standard deviations are wildly different we should be concerned that the population standard deviations are different too.

In this case, since there is not a single σ_x, it is not reasonable to combine the samples for a single pooled estimate, s_P. As Figure 9.9 indicates,

$$t_c = \frac{(\bar{X}_1 - \bar{X}_2) - (\mu_1 - \mu_2)}{s_{\bar{X}_1 - \bar{X}_2}} = \frac{(\bar{X}_1 - \bar{X}_2) - 0}{s_{\bar{X}_1 - \bar{X}_2}} = \frac{(\bar{X}_1 - \bar{X}_2)}{s_{\bar{X}_1 - \bar{X}_2}}$$

$$\text{where } s_{\bar{X}_1 - \bar{X}_2} = \sqrt{\frac{s_1^2}{n_1} + \frac{s_2^2}{n_2}}$$

$$\text{and df} = \frac{(s_1^2/n_1 + s_2^2/n_2)^2}{(s_1^2/n_1)^2/(n_1 - 1) + (s_2^2/n_2)^2/(n_2 - 1)}$$

Figure 9.9 Two-sample test of means: The formulas, σ_Xs unknown and unequal.

we now use the individual sample standard deviations, s_1 and s_2, as our estimates for the population ones. In order to compensate, though, we need to calculate the degrees of freedom differently. Figure 9.9 shows that as well. The degrees of freedom now depend in a fairly complicated way on the relative sizes and standard deviations of the two samples.

The formula for degrees of freedom is far from intuitive; Figure 9.10 shows how the value changes as relative sample sizes and standard deviations change. All the numbers assume that the sample size and standard deviation for sample 1 are fixed at $n_1 = 10$, and $s_1 = 10$, while those for sample 2—n_2 and/or s_2—change.

Start with the upper left. When $n_1 = n_2 = 10$, and $s_1 = s_2 = 10$, the degrees of freedom = 18, the same as if we had assumed the populations were equal, and used a pooled estimate. This seems reasonable since $s_1 = s_2$. What seems surprising is what happens as we increase the size of sample two. Read down the first column. As n_2 increases from 10 to 25, 100, 250, and 1000, the degrees of freedom *decrease!* Now this does not mean that additional information is bad; the larger values for n_2 go into the denominator of the estimated standard error, $s_{\bar{X}_1 - \bar{X}_2}$. The estimated standard error gets smaller and t_c gets larger. But it is true that the degrees of freedom also fall, meaning that the criterion for rejection rises. In the limit, the degrees of freedom fall to $n_x - 1 = 9$, the degrees of freedom for the smaller sample.

What is the intuition here? Remember, s_1 and s_2 are estimates, with their own sampling distributions; though their expected values equal σ_1 and σ_2, from sample to sample they will tend to be too high or too low. Still, some of the time they will be too high or low in opposite directions; the combined error can be lower than their individual errors. As the second sample gets large, though, sample variation in s_2 becomes less and

		Value of s_2				
		10	25	100	250	1000
Value of n_2	10	18.00	11.81	9.18	9.03	9.00
	25	16.64	32.97	25.17	24.19	24.01
	100	10.88	22.95	107.92	101.91	99.20
	250	9.73	14.03	142.56	257.90	250.20
	1000	9.18	10.16	35.68	349.92	1007.89

Figure 9.10 Degrees of freedom for various s_2 and n_2, given $s_1 = 10$ and $n_1 = 10$.

less. In the limit, all the variability in our estimate of the standard error, $s_{\bar{X}_1-\bar{X}_2}$, is due to the smaller sample. And thus, the relevant degrees of freedom are $n_1 - 1$, the degrees of freedom of that sample.

We can also look across the first or second row to see the effect of greater and greater relative values of s_2. Again, the degrees of freedom fall—this time to $n_2 - 1$—the degrees of freedom of the sample with the larger standard deviation. The intuition is similar to that above. As s_2 increases, sample two becomes more and more dominant in determining the variation in $s_{\bar{X}_1-\bar{X}_2}$. In the limit, all the variability in our estimate of the standard error, $s_{\bar{X}_1-\bar{X}_2}$, is due to the sample with the larger standard deviation. Hence, the relevant degrees of freedom are $n_2 - 1$, the degrees of freedom of that sample.

Finally, read down the diagonal—18.00, 32.97, 107.92, 257.90, and 1007.89. Each of these numbers (almost) equals $n_1 + n_2 - 2$, the degrees of freedom for the pooled test. And each is the highest possible value for that particular combination of standard deviations. That is, with $s_1 = 10$; $s_2 = 10$, the highest degrees of freedom are 18, which we have if $n_1 = 10$; $n_2 = 10$. However, with $s_1 = 10$; $s_2 = 25$, the highest degrees of freedom are (almost) 33, which we have if $n_1 = 10$; $n_2 = 25$. Generally, the degrees of freedom are highest, and the test most powerful, if the ratio of sample sizes is the same as the ratio of sample standard deviations.

Of course, you will not know exact sample standard deviations until you have taken your samples. But, if you have reason to think that the standard deviation of group one is likely to be twice that of group two, it would make sense to take twice as large a sample from group one.

Consider the following example:

> Reconsider the earlier example comparing the "expected returns" of two stocks. We might theorize that a higher expected return on a stock would be compensation for higher risk. If so, we might not want to assume that the risks are equal. The table below repeats the summary statistics. Based on these statistics, is there a significant difference in the "expected return" of these two stocks? Assume the true risks are unequal. Use $\alpha = 0.05$. Use $\alpha = 0.01$. What is the *p*-value of this test?

	ABC Inc.	XYZ Corp.
Expected return (mean)	9.50	6.40
Risk (standard deviation)	3.90	2.60
Number of periods	10	14

The hypotheses are unchanged.

1. H_o: $\mu_{ABC} - \mu_{XYZ} = 0$ $\quad$ H_o: $\mu_{ABC} = \mu_{XYZ}$
   H_a: $\mu_{ABC} - \mu_{XYZ} \neq 0$ $\quad$ H_a: $\mu_{ABC} \neq \mu_{XYZ}$.

Now, though, we have fewer degrees of freedom, and a higher criterion.

2. $$\text{df} = \frac{(3.90^2/10 + 2.60^2/14)^2}{(3.90^2/10)^2/9 + (2.60^2/14)^2/13}$$

$$= \frac{(1.521 + 0.4829)^2}{(1.521)^2/9 + (0.4829)^2/13} = 14.60$$

Reject H_o if $|t_c| > 2.131$ (df = 15; $\alpha = 0.05$)   Reject H_o if $|t_c| > 2.947$ (df = 15; $\alpha = 0.01$).

And the formula for the standard error is different.

3. $$s_{\bar{X}_{ABC}-\bar{X}_{XYZ}} = \sqrt{\frac{3.90^2}{10}+\frac{2.60^2}{14}} = \sqrt{1.521+.4829} = 1.416;$$

$$t_c = \frac{9.5-6.4}{1.416} = 2.190.$$

4. ∴ Reject H_o ∴ Fail to Reject H_o.

We can still reject with an α of 0.05, but not with an α of 0.01. The *p*-value is still between 0.05 and 0.02. However it has risen to =TDIST(2.190,15,2) = 0.0447.

Consider the following example:

In a market study for Walton's, a local department store, you select a sample of 60 actual and potential patrons to interview. Among the questions you wish to answer is whether the patrons and nonpatrons differ in their incomes. The table below gives summary statistics. Noting the rather large difference in sample standard deviations, you decide that you must assume that the population standard deviations are unequal. Can you conclude that there is a difference in the mean incomes of patrons and nonpatrons? Use $\alpha = 0.05$. Use $\alpha = 0.01$. What is the *p*-value of this test?

	Patrons	Nonpatrons
Mean income (in $ 1000s)	58.7	50.4
Standard deviation (in $ 1000s)	16.8	9.8
Number	27	33

Nothing in the problem suggests a one-tailed alternative. Hence

1. H_o: $\mu_{Patron} - \mu_{Nonpatron} = 0$ H_o: $\mu_{Patron} - \mu_{Nonpatron}$
H_a: $\mu_{Patron} - \mu_{Nonpatron} \neq 0$ H_a: $\mu_{Patron} \neq \mu_{Nonpatron}$.

2. $$df = \frac{(16.8^2/27+9.8^2/33)^2}{(16.8^2/27)^2/26+(9.8^2/33)^2/32}$$

$$= \frac{(10.453+2.910)^2}{(10.453)^2/26+(2.910)^2/32} = 40.0$$

Reject H_o if $|t_c| > 2.021$ (df = 40; $\alpha = 0.05$)   Reject H_o if $|t_c| > 2.704$ (df = 40; $\alpha = 0.01$).

3. $$s_{\bar{X}_{Patron}-\bar{X}_{Nonpatron}} = \sqrt{\frac{16.8^2}{27}+\frac{9.8^2}{33}} = \sqrt{10.453+2.910} = 3.656;$$

$$t_c = \frac{58.7-50.4}{3.656} = 2.270.$$

4. ∴ Reject H_o ∴ Fail to Reject H_o.

We can reject H_o with an α of 0.05, but not with an α of 0.01. The p-value of this test is =TDIST(2.270,40,2) = 0.0286. We can reject the null hypothesis and conclude that there is a difference if we are willing to accept a probability of 0.0286 or more that we are making a type I error.

9.7 A Difference in Means: Using Paired Data*

So far, the comparison of proportions and means have all assumed "unmatched" samples. For example, our sample of student's sex, major, and GPA included 30 women and 20 men. We could sort them into two groups by sex, and see whether there was a difference between the two sexes in their choice of major or their GPA. There was no attempt, though, to "match up" the men and women. Indeed that would have been impossible since there were unequal numbers. We simply found the appropriate summary statistic for each group, and compared them.

There are cases, though, in which we can "match" or "pair" the data. Consider the data below on the weights of a sample of people who have been on a new weight-loss diet. Suppose we want to know whether we can conclude that the diet works. That is, can we conclude that $\mu_{Before} > \mu_{After}$? Figure 9.11 illustrates.

The hypotheses look very much like those we have been testing.

1. H_o: $\mu_{Before} - \mu_{After} \leq 0$ $\quad$ H_o: $\mu_{Before} \leq \mu_{After}$
   H_a: $\mu_{Before} - \mu_{After} > 0$ $\quad$ H_a: $\mu_{Before} > \mu_{After}$.

And we *could* do this as we have done the problems of the last two sections. That is, we could calculate the sample mean weights, $\bar{X}_{Before}$ and $\bar{X}_{After}$, find the difference, and calculate t_c using the formulas of Figure 9.8 or Figure 9.9, depending on our assumption about the population standard deviations. But this would be terribly inefficient. It would be ignoring the fact that the person who weighed 182 lbs after the diet is the same one who weighed 200 lbs before; the person who weighed 104 lbs after the diet is same one who weighed 110 lbs before; and so on. Instead of finding

ID	Before	After	Diff
1	200	182	18
2	155	142	13
3	135	122	13
4	130	126	4
5	184	170	14
6	127	117	10
7	155	156	−1
8	110	104	6
9	150	150	0
10	120	124	−4
11	110	113	−3
12	135	133	2
$\bar{X}$ =	142.583	136.583	6.000
s =	27.839	23.861	7.435

Figure 9.11 The weights of 12 people before and after a diet.

$$t_c = \frac{\overline{\text{Diff}} - \mu_{\text{Diff}}}{s_{\overline{\text{Diff}}}} = \frac{\overline{\text{Diff}} - 0}{s_{\overline{\text{Diff}}}} = \frac{\overline{\text{Diff}}}{s_{\overline{\text{Diff}}}}$$

$$\text{where } s_{\overline{\text{Diff}}} = \frac{s_{\text{Diff}}}{\sqrt{n}}$$

Figure 9.12 Two-sample test of means: The formulas, paired samples, σ_X unknown.

the averages first and then their difference, we should find the individual differences first and then average them. The average of the differences is the same as difference in the averages. That is, 142.583 – 136.583 = 6.000. However, the standard deviation is much less. If we were to assume equal population standard deviations and calculate a pooled estimate for the before and after samples, we would get $s_P = 25.926$. By contrast, the standard deviation of the differences is just $s_{\text{Diff}} = 7.435$. The variation in *weight loss* is much less than the variation in *weight*.

In essence, by taking the individual differences first, we turn this two-sample problem into a one-sample one, and the formulas in Figure 9.12 are just the one-sample formulas from Chapter 8 applied to the differences. There are 12 differences, so we have $n - 1 = 11$ degrees of freedom:

2. Reject H_o if $t_c > 1.796$ (df = 11; $\alpha = 0.05$) | Reject H_o if $t_c > 2.718$ (df = 11; $\alpha = 0.01$).

3. $s_{\overline{\text{Diff}}} = \dfrac{7.435}{\sqrt{12}} = 2.146, \quad t_c = \dfrac{6.000}{2.146} = 2.796.$

4. ∴ Reject H_o | ∴ Reject H_o.

The *p*-value of the test is =TDIST(2.796,11,1) = 0.0087. We can reject the null hypothesis and conclude that there is a difference if we are willing to accept a probability of 0.0087 or more that we are being misled.

Contrast this result with what we would have found if we had not taken advantage of the pairing. As noted above, the pooled estimate of the standard deviation would have been 25.926, not 7.435. The estimated standard error, then, would have been 10.584, not 2.146, and t_c would have been 0.567, not 2.796. The *p*-value of the test would have been =TDIST(0.567,22,1)=0.2883. We would not have come close to rejecting the null hypothesis. By taking advantage of the pairing, we dramatically increased the power of the test.

Consider the following example:

Suppose you are testing two different techniques for manufacturing your product. You train 10 workers equally in each technique; you randomly assign five workers to each technique for a week; then you reverse them for a week. The results, in Figure 9.13, are output measures for each of the 10 workers using each of the two techniques. Can you conclude that there is a difference in average output using the two techniques? Use $\alpha = 0.05$. Use $\alpha = 0.01$. What is the *p*-value of this test?

1. H_o: $\mu_A - \mu_B = 0$ | H_o: $\mu_A = \mu_B$
   H_a: $\mu_A - \mu_B \neq 0$ | H_a: $\mu_A \neq \mu_B$.

Worker	Output A	Output B	Diff
1	65	63	2
2	47	51	−4
3	41	47	−6
4	43	46	−3
5	57	56	1
6	39	41	−2
7	52	56	−4
8	35	38	−3
9	50	54	−4
10	41	40	1
$\bar{X}$	47.000	49.200	−2.200
$s =$	9.153	8.176	2.658

Figure 9.13 Output per day of 10 workers using techniques A and B.

2. Reject H_0 if $|t_c| > 2.262$ (df = 9; $\alpha = 0.05$) Reject H_0 if $|t_c| > 3.250$ (df = 9; $\alpha = 0.01$).
3. $s_{\overline{\text{Diff}}} = \dfrac{2.658}{\sqrt{10}} = 0.841,\quad t_c = \dfrac{-2.2}{.841} = -2.617$
4. ∴ Reject H_0 ∴ Fail to Reject H_0.

You can reject H_0 with an α of 0.05 but not with an α of 0.01. The *p*-value of the test is =TDIST(−2.617,9,2) = 0.0279. You can reject the null hypothesis and conclude that there is a difference if you are willing to accept a probability of 0.0279 or more that you are being misled.

Using the same workers with each technique was very important here. It allowed you to use the paired-data test. And again, the variability in the differences is much smaller than the variability in either of the output samples themselves. Worker 1 seems to be much more productive than the others, increasing the variability in both samples. However, since he is more productive in *both* he adds very little to the variability in the *difference*. Hence, the paired data test is much more efficient.

If you had used 10 different workers with each technique, you would probably have used the test of Section 9.5. Suppose your sample results were the same as the two samples in Figure 9.13. Your pooled standard deviation would have been $s_P = 8.678$; your $t_c = -0.567$ and you would not have been able to reject H_0 at any reasonable α.

In business and the social sciences, we do not generate a lot of our data through experiments. Indeed, a lot of the questions we address do not lend themselves to experiments. As the preceding example shows, though, when we have the opportunity, it is important to stop and think through that experiment. Generating the matched samples above would not have been any more expensive than generating unmatched ones, and they are much more informative.

9.8 A Final Word on Two-Sample Tests

In this chapter, we have examined two-sample tests of the proportion and of the mean. The language of two-sample tests tends toward looking for "differences between samples." However, we can also think of these tests as looking for "relationships" in which an "explanatory" variable takes on just two values. Hence, a student's sex might help explain his or her choice of a Natural Science major or GPA. Location—Chicago or Downstate—might help explain support for a political candidate. Shift—day or night—might help explain how long a task takes.

For proportions, there is just one case. We compare the difference in sample proportions with the hypothesized population difference (zero) in terms of the estimated standard error, $s_{p_1-p_2}$. If the difference in sample proportions is unlikely enough given the null hypothesis, we reject the null hypothesis and conclude that the population difference must not be zero. And, of course, this means the variable that defines the two groups—sex or location—can help us explain the variable we have summarized in proportions—choice of Natural Science major or preference for a particular candidate.

For means, the logic is exactly the same, but there are four different cases. The first, which we disposed of quickly, assumes that we know the two population standard deviations. It is hard to imagine a case in which this would be so. The second case assumes that the population standard deviations, while unknown, are equal. This assumption is often very plausible, and should always be considered. It allows a pooled estimate of the unknown population standard deviation. The third case is the one for cases in which the population standard deviations cannot plausibly be assumed equal. It is a weaker test, but it is sometimes necessary. And the fourth case is the one that takes advantage of paired data. In instances for which we have data matched case by case, we can work with the differences, and greatly increase the power of the test.

A limitation of all these tests is that the explanatory variable must be a dummy, "yes–no" variable. If it divides the data into three or more categories instead of just two, these tests all fail. Dealing with more general cases will be the subject of the next several chapters.

9.9 Exercises

9.1 Suppose you expect that hourly-wage workers are less likely than other (primarily salaried and self-employed) workers to have at least some college education. Using a random sample of 720 workers from the National Longitudinal Survey of Youth (NLSY), you get the following frequencies. Do these results support your expectations? Use $\alpha = 0.05$. Use $\alpha = 0.01$. What is the p-value of this test?

Hourly-Wage Worker	At Least Some College	
	Yes	No
Yes	102	244
No	216	158

9.2 Suppose that you expect college freshmen who live on campus will do better academically than those who do not. You select a random sample of 75 freshmen, group them according to where they live, and calculate the following summary statistics. Do these results support your expectations? Assume that the true population standard deviations are equal. Use $\alpha = 0.10$. Use $\alpha = 0.05$. What is the p-value of this test?

GPA	On Campus	Off Campus
Mean	2.893	2.714
Standard deviation	0.5055	0.5377
Count	53	22

9.3 In Exercises 8.7 using NLSY1.xls, you tested separate hypotheses about the heights of young adult men and women. Return to these data.

a. Are the mean heights of young adult men and women significantly different from each other? Assume population standard deviations are equal. Use $\alpha = 0.05$. Use $\alpha = 0.01$. What is the p-value of this test?

b. Did the assumption of equal population standard deviations seem tenable in this case?

*c. Suppose you were unwilling to assume equal population standard deviations above. How would your results change?

9.4 Continue with these same data.

a. Are the mean weights of young adult men and women significantly different from each other? Assume population standard deviations are equal. Use $\alpha = 0.05$. Use $\alpha = 0.01$. What is the p-value of this test?

b. Did the assumption of equal population standard deviations seem tenable in this case?

*c. Suppose you were unwilling to assume equal population standard deviations above. How would your results change?

9.5 In Exercises 8.12 and 8.13, using Nickelsl.xls, you tested separate hypotheses about Nickels customers and non-customers. Return to those data. Are the two groups significantly different with respect to each of the following? Where necessary, assume population standard deviations are equal. Use $\alpha = 0.05$. Use $\alpha = 0.01$. What is the p-value of each test?

a. Proportions who are female?

b. Mean age?

c. Mean income?

d. Proportions whose primary source of market information is the newspaper?

9.6 Continue with these same data.

a. For which cases did you need to assume population standard deviations equal? Did the assumption seem tenable in these cases?

*b. Suppose you were unwilling to assume equal population standard deviations. How would your results change?

9.7 In Exercise 8.14 using Employees2.xls, you tested a number of hypotheses about employees of your firm. Return to these data. Are your female and male employees significantly different with respect to each of the following? Where necessary, assume population standard deviations are equal. Use $\alpha = 0.10$. Use $\alpha = 0.05$. What is the p-value of each test?

a. Mean education?
b. Mean job experience?
c. Proportions employed in management?
d. Mean salary?

9.8 Continue with these same data.

a. For which cases did you need to assume population standard deviations equal? Did the assumption seem tenable in these cases?
*b. Suppose you were unwilling to assume equal population standard deviations. How would your results change?

9.9 In Chapter 3, using Students2.xls, you calculated summary statistics separately for male and female students. Return to these data. Are the two groups significantly different with respect to each of the following? Where necessary, assume population standard deviations are equal. Use $\alpha = 0.05$. Use $\alpha = 0.01$. What is the p-value of each test?

a. Mean height.
b. Mean weight.
c. Proportion economics major.
d. Mean financial aid.
e. Proportion holding a job.
f. Proportion participating in a varsity sport.
g. Proportion participating in a music ensemble.
h. Proportion belonging to a fraternity or sorority.
i. Mean entertainment expenditure.
j. Mean study time.
k. Mean college GPA.

9.10 Continue with these same data.

a. For which of the tests above did you need to assume population standard deviations were equal? Did the assumption seem tenable in these cases?
*b. Suppose you were unwilling to assume equal population standard deviations. How would your results change?

9.11 Continue with these same data. In Chapter 3, you also calculated summary statistics separately for athletes and nonathletes. Are the two groups significantly different with respect to each of the following? Where necessary, assume population standard deviations are equal. Use $\alpha = 0.05$. Use $\alpha = 0.01$. What is the p-value of each test?

a. Proportion female.
b. Mean height.
c. Mean weight.
d. Proportion economics major.
e. Mean financial aid.

f. Proportion holding a job.
g. Proportion participating in a music ensemble.
h. Proportion belonging to a fraternity or sorority.
i. Mean entertainment expenditure.
j. Mean study time.
k. Mean college GPA.

9.12 Continue with these same data.

a. For which of the tests above did you need to assume population standard deviations were equal? Did the assumption seem tenable in these cases?

*b. Suppose you were unwilling to assume equal population standard deviations. How would your results change?

9.13 The data file Dietl.xls contains data from another trial of a weight-loss diet. The data are Before and After weights for 25 individuals following this diet. Can you conclude that the diet works? Use $\alpha = 0.05$. Use $\alpha = 0.01$. What is the p-value of this test?

10

Tests of Hypotheses: Contingency and Goodness-of-Fit

Chapter 9 introduced a test for the difference between two proportions. The test had the advantage that it was a straightforward extension of the one-sample test of a proportion, and used the by-now-familiar normal distribution. However, the test has limitations. Most importantly, it cannot be extended to more than two samples. It is fine if we have two assembly lines and want to know if they differ in the proportion of defectives they turn out. But what if we have three assembly lines? It is fine if we want to know whether the proportion of Natural Science majors differs by sex. But what if we want to know whether the whole distribution of majors differs by sex?

Our new approach will be to create a **contingency table**—really just a two-way frequency table of our data—and then compare this whole set of observed frequencies with those we would have expected according to our null hypothesis. If our observed frequencies are so different from the expected frequencies that they would arise only rarely when the null hypothesis is true, we will conclude that the null hypothesis is not true and reject it.

The general reasoning, then, is exactly the same as that we have been using in the last two chapters. We will specify null and alternative hypotheses; we will specify a rejection criterion based on the probability we are willing to accept of making a type I error. We will calculate a test statistic. And we will either reject or fail to reject the null hypothesis based on whether our test statistic meets or does not meet our criterion.

What is new is our test statistic, and its sampling distribution. We will calculate a statistic that follows the **chi-square (χ^2) distribution**.

As it turns out, this same approach is good for answering a different sort of problem as well. Suppose we have sample values for a single variable, and wish to test whether its population distribution follows a particular theoretical distribution. Perhaps we want to know if the heights of young adult men really are normally distributed. Perhaps we want to know if that die you were rolling back in Section 1.1 really is fair. In these cases our data can be organized in an ordinary one-way frequency table. But again, we can compare this whole set of observed frequencies with those we would have expected if the population distribution were as hypothesized. And, again, if our observed frequencies are so different from the expected frequencies that they would arise only rarely when the null hypothesis is true, we will conclude that the null hypothesis is not true, and reject it. Such tests are called **goodness-of-fit** tests.

10.1 A Difference in Proportions: An Alternate Approach

10.1.1 The Null and Alternative Hypotheses

In the last chapter, we tested to see whether there was a difference in the proportion of men and women who were Natural Science majors. Our hypotheses were

1. H_o: $\pi_{NS\text{–Female}} - \pi_{NS\text{–Male}} = 0$ $\quad$ H_o: $\pi_{NS\text{–Female}} = \pi_{NS\text{–Male}}$
   H_a: $\pi_{NS\text{–Female}} - \pi_{NS\text{–Male}} \neq 0$ $\quad$ H_a: $\pi_{NS\text{–Female}} \neq \pi_{NS\text{–Male}}$.

An alternative way of expressing these hypotheses is

1. H_o: Majoring in Natural Science is independent of sex
   H_a: Majoring in Natural Science is not independent of sex.

That is, saying that the true proportions are equal is the same as saying that knowing a student's sex tells us nothing about his or her chance of being a Natural Science major. These are unrelated, or independent events. Putting the hypotheses in this second form emphasizes that we have categorized these students in two dimensions—by their sex and by their major—and we are looking to see if one can help explain the other. We are looking for a relationship. This second form is also less awkward as we start to have multiple categories instead of just two.

10.1.2 The Decision Criterion

The statistic we will calculate in testing the hypothesis above follows a new sampling distribution, the **chi square (χ^2) distribution.** Like the *t* distribution, it has a degrees of freedom parameter associated with it. Figure 10.1 shows it for 1, 5, and 9 degrees of freedom. It differs from the normal and *t* distributions in several ways. First, clearly it is not symmetric, though it becomes more so as the degrees of freedom increase. Second, it must be positive. Under the null hypothesis, the expected value for the statistic is zero, just as it was for the normal distribution. In this case, though, the calculations comparing actual and expected values under the null hypothesis involve finding differences and then *squaring* them. Regardless of the direction of the difference, then, the measure is always positive. We will always be looking for a value that is too large positive. We will only be concerned with the upper tail.

Like the normal and *t* distributions, the χ^2 is tabulated and the table is widely available. Table 5, in Appendix C, shows how it is typically laid out. Find it now. It looks very much like the *t* table, with upper-tail probabilities across the top and degrees of freedom down the side.

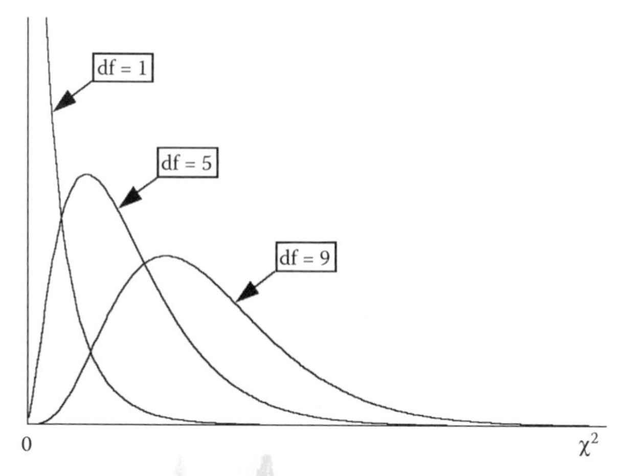

Figure 10.1 The χ^2 distribution for one, five, and nine degrees of freedom.

As with the normal and *t*, there are also special spreadsheet functions. Figure 10.2 shows the special functions for several of the most common spreadsheets.

An important difference between this table and the *t* table deserves emphasizing immediately. In the *t* table, if we use just the upper tail, it is in order to test a one-tailed alternative; we are predicting a positive rather than a negative value for the t_c statistic. If we want a two-tailed alternative, we divide α in half to look it up and then ignore the sign. In the χ^2 table there is no negative tail. Deviations of actual from expected values will always lead to a positive result. Hence, the probability in the upper tail of the χ^2 is really the probability associated with what we have so far called a two-tailed test. The χ^2 test is naturally a test of differences.

Finally, a word on degrees of freedom. They are figured quite differently for χ^2 tests than they are for *t* tests. For this particular example, testing whether Natural Science major is independent of sex, there is just one degree of freedom. It will be easier to explain why after we look at the calculations.

2. Reject H_o if $\chi^2_c > 3.841$ (df = 1; $\alpha = 0.05$).

10.1.3 The Calculations

Panel A of Figure 10.3 shows our data organized into a contingency table—really just a two-way frequency table. Among the women, there were three Natural Science majors and 27 others. Among the men, there

	Formula	Example
From χ^2_o to a probability:	=CHIDIST(χ^2_o, df)	=CHIDIST(3.841,1) = 0.05
From a probability to χ^2_o:	=CHIINV(*p*, df)	=CHIINV(0.05, 1) = 3.841
(For Lotus/Quattro Pro, replace "=" with "@")		

Figure 10.2 The χ^2 distribution special functions in common spreadsheet programs.

Panel A

	NS	Not	
Female	3	27	30
Male	5	15	20
	8	42	50
	0.16	0.84	

Panel B

	NS	Not	
Female	4.8	25.2	30
Male	3.2	16.8	20
	8.0	42.0	50

Panel C

f_o	f_e			
3	4.8	−1.8	3.24	0.675
5	3.2	1.8	3.24	1.012
27	25.2	1.8	3.24	0.129
15	16.8	−1.8	3.24	0.193
50	50.0			2.009

Figure 10.3 Contingency tables and the χ_c^2 statistic: An example.

were five Natural Science majors and 15 others. In total, there were eight Natural Science majors and 42 others.

The proportions below the table in Panel A are the overall proportions of Natural Science and other majors. That is, overall, 8/50 = 0.16 is the overall proportion of Natural Science majors and 42/50 = 0.84 is the overall proportion of other majors. Since our null hypothesis is that choice of major does not depend on sex, we should expect these same proportions of Natural Science and other majors among both women and men. Panel B, then, simply applies these proportions to the 30 women and 20 men, to find the expected frequency of each. That is, since 16% of the total are science majors, $0.16 \times 30 = 4.8$ women and $0.16 \times 20 = 3.2$ men should be Natural Science majors. And since 84% of the total are other majors, $0.84 \times 30 = 25.2$ women and $0.84 \times 20 = 16.8$ men should be other majors.

It does not matter which variable is in the rows and which is in columns. It would have been just as correct to note that 60% of the sample are women while 40% are men. Hence, $0.60 \times 8 = 4.8$ Natural Science majors and $0.60 \times 42 = 25.2$ other majors should be women, while $0.40 \times 8 = 3.2$ Natural Science majors and $0.40 \times 42 = 16.8$ other majors should be men. Either way, we are calculating the category proportions for one variable and applying them to the category totals of the other variable. And the results are the same.

Notice that all the totals in Panel B correspond to the totals in Panel A. We still have 30 women and 20 men; we still have eight Natural Science and 42 other majors. This means that we did not really need to do all four of the calculations above. Once we had calculated the first number, $0.16 \times 30 = 4.8$, we had really determined all four. The number below it had to be 3.2 in order for their sum to equal eight; the number to the right had to be 25.2 in order for their sum to equal 30; the remaining number had to be 16.8 in order for the sums down and across to be 42 and 20. This is where our one degree of freedom comes from.

In general, once you have all but the last row, you really have the last row too, since you know what the column totals are. And once you have all but the last column, you really have the last column too, since you know what the row totals are. The general rule for degrees of freedom in a contingency table is $\mathbf{df} = (\boldsymbol{r} - \mathbf{1}) \times (\boldsymbol{c} - \mathbf{1})$, where $\boldsymbol{r}$ is the number of row categories and $\boldsymbol{c}$ is the number of column categories. In a 2×2 table like ours, that is, $\text{df} = (2-1) \times (2-1) = 1 \times 1 = 1$.

Panel B, then, contains our expected frequencies. But of course we would not really expect to get these values exactly; indeed, we cannot since the actual frequencies must be integers. As always, the question is—are our observed values so unlikely, given our expected ones based on the null hypothesis, that the null hypothesis must be wrong? The χ_c^2 statistic compares these two sets of frequencies—observed and expected—to decide if they are too different.

Panel C shows the calculation of the χ_c^2 statistic. The first two columns simply repeat the observed and expected frequencies. They can be listed in any order; just make sure to list them both in the same order. Column 3 finds the differences. These will often show a pattern. And of course, since the observed and expected frequencies always sum to the same amount, their differences always sum to zero. We want a measure in which big positive differences do not just get canceled out by big negative ones. This is a problem we have dealt with before, starting with the calculation of the standard deviation. We squared all the deviations, to make them all positive. We do so again here in Column 4.

Column 5, then, divides the squared differences by the expected frequencies. Intuitively, whether a squared difference like 3.24 seems large depends on the size of the frequencies with which we are working. Dividing through by the expected frequencies converts the squared differences to relative measures.

Finally, the sum of column 5 is our χ_c^2 statistic.

3. $\chi_c^2 = \sum \frac{(f_o - f_e)^2}{f_e} = 2.009.$

10.1.4 The Conclusion

Our χ_c^2 is not greater than 3.841. Our conclusion is

4. $\therefore$ Fail to Reject H_o.

These observed frequencies are not different enough from the expected ones under the null hypothesis to reject the null hypothesis with an α of 0.05. They could too easily have come from a population in which the choice of a Natural Science major is independent of sex. How easily? Looking across the χ^2 table at one degree of freedom, the p-value of this test is between 0.1000 and 0.2500. More precisely =CHIDIST(2.009,1) = 0.1564. We reach the same conclusion we did using the two-sample test of proportions. Indeed, the p-values are the same except for rounding. The two tests are mathematically equivalent. χ^2 with one degree of freedom is the square of z. The strength of this one is that it can deal with variables that have more than two categories.

10.2 Contingency Tables with Several Rows and/or Columns

In testing whether the choice of a Natural Science major depended on sex, we simply lumped together all the other majors. I hope you found that rather artificial. A much more natural question would have been whether choice of major depends on sex. Our sample has students' majors grouped into six categories; why single out just Natural Science? While there might be some good reason, in truth it is because we had no way of dealing with more-than-two-category, categorical variables until now. Now we do.

1. H_o: Choice of major is independent of sex
   H_a: Choice of major is not independent of sex.

Panel A below shows our data organized into a 6×2 contingency table. This means that our degrees of freedom are $(6-1) \times (2-1) = 5 \times 1 = 5$. If we choose an α of 0.05, our rejection criterion is as follows:

2. Reject H_o if $\chi^2 > 11.070$ (df = 5; $\alpha = 0.05$).

Figure 10.4 shows the calculations. Again, Panel A displays our observed frequencies. Panel B calculates the expected frequency in each cell. Since 60% of the students are female, 60% of each major should be female. Hence, $0.60 \times 8 = 4.8$, $0.60 \times 9 = 5.4$, and so on. Since 40% of the students are male, 40% of each major should be male. Hence, $0.40 \times 8 = 3.2$, $0.40 \times 9 = 3.6$, and so on.

Finally, Panel C compares these two sets of numbers. Columns 1 and 2 just list the two sets of frequencies for convenience. Column 3 finds the 12 differences, column 4 squares these differences, and column 5 divides each squared difference by its expected frequency. The sum of column 5 is our χ_c^2.

3. $\chi_c^2 = \sum \frac{(f_o - f_e)^2}{f_e} = 6.923.$

Our χ_c^2 is not greater than 11.070.

4. $\therefore$ Fail to Reject H_o.

Again, these observed frequencies are not different enough from the expected ones to reject the null hypothesis. Looking across the χ^2 table

Panel A

	F	M	
Natural Science	3	5	8
Social Science	5	4	9
Humanities	7	2	9
Fine Arts	5	2	7
Business	4	6	10
Nursing	6	1	7
	30	20	50
	0.60	0.40	

Panel B

	F	M	
Natural Science	4.8	3.2	8
Social Science	5.4	3.6	9
Humanities	5.4	3.6	9
Fine Arts	4.2	2.8	7
Business	6.0	4.0	10
Nursing	4.2	2.8	7
	30.0	20.0	50

Panel C

f_o	f_e			
3	4.8	−1.8	3.24	0.675
5	5.4	−0.4	0.16	0.030
7	5.4	1.6	2.56	0.474
5	4.2	0.8	0.64	0.152
4	6.0	−2.0	4.00	0.667
6	4.2	1.8	3.24	0.771
5	3.2	1.8	3.24	1.013
4	3.6	0.4	0.16	0.044
2	3.6	−1.6	2.56	0.711
2	2.8	−0.8	0.64	0.229
6	4.0	2.0	4.00	1.000
1	2.8	−1.8	3.24	1.157
50	50.0			6.923

Figure 10.4 Contingency tables and the χ_c^2 statistic: Another example (a).

at five degrees of freedom, we could barely reject at an α of 0.2500. The p-value of the test is =CHIDIST(6.923,5) = 0.2264. We would get departures from expected frequencies this large nearly 23% of the time when the null hypothesis is true. Clearly, then, we do not have convincing evidence against the null hypothesis.

A word is in order on sample size here. I have stuck with this sample of 50 students because it is small enough to list (Figures 2.1 and 9.1) and to absorb easily. For a serious test of these hypotheses, though, it is not large enough. Many authors suggest that expected frequencies be no smaller than five and, especially in this second example, most of our expected frequencies are smaller than that. One might try to get around this by combining major categories further, though this might also reduce the observed departures from the expected numbers. Preferably, we could collect a larger sample. Suppose we doubled the sample and each of the observed frequencies turned out exactly twice as big. That would exactly double the χ^2_c statistic and we would be able to reject the null hypothesis.

Consider the following example:

Three national brands of tires are tested to compare the percentages failing to last 50,000 miles. If 200 of each brand are tested with the following results, can you conclude that there is a significant difference among brands in the percentage failing to last 50,000 miles? Use $\alpha = 0.05$. Use $\alpha = 0.01$. What is the p-value of this test?

Brand	Percentage Failing
X	15%
Y	20%
Z	10%

If there had been just two brands, it would probably have seemed natural to attack this problem as a comparison of proportions and that would have been fine. Because there are three brands, though, we need to turn it into a contingency table problem. First, it is important that the contingency table frequencies are ordinary frequencies and not percentages or proportions. So, to find the actual number of times that failed, multiply each of their proportions by 200. That is, Brand X had $0.15 \times 200 = 30$ failures, and so on. Second, do not forget that the others did not fail. That is, there is a whole other column that is only implied. Brand *X* had $0.85 \times 200 = 170$ that did not fail, and so on. Panel A in Figure 10.5 shows the data in the problem transformed into a contingency table. We have all 600 tires categorized both by their brand and by whether or not they failed. Our null hypothesis, then, is that there is no relationship between these two variables.

1. Failure is independent of brand.
   Failure is not independent of brand.

Since this is a 3 × 2 table, the degrees of freedom are
$\text{df} = (3 - 1) \times (2 - 1) = 2 \times 1 = 2.$

2. Reject H_o if $\chi^2_c > 5.991$ (df = 2; $\alpha = 0.05$)   Reject H_o if $\chi^2_c > 9.210$ (df = 2; $\alpha = 0.01$).

Panel A

	Fail	Not	
X	30	170	200
Y	40	160	200
Z	20	180	200
	90	510	600
	0.15	0.85	

Panel B

	Fail	Not	
X	30	170	200
Y	30	170	200
Z	30	170	200
	90	510	600

Panel C

f_o	f_e			
30	30	0	0	0
40	30	10	100	3.3333
20	30	−10	100	3.3333
170	170	0	0	0
160	170	−10	100	0.5888
180	170	10	100	0.5888
600	600			7.8433

Figure 10.5 Contingency tables and the χ_c^2 statistic: Another example (b).

Panel B shows the expected frequencies; since 15% failed overall, we should expect 15% of each brand to fail according to our null hypothesis. Panel C compares the two sets of numbers.

3. $\chi_c^2 = \sum \frac{(f_o - f_e)^2}{f_e} = 7.843.$

   Our χ_c^2 is greater than 5.991 but not greater than 9.210.

4. ∴ Reject H_o ∴ Fail to Reject H_o.

The *p*-value of the test is =CHIDIST(7.843,2) = 0.0198. There is about a 0.02 probability of getting a χ_c^2 statistic this large if the null hypothesis is true. If we are willing to accept that large a probability of a type I error, we can reject the null hypothesis and conclude that there is a difference among brands.

Consider the following example:

Employees of a large firm are able to choose any one of three health care plans. To explore whether there is a pattern to employee choice, you select a random sample of 200 employees and record their choice along with other characteristics. Panel A in Figure 10.6 shows a contingency table of their choice versus their job category. Can you conclude that workers in the various job categories differ in their choice? Use $\alpha = 0.05$. Use $\alpha = 0.01$. What is the *p*-value of this test?

1. Choice is independent of job category.
   Choice is not independent of job category.
   Since this is a 3×3 table, the degrees of freedom are $df = (3–1) \times (3–1) = 2 \times 2 = 4$.

2. Reject H_o if $\chi_c^2 > 9.488$ ($df = 4$; $\alpha = 0.05$)   Reject H_o if $\chi_c^2 > 13.277$ ($df = 4$; $\alpha = 0.01$).

   Panel B shows the expected frequencies; panel C compares the two sets of numbers.

3. $\chi_c^2 = \sum \frac{(f_o - f_e)^2}{f_e} = 13.333.$

   Our χ_c^2 is greater than both 9.488 and 13.277 (barely).

4. ∴ Reject H_o ∴ Reject H_o.

Panel A

	Plan 1	Plan 2	Plan 3	
Administration	6	10	4	20
Office	14	10	6	30
Line	40	40	70	150
	60	60	80	200
	0.3	0.3	0.4	

Panel B

	Plan 1	Plan 2	Plan 3	
Administration	6	6	8	20
Office	9	9	12	30
Line	45	45	60	150
	60	60	80	200

Panel C

f_o	f_e			
6	6	0	0	0.000
14	9	5	25	2.778
40	45	−5	25	0.556
10	6	4	16	2.667
10	9	1	1	0.111
40	45	−5	25	0.556
4	8	−4	16	2.000
6	12	−6	36	3.000
70	60	10	100	1.667
200	200			13.333

Figure 10.6 Contingency tables and the χ_c^2 statistic: Another example (c).

The *p*-value of the test is =CHIDIST(13.333,4) = 0.0098. We can conclude, with an α of a bit less than 0.01 that the employee groups differ in their preference for plans.

10.3 A Final Word on Contingency Tables

As we have seen, we can use contingency tables and the chi-square test statistic to test whether there is a relationship between two categorical variables. This test is equivalent to the test for a difference in proportions, for the case in which each variable has just two categories; its advantage is that it can be extended to cases in which one or both of the variables has several categories.

What this test does not do particularly well is characterize the nature of the relationship. For example, look back at Figure 10.4. There was not a significant relationship between sex and choice of major, but suppose there had been. How might we describe the relationship? We might note that there are fewer than the expected number of women in Natural Science, Social Science, and Business and more than the expected number in the Humanities, Fine Arts, and Nursing. But we do not know that each of these differences from the expected number is significant; only that together they add up to a pattern that we should not have seen according to the null hypothesis.

It would be nice, as well, to have some relative measure of how well we have done in explaining a student's choice of major. A zero might mean that knowing the student's sex tells us nothing about his or her choice of major; a one might mean that knowing the student's sex tells us for certain his or her choice of major. Numbers in between would tell us the proportion of the variation among majors that the student's sex explains. The chi-square test statistic is not such a measure.

These limitations are due largely to the type of variables with which we are working: categorical variables. Female is neither larger nor smaller than male; the majors are simply categories which can be listed in any order with equal validity. Those who work extensively with categorical data have developed measures that attempt to go beyond the chi-square statistic, to address some of these limitations. However, these measures are beyond the scope of this text.

10.4 Testing for Goodness-of-Fit

The test above was a test for a relationship. We classified each case in two different ways—sex and major, for example—creating a two-way frequency table. Then we compared the observed frequencies with the expected ones according to the null hypothesis that there was no relationship. Now that we know the chi-square test statistic, though, we can also use it to test a very different sort of hypothesis. We can test whether a single random variable follows a hypothesized true probability distribution.

In the first example in Section 1.1 of this text, I offered this scenario.

> Imagine you are playing a board game with some friends and a die seems to be coming up "1" too often. Perhaps this is just chance; perhaps the die is unbalanced. How do you decide?

Giving you the benefit of the doubt, I suggested that you would probably roll the die a number of times and keep track of the outcomes—much as in panel A of Figure 10.7, which simply repeats Figure 1.1. These are your observed frequencies. If the die is fair, the expected frequencies would be uniform. The chi-square test statistic gives us a measure of how far these observed frequencies are from the expected ones, and a criterion for deciding if they are too different.

10.4.1 The Null and Alternative Hypotheses

The null hypothesis is that the die is fair.

1. H_0: Observations fit a uniform distribution (die is fair).
   H_a: Observations do not fit a uniform distribution (die is not fair).

Panel A

Result	Frequency
1	~~////~~ ~~////~~ ~~////~~ ///
2	~~////~~ ~~////~~ //
3	~~////~~ ////
4	~~////~~ ////
5	~~////~~ /
6	~~////~~ /

Panel B

f_o	f_e			
18	10	8	64	6.4
12	10	2	4	0.4
9	10	−1	1	0.1
9	10	−1	1	0.1
6	10	−4	16	1.6
6	10	−4	16	1.6
60	60			10.2

Figure 10.7 Goodness-of-Fit and the χ_c^2 statistic: An example.

10.4.2 The Decision Criterion

We know our criterion is a chi-square. For that we need to know the degrees of freedom. The general rule for goodness-of-fit tests is **df = # of categories – # of constraints**. We start with the number of categories—six for a six-sided die—and then subtract one for each way in which we constrain the expected frequencies to be like the observed.

In this case, our null hypothesis implies that the expected values are equal; it does not imply that they equal 10. We get the 10 from the fact that the observed values sum to 60. For a fair comparison, then, we constrain the expected numbers to sum to 60 as well. We always do this, so we always lose at least one degree of freedom. In this example, this is the only way in which we constrain the expected frequencies to be like the observed, so the degrees of freedom is 6 – 1 = 5.

In this section, all our examples will be ones like this, so df = # of categories – 1. You should understand, though, that this is not completely general. In some cases, we would want to constrain the expected frequencies in additional ways. If, for example, we were testing whether a sample of observations fit a normal distribution, we would want to compare it to the normal distribution that had the same mean and standard deviation. In doing so, we would lose two additional degrees of freedom. The final section will say a little more about such cases.

2. Reject H_o if $\chi_c^2 > 11.070$ (df = 5; $\alpha = 0.05$).

10.4.3 The Calculations

The expected frequencies are each 1/6th of the total. The f_o and f_e are compared just as before. We find the six differences; square each; divide each squared difference by its f_e; and sum.

3. $\chi_c^2 = \sum \frac{(f_o - f_e)^2}{f_e} = 10.2.$

10.4.4 The Conclusion

Our χ_c^2 does not exceed 11.070. Thus,

4. ∴ Fail to Reject H_o.

Sets of observed frequencies this far from the expected frequencies would arise randomly more than 5% of the time when the die is fair. Therefore we cannot conclude with an α of 0.05 that the die is unfair. Notice, though, that we could have rejected with an α of 0.10. The p-value of the test is =CHIDIST(10.2,5) = 0.0698.

Consider the following example:

> Suppose you suspect that workers at your firm have begun using their allotted sick days as personal days, calling in sick when they are not so they can do something else that day. A possible indicator of this would be a disproportionate use of sick days on Mondays and Fridays. You collect information on a sample of sick days taken

Panel A

	f_o	f_e			
Monday	22	18.4	3.6	12.96	0.704
Tuesday	15	18.4	−3.4	11.56	0.628
Wednesday	14	18.4	−4.4	19.36	1.052
Thursday	16	18.4	−2.4	5.76	0.313
Friday	25	18.4	6.6	43.56	2.367
	92	92.0			5.065

Panel B

	f_o	f_e			
Monday, Friday	47	36.8	10.2	104.04	2.827
Tuesday–Thursday	45	55.2	−10.2	104.04	1.885
	92	92.0			4.712

Figure 10.8 Goodness-of-Fit and the χ_c^2 statistic: Another example (a).

over the last year. The data, arranged in a frequency distribution, are in the first two columns of Figure 10.8, Panel A. Do they support your suspicion? Use $\alpha = 0.05$.

1. H_o: Observations fit a uniform distribution.
   H_a: Observations do not fit a uniform distribution.

There are five workdays, hence five categories, so df = 5 – 1 = 4.

2. Reject H_o if $\chi_c^2 > 9.488$ (df = 4; $\alpha = 0.05$).
   The calculations are shown in Figure 10.8, Panel A. The expected frequencies are 92/5 = 18.4. As always we find each difference; square each; divide each squared difference by its f_e; and sum.

3. $\chi_c^2 = \sum \frac{(f_o - f_e)^2}{f_e} = 5.065.$
   Our χ_c^2 does not exceed 9.488.

4. ∴ Fail to Reject H_o.

Sets of observed frequencies this far from the expected frequencies would arise randomly more than 5% of the time when the true distribution is uniform. Therefore we cannot conclude, with an α of 0.05, that the distribution is nonuniform.

Notice, though, that your suspicion was not simply that the distribution is nonuniform. Your suspicion was that it was nonuniform in a particular way; your suspicion was that the {Monday, Friday} frequencies were too high relative to the {Tuesday through Thursday} ones. A more focused test, then, would divide the days into these groupings, with expected proportions of 0.40 and 0.60.

1. H_o: The {Mon, Fri}/{Tues–Thurs} distribution is 0.40/0.60.
   H_a: The {Mon, Fri}/{Tues–Thurs} distribution is not 0.40/0.60.
   We now have just two categories, so df = 2 – 1 = 1.

2. Reject H_o if $\chi_c^2 > 3.841$ (df = 1; $\alpha = 0.05$).
   The calculations are shown in Figure 10.8, Panel B. The expected frequencies are no longer equal; they are $0.40 \times 92 = 36.8$ and $0.60 \times 92 = 55.2$.

3. $\chi_c^2 = \sum \frac{(f_o - f_e)^2}{f_e} = 4.712.$

   Our χ_c^2 exceeds 3.841. Thus,

4. ∴ Reject H_o.

The true proportions are not 0.40/0.60.

Finally, recall that you predicted more than just a "difference;" you predicted the direction. Check to make sure the observed frequency for {Monday, Friday} is too high rather than too low. It is. With this more focused test, you do find support for your suspicion.

Consider the following example:

A large batch of mixed nuts is supposed to contain (by count) 30% cashews, 20% Brazil nuts, and 50% peanuts. If a random sample of 400 nuts gives the following results, test the hypothesis that the mixture is correct. Use $\alpha = 0.05$. Use $\alpha = 0.01$. What is the p-value of this test?

Type	Number
Cashews	100
Brazil nuts	75
Peanuts	225

As in the second version of the sick-leave example, we do not expect a uniform distribution, but it is clear what we do expect

1. H_o: The cashew/brazil nut/peanut distribution is 0.30/0.20/0.50.
   H_a: The cashew/brazil nut/peanut distribution is not 0.30/0.20/ 0.50.
   We have three categories, so df = 3 − 1 = 2.
2. Reject H_o if $\chi_c^2 > 5.991$ (df = 2; $\alpha = 0.05$)   Reject H_o $\chi_c^2 > 9.210$ (df = 2; $\alpha = 0.01$).

   The calculations are shown in Figure 10.9. The expected frequencies are $0.30 \times 400 = 120$, $0.20 \times 400 = 80$, and $0.50 \times 400 = 200$.

3. $\chi_c^2 = \sum \frac{(f_o - f_e)^2}{f_e} = 6.771\,.$

4. ∴ Reject H_o ∴ Fail to Reject H_o.

We are able to reject the null hypothesis with an α of 0.05, but not with an α of 0.01. In the first case, either we have found an error in the

	f_o	f_e			
Cashews	100	120	−20	400	3.333
Brazil nuts	75	80	−5	25	0.313
Peanuts	225	200	25	625	3.125
	400	400			6.771

Figure 10.9 Goodness-of-Fit and the χ_c^2 statistic: Another example (b).

mixture or we are making a type I error. In the second case, either the mixture is fine or we are making a type II error.

The *p*-value of the test is = CHIDIST(6.771,2) = 0.0339. If we are willing to accept a probability this high of making a type I error, we should reject the null hypothesis; otherwise we should not.

10.5 A Final Example on Testing for Goodness-of-Fit

I indicated at the beginning of the last section that not all goodness-of-fit tests will have degrees of freedom of df = # of categories – 1, since some require constraining the expected frequencies to be like the observed frequencies in more than just one way—their sum. While such tests are generally beyond the scope of this text, the following example gives a sense of what is involved.

Suppose we wanted to test whether the heights of young adult men are normally distributed. We could take a random sample and create a frequency distribution for the heights of those in our sample. Indeed, the sample of 281 young adults from the National Longitudinal Survey of Youth (NLSY) that we have used several times contains data on the heights of 135 young men. Figure 10.10 shows a frequency distribution and histogram of these heights. (Since heights are measured to the nearest inch, the category boundaries occur at the half inch.)

However, there are an infinite number of normal distributions; one for every possible combination of mean and standard deviation. We need to decide what normal distribution we are going to use for comparison. Perhaps we have a preconceived notion of the mean and standard deviation.

1. H_o: The distribution is normal with $\mu = 81$ and $\sigma = 3$
   H_a: The distribution is not normal with $\mu = 81$ and $\sigma = 3$.

Since μ and σ are specified in the null hypothesis, we would not worry about whether they matched the sample $\overline{X}$ and s. We would just make sure the expected frequencies summed to 135, the sample sum. And degrees of freedom would still be df = # of categories – 1. But notice that this is not a test just of shape; it is a test of mean, standard deviation, and shape. If we ended up rejecting H_o, it might not be because the distribution is nonnormal; it might be because our hypothesized μ and/or σ were wrong.

Figure 10.10 shows two different normal distributions fitted to our histogram. Clearly, the one that peaks around 70.4 fits the data a whole lot better than the one that peaks around 81, and the difference has nothing to do with shape.

Instead, then, we would ordinarily specify our hypotheses as simply

1. H_o: The distribution is normal
   H_a: The distribution is not normal.

The sensible μ and σ to assume are those that match $\overline{X}$ and s for our sample. We would calculate $\overline{X} = 70.415$ and $s = 2.768$ (from the original data not the frequency distribution), find the expected frequencies for a

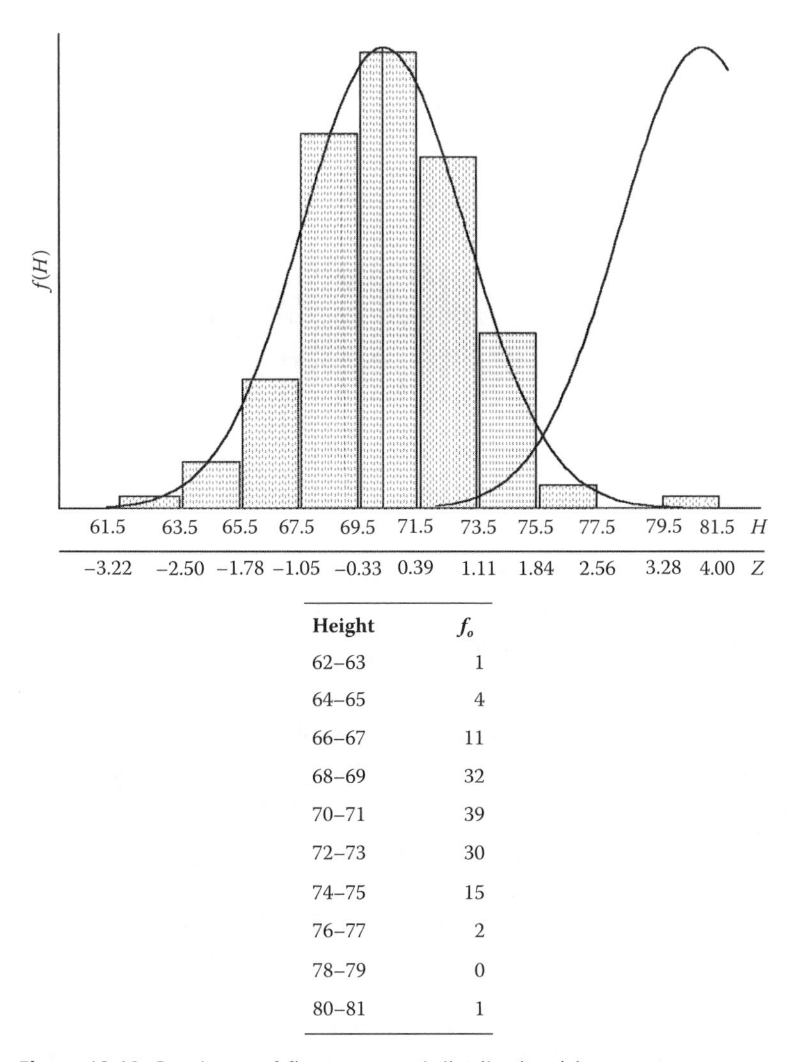

Height	f_o
62–63	1
64–65	4
66–67	11
68–69	32
70–71	39
72–73	30
74–75	15
76–77	2
78–79	0
80–81	1

Figure 10.10 Goodness-of-fit to a normal distribution (a).

normal with the same mean and standard deviation, and compare our observed frequencies to these.

Notice, though, that we have now constrained the expected frequencies to agree with our observed frequencies in three ways that are not part of the null hypothesis—they have the same mean, the same standard deviation, and the same count. Thus, in this case, our degrees of freedom will be df = # of categories – 3.

The rest of this problem, while messier, is not very different from all the previous ones. One difference is that, since the normal distribution is continuous, the number of categories is somewhat arbitrary. I have broken the heights down into two inch increments. This gives 10 observed categories. Since the normal distribution goes to infinity in both directions, it actually gives twp additional expected categories—below 61.5 and above 81.5. However, some of these categories may turn out to be too small. The rule of thumb that expected frequencies should be at least

Panel A

Height	Z	$P(Z)$
61.5	−3.22	0.4994
63.5	−2.50	0.4938
65.5	−1.78	0.4625
67.5	−1.05	0.3531
69.5	−0.33	0.1293
71.5	0.39	0.1517
73.5	1.11	0.3665
75.5	1.84	0.4671
77.5	2.56	0.4948
79.5	3.28	0.4995
81.5	4.00	0.5000

Panel B

Range	Probability	f_e	f_e	f_o			
−61.5	0.0006	0.081					
61.5–63.5	0.0056	0.756					
63.5–65.5	0.0313	4.226	5.063	5	−0.062	0.004	0.001
65.5–67.5	0.1094	14.769	14.769	11	−3.769	14.205	0.962
67.5–69.5	0.2238	30.213	30.213	32	1.787	3.193	0.106
69.5–71.5	0.2810	37.935	37.935	39	1.065	1.134	0.030
71.5–73.5	0.2148	28.998	28.998	30	1.002	1.004	0.035
73.5–75.5	0.1006	13.581	13.581	15	1.419	2.014	0.148
75.5–77.5	0.0277	3.740	4.442	3	−1.442	2.078	0.468
77.5–79.5	0.0047	0.634					
79.5–81.5	0.0005	0.068					
81.5–	0.0000	0.000					
	1.0000	135.000	135.000	135			1.749

Figure 10.11 Goodness-of-fit to a normal distribution (b).

five applies; we may end up having to combine some categories before we are done. Thus, we will need to wait to see how many categories we have left to determine our final number of categories and degrees of freedom.

The calculations are in Figure 10.11. To get the expected frequency in each category, we need to find the normal probability in that category. To do so, we first find the normal probability between each category boundary and the mean. This is what we do in Panel A. We convert each category boundary to its z-value equivalent, by subtracting the mean and dividing by the standard deviation. Looking up these z values in the standard normal table gives us the probability between each boundary and the mean.

Panel B then converts these to the probabilities in each category. To find the probability below 61.5, we note that if 0.4994 is the probability from 61.5 up to the mean, 0.5000 – 0.4994 = 0.0006 must be the probability below 61.5. To find the probability in the 61.5–63.5 range, we note that they are both below the mean. The probabilities from 61.5 and 63.5 up to the mean are 0.4994 and 0.4938, respectively. The probability between them, then, must be 0.4994 – 0.4938 = 0.0056.

We find most of the remaining probabilities in the same way. The 69.5–71.5 range is the exception because it contains the mean. Hence 0.1293 is the probability from 69.5 *up* to the mean and 0.1517 is the probability from 71.5 *down* to the mean; we need to sum these probabilities—0.1293 + 0.1517 = 0.2810—instead of finding their difference.

Notice that the sum of the probabilities is 1.0000, while we want our expected frequencies to equal 135. Thus, the next step is simply to multiply through by 135.

Finally, note that some of these expected frequencies are very small; we want them all to be at least five. Hence, we begin combining ranges until the expected frequency reaches that size. In this case, we need to combine the three shortest and four tallest ranges. Actually, the four tallest still have an expected frequency of only 4.442, but I decided that this was close enough. It would have made very little difference in our result. We are left with seven categories.

We can finally find our degrees of freedom and state our rejection criterion. df = 7 − 3 = 4.

2. Reject H_o if $\chi_c^2 > 9.488$ (df = 4; $\alpha = 0.05$).
   We need to combine observed categories to match the expected ones. The final calculations, then, are the same as for any other chi square.

3. $\chi_c^2 = \sum \frac{(f_o - f_e)^2}{f_e} = 1.749.$

4. ∴ Fail to Reject H_o.

We are unable to reject, with an α of 0.05, the null hypothesis that the heights of young adult men are normally distributed. The *p*-value of the test is 0.7818.

While none of the steps in this example are that difficult, there are a lot of steps and it is easy to get lost. The point in showing it is not that you should be doing something like this by hand. Indeed, the test is less important than it might at first seem. Only our small-sample tests require that a population be normally distributed and this test requires a fairly large sample to detect deviations from normality.

Still, the hypothesis—that a variable fits a normal distribution—is one you can understand. Hopefully you understand, too, how and why we have put additional constraints on the expected frequencies—constraints that cost additional degrees of freedom.

10.6 Exercises

10.1 If four candidates get the following votes in a random poll of 120 voters, can you reject the hypothesis that they are equally popular? Use $\alpha = 0.05$. Use $\alpha = 0.01$. What is the *p*-value of this test?

Adams	33
Baker	17
Clark	40
Davis	30

10.2 a. The producers of Brand X conduct a survey of consumer preferences to see if there is a difference in overall preference among the three major brands. Based on the data below, what can they conclude? Use $\alpha = 0.05$. Use $\alpha = 0.01$. What is the *p*-value of this test?

Brand	**X**	**Y**	**Z**
Number preferring	40	60	80

b. They now initiate an advertising campaign designed to increase their market share. (Naturally, their competitors respond in kind.) A year later they conduct a second survey to see if there has been a change in overall preference among the three major brands. Based on the data above and below, what can they conclude? Use $\alpha = 0.05$. Use $\alpha = 0.01$. What is the p-value of this test?

Brand	X	Y	Z
Number preferring	50	30	40

10.3 In Exercise 9.5 you tested a number of hypotheses about Nickels customers and noncustomers. You tested, for example, to see whether customers and noncustomers differed significantly in the proportion female and in the proportion whose primary source of market information is the newspaper. Return to these data (Nickels1.xls).

a. Test the hypothesis that patronage of Nickels is independent of sex. Use $\alpha = 0.05$. Use $\alpha = 0.01$. What is the p-value of this test?
b. How does this test relate to the test in Chapter 9 for a difference in proportion female?
c. Test the hypothesis that patronage of Nickels is independent of primary source of market information. Use $\alpha = 0.05$. Use $\alpha = 0.01$. What is the p-value of this test?
d. How does this test relate to the test in Chapter 9 for a difference in proportion whose primary source of market information is the newspaper?

10.4 In Exercise 9.7 you tested a number of hypotheses about employees of your firm. You tested, for example, to see whether your female and male employees differed significantly in the proportions employed in management. Return to these data (Employees2.xls).

a. Test the hypothesis that being employed in management is independent of sex. Use $\alpha = 0.05$. Use $\alpha = 0.01$. What is the p-value of this test?
b. How does this test relate to the test in Chapter 9 for a difference between female and male employees in the proportion employed in management?
c. Test the hypothesis that employee job type is independent of sex. Use $\alpha = 0.05$. Use $\alpha = 0.01$. What is the p-value of this test?
d. How does this test relate to the test in Chapter 9 for a difference between female and male employees in the proportion employed in management?
e. A management consultant suggests that, in a normal, well-run manufacturing firm, the proportions of employees in the line, office, and management categories would be 0.75, 0.15, and 0.10. Test the hypothesis that your firm matches this norm. Use $\alpha = 0.05$. Use $\alpha = 0.01$. What is the p-value of this test?

10.5 In Exercise 9.9 you tested a number of hypotheses about students at a large university. You tested, for example, to see if female and male students differed significantly in the proportions majoring in economics. Return to these data (Students2.xls).

a. Test the hypothesis that majoring in economics is independent of sex. Use $\alpha = 0.05$. Use $\alpha = 0.01$. What is the p-value of this test?
b. How does this test relate to the test in Exercise 9.9c for a difference between female and male students in the proportion majoring in economics?
c. Test the hypothesis that major is independent of sex. Use $\alpha = 0.05$. Use $\alpha = 0.01$. What is the p-value of this test?
d. How does this test relate to the test in Exercise 9.9c, for a difference between female and male students in the proportion majoring in economics?
e. Which other tests from Exercise 9.9c can you carry out as χ^2 tests of independence? Confirm that they give the same result.

10.6 Continue with these same data. In Exercise 9.11, you tested to see if athletes and nonathletes differed significantly in the proportions majoring in economics.

a. Test the hypothesis that being majoring in economics is independent of athletic status. Use $\alpha = 0.05$. Use $\alpha = 0.01$. What is the p-value of this test?
b. How does this test relate to the test in Exercise 9.11d for a difference between athletes and nonathletes in the proportion majoring in economics?
c. Test the hypothesis that major is independent of athletic status. Use $\alpha = 0.05$. Use $\alpha = 0.01$. What is the p-value of this test?
d. How does this test relate to the test in Exercise 9.11d for a difference between athletes and nonathletes in the proportion majoring in economics?
e. Which other tests from Exercise 9.11 can you carry out as χ^2 tests of independence? Confirm that they give the same result.

10.7 Continue with these same data. Test the hypothesis that the six majors are equally popular. Use $\alpha = 0.05$. Use $\alpha = 0.01$. What is the p-value of this test?

10.8 A bank audits its four branches, looking at 500 randomly chosen accounts at each. Given the following percentages with errors, test the hypothesis that the branches are equally accurate. Use $\alpha = 0.05$. Use $\alpha = 0.01$. What is the p-value of this test?

Branch	**A**	**B**	**C**	**D**
Percentage Error	1.0%	0.8%	3.0%	2.0%

10.9 Company executives are considering three possible benefit packages.

a. A random sample of office employees shows the following preferences for the three packages. Test the hypothesis that the three packages are equally popular with their office employees. Use $\alpha = 0.05$. Use $\alpha = 0.01$. What is the p-value of this test?

Package	**A**	**B**	**C**
Number preferring	32	17	35

b. A random sample of line employees shows the following preferences for the three packages. Test the hypothesis that the preferences of office and line employees are the same. Use $\alpha = 0.05$. Use $\alpha = 0.01$. What is the p-value of this test?

Package	**A**	**B**	**C**
Number preferring	48	13	55

11

Tests of Hypotheses: ANOVA and Tests of Variances

The main thrust of this chapter parallels that of the last. Chapter 10 introduced a new approach to comparing proportions that gave the same result for the two-sample case, but could be extended to more than two samples or groups. This chapter introduces a new approach to comparing means that gives the same result for the two-sample case, standard deviations unknown but equal, but can be extended to more than two samples or groups.

Our approach will be to develop a measure of the variance between or among samples or groups based on their means that we can then compare to the variance within the samples or groups. The basic notion is that, if the population means are all equal, the between-sample variance should be small in comparison to the variance within the samples. If the between-sample variance is so large that it would arise only rarely when the null hypothesis is true, we will conclude that the null hypothesis is not true and reject it. The approach is called **Analysis of Variance**, or simply **ANOVA**.

The general reasoning is exactly the same as that we have been using in the last three chapters. We will specify null and alternative hypotheses; we will specify a rejection criterion based on the probability we are willing to accept of making a type I error. We will calculate a test statistic. And we will either reject or fail to reject the null hypothesis based on whether our test statistic meets or does not meet our criterion.

What is new is our test statistic, and its sampling distribution. We will calculate a statistic that follows the ***F* distribution**.

The F is a ratio of variances; hence it is good for answering another sort of problem as well. Back in Chapter 9 we had two different versions of the t test, depending on whether or not we could assume that the two unknown population standard deviations or variances were equal. At the time we deferred discussion of a formal test for the equality of standard deviations or variances because it required a new sampling distribution. We will now be able to present that test as well.

11.1 A Difference in Means: An Alternate Approach

11.1.1 The Null and Alternative Hypotheses

In Chapter 9 we tested to see whether there was a difference in the mean GPAs of female and male students. Our hypotheses were

1. H_o: $\mu_{GPA-Female} - \mu_{GPA-Male} = 0$ H_o: $\mu_{GPA-Female} = \mu_{GPA-Male}$
   H_a: $\mu_{GPA-Female} - \mu_{GPA-Male} \neq 0$ H_a: $\mu_{GPA-Female} \neq \mu_{GPA-Male}$.

An alternative way of expressing these hypotheses is

1. H_o: GPA is independent of sex.
   H_a: GPA is not independent of sex.

That is, saying that the true means are equal is the same as saying that knowing a student's sex tells us nothing about his or her GPA. You should recognize this shift in expressing our hypotheses as analogous to the shift we made with proportions in the last chapter. It emphasizes that we are looking to see whether one of these variables—the student's sex—can help to explain the other—the student's GPA. We are looking for a relationship. This second form is also less awkward as we start to have multiple categories instead of just two.

11.1.2 The Decision Criterion

The statistic we will calculate in testing the hypothesis above follows another new sampling distribution, the ***F* distribution**. The *F* distribution is a bit more complicated than those we have dealt with so far. It has two degrees of freedom—the first one associated with the numerator and the second one associated with the denominator. Still, it is related to χ^2 and looks similar. The *F* like χ^2 is nonsymmetrical and since it is a ratio of variances, *F* like χ^2 must be positive.

The variance in the numerator measures variation between or among samples, based on their means; that in the denominator measures variation within the samples. Under the null hypothesis, the expected value for the *F* statistic is one. Of course, due to sampling error it will not be exactly that. We will be looking for evidence that the variation between or among samples is too large; hence, we will be looking for an *F* statistic that is too much greater than one. We will be concerned just with the upper tail. Still, the probability in the upper tail is equivalent to both tails of the *t*. *F* like χ^2 is naturally a test of *differences*.

Like the normal, *t* and χ^2, the *F* is tabulated and the table is widely available. Table 4, in Appendix C, shows how it might be laid out. Find it now. The first degrees of freedom (df_n) associated with the numerator is across the top; the second degrees of freedom (df_d) associated with the denominator is down the side. Often, *F* tables are laid out as a series of subtables on a series of facing pages, one for each value of α. You would first find the subtable for the α you wanted and then you would find your combination of df_n and df_d.

	Formula	Example
From F to a probability:	=FDIST(F,df$_n$,df$_d$)	=FDIST(2.922,3,30) = 0.05
From a probability to F:	=FINV(p,df$_n$,df$_d$)	=FINV(0.05,3,30) = 2.922
(For Lotus/Quattro Pro, replace " = " with "@"		

Figure 11.1 The F distribution special functions in common spreadsheet programs.

Instead, I have organized it as a single table that continues on for four sets of facing pages. Each set of facing pages contains a full set of columns and five different values for α; successive sets of facing pages just continue those columns. For example, suppose you had (3,30) degrees of freedom. You would read across until you found the column for 3, and read down (turning the page twice) to find the row for 30. There, you find criteria for five different values for α. If you were using an α of 0.05, your criterion would be 2.922.

As with the normal, t, and χ^2, there are also special spreadsheet functions. Figure 11.1 shows the special functions for several of the most common spreadsheets.

Finally, the two degrees of freedom are calculated as follows:

$$df_n = \text{\# of samples} - 1 = 2 - 1 = 1$$
$$df_d = \text{combined sample } n - \text{\# of samples} = 50 - 2 = 48.$$

The denominator degrees of freedom should look familiar; it was the degrees of freedom for our two-sample t test with standard deviations unknown but equal. Indeed, the measure of variance within samples that we will use in the denominator of F_c is just the square of our pooled standard deviation measure in that t test. In essence, we have $n - 1$ degrees of freedom for each sample.

The numerator degrees of freedom requires a little justification. The measure of variance between or among samples that we will use in the numerator of F_c, treats each sample in essence as a single value, $\bar{X}$. Hence, for the numerator "# of samples – 1" is in this sense "n – 1."

Suppose we choose an α of 0.05. The F table in the back of the book jumps from {1,45} to {1,50} degrees of freedom. We could use either 4.057 or 4.034; since the F changes so slowly it is unlikely to affect our decision. Or we could use the special spreadsheet function =FINV(0.05,1,48) = 4.043.

2. Reject H_0 if $F_c > 4.043$ ($df_n = 1$; $df_d = 48$; $\alpha = 0.05$).

11.1.3 The Calculations

For the basic intuition of what we are going to do, consider the two panels of Figure 11.2. Each has the six numbers from three to eight arranged in two samples. In Panel A, the two samples are relatively spread out. The sample averages of five and six are not very different, given this much variation within the samples. This pattern would be consistent with our

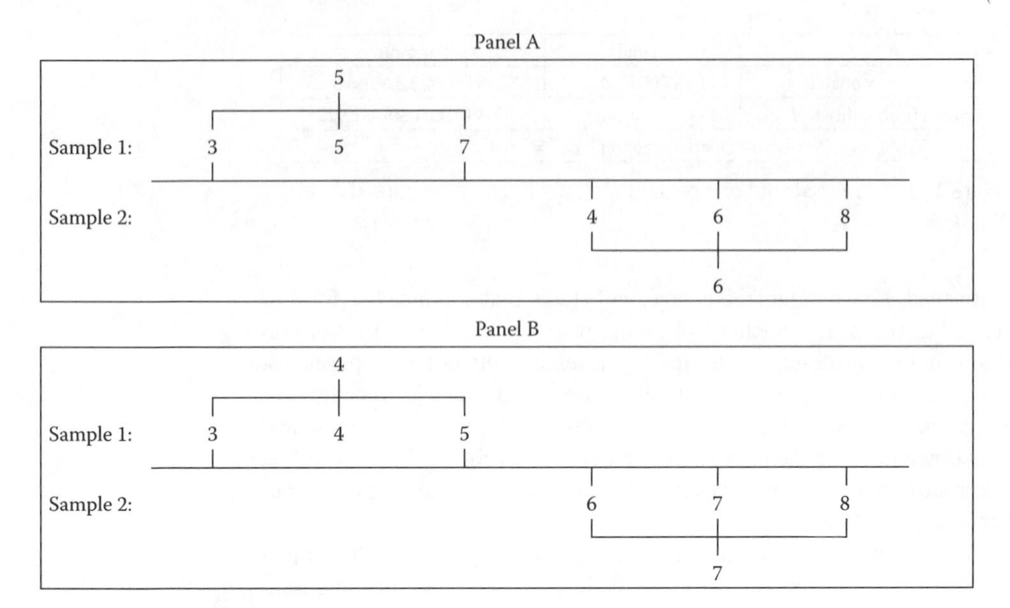

Figure 11.2 ANOVA: The basic intuition.

null hypothesis; these samples could easily be two samples from the same population. In Panel B, by contrast, the two samples are rather tightly grouped. And the sample averages of four and seven are quite different given so little variation within the samples. This pattern seems at odds with our null hypothesis; these samples are so different that they are unlikely to be two samples from the same population.

Figure 11.3 summarizes the formulas. You should recognize the denominator as the pooled variance; if you do not, turn back to Figure 9.8 (page 216) where we calculated the pooled standard deviation—a sort of weighted average of the two sample standard deviations. Since F is a ratio of variances, we do not take the square root as we did then. I have written the formula for k samples, not just two, because this approach will work for more than just two samples.

The numerator is new. We calculate an overall, pooled average, $\overline{X}_p$. We find the deviations of each of the sample means from this overall mean. Some will be positive and some will be negative; as you should expect by this point in the course, we square them to make them all positive. Each of these squared deviations represents a sample of size n_j, so we weight each by its sample size. And we sum.

We are now able to use our sample of 30 women and 20 men to test the hypothesis that mean GPA is independent of sex. Retrieving the summary data from Chapter 9:

$$\overline{X}_F = 2.9076 \qquad \overline{X}_M = 2.9634$$

3. $$s_F = 0.5277 \qquad s_M = 0.6470$$

$$n_F = 30 \qquad n_M = 20.$$

The pooled mean $\overline{X}_p = (2.9076 \times 30 + 2.9634 \times 20)/50 = 2.9299.$

$$F_c = \frac{\dfrac{\text{Between-Sample Variation}}{\text{\# of Samples} - 1}}{\dfrac{\text{Within-Sample Variation}}{\text{Combined } n - \text{\# of Samples}}} = \frac{\dfrac{\sum_{j=1}^{j=k} n_j(\bar{X}_j - \bar{X}_p)^2}{k-1}}{\dfrac{\sum_{j=1}^{j=k}\sum_{i=1}^{i=n_j}(X_{ij} - \bar{X}_j)^2}{\sum_{j=1}^{j=k} n_j - k}} = \frac{\dfrac{\sum_{j=1}^{j=k} n_j(\bar{X}_j - \bar{X}_p)^2}{k-1}}{\dfrac{\sum_{j=1}^{j=k} s_j^2(n_j - 1)}{\sum_{j=1}^{j=k} n_j - k}}$$

where k = # of samples.

Figure 11.3 ANOVA: The formulas.

In finding the between-sample variation, a worksheet helps:

Sample	$\bar{X}$	$\bar{X} - \bar{X}_p$	$(\bar{X} - \bar{X}_p)^2$	$n(\bar{X} - \bar{X}_p)^2$
Women	2.9076	–0.0223	0.0005	0.0149
Men	2.9634	0.0335	0.0011	0.0224
				0.0373

— numerator — $\frac{0.0373}{k-1} = \frac{0.0373}{2-1}$

Finding the within-sample variation

Sample		
Women	$s^2(n-1) = 0.5277^2(30-1) =$	8.0754
Men	$s^2(n-1) = 0.6470^2(20-1) =$	7.9527
		16.0282

— denominator — $\frac{16.0282}{n-k} = \frac{16.0282}{50-2}$

Finally, putting it all together:

$$F_c = \frac{\dfrac{\sum_{j=1}^{j=k} n_j\left(\bar{X}_j - \bar{X}_p\right)^2}{k-1}}{\dfrac{\sum_{j=1}^{j=k} s_j^2\left(n_j - 1\right)}{\sum_{j=1}^{j=k} n_j - k}} = \frac{\dfrac{0.0373}{2-1}}{\dfrac{16.0282}{50-2}} = \frac{0.0373}{0.3339} = 0.1118.$$

11.1.4 The Conclusion

The F_c is not greater than 4.043.

4. ∴ Fail to Reject H_0.

As far as we can tell, mean GPA is independent of sex. This is the same conclusion we reached with the two-sample t test with σs unknown but equal. Indeed, the two tests are mathematically equivalent. F with df = {1,ν} is just the square of t with df = {ν}. The strength of this test is that it can deal with explanatory variables that have more than two categories.

11.2 ANOVA with Several Categories

Suppose, instead of sex, we think mean GPA may depend on major. Our sample has students' majors grouped into six categories. The *t* test, based as it is on a difference, no longer works; we cannot find a difference between six things. However the *F* test generalizes easily.

1. H_o: GPA is independent of major
   H_a: GPA is not independent of major.

Since our combined sample of 50 students is grouped into $k = 6$ major groups, our degrees of freedom are $df_n = 6 - 1 = 5$ and $df_d = 50 - 6 = 44$. Since {5,44} is not in the table at the back, we would use {5,45}; or we could use the special spreadsheet function. If we choose an α of 0.05, =FINV(0.05,5,44) = 2.427.

2. Reject H_o if $F_c > 2.427$ ($df_n = 5$, $df_d = 44$; $\alpha = 0.05$).

Summary statistics by major are as follows:

Group	Natural Science	Social Science	Humanities	Fine Arts	Business	Nursing	Overall
$\bar{X}$	2.9208	3.0863	2.8460	3.0536	2.8456	2.8441	2.9299
s	0.5774	0.5155	0.5645	0.5335	0.6909	0.6504	
n	8	9	9	7	10	7	50

3. The overall mean could be calculated as the weighted average of the group means

$$\bar{X}_p = \frac{2.9208 \times 8 + 3.0863 \times 9 + 2.8460 \times 9 + \ldots + 2.8441 \times 7}{50} = 2.9299,$$

though since we have the original data in a spreadsheet, it would be easier just to calculate it directly as the mean of the 50 GPAs.

Finding the between-group variation

Group	$\bar{X}$	$\bar{X} - \bar{X}_p$	$(\bar{X} - \bar{X}_p)^2$	$n(\bar{X} - \bar{X}_p)^2$
Natural Science	2.9208	−0.0092	0.0001	0.0007
Social Science	3.0863	0.1564	0.0245	0.2201
Humanities	2.8460	−0.0839	0.0070	0.0634
Fine Arts	3.0536	0.1236	0.0153	0.1071
Business	2.8456	−0.0843	0.0071	0.0711
Nursing	2.8441	−0.0858	0.0074	0.0515
				0.5139

Finding the within-group variation

Group		
Natural Science	$s^2(n-1) = 0.5774^2(8-1) =$	2.3338
Social Science	$s^2(n-1) = 0.5155^2(9-1) =$	2.1263
Humanities	$s^2(n-1) = 0.5645^2(9-1) =$	2.5492
Fine Arts	$s^2(n-1) = 0.5335^2(7-1) =$	1.7080
Business	$s^2(n-1) = 0.6909^2(10-1) =$	4.2963
Nursing	$s^2(n-1) = 0.6504^2(7-1) =$	2.5379
		15.5516

Putting it all together

$$F_c = \frac{\dfrac{\sum_{j=1}^{j=k} n_j\left(\bar{X}_j - \bar{X}_p\right)^2}{k-1}}{\dfrac{\sum_{j=1}^{j=k} s_j^2\left(n_j - 1\right)}{\sum_{j=1}^{j=k} n_j - k}} = \frac{\dfrac{0.5139}{6-1}}{\dfrac{15.5516}{50-6}} = \frac{0.1028}{0.3534} = 0.2908.$$

Our F_c is not greater than 2.427. Hence, our conclusion is

4. ∴ Fail to Reject H_o.

These observed group means are not different enough to reject the null hypothesis. The p-value of the test is =FDIST(0.2908,5,44) = 0.9155. We would almost always get an F_c statistic this large, just randomly, even when the null hypothesis is true.

Consider the following example:

A consumer research organization tested random samples of three automobile models, to see if they differ in their mean highway gas mileage. Based on the following sample statistics, what can you conclude? Use $\alpha = 0.05$. Use $\alpha = 0.01$. What is the p-value for this test?

	Model A	Model T	Model Z
$\bar{X}$	26.5	24.5	29.0
s	2.3805	2.4290	2.8284
n	4	6	7

1. H_o: Mileage is independent of model
   H_a: Mileage is not independent of model.
   Our degrees of freedom are $df_n = 3 - 1 = 2$ and $df_d = 17 - 3 = 14$.
2. Reject H_o if $F_c > 3.739$ ($df_n = 2$, $df_d = 14$; $\alpha = 0.05$)   Reject H_o if $F_c > 6.515$ ($df_n = 2$, $df_d = 14$; $\alpha = 0.01$).
3. The pooled mean is $\bar{X}_p = (26.5 \times 4 + 24.5 \times 6 + 29.0 \times 7)/17 = 26.8235$.

Finding the between-group variation

Model	$\bar{X}$	$\bar{X} - \bar{X}_p$	$(\bar{X} - \bar{X}_p)^2$	$n(\bar{X} - \bar{X}_p)^2$
Model A	26.5	−0.3235	0.1047	0.4187
Model T	24.5	−2.3235	5.3988	32.3927
Model Z	29.0	2.1765	4.7370	33.1592
				65.9706

Finding the within-group variation

Model		
Model A	$s^2(n-1) = 2.3805^2(4-1) =$	17.0003
Model T	$s^2(n-1) = 2.4290^2(6-1) =$	29.5002
Model Z	$s^2(n-1) = 2.8284^2(7-1) =$	47.9991
		94.4996

Finally, putting it all together

$$F_c = \frac{\dfrac{\sum_{j=1}^{j=k} n_j(\bar{X}_j - \bar{X}_p)^2}{k-1}}{\dfrac{\sum_{j=1}^{j=k} s_j^2(n_j-1)}{\sum_{j=1}^{j=k} n_j - k}} = \frac{\dfrac{65.9706}{3-1}}{\dfrac{94.4996}{17-3}} = \frac{32.9853}{6.7500} = 4.8867.$$

4. ∴ Reject H_o ∴ Fail to Reject H_o.

We can reject the null hypothesis with an α of 0.05 but not with an α of 0.01. The *p*-value of the test is =FDIST(4.8867,2,14) = 0.0246.

Consider the following example:

As marketing director for your company, you are considering new packaging for your product. To test the appeal of various alternatives, you place each in several stores that you consider equally favorable locations and record their sales (in units). Based on the following sample statistics, what can you conclude? Use $\alpha = 0.05$. Use $\alpha = 0.01$. What is the *p*-value for this test?

	Current Packaging	New Version A	New Version B	New Version C
$\bar{X}$	186.0	192.0	145.5	154.0
s	22.521	8.367	12.152	18.908
n	6	5	4	5

1. H_o: Sales are independent of packaging.
   H_a: Sales are not independent of packaging.

Our degrees of freedom are $df_n = 4 - 1 = 3$ and $df_d = 20 - 4 = 16$. Hence,

2. Reject H_o if $F_c > 3.239$ ($df_n = 3$, $df_d = 16$; $\alpha = 0.05$)   Reject H_o if $F_c > 5.292$ ($df_n = 3$, $df_d = 16$; $\alpha = 0.01$).

3. The pooled mean is

$$\bar{X}_p = \frac{186.0 \times 6 + 192.0 \times 5 + 145.5 \times 4 + 154.0 \times 5}{20} = 171.4$$

Finding the between-group variation

Version	$\bar{X}$	$\bar{X} - \bar{X}_p$	$(\bar{X} - \bar{X}_p)^2$	$n(\bar{X} - \bar{X}_p)^2$
Current	186.0	14.600	213.160	1278.960
New A	192.0	20.600	424.360	2121.800
New B	145.5	–25.900	670.810	2683.240
New C	154.0	–17.400	302.760	1513.800
				7597.800

Finding the within-group variation

Version		
Current	$s^2(n-1) = 22.521^2(6-1) =$	2535.977
New A	$s^2(n-1) = 8.367^2(5-1) =$	280.027
New B	$s^2(n-1) = 12.152^2(4-1) =$	443.013
New C	$s^2(n-1) = 18.908^2(5-1) =$	1430.050
		4689.067

Finally, putting it all together

$$F_c = \frac{\dfrac{\sum_{j=1}^{j=k} n_j(\bar{X}_j - \bar{X}_p)^2}{k-1}}{\dfrac{\sum_{j=1}^{j=k} s_j^2(n_j - 1)}{\sum_{j=1}^{j=k} n_j - k}} = \frac{\dfrac{7597.800}{4-1}}{\dfrac{4689.067}{20-4}} = \frac{2532.6}{293.067} = 8.642.$$

4. $\therefore$ Reject H_o $\therefore$ Reject H_o.

We can reject the null hypothesis with an α of less than 0.01. The p-value of the test is =FDIST(8.642,3,16) = 0.0012. Sales are almost certainly dependent on packaging.

11.3 A Final Word on ANOVA

As we have seen, we can use ANOVA and the F statistic to test whether there is a relationship between a numerical and a categorical variable. This test is equivalent to the test for a difference in means, standard deviations unknown but equal, for the case in which the categorical variable has just two categories. Its advantage is that it can be extended to cases in which the categorical variable has several categories.

ANOVA does share some of the limitations of contingency tables when it comes to characterizing the nature of the relationship. For example, look back at the last two examples. We found that car model affected gas mileage, and we found that packaging affected sales. Beyond that, how might we describe the relationships? We might note that Model T seems to have the worst mileage and Model Z the best, with Model A somewhere in the middle. But we do not know that each of these differences is significant; only that together they add up to a pattern that we should not have seen according to the null hypothesis. Likewise, we might note that only New Version A seems to hold any promise of being more effective than the Current packaging. But we have not shown that the difference between New Version A and the Current packaging is significant, only that the whole set of four possibilities are too different to accord with our null hypothesis.

I noted in the last chapter that it would be nice to have some relative measure of how well we have done in explaining our dependent variable. That is, how much better do we understand the variation in gas mileage taking model into account? Or how much better do we understand the variation in our sales taking packaging into account? The F statistic is not such a measure, but ANOVA does provide one. It is **R^2, the Coefficient of Determination**. It is simply the proportion of the total variation that our model can explain. Moreover, we can consider the F test as a test of whether this R^2 is significantly greater than zero. Rejecting H_o means that our R^2 is too high to be just randomly greater than zero.

The between-sample variation and within-sample variation add up to the total variation in the combined samples.

$$\begin{array}{ccccc} \text{Between-Sample} & & \text{Within-Sample} & & \text{Total} \\ & + & & = & \\ \text{Variation} & & \text{Variation} & & \text{Variation} \end{array}$$

$$\sum_{j=1}^{j=k} n_j(\bar{X}_j - \bar{X}_p)^2 + \sum_{j=1}^{j=k}\sum_{i=1}^{i=n_j}(X_{ij} - \bar{X}_j)^2 = \sum_{j=1}^{j=k}\sum_{i=1}^{i=n_j}(X_{ij} - \bar{X}_p)^2.$$

For the automobile mileage example:

$$65.9706 + 94.4996 = 160.4702.$$

That is, if we had lumped together the mileages of all 17 automobiles, regardless of model, the sum of the squared deviations around their pooled mean would have equaled 160.4702. This is a measure of the difference in mileage among cars that we would like to explain. Why do some of these cars get better mileage than others? Dividing the cars by model, we find that 65.9706 can be explained as differences among models. Taking model into account, we have explained $R^2 = 65.9706/160.4702 = 0.4111$, or 41.11% of the variation in mileage. And according to our earlier F test, this is significantly greater than zero with a p-value = 0.0246.

For the sales/packaging example:

$$7597.800 + 4689.067 = 12286.867.$$

If we had lumped together the sales of our product at all 20 stores, regardless of packaging, the sum of the squared deviations around their pooled mean would have equaled 12,286.867. This is a measure of the difference in sales of our product among stores that we would like to explain. Why do some stores have greater sales than others? Dividing the stores according to the packaging type displayed there, we find that 7,597.800 can be explained as differences among packaging types. Taking packaging into account, we have explained $R^2 = 7{,}597.800/12{,}286.867 = 0.6184$, or 61.84% of the variation in sales. And according to our earlier F test, this is significantly greater than zero with a p value of 0.0012.

Finally, you should be aware that there is a great deal more to ANOVA; indeed, we have dealt only with the simplest possible case. Common complications involve multiple explanatory variables. In the automobile mileage example, the cars might be categorized in more than one dimension. Perhaps some have large engines and some do not; perhaps some are designed for off-road use and some are not; perhaps some have air conditioning and some do not. We could ask which of these characteristics help explain the variation in automobile mileage and how much of the explained variation can be attributed to each.

Interest in such questions often leads to questions of experimental design. If you want to test the effects on crop yield of fertilizer use, pesticide use, and hybrid variety, you can set up various agricultural plots with various combinations of these characteristics. How would you decide on the distribution of these characteristics among plots? These complications are beyond the scope of this text. However you should know they exist.

11.4 A Difference in Population Variances

Since F is a ratio of variances, we can also use it to test another sort of hypothesis—the hypothesis that two standard deviations or variances are equal. Such a test can be useful in at least two situations.

First, as you know, both the two-sample t test, standard deviations unknown but equal, and the comparable F test, make the assumption that the population standard deviations or variances are equal. While this is often a reasonable assumption, there are instances in which it is not; indeed, we looked at an alternative t test for such cases. It is helpful, then, to have a formal test, especially when the sample standard deviations or variances seem rather different.

Second, the standard deviation or variance is sometimes of interest for its own sake. We have looked at quality-control problems in which we were looking for parts to be the correct size on average. Hopefully it occurred to you that it might not be good enough that a piece is the correct size, on average, if the standard deviation or variance is so large that many individual parts are too large or too small. Other things equal, a process with a smaller variance is better.

11.4.1 The Null and Alternative Hypotheses

In Chapter 9 and then again in this chapter we used our sample of 30 female and 20 male students to test whether the population mean GPAs of female and male students were equal. In doing so, we assumed that the population standard deviations were equal. We now test that assumption.

1. $H_o: \sigma_F = \sigma_M$ $\quad H_o: \sigma_F^2 = \sigma_M^2$

   $H_a: \sigma_F \neq \sigma_M$ $\quad H_a: \sigma_F^2 \neq \sigma_M^2$.

I have written these both in terms of standard deviations and in terms of variances; clearly, if the σs are equal, so are the σ^2s. I have written them as two-tailed tests, since there is no obvious reason to have expected a particular sex to have the larger standard deviation. (Recall that you cannot decide on a one-tailed test by looking at the data.) However, this particular F test can be one- or two-tailed.

11.4.2 The Decision Criterion

The statistic we will calculate in testing the hypothesis above is again an F statistic. In this case it is not nearly as complicated. It is just the ratio of the two sample variances. Again, under the null hypothesis, the expected value of this statistic is one. That is, under the null hypothesis, s_F^2 and s_M^2 are both estimates of the same thing; they should be similar in size; their ratio should be approximately one. If the ratio we get is too much greater or too much less than one, we will reject the null hypothesis.

This test is two-tailed in the same sense that the normal and t were two-tailed. Recall for the t the statistic could be too far from zero, positive, or negative. However, the t table included just the positive tail. Thus, for a two-tailed test with an α of 0.05, we looked up 0.025 for the positive tail and then included the negative tail as well by taking the absolute value of t_c. For the F the statistic cannot be negative, but it can be too far from one, high, or low. Again, the F table includes just the upper tail. Thus, for a two-tailed test with an α of 0.05, we look up 0.025 for the upper tail and then include the lower tail by always putting the larger sample variance on top.

Notice that we cannot specify our two-tailed criterion without looking back at our data to see which sample variance will end up on top, since this will determine the order of the degrees of freedom. The summary statistics are repeated below for convenience. The sample standard deviation (hence the sample variance) is greater for men. Hence, the degrees of freedom are $df_n = 20 - 1 = 19$; and $df_d = 30 - 1 = 29$. Since {19, 29} is not in the table in the back, we would use {20, 29}; or we could use the spreadsheet function =FINV(0.025,19,29) = 2.231.

2. Reject H_o if $F_c > 2.231$ ($df_n = 19$, $df_d = 29$; $\alpha = 0.05$).

What if our test had been a one-tailed test? Again, the analogy to the t may help. For a one-tailed t test with an α of 0.05, we looked up 0.05 in the t table *and also specified the sign, based on the alternative hypothesis.* And, of course, we did not reject the null hypothesis if we got the wrong sign, no matter how large t_c was. For a one-tailed F test with an α of 0.05, we would look up 0.05 in the F table *and also specify which sample variance goes in the numerator, based on the alternative hypothesis of which should be larger.* And we would not reject the null hypothesis if that one turned out smaller instead, no matter how much smaller it was.

11.4.3 The Calculations

Retrieving the summary data from earlier

3. $\bar{X}_F = 2.9076 \qquad \bar{X}_M = 2.9634$

$s_F = 0.5277 \qquad s_M = 0.6470$

$n_F = 30 \qquad n_M = 20,$

$$F_c = \frac{s_M^2}{s_F^2} = \frac{0.6470^2}{0.5277^2} = 1.5031\,.$$

Note that, though we can think of this as a test of standard deviations or variances, the F_c statistic is always a ratio of *variances.*

11.4.4 The Conclusion

The F_c is not greater than 2.231.

4. ∴ Fail to Reject H_o.

As far as we can tell, there is no difference in the standard deviations or variances. The p-value of the test is =FDIST(1.5031,19,29) × 2 = 0.1572 × 2 = 0.3145. Note that, since the spreadsheet function like the table includes only the upper tail, we need to double the probability it returns. This is just the reverse of what we did initially when we divided our α in half to look it up.

Consider the following example:

> In Section 9.6 you were given information on patrons and nonpatrons of Walton's department store and asked whether you conclude that their mean incomes differed. The summary statistics are repeated below for convenience. Noting the rather large difference in sample standard deviations, you decided that you must assume that the population standard deviations are unequal. Now that you have a formal test, were you right? Use $\alpha = 0.05$. Use $\alpha = 0.01$. What is the p-value of this test?

	Patrons	Nonpatrons
Mean income (in $ 1000s)	58.7	50.4
Standard deviation (in $1000s)	16.8	9.8
Number	27	33

Nothing in the problem suggests a one-tailed alternative. Hence,

1. H_o. $\sigma_{\text{patron}} = \sigma_{\text{Nonpatron}}$
   H_a: $\sigma_{\text{patron}} \neq \sigma_{\text{Nonpatron}}$.

Since this is a two-tailed test, we divide α in half to look it up. Since s^2_{Patrons} is larger, the F_c ratio will be $s^2_{\text{Patrons}} / s^2_{\text{Nonpatrons}}$. Thus, degrees of freedom are {26,32}. Since {26,32} is not in the table in the back we would use {24,32}; or we could use the spreadsheet function =FINV(0.025,26,32) = 2.080 and =FINV(0.005,26,32) = 2.632.

2. Reject H_o if $F_c > 2.080$ $(\text{df}_n = 26, \text{df}_d = 32; \alpha = 0.05)$ Reject H_o if $F_c > 2.632$ $(\text{df}_n = 26, \text{df}_d = 32; \alpha = 0.01)$

3. $$F_c = \frac{s^2_M}{s^2_F} = \frac{16.8^2}{9.8^2} = 2.9388\cdot$$

4. ∴ Reject H_o ∴ Reject H_o.

The p-value of the test is =FDIST(2.9388,26,32) × 2 = 0.0021 × 2 = 0.0042. There is very little chance of a difference this large arising just randomly when the population values are the same. Hence, we were right to assume unequal standard deviations.

Consider the following example:

Your current supplier provides you with parts of a required size, on average; because of random variation from part to part, though, some of the parts are too large or small. A potential new supplier claims that it can provide you with parts with a smaller standard deviation in size. You examine random samples of parts from each supplier. Based on the results, below, can you conclude that the claim is true? Use $\alpha = 0.05$. Use $\alpha = 0.01$. What is the p-value of this test?

$$s_{\text{Current}} = 0.21 \text{ inches} \qquad s_{\text{New}} = 0.15 \text{ inches}$$

$$n_{\text{Current}} = 25 \text{ parts} \qquad n_{\text{New}} = 30 \text{ parts}$$

This is a one-tailed claim. And to conclude that the claim is true, it needs to be the alternative hypothesis. Hence,

1. H_o: $\sigma_{\text{New}} \geq \sigma_{\text{Current}}$
   H_a: $\sigma_{\text{New}} < \sigma_{\text{Current}}$.

The F_c ratio will be $s^2_{\text{Curremt}} / s^2_{\text{New}}$, not because $s^2_{Current}$ is larger, but because of the one-tailed alternative. Only if the ratio is large enough, this way around, will we have evidence in favor of this alternative. Thus, the degrees of freedom are {24,29}. And, because this is a one-tailed test, we do *not* divide α in half in looking up our criterion.

2. Reject H_o if $F_c > 1.901$ $(\text{df}_n = 24, \text{df}_d = 29; \alpha = 0.05)$ Reject H_o if $F_c > 2.495$ $(\text{df}_n = 24, \text{df}_d = 29; \alpha = 0.01)$

3. $$F_c = \frac{s^2_{Current}}{s^2_{New}} = \frac{0.21^2}{0.15^2} = 1.960\cdot$$

4. ∴ Reject H_o ∴ Fail to Reject H_o.

We can reject the null hypothesis—and conclude that the claim is true—with an α of 0.05 but not with an α of 0.01. The p-value of the test is =FDIST(1.960,24,29) = 0.0425.

11.5 Exercises

11.1 In an example in Chapter 9, we decided that a higher expected return (mean) on a stock might be compensation for higher risk (standard deviation). If so, we might not want to assume that risks (standard deviations) are equal. The table below repeats the summary statistics. Based on these statistics, is there a significant difference in the risk of these stocks? Use $\alpha = 0.10$. Use $\alpha = 0.05$. What is the p-value of this test?

	ABC Inc.	XYZ Corp.
Expected return (mean)	9.50	6.40
Risk (standard deviation)	3.90	2.60
Number of periods	10	14

11.2 In Exercise 9.3 you tested hypotheses about the heights of young adults. Return to these data (NLSY1.xls).

a. Test the hypothesis that the heights of young adults are independent of sex. Assume population standard deviations are equal. Use $\alpha = 0.05$. Use $\alpha = 0.01$. What is the p-value of this test? What proportion of the variation in heights can be explained by sex?

b. How does this test relate to the test in Chapter 9?

c. In both these tests, you assumed population standard deviations are equal. Test this assumption. Use $\alpha = 0.05$. Use $\alpha = 0.01$. What is the p-value of this test?

11.3 Continue with these same data.

a. Test the hypothesis that the weights of young adults are independent of sex. Assume population standard deviations are equal. Use $\alpha = 0.05$. Use $\alpha = 0.01$. What is the p-value of this test? What proportion of the variation in weights can be explained by sex?

b. How does this test relate to the test in Chapter 9?

c. In both these tests you assumed population standard deviations are equal. Test this assumption. Use $\alpha = 0.05$. Use $\alpha = 0.01$. What is the p-value of this test?

11.4 In Exercise 9.5 you tested a number of hypotheses about Nickels customers and noncustomers. You tested, for example, whether customers and noncustomers differed significantly in average age and income. Return to these data (Nickels1.xls).

a. Can you conclude, using the techniques of this chapter, that customers and noncustomers differ in their average age? Assume population standard deviations are equal. Use $\alpha = 0.05$. Use $\alpha = 0.01$. What is the p-value of this test? What proportion of the variation in ages can be explained by their patronage?

b. Using the techniques of this chapter, can you conclude that customers and noncustomers differ in their average income? Assume population standard deviations are equal. Use $\alpha = 0.05$. Use $\alpha = 0.01$. What is the p-value of this test? What proportion of the variation in incomes can be explained by their patronage?

c. How do these tests relate to the comparable tests in Chapter 9?

d. In all these tests, you assumed population standard deviations are equal. Test this assumption. Use $\alpha = 0.05$. Use $\alpha = 0.01$. What is the p-value of this test?

11.5 Continue with these same data.

a. Can you conclude that consumers who get their market information from different sources differ in age? Assume population standard deviations are equal. Use $\alpha = 0.05$. Use $\alpha = 0.01$. What is the p-value of this test? What proportion of the variation in ages can be explained by information source?

b. Can you conclude that consumers who get their market information from different sources differ in income? Assume population standard deviations are equal. Use $\alpha = 0.05$. Use $\alpha = 0.01$. What is the p-value of this test? What proportion of the variation in incomes can be explained by information source?

11.6 In Exercise 9.7 you tested a number of hypotheses about employees of your firm. Return to these data (Employees2.xls).

a. Test the hypothesis that your employees' salaries are independent of sex. Assume population standard deviations are equal. Use $\alpha = 0.10$. Use $\alpha = 0.05$. What is the p-value of this test? What proportion of the variation in salaries can be explained by sex?

b. How does this test relate to the comparable test in Chapter 9?

c. In both these tests, you assumed population standard deviations are equal. Test this assumption. Use $\alpha = 0.10$. Use $\alpha = 0.05$. What is the p-value of this test?

11.7 Continue with these same data. Test the hypothesis that your employees' salaries are independent of job type? Assume population standard deviations are equal. Use $\alpha = 0.10$. Use $\alpha = 0.05$. What is the p-value of this test? What proportion of the variation in salaries can be explained by job type?

11.8 In Exercise 9.9 you tested a number of hypotheses about students at a large university. Return to these data (Students2.xls). Can you conclude, using the techniques of this chapter, that male and female students differ with respect to each of the following? Assume population standard deviations are equal. Use $\alpha = 0.05$. Use $\alpha = 0.01$. For each, what is the p-value of this test? For each, what proportion of the variation can be explained by sex?

a. Mean height
b. Mean weight
c. Mean financial aid
d. Mean entertainment expenditure
e. Mean study time
f. Mean college GPA

11.9 Continue with these same data. In each of these tests you assumed population standard deviations are equal. Test this assumption for each. Use $\alpha = 0.05$. Use $\alpha = 0.01$. What is the p-value of each test?

11.10 Continue with these same data.
Can you conclude that students in different majors differ with respect to each of the following? Assume population standard deviations are equal. Use $\alpha = 0.05$. Use $\alpha = 0.01$. For each, what is the p-value of this test? For each, what proportion of the variation can be explained by major?

a. Mean financial aid
b. Mean study time
c. Mean college GPA

12

Simple Regression and Correlation

In the last three chapters, we have explored relationships between variables. We have asked whether sex could help explain a student's choice of major, or whether choice of major could help explain a student's GPA. We have tested whether sex or job type could help explain an employee's salary. In each case, we started with a null hypothesis that the variables were independent; that one could not help explain the other. We calculated a statistic—Z_c, t_c, χ_c^2, or F_c—based on our sample values of these variables. And then, if our statistic was so large that it was unlikely enough, given our null hypothesis, we rejected our null hypothesis and concluded that the variables were not independent; that one could in fact help explain the other.

In these examples, the dependent variable—the one we were trying to explain—could be categorical or numerical. If categorical, we calculated a Z_c or χ_c^2 statistic; if numerical, we calculated a t_c or F_c statistic. However, the explanatory variable was always categorical. This was because we always calculated proportions, means, counts, or squared deviations within the categories. However, there are many cases in which our explanatory variable might be numerical. Certainly, course load or hours spent studying might help explain a student's GPA. And years of education or experience might help explain an employee's salary.

In this chapter we look at such relationships. We restrict ourselves to linear relationships between two numeric variables for now. This is the sense in which the topic is "simple" regression. Chapter 13, then, will look at relationships that involve more than two variables, or that are nonlinear.

We actually discussed numerical explanatory variables back in Chapters 2 and 3. In Section 2.2.2 we used scattergrams to show a relationship between numerical variables graphically. In Section 3.4.2 we estimated the equation of the line relating them. Since we have been away from this material for a while, Section 12.2, below, offers a fairly complete review. First though, Section 12.1 gives a further introduction to our goals and assumptions. Chapters 2 and 3 dealt with description; now we are interested in inference about a population relationship.

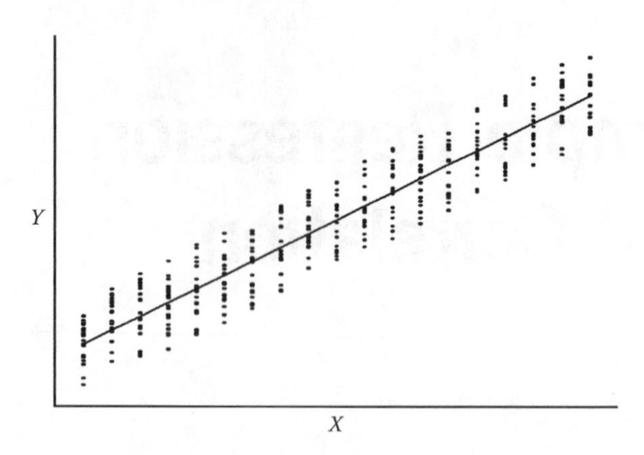

Figure 12.1 The population regression line (a).

12.1 The Population Regression Line

When we estimate a sample regression line between two variables, it is because we have reason to think that there really is a relationship between these variables in the population. This population regression line is assumed to look something like that in Figure 12.1, above. We can characterize this assumed relationship in a couple of ways, as suggested in Figure 12.2, below. Version 1 emphasizes that the *expected*, or *average*, value of Y depends on X. Version 2 emphasizes that the *actual* value of Y—even in the population—is the sum of a deterministic part that depends on X and a random part that does not.

Of course, we do not have access to the whole population; hence we will be taking a sample and estimating the line from just this sample. But we want this sample line to be a good estimate of the unseen population line. We want to be able to use our estimate to make good predictions about average or individual values of Y for new values of X. Moreover, we could be wrong about there being any relationship between these variables in the population at all. Hence, we also want to test hypotheses about this relationship, just as we tested hypotheses about means, proportions, and the like.

If it were not for the random variation around the population regression line, none of this would present a problem. Our sample points would all be on the population regression line; we would get an exact estimate of the population regression line. Our predictions would be exactly right. And as

1. $E(Y_i|X_i) = a + \beta X_i$

or

2. $Y_i|X_i = a + \beta X_i + \varepsilon_i$

where ε_i is a normally distributed random variable with zero mean and constant variance. which is uncorrelated with X or itself.

Figure 12.2 The population regression line (b).

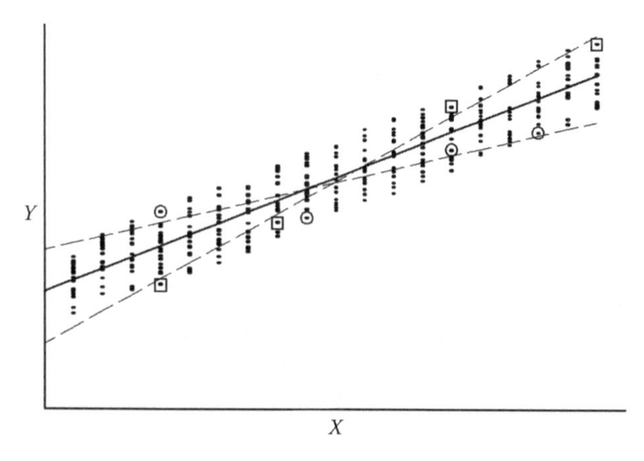

Figure 12.3 Two sample regression lines from the same population.

long as the slope were not zero, we would conclude that we had been right to think the two variables were related.

But because of the random variation around the population regression line, our sample points will almost certainly not be on the population regression line. In Figure 12.3, above, I have circled four points and boxed four others. If our sample consists of the circled points and we put the best possible line through them, we will get a sample regression line that is flatter than the true line. If our sample consists of the boxed points and we put the best possible line through them, we will get a sample regression line that is steeper than the true one. There is a whole distribution of sample slopes (and intercepts) that we could get from this population. There is a whole distribution of predictions about average or individual values of Y for new values of X. We will need confidence intervals again.

Moreover, suppose the population regression slope were actually zero; that the value for Y did not really depend on X at all. There is still a whole distribution of sample slopes we could get, most of which are not zero. Hence, we will need to test hypotheses about slopes, just as we tested hypotheses about means, proportions, and the like. Is our sample slope so different from zero, that we can conclude the true population slope is not zero with an acceptable probability of being wrong? This is where we are headed.

12.2 The Sample Regression Line

12.2.1 The Best Sample Line: Ordinary Least Squares

The meaning and calculation of the best sample regression line was covered briefly in Chapter 3, but it has been a while so we will review. Actually, there is more than one technique; the one we will use is the most common and is called **Ordinary Least Squares**, or just **OLS**. The "ordinary" just indicates that there are other variations on the theme. But if the population is as assumed in Figure 12.2, OLS estimates are best. Of course, this does not mean they are necessarily good. Both the sample lines in Figure 12.3 were OLS estimates. But because the samples were small and somewhat unrepresentative, these sample lines were quite far from the mark as estimates of the population line. Still, OLS gives unbiased estimates; they have no tendency to be too high, low, steep, or flat. And they are efficient. The distribution of different possible sample lines around the population line is as small as possible.

12.2.2 Finding the Intercept and Slope

Figure 12.4, repeated from Chapter 3, shows a scattergram for a small sample of two variables. If you did not know the value of *X*, you could do no better than predict $\overline{Y}$ for each case. This is the horizontal line in the figure. Now, though, suppose you know the value of *X*. It certainly appears that higher values of *X* should lead you to predict higher values of *Y*, and lower values of *X* should lead you to predict lower values of *Y*. The upward sloping **sample regression line** in the figure is the OLS estimate for *Y*, given *X*.

How is the line calculated? Any straight line can be represented as $\hat{Y} = a + bX$, where *a* is the *Y* intercept, and *b* is the slope. The $\hat{Y}$ ("y hat") is the value of *Y* predicted by the equation, as opposed to the actual data value, *Y*. Clearly, we want the prediction errors—the $(Y - \hat{Y})$s—to be small. But we cannot simply minimize the sum of these distances since, as always, the positive and negative values would cancel out. So—as you should expect by now—we square these errors, and minimize the sum of the squared errors, $\Sigma(Y_i - \hat{Y}_i)^2$. Imagine shifting or rotating the line in Figure 12.4. What you are doing is changing the intercept or slope. As you do, the line will get closer to some points and further from others. So, some of the squared errors will get smaller and some of the squared errors will get larger. According to the OLS criterion, our best line is the one with the combination of intercept and slope that minimizes the sum of squared errors. Figure 12.5 gives the formulas for calculating this combination of intercept and slope; and Figure 12.6 shows the spreadsheet calculations for this example.

Columns B and C in Figure 12.6 give the data. That is, the first point is three across and eight up. And there are 10 such points. The means are calculated in both directions. The mean in the *X* direction—$\overline{X}$—is 5; the mean in the *Y* direction—$\overline{Y}$—is 17. On the graph, these are the vertical and horizontal lines. Columns E and F find the deviations from

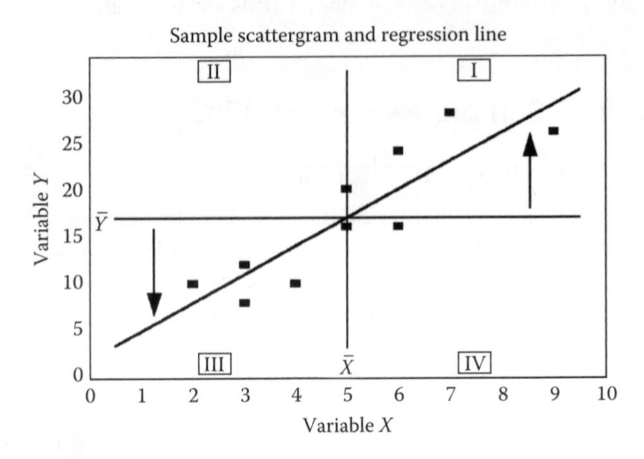

Figure 12.4 The sample regression line.

$$\hat{Y}_i = a + bX_i$$

where the slope is $b = \dfrac{\sum (X_i - \bar{X})(Y_i - \bar{Y})}{\sum (X_i - \bar{X})^2} = \dfrac{SCD_{XY}}{SSD_X}$

and the intercept is $a = \bar{Y} - b\bar{X}$

Figure 12.5 The formulas for the slope and intercept.

the mean in each direction. Of course, as always, these deviations in each direction sum to zero, because positive and negative deviations from the mean always just cancel out. To prevent this we do exactly what we did for the variance—*square* the deviations. The calculations shown in columns G and H are exactly like those for the variance. Indeed, if I had divided SSD_X by $(n - 1)$, I would have found the variance of X. And if I had divided SSD_Y by $(n - 1)$, I would have found the variance of Y.

But now we want to tie together the variation in the two variables. We can do this by multiplying $(X - \bar{X})$ by $(Y - \bar{Y})$ for each point. This is done in column I. "The sum of this column is the Sum of Cross Deviations from the means for the pair of variables, X and Y." The slope of the line, then, is just $SCD_{XY}/SSD_X = 120/40 = 3$.

Unlike the numbers in columns G and H, the numbers in column I are not squares, and do not need to be positive. In this example, most of the points are in quadrants I and III of the graph. That is, most of the points that are above $\bar{X}$ in the X direction are also above $\bar{Y}$ in the Y direction, and most of the points that are below $\bar{X}$ in the X direction are also below $\bar{Y}$ in the Y direction. This causes the cross-products to be positive. Their sum, SCD_{XY}, is positive. And when we calculate the slope, SCD_{XY}/SSD_X, it is positive as well. We have an upward sloping line. But

	A	B	C	D	E	F	G	H	I
1	Case	X	Y		$X - \bar{X}$	$Y - \bar{Y}$	$(X - \bar{X})^2$	$(Y - \bar{Y})^2$	$(X - \bar{X})(Y - \bar{Y})$
2	1	3	8		−2	−9	4	81	18
3	2	5	20		0	3	0	9	0
4	3	9	26		4	9	16	81	36
5	4	6	24		1	7	1	49	7
6	5	3	12		−2	−5	4	25	10
7	6	2	10		−3	−7	9	49	21
8	7	5	16		0	−1	0	1	0
9	8	4	10		−1	−7	1	49	7
10	9	7	28		2	11	4	121	22
11	10	6	16		1	−1	1	1	−1
12									
13	Means →	5	17			Sums	→ 40	466	120
14							↑	↑	↑
15	Slope →	3	← = I13/G13				SSD_X	SSD_Y	SCD_{XY}
16	Intercept →	2	← = C13–B15*B13						

Figure 12.6 Calculating the sample regression line.

imagine a different example in which the points had been in quadrants II and IV. The positive deviations would have been paired with negative deviations and the cross-products would have been negative. Their sum, SCD_{XY}, would have been negative. And when we calculated the slope, SCD_{XY}/SSD_X, it would have been negative as well. We would have had a downward sloping line.

Notice, on the graph, that the line goes through the point $(\bar{X}, \bar{Y})$. This is not a coincidence; the point $(\bar{X}, \bar{Y})$ is always on the line. Thus, we can write $\bar{Y} = a + b\bar{X}$ and—since we know $\bar{X}$, $\bar{Y}$ and b—we can solve for a. In this example, $a = \bar{Y} - b\bar{X} = 17 - 3 \times 5 = 2$. So, in this example, the sample regression equation is $\hat{Y} = 2 + 3X$.

12.2.3 Interpreting the Intercept and Slope

The Y intercept—2 in this example—is our estimate for the mean value of Y if $X = 0$. It is also our estimate for any individual case for which $X = 0$. It is in the same units as Y. In practice, it is often a fairly meaningless number. If Y is salary, in dollars, and X is years of education, the Y intercept would be our estimate for the salary (in dollars) of someone with zero years of education. However, we are not likely to hire people with zero years of education. Still, it is a necessary part of our estimate. Most statistical programs allow you to estimate the line without an intercept, but this is almost never a good idea.

The slope—3 in this example—is the more interesting number. It is our estimate for the change in the mean value of Y with an increase of one unit in X. It is also our estimate for the change in an individual case with an increase of one unit in X. Its units are those of Y over those of X. If Y is salary, in dollars, and X is years of education, the slope is our estimate of the salary increase, in dollars, for an additional year of education.

The slope is more interesting because it captures the relationship between the variables. If the true population slope is zero, we were wrong to think that Y depended on X. The sample slope is not the population slope, but it is our estimate and we will be able to use it to test the hypothesis that the population slope is zero. If we can reject that hypothesis, we can conclude (with some known probability of being wrong) that Y truly does depend on X.

12.3 Evaluating the Sample Regression Line

12.3.1 The Sum of Squared Errors

We have found the best sample regression; but that does not mean that it is especially good. We need to evaluate it. Since we minimized the sum of squared errors, a fairly obvious first step might be to calculate what that sum of squared errors turned out to be. Figure 12.7 adds these calculations to those of Figure 12.6. Column J, calculates the predicted value of the regression for each of the actual X-values in the data. By hand this would be tedious, but in a spreadsheet it is not. Cell J2 contains the formula = $B$16 + $B$15*B2; the references to the intercept and slope are absolute references; the reference to the first value of X is relative. Copying this formula down, then, finds the predicted values of all the others. Column K, then just finds the

	A	B	C	//	G	H	I	J	K	L
1	Case	X	Y		$(X-\bar{X})^2$	$(Y-\bar{Y})^2$	$(X-\bar{X})(Y-\bar{Y})$	$\hat{Y}$	$Y-\hat{Y}$	$(Y-\hat{Y})^2$
2	1	3	8		4	81	18	11	−3	9
3	2	5	20		0	9	0	17	3	9
4	3	9	26		16	81	36	29	−3	9
5	4	6	24		1	49	7	20	4	16
6	5	3	12		4	25	10	11	1	1
7	6	2	10		9	49	21	8	2	4
8	7	5	16		0	1	0	17	−1	1
9	8	4	10		1	49	7	14	−4	16
10	9	7	28		4	121	22	23	5	25
11	10	6	16		1	1	−1	20	−4	16
12										
13	Means	→ 5	17		40	466	120			106
14					↑	↑	↑			↑
15	Slope	→ 3			SSD_X	SSD_Y	SCD_{XY}			SSE
16	Intercept	→ 2								

Figure 12.7 Calculating the sum of squared errors.

difference between columns C and J; column L squares those differences; the sum of these squared differences is the **Sum of Squared Errors, SSE.** $\Sigma(Y-\hat{Y})^2 = \text{SSE} = 106$. While the SSE is not that easy to interpret in isolation, it is the basis for other measures that are.

12.3.2 The Mean Square Error and Standard Error of the Estimate

Recall that, by assumption, the population points vary around the population regression line with a constant variance (hence constant standard deviation or error). Now that we have our sample estimate of the population regression line, it would be nice to have sample estimates of that variance and standard error around the line. These estimates are s_e^2, often called the **mean square error,** and its square root, s_e, **the standard error of the estimate**. Figure 12.8 gives their formulas.

Why divide by $(n-2)$? As always, we want our sample statistics to be the best estimates of the population parameters. This led us to calculate the sample variance and standard deviation with $(n-1)$ in the denominator instead of n. One rationale was that because we were using $\bar{X}$, calculated from the same sample, instead of μ, the last squared deviation contained

$$s_e^2 = \frac{\sum\left(Y_i-\hat{Y}_i\right)^2}{n-2} = \frac{\text{SSE}}{n-2} = \text{MSE}$$

$$s_e = \sqrt{\frac{\sum(Y_i-\hat{Y}_i)^2}{n-2}} = \sqrt{\frac{\text{SSE}}{n-2}} = \sqrt{\text{MSE}}$$

Figure 12.8 The mean square error and standard error of the estimate.

no new information. Tell me the first $(n-1)$ X-values plus $\bar{X}$, and I can figure out what the last value must be. The upshot was that, for s_X to be an unbiased estimator of σ_X, we needed to divide by $(n-1)$. In the present case, we are using a and b, calculated from the same sample, instead of α and β, so the last two squared deviations contain no new information. Tell me the first $(n-2)$ (X, Y)-values plus a and b and I can figure out what the last two values must be. The upshot is that, for s_e^2 and s_e to be unbiased estimators of σ_e^2 and σ_e, we need to divide by $(n-2)$. More generally, we divide by the degrees of freedom which, for regression, will be n minus the number of coefficients estimated. For simple regression, this is $n-2$.

In this example, then, the mean square error = 106/8 = 13.25 and the standard error of the estimate $s_e = \sqrt{\text{MSE}} = \sqrt{13.25} = 3.6401$.

We will encounter the mean square error again shortly when we do an analysis of variance of our regression results. As for the standard error of the estimate, we can think of s_e as a sort of average deviation of the points from the line. It is in the same units as Y itself. It is also a basic input into formulas for a number of other standard errors. For example, in order to test whether our slope, b, is large enough that we can conclude that the true slope, β, is not zero, we will need a standard error of the slope. To create confidence intervals for our predictions of mean or individual values of Y given X, we will need standard errors for mean and individual predictions of Y given X. All of these standard errors are based on the standard error of the estimate, s_e.

12.3.3 R^2: The Coefficient of Determination

The coefficient of determination, R^2, was introduced in the last chapter, on analysis of variance (ANOVA). It measures the proportion of the variation in a dependent variable that we are able to explain with our explanatory variable. Its application in the regression context is quite straightforward. Indeed, we have already calculated just about everything we need. Our dependent variable, Y, has a certain amount of variation; we already have a measure of that variation in SSD_Y which, in this context, is usually called the **Total Sum of Squares, SST**. And this variation can be partitioned, just as in ANOVA, into that which we can explain, here called the **Regression Sum of Squares, SSR**, and that which we cannot, here called the **Sum of Squared Errors, SSE**.

Figure 12.9 summarizes. The total variation is the sum of the variation of the individual Y-values from their mean (squared). More concretely, in case 1 of our example (Figure 12.7), why does $Y = 8$ instead of 17, the mean for the Ys? The sum of this variation from the mean (squared) for each point is what there is to explain. According to our regression, because $X = 3$, we should not have expected 17, we should have expected $\hat{Y} = 2 + 3 \times 3 = 11$. So our equation explains why Y is 11 instead of 17; it does not explain the remaining discrepancy between 11 and the actual value of 8.

In this example, then

Total Variation	=	Variation Explained	+	Remaining Variation
SST		SSR		SSE
$\Sigma(Y-\bar{Y})^2$	=	$\Sigma(\hat{Y}_i-\bar{Y})^2$	+	$\Sigma(Y_i-\hat{Y}_i)^2$
466	=	360	+	106

Total variation	=	variation Explained	+	Remaining Variation
SST	=	SSR	+	SSE
$\sum(Y_i - \bar{Y})^2$	=	$\sum(\hat{Y}_i - \bar{Y})^2$	+	$\sum(Y_i - \hat{Y}_i)^2$

$$R^2 = \frac{\text{SSR}}{\text{SST}} = 1 - \frac{\text{SSE}}{\text{SST}}$$

Figure 12.9 Partitioning the total sum of squares and finding R^2.

Note that since we already know SST and SSE, we can find SSR as simply the difference: SSR = SST – SSE = 466 – 106 = 360. And in this example the coefficient of determination: $R^2 = 360/466 = 1 - 106/466 = 0.7725$. Our regression equation can explain 77.25% of the variation in Y.

12.3.4 Testing the Sample Regression Line

You will recall that in the last chapter, on ANOVA, we did not just estimate the amount of variation in one variable that could be explained by the other. We did formal inference, testing the null hypothesis that the dependent variable was actually independent of the explanatory variable; that there was really no relationship. We calculated an F_c statistic. And if this F_c statistic was large enough, we rejected the null hypothesis and concluded (with a known probability of being wrong) that there really was a relationship. We can do the same with regression, to see whether this regression explains more than a random amount of the variation in Y.

As in the last chapter, our null hypothesis is that there is no relationship; that Y is actually independent of X.

1. H_o: Y is independent of X
   H_a: Y is not independent of X.

Since we are doing ANOVA, our statistic is an F. Figure 12.10 summarizes. Any standard statistics package will give you this information automatically. Indeed, it will probably give you the p-value of the test as well.

As always, F_c is the ratio of two variances. You should recognize the variance in the denominator. It is just the mean square error. Recall that it is calculated so as to be an unbiased estimator of σ_e^2, the constant variance of the points around the regression line. What is much less obvious is that, under the null hypothesis, the numerator is also an estimator of σ_e^2. Under the null hypothesis, there is no real relationship between X and Y. Still, we would expect essentially *any* variable to have some small, random correlation with Y. Hence, the SSR will not be zero. Rather, the SSR/df_n—where the

	Sum of Squares	df	Mean Square	F_c
Regression	SSR	$2-1$	SSR/df_n	$\frac{\text{SSR}/df_n}{\text{SSE}/df_d}$
Error	SSE	$n-2$	SSE/df_d	
Total	SST	$n-1$		

Figure 12.10 The regression ANOVA table.

degrees of freedom equal the number of coefficients minus one—will have an expected value of σ_e^2. The result is that, under the null hypothesis, the F_c statistic has an expected value of one. Of course, if there *is* a relationship between X and Y, we would expect to explain more than just a random amount of variation in Y. SSR/df_n will be larger; hence, F_c will be larger. We are looking for a value of F_c so large that we can conclude (with an acceptable probability of being wrong) that there is, in fact, a relationship.

2. Reject H_0 if $F_c > 5.318$ ($df_n = 1$, $df_d = 8$; $\alpha = 0.05$).
3. We already have all the information for the ANOVA table.

	Sum of Squares	df	Mean Square	F_c
Regression	360	1	360.000	27.1698
Error	106	8	13.250	
Total	466	9		

4. ∴ Reject H_0.

The p-value of the test is less than 0.005: =FDIST(27.1698,1,8) = 0.0008.

12.4 Evaluating the Sample Regression Slope

As we saw in the last section, ANOVA offers one approach to evaluating our sample regression. An alternative is to look at the slope. The null hypothesis that β, the population slope, equals zero is logically equivalent to the null hypothesis that Y is independent of X.

Suppose, as in Figure 12.11, there is really no relationship between X and Y in the population. In this case, the population slope would be zero. The population line would be just a horizontal line at $\overline{Y}$. The value of X would not matter. Still, because of the random variation around $\overline{Y}$, our sample Y-values will almost certainly not all be equal. In Figure 12.11, I have circled four points and boxed four others. If our sample consists of the circled points, our best sample line will have a negative slope. If our sample consists of the boxed points, our best sample line will have a positive slope. There is a whole distribution of sample slopes we could get when, in fact, there is no relationship in the population.

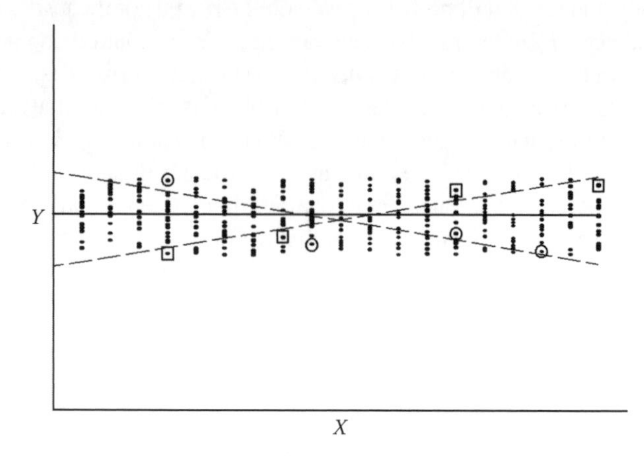

Figure 12.11 Two sample regression lines from a population with no relationship.

Fortunately, the sample slopes we could get follow the familiar t distribution, with a mean of β, a standard error of s_b, and $n - 2$ degrees of freedom. Hence, we can apply what we learned about t tests for the mean in Chapters 8 and 9. In this context, we begin with a null hypothesis that the true slope is indeed zero. Then, we calculate t_c based on the difference between our sample slope and our hypothesized true slope. Finally, if the t_c we get is so large that it is unlikely enough, given the null hypothesis, we reject the null hypothesis and conclude (with a known probability of being wrong) that the true slope is not zero. And, of course, if the slope is not zero, there is indeed a relationship.

A t test also opens up the possibility of either a two- or a one-tailed alternative. If you are just looking for a relationship, you would choose a two-tailed test. If you have reason (before looking at the data!) to expect a positive or negative relationship, you would choose a one-tailed test.

In our example, since the variables are just X and Y, we have no reason to expect a particular sign; you cannot use the graph of the data to decide. Hence, it is a two-tailed test.

1. H_o: $\beta = 0$
   H_a: $\beta \neq 0$.

Our test statistic follows the t distribution. The degrees of freedom equal n minus the number of coefficients estimated; $n - 2$ for simple regression. And, because this is a two-tailed test, we need to remember to divide α in half to look it up. For an α of 0.05, then, our criterion is:

2. Reject H_o if $|t_c| > 2.306$ (df = 8; $\alpha = 0.05$).

Our test statistic is calculated just as you should expect given what you know about the case with averages. The only thing new is the calculation of s_b, the standard error of the slopes we could get. Figure 12.12 summarizes. As mentioned earlier, s_b depends on s_e, the standard error of the estimate. The greater the variation of the data around the population regression line (of which s_e is an estimate), the wider the range of sample slopes we might reasonably get. However, notice that s_b also depends on SSD_X, the sum of squared deviations in the X direction. The wider the spread of the data in the X direction, the narrower the range of sample slopes we might reasonably get.

3. For our example, then, we already have everything necessary: the sample slope, $b = 3$ (Section 12.2.2), the standard error of the estimate, $s_e = 3.6401$ (Section 12.3.2), and the sum of squared deviations in the X direction, $SSD_X = 40$ (G13 in the spreadsheet, Section 12.2.2). Hence:

$$s_b = \frac{s_e}{\sqrt{SSD_x}} = \frac{3.6401}{\sqrt{40}} = 0.5755$$

$$t = \frac{b - \beta}{s_b} = \frac{b - 0}{s_b} = \frac{b}{s_b}$$

$$\text{where } s_b = \frac{s_e}{\sqrt{SSD_x}}$$

Figure 12.12 The t test for $\beta = 0$.

$$t = \frac{3-0}{0.5755} = 5.2125.$$

4. $\therefore$ Reject H_o.

The p-value of the test is less than $0.0005 \times 2 = 0.0010$, = TDIST (5.2125,8,2) = 0.0008.

12.5 The Relationship of *F* and *t*: Here and Beyond

Recall from Chapter 11 on ANOVA, that the F distribution with 1 and $n-2$ degrees of freedom was the square of the t distribution with $n-2$ degrees of freedom, and that the ANOVA F test for the two sample case was mathematically equivalent to the t test for the two sample case with standard deviations unknown but assumed equal. You should be wondering, then, if the F and t tests we have just done in the last two sections are also mathematically equivalent. Or perhaps you have already noticed the fact that both tests rejected the null hypothesis with p-values of 0.0008. Coincidence? Hardly. The criterion for the F test with 1 and 8 degrees of freedom was 5.318, the square of 2.306, the criterion for the t test with 8 degrees of freedom. And the F_c test statistic was 27.1698, the square of 5.2125, the t_c test statistic. Hence, the two must always agree.

Why, then, do we need both tests? If we were stopping with this chapter on simple regression, we would not. But when we get to multiple regression, in Chapter 13, the two tests will diverge. Think of the F test as testing whether the dependent variable really depends on *the whole set of explanatory variables, taken together.* It is a test of the entire regression. And think of the t test as testing whether the dependent variable really depends on *the particular variable for which b is the slope.* It is a test of the effect of a particular variable. Of course, in this chapter, we are confining our "set" of explanatory variables to one. If the regression taken as a whole explains something, and there is only one variable in the regression, then that variable must be doing the explaining. And that must mean that its slope is not zero. But once we get to multiple regression, it will be perfectly possible to conclude that the dependent variable depends on some explanatory variables but fail to conclude that it depends on others. It will even be possible to conclude that the dependent variable depends on some set of explanatory variables, while failing to conclude that it depends on any one of them in particular.

12.6 Predictions Using the Regression Line

12.6.1 Using the Regression Line to Predict *Y* Given *X*

Perhaps the most straightforward use of the sample regression line is to predict mean or individual values of Y given X. We have waited until now to do so for two reasons. First, it would make no sense to start using the

1. The mean Y given X –	2. An individual Y given X –
$\hat{Y} \pm t_{(n-2)} S_{\bar{Y}\|X}$	$\hat{Y} \pm t_{(n-2)} S_{\hat{Y}\|X}$
where $S_{\bar{Y}\|X} = S_e \sqrt{0 + \frac{1}{n} + \frac{(X^* - \bar{X})^2}{\text{SSD}_X}}$	where $S_{\hat{Y}\|X} = S_e \sqrt{1 + \frac{1}{n} + \frac{(X^* - \bar{X})^2}{\text{SSD}_X}}$

Figure 12.13 Confidence intervals for Y given X: The formulas.

regression until we have established that it is not just a random relationship. We want to satisfy ourselves first that Y really does depend on X. And second, when we make our predictions, we are going to want to put confidence intervals around them. And these confidence intervals involve some rather messy standard errors.

In our example, if $X = 8$, $\hat{Y} = 2 + 3 \times 8 = 26$. We can think of $\hat{Y}$ as an estimate of the mean—$E(Y|X = 8)$—the expected value of Y given that $X = 8$. Or, we can think of it as a prediction for an individual case, given that $X = 8$. The predictions are the same—26—but the confidence intervals around them are different. Predictions of individual cases are subject to larger errors. Hence it is important to be clear what we are predicting.

12.6.2 Confidence Intervals for *Y* Given *X*

Figure 12.13 summarizes confidence intervals for Y given X. Whether we are predicting a mean or an individual case, we get our point prediction by just plugging our specific X—call it X^*—into our regression equation to get $\hat{Y}$. Our confidence interval, then, is $t_{(n-2)}$ standard errors around $\hat{Y}$. The only difference is in the standard errors. And the only difference there is in the first figure under the radical. For $s_{\bar{Y}|X}$, it is zero; for $s_{\hat{Y}|X}$, it is one.

To understand this difference, consider Figure 12.14, which simply repeats Figure 12.3, showing the population regression line and two possible sample regression lines. To start, suppose we actually knew the population regression line. In that case, our predicted mean Y given X-values would be absolutely correct; our errors would be zero. Our predicted individual Y

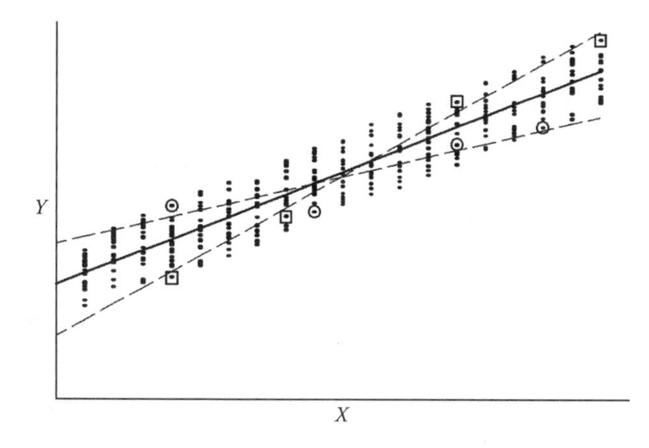

Figure 12.14 The population and two sample regression lines.

given X-values would not be absolutely correct, though, since the individual values are scattered about the population line. And recall, s_e is our estimate of the standard error around the population line. If we knew the population regression line, $s_{\bar{Y}|X}$ would equal zero and $s_{\hat{Y}|X}$ would equal s_e.

Now notice that, if the second and third parts under the radical were not there, the equations in Figure 12.13 would give these population values. That is, $s_{\bar{Y}|X} = s_e\sqrt{0} = 0$, and $s_{\hat{Y}|X} = s_e\sqrt{1} = s_e$. The actual values are somewhat larger because of the second and third parts under the radical. These reflect the fact that we are not working from the population regression line. We are working from a sample estimate, and the sample estimate is almost certainly somewhat wrong.

The second part under the radical—$1/n$—reflects the fact that sample regression lines calculated from small samples are likely to be worse estimates of the population line. However, as n increases sample lines converge to the population line. As a practical matter $1/n$ is pretty small even for a moderate size sample.

The third part under the radical—$(X^* - \bar{X})^2/\mathrm{SSD}_x$—is more important. The X^* is the specific value of X that we plugged in to get $\hat{Y}$. If $X^* = \bar{X}$, this term drops out entirely; as X^* gets further from $\bar{X}$, this term grows more and more rapidly. Because of this term, both standard errors—thus both confidence intervals—are smallest at the mean of the data and get larger and larger as we get away from the mean.

What is the reason for the confidence intervals widening as we get away from the mean of the data? A glance at Figure 12.14 should suggest the answer. The two sample regression lines have rather different slopes. Still, at the means of the data they give similar predictions. It is only as we get away from the mean that the different slopes cause the predictions to be very different.

In our example, we plugged in $X^* = 8$ to get $\hat{Y} = 2 + 3 \times 8 = 26$. For a 95% confidence interval with $\mathrm{df} = 10 - 2 = 8$, $t = 2.306$. And the standard errors are

$$\begin{aligned} s_{\bar{Y}|X} &= s_e\sqrt{0 + \frac{1}{n} + \frac{(X^* - \bar{X})^2}{\mathrm{SSD}_X}} \\ &= 3.6401\sqrt{0 + \frac{1}{10} + \frac{(8-5)^2}{40}} \\ &= 3.6401\sqrt{0 + 0.10 + 0.225} \\ &= 3.6401\sqrt{0.325} \\ &= 3.6401 \times 0.5701 = 2.0752 \end{aligned}$$

$$\begin{aligned} s_{\hat{Y}|X} &= s_e\sqrt{1 + \frac{1}{n} + \frac{(X^* - \bar{X})^2}{\mathrm{SSD}_X}} \\ &= 3.6401\sqrt{1 + \frac{1}{10} + \frac{(8-5)^2}{40}} \\ &= 3.6401\sqrt{1 + 0.10 + 0.225} \\ &= 3.6401\sqrt{1.325} \\ &= 3.6401 \times 1.1511 = 4.1900. \end{aligned}$$

So, the 95% confidence intervals are:

95% CI: $26 \pm 2.306 \times 2.0752$ 95% CI: $26 \pm 2.306 \times 4.1900$
95% CI: 26 ± 4.785 95% CI: 26 ± 9.662.

The confidence interval for an individual Y given X is more than twice as wide.

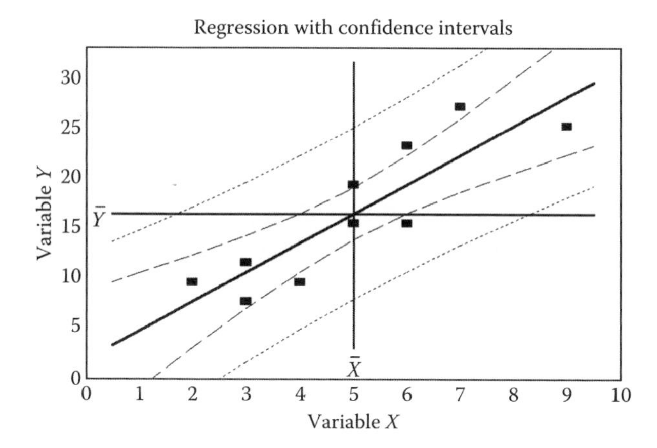

Figure 12.15 Confidence intervals around the sample regression line.

Remember that these confidence intervals are for a particular value of *X*. Those closer to the mean will be narrower; those further from the mean will be wider. Figure 12.15 repeats the sample scattergram and regression graph with the two sets of confidence intervals drawn in. The narrower, dashed intervals represent 95% confidence intervals for the mean *Y* given *X*; the wider, dotted lines represent 95% confidence intervals for an individual *Y* given *X*.

12.7 Regression and Correlation

Regression analysis is a powerful tool for exploring relationships between variables, especially in cases in which you believe they are causally related. We actually estimate that relationship. And, while regression cannot prove causation, if the relationship you find is a strong one, it certainly supports your belief. Sometimes, though, we may want to measure the "relatedness" of two variables without positing causation. We just want to measure the extent to which they vary together. In such cases we can use correlation instead of regression analysis. The common correlation coefficient, sometimes called the **Pearson Correlation Coefficient** to distinguish it from other varieties, is actually related to the regression analysis we have been doing. But it treats the variables symmetrically, instead of treating one as dependent on the other.

12.7.1 Finding a Sample Correlation

Figure 12.16 gives the basic formula. Notice that it makes use of sums of squares that we have already computed in order to do our regression analysis. Looking back to our initial calculations, $SSD_X = 40$, $SSD_Y = 466$, and $SCD_{XY} = 120$. This is all we need. The sample correlation is

$$r = \frac{SCD_{XY}}{\sqrt{SSD_X \times SSD_Y}} = \frac{120}{\sqrt{40 \times 466}} = \frac{120}{136.5284} = 0.8789.$$

The correlation coefficient

$$r = \frac{\text{SCD}_{XY}}{\sqrt{\text{SSD}_X \text{SSD}_Y}}$$

The t test for $\rho = 0$

$$t = \frac{r-\rho}{S_r} = \frac{r-0}{S_r} = \frac{r}{S_r}$$

$$\text{where } s_r = \sqrt{\frac{1-r^2}{n-2}}$$

Figure 12.16 Finding and testing the correlation coefficient.

12.7.2 Interpreting a Sample Correlation

The correlation coefficient is defined from –1, for a perfect negative linear relationship, to + 1 for a perfect positive linear relationship. A correlation of 0 means that there is no linear relationship between the variables.

Except at these three points, though, there is no simple interpretation. Certainly, a 0.80 correlation is stronger than a 0.40 correlation, but there is no meaningful sense that it is twice as strong. How large a correlation needs to be to be "strong" will vary with the context.

12.7.3 Testing a Sample Correlation

We can test whether a correlation is significantly different from zero. Our null hypothesis would be that the true population correlation, ρ, is zero. The test can have either a two-tailed or a one-tailed alternative. If you have reason (before looking at the data!) to expect a positive or negative relationship, you would choose a one-tailed test. In our example, since the variables are just X and Y, we have no reason to expect a particular sign; you cannot use the graph of the data to decide. Hence it is a two-tailed test.

1. H_o: $\rho = 0$
   H_a: $\rho \neq 0$.

Our test statistic follows the t distribution. As with simple regression, the degrees of freedom equals $n - 2$. And, because this is a two-tailed test, we need to remember to divide α in half to look it up. For an α of 0.05, then, our criterion is

2. Reject H_o if $|t_c| > 2.306$ (df = 8; α = 0.05).

Our test statistic is calculated just as you should expect given what you know about the case with slopes. The only thing new is the calculation of s_r, the standard error of the correlations we could get.

3. $$s_r = \sqrt{\frac{1-r^2}{n-2}} = \sqrt{\frac{1-0.7725}{10-2}} = \sqrt{\frac{0.2275}{8}} = 0.1686$$
   $$t = \frac{0.8789-0}{0.1686} = 5.2125.$$

4. ∴ Reject H_o.

The p-value of the test is less than $0.0005 \times 2 = 0.0010$, =TDIST (5.2125,8,2) = 0.0008.

12.7.4 The Relationship of Regression and Correlation

Notice that the criterion and test statistic for the correlation, above, are both the same as for the slope. We again get a p-value of 0.0008. Again, this is not coincidence. For simple regression, the ANOVA F test, the t test for the slope, and the t test for the correlation are all mathematically equivalent. Indeed, simple regression and correlation are closely related. Two relationships are worth highlighting.

First, compare the equations for the slope, b, and the correlation coefficient, r.

$$b = \frac{\text{SCD}_{XY}}{\text{SSD}_X} \qquad r = \frac{\text{SCD}_{XY}}{\sqrt{\text{SSD}_X \text{SSD}_Y}}$$

You can think of r as a sort of average (literally, a geometric average) of the regression slope and the slope we would have found with X and Y reversed. If the line is steep, meaning a large b, and we reverse X and Y, the new line will be flat, meaning a small b. Multiplying these two bs and taking the square root gives us r. As an average of a steep and a flat slope, it is not itself a slope. Rather, it is a measure of "relatedness" that treats the two variables symmetrically.

Second, the square of the correlation coefficient, r^2, is the regression coefficient of determination, R^2. Hence it was perhaps unfair to say, earlier, that the correlation coefficient has no simple interpretation. After all, R^2 has a fairly simple interpretation. It is the proportion of the variation in Y that can be explained by the variation in X. So, in our example, $r = 0.8789$, means that $r^2 = R^2 = 0.8789^2 = 0.7725$. Variation in X is capable of explaining 77.25% of the variation in Y. And, symmetrically, variation in Y is also capable of explaining 77.25% of the variation in X.

12.8 Another Example

We have covered a great deal so far in this chapter, all with just one simple example. I have done it this way to bring out the interconnectedness of the various components of regression analysis. I wanted you to see the same SSD_X-value coming up repeatedly. I wanted you to notice, hopefully even before I pointed it out, that the p-values of all the tests were identical. But now it is time to work another, more interesting example.

Consider the following example:

In Chapters 2 and 3, we used data on a sample of 50 employees to explore whether education might help explain differences in employee salaries. Employees2.xls contains the data.

a. Create a scattergram of salary and education.
b. Calculate the OLS sample regression line relating salary to education.

c. Interpret your coefficients in words.
d. Calculate the sum of squared errors, mean square error, and standard error of the estimate.
e. Create an ANOVA table. Can you conclude that salary depends on education in the population? Use $\alpha = 0.05$. What is the p-value of this test?
f. How much of the variation in salary can be explained by education?
g. Can you conclude that the true population slope coefficient is different from zero? Use $\alpha = 0.05$. What is the p-value of this test?
h. Find a 95% confidence interval estimate for the mean salary of employees with 10 years of education.
i. Find a 95% confidence interval estimate for the salary of an individual employee with 10 years of education.

a. Figure 12.17 shows the scattergram. It does look like salary rises with education.
b. The first step is to calculate SSD_X, SSD_Y, and SCD_{XY}. Figure 12.18 shows the data and calculations for the first and last four cases of Ed and Salary, along with the means or sums for all 50 cases. Thus, the first value of $(\text{Ed} - \overline{\text{Ed}}) = (\text{B2} - \$\text{B}\$53) = 3.46$; the first value of $(\text{Salary} - \overline{\text{Salary}}) = (\text{F2} - \$\text{F}\$53) = 22.734$. Columns J, K, and L find the squared deviations in the X direction, the squared deviations in the Y direction, and the cross deviations. For the first case, (H2^2) = 11.972, (I^2) = 516.835, and (H2*I2) = 78.660. Copying these down, we get the other 49 cases. The sums of these columns are SSD_X, SSD_Y, and SCD_{xy}. You should verify these.

   The slope, then, is just $SCD_{XY}/SSD_X = 765.718/282.420 = 2.711$. Using this slope, and the $(\overline{X}, \overline{Y})$ point, the intercept is just $a = \overline{Y} - b\overline{X} = 42.266 - 2.711 \times 13.54 = 5.555$. The OLS sample regression line is: Salary = 5.555 + 2.711 × Ed.
c. Interpreting these results, 5.555 is our estimated salary (in $ thousands) for an employee with zero years of education,

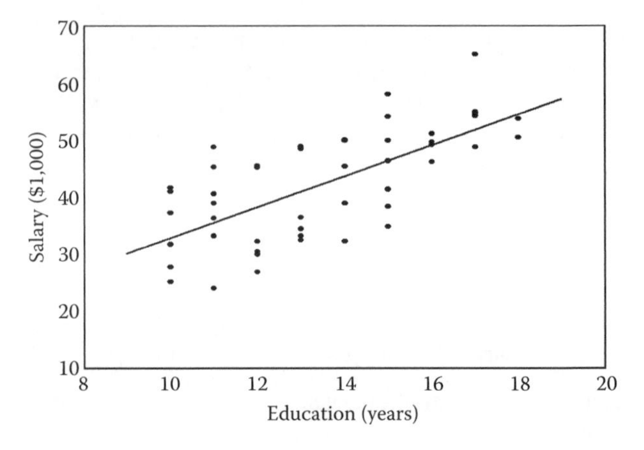

Figure 12.17 Another scattergram and sample regression line.

	B	//	F	G	H	I	J	K	L
1	Ed		Salary		$X-\bar{X}$	$Y-\bar{Y}$	$(X-\bar{X})^2$	$(Y-\bar{Y})^2$	$(X-\bar{X})(Y-\bar{Y})$
2	17		65.0		3.46	22.734	11.972	516.835	78.660
3	11		48.7		−2.54	6.434	6.452	41.396	−16.342
4	15		58.0		1.46	15.734	2.132	247.559	22.972
5	14		49.9		0.46	7.634	0.212	58.278	3.512
:	:		:		:	:	:	:	:
48	10		25.1		−3.54	−17.166	12.532	294.672	60.768
49	12		29.9		−1.54	−12.366	2.372	152.918	19.044
50	14		32.2		0.46	−10.066	0.212	101.324	−4.630
51	13		32.4		−0.54	−9.866	0.292	97.338	5.328
52									
53	13.54		42.266		← Means	Sums →	282.420	4511.012	765.718
54							↑	↑	↑
55	Slope →		2.711				SSD_X	SSD_Y	SCD_{XY}
56	Intercept →		5.555						

Figure 12.18 Calculating the sample regression line.

	B	//	F	//	J	K	L	M	N	O
1	Ed		Salary		$(X-\bar{X})^2$	$(Y-\bar{Y})^2$	$(X-\bar{X})(Y-\bar{Y})$	$\hat{Y}$	$Y-\hat{Y}$	$(Y-\hat{Y})^2$
2	17		65		11.972	516.835	78.660	51.647	13.353	178.302
3	11		48.7		6.452	41.396	−16.342	35.379	13.321	177.439
4	15		58.0		2.132	247.559	22.972	46.224	11.776	138.663
5	14		49.9		0.212	58.278	3.512	43.513	6.387	40.791
:	:		:		:	:	:	:	:	:
48	10		25.1		12.532	294.672	60.768	32.668	−7.568	57.276
49	12		29.9		2.372	152.918	19.044	38.091	−8.191	67.087
50	14		32.2		0.212	101.324	−4.630	43.513	−11.313	127.988
51	13		32.4		0.292	97.338	5.328	40.802	−8.402	70.592
52										
53	13.54		42.266		282.420	4511.012	765.718			2434.941
54					↑	↑	↑			↑
55	Slope →		2.711		SSD_X	SSD_Y	SCD_{XY}			SSE
56	Intercept →		5.555							

Figure 12.19 Calculating the sum of squared errors.

though we probably have no such employees. More interestingly, 2.711 is our estimated increase in salary (in $ thousands) for each additional year of education.

d. The rest of the question is largely about evaluating how well we have done, and requires finding SSE, the sum of squared errors. Figure 12.19 show this step. Again, it shows the data and calculations for the first and last four cases, along with means or sums for all 50. Thus, the first $\hat{Y}$ = $F$56 + $F$55 × B2 = 51.647; the first error $(Y - \hat{Y})$ = (B2 − M2) = 13.353; and the first squared error, $(Y - \hat{Y})^2$ = (N2^2) = 178.302. Copying down, we get the other 49 cases. The sum of the squared errors, SSE = 2434.941.

From this, the mean square error MSE = SSE/(n − 2) = 2434.941/48 = 50.728. And the standard error of the estimate, $s_e = \sqrt{\text{MSE}} = \sqrt{50.728} = 7.1224$.

e. Knowing SST (= SSD_Y) and SSE, we can set up the ANOVA table for the F test.

1. H_o: Salary is independent of education
   H_a: Salary is not independent of education.
2. Reject H_o if $F_c > 4.034$ ($df_n = 1$, $df_d = 48$; $\alpha = 0.05$).
3.

	Sum of Squares	df	Mean Squares	F_c
Regression	2076.071	1	2076.0713	40.9256
Error	2434.941	48	50.7279	
Total	4511.012	49		

   $F_c = 40.9256$.
4. ∴ Reject H_o.
   The p-value of the test is less than 0.005. =FDIST(40.9256, 1,48) = 0.00000006. There is almost no chance that an F_c this large could have arisen just randomly.

f. From the ANOVA table, we can also calculate R^2. R^2 = 2076.071/4511.012 = 0.4602. We can explain about 46% of the variation in Salary by taking Ed into account.

g. Different from zero implies a two-tailed test; in this case, though, we might argue for a one-tailed test on the grounds that economic theory strongly suggests a positive relationship.

1. $H_o: \beta = 0$ $\qquad H_o: \beta = 0$
   $H_a: \beta \neq 0$ $\qquad H_a: \beta > 0$.
2. Reject H_o if $|t_c| > 2.009$ $\qquad$ Reject H_o if $t_c > 1.676$
   (df = 48; $\alpha = 0.05$) $\qquad$ (df = 48; $\alpha = 0.05$).
3.

$$s_b = \frac{s_e}{\sqrt{\text{SSD}_x}} = \frac{7.1224}{\sqrt{282.420}} = 0.4238$$

$$t = \frac{2.711 - 0}{0.4238} = 6.3973.$$

4. ∴ Reject H_o $\qquad$ ∴ Reject H_o.
   The p-value for the two-tailed test is less than 0.0005 × 2 = 0.001. =TDIST(6.3973,48,2) = 0.00000006, exactly the same as for the F test. For the one-tailed test, it is half of that. The true population slope is almost certainly not zero.

   The predicted Salary for the mean or individual with 10 years of education is

$$\text{Salary} = 5.555 + 2.711(10) = 32.668.$$

The standard errors are

h. $$s_{\bar{Y}|X} = s_e\sqrt{0+\frac{1}{n}+\frac{(X^*-\bar{X})^2}{\text{SSD}_X}}$$

$$= 7.1224\sqrt{0+\frac{1}{50}+\frac{(10-13.54)^2}{282.420}}$$

$$= 7.1224\sqrt{0+0.02+0.04437}$$

$$= 7.1224\sqrt{0.06437}$$

$$= 7.1224\times 0.25372 = 1.8071$$

i. $$s_{\hat{Y}|X} = s_e\sqrt{1+\frac{1}{n}+\frac{(X^*-\bar{X})^2}{\text{SSD}_X}}$$

$$= 7.1224\sqrt{1+\frac{1}{50}+\frac{(10-13.54)^2}{282.420}}$$

$$= 7.1224\sqrt{1+0.02+0.04437}$$

$$= 7.1224\sqrt{1.06437}$$

$$= 7.1224\times 1.03168 = 7.3480.$$

The 95% CI for the mean Salary of employees with 10 years education

95% CI: 32.668 ± 2.009 × 1.8071
95% CI: 32.668 ± 3.630

The 95% CI for the Salary of an individual employee with 10 years education

95% CI: 32.668 ± 2.009 × 7.3480
95% CI: 32.668 ± 14.762.

As before, we are a lot more certain about the mean value along the regression line than we are about individual cases, which are scattered about the regression line even in the population.

12.9 Dummy Explanatory Variables

So far, we have limited our discussion to numerical explanatory variables. Indeed, that is what has been new in this chapter. But can we also use regression when the explanatory variable is categorical? Can we use regression to see whether salary depends on job type or sex? In general, no. Consider Type: line = 1, office = 2, and management = 3. One would need to be able to assume that the intervals from 1 to 2 and 2 to 3 are somehow equal. And that is unlikely. In Chapter 13, we will examine a technique for transforming Type so that it is usable.

The exception—the case in which a categorical variable works just fine—is a dummy, yes–no variable. Consider Female: male = 0 and female = 1. There is only one difference here—between zero and one—and one can interpret the "slope" coefficient as the effect of being female instead of male. And, if that slope is significant, then we can conclude that salary depends on sex.

Consider the following example:

In end-of-chapter Exercises 9.7 and 11.6, you used these same Employees2.xls data to explore whether sex might help explain differences in employee salaries.

a. Create a scattergram for salary and sex.
b. Calculate the OLS sample regression line relating salary to sex. Interpret your coefficients in words.

c. Calculate the sum of squared errors, mean square error, and standard error of the estimate.
d. Create an ANOVA table. Can you conclude that salary depends on sex in the population? Use $\alpha = 0.05$. What is the p-value of this test?
e. How much of the variation in salary can be explained by sex?
f. Can you conclude that the true population slope coefficient is different from zero? Use $\alpha = 0.05$. What is the p-value of this test?

a. Figure 12.20 shows a scattergram and sample regression line.
b. Again, the first step is to calculate SSD_X, SSD_Y, and SCD_{XY}. Figure 12.21 shows the data and calculations for the first and last four cases of Female and Salary, along with the means or sums for all 50 cases. The computations are all the same as in the last example. The OLS sample regression line is: Salary = 43.406 − 3.167 × Female.
c. Interpreting these results, since Female = 0 for men, the intercept, 43.406, is our estimate for the mean salary of men; since Female = 1 for women, 43.406 − 3.167 × 1 = 40.239 is our estimate for the mean salary of women. The slope, −3.167, is our estimate of the difference.
d. Again, the rest of the question is largely about evaluating how well we have done, and requires finding SSE, the sum of squared errors. Figure 12.22 shows this step. Again, it shows the data and calculations for the first and last four cases, along with means or sums for all 50. Again, the computations are all the same as in the last example. The sum of the squared errors, SSE = 4395.442.

   From this, the mean square error, $MSE = SSE/(n - 2) = 4395.442/48 = 91.572$. And the standard error of the estimate, $s_e = \sqrt{MSE} = \sqrt{91.572} = 9.5693$.
e. Knowing SST ($= SSD_Y$) and SSE, we can set up the ANOVA table for the F test:

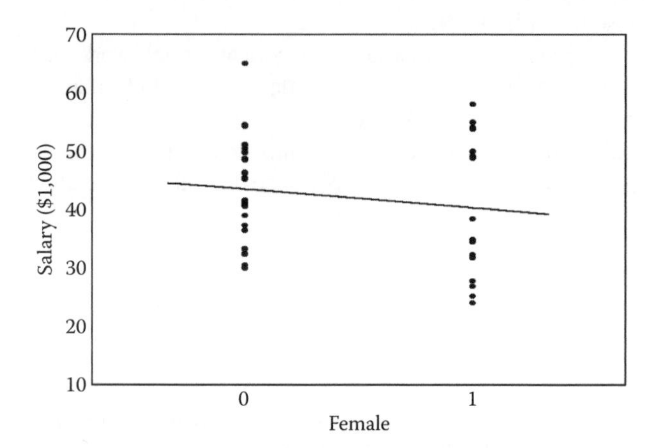

Figure 12.20 Another scattergram and sample regression line.

	//	E	F	G	H	I	J	K	L
1		Female	Salary		$X-\bar{X}$	$Y-\bar{Y}$	$(X-\bar{X})^2$	$(Y-\bar{Y})^2$	$(X-\bar{X})(Y-\bar{Y})$
2		0	65.0		−0.36	22.734	0.130	516.835	−8.184
3		0	48.7		−0.36	6.434	0.130	41.396	−2.316
4		1	58.0		0.64	15.734	0.410	247.559	10.070
5		1	49.9		0.64	7.634	0.410	58.278	4.886
:		:	:		:	:	:	:	:
48		1	25.1		0.64	−17.166	0.410	294.672	−10.986
49		0	29.9		−0.36	−12.366	0.130	152.918	4.452
50		1	32.2		0.64	−10.066	0.410	101.324	−6.442
51		0	32.4		−0.36	−9.866	0.130	97.338	3.552
52									
53		0.36	42.266	← Means		Sums →	11.520	4511.012	−36.488
54							↑	↑	↑
55	Slope →		−3.167				SSD_X	SSD_Y	SCD_{XY}
56	Intercept →		43.406						

Figure 12.21 Calculating the sample regression line.

	//	E	F	//	J	K	L	M	N	O
1		Female	Salary		$(X-\bar{X})^2$	$(Y-\bar{Y})^2$	$(X-\bar{X})(Y-\bar{Y})$	$\hat{Y}$	$Y-\hat{Y}$	$(Y-\hat{Y})^2$
2		0	65.0		0.130	516.835	−8.184	43.406	21.594	466.290
3		0	48.7		0.130	41.396	−2.316	43.406	5.294	28.024
4		1	58.0		0.410	247.559	10.070	40.239	17.761	315.457
5		1	49.9		0.410	58.278	4.886	40.239	9.661	93.337
:		:	:		:	:	:	:	:	:
48		1	25.1		0.410	294.672	−10.986	40.239	−15.139	229.186
49		0	29.9		0.130	152.918	4.452	43.406	−13.506	182.419
50		1	32.2		0.410	101.324	−6.442	40.239	−8.039	64.624
51		0	32.4		0.130	97.338	3.552	43.406	−11.006	121.138
52										
53		0.36	42.266		11.520	4511.012	−36.488			4395.442
54					↑	↑	↑			↑
55	Slope →		−3.167		SSD_X	SSD_Y	SCD_{XY}			SSE
56	Intercept →		43.406							

Figure 12.22 Calculating the sum of squared errors.

1. H_o: Salary is independent of sex
   H_a: Salary is not independent of sex.
2. Reject H_o if $F_c > 4.034$ ($df_n = 1$, $df_d = 48$; $\alpha = 0.05$).
3.

	Sum of Squares	df	Mean Squares	F_c
Regression	115.571	1	115.5707	1.2621
Error	4395.442	48	91.5717	
Total	4511.012	49		

4. ∴ Fail to Reject H_o.

The p-value of the test is greater than 0.1, =FDIST (1.2621,1,48) = 0.2668. There is a 0.2668 chance that an F_c this large could have arisen just randomly.

f. From the ANOVA table, we can also calculate R^2. $R^2 = 115.571/4511.012 = 0.0256$. We can explain only 2.56% of the variation in Salary by taking sex into account.

g. Different from zero implies a two-tailed test; in this case, though, we might argue for a one-tailed test on the grounds that, historically, women have earned less than men.

1. $H_o: \beta = 0$ $\quad\quad$ $H_o: \beta = 0$
   $H_a: \beta \neq 0$ $\quad\quad$ $H_a: \beta < 0.$
2. Reject H_o if $|t_c| > 2.009$ $\quad$ Reject H_o if $t_c < -1.676$
   (df = 48; $\alpha = 0.05$) $\quad\quad$ (df = 48; $\alpha = 0.05$).

$$s_b = \frac{s_e}{\sqrt{\text{SSD}_x}} = \frac{9.5693}{\sqrt{282.420}} = 2.8194$$

$$t = \frac{-3.167 - 0}{2.8194} = -1.1234.$$

4. $\therefore$ Fail to Reject H_o $\quad\quad$ $\therefore$ Fail to Reject H_o.

The p-value for the two-tailed test is between $0.2500 \times 2 = 0.500$ and $0.1000 \times 2 = 0.2000$. =TDIST(1.1234,48,2) = 0.2668, exactly the same as for the F test. For the one-tailed test, it is half of that. We cannot conclude that salary depends on sex.

I introduced this problem by noting that we had addressed it twice already. In Exercise 9.7, we ran a t test of whether the difference in true mean salaries equaled zero. We found the sample means for each sex—43.406 for men and 40.239 for women—and found a t_c statistic equal to the difference in sample means divided by the standard error of the differences we could have gotten.

We got

$$t_c = \frac{43.406 - 40.239}{2.8194} = \frac{-3.167}{2.8194} = -1.1234.$$

This is exactly the same result we just got, testing whether the slope equaled zero. These are mathematically equivalent tests. In Exercise 11.6, we ran an ANOVA test of whether salaries were independent of sex. We found the between-sample sum of squares to be 115.5707, the within-sample sum of squares to be 4395.4415, and our F_c statistic to be

$$F_c = \frac{115.5707/1}{4395.4415/48} = \frac{115.5707}{91.5717} = 1.2621\cdot$$

Again, this is exactly the same result we just got using simple regression. And, of course, it is exactly the square of the t_c. These are all mathematically equivalent tests. The advantage of the regression approach is that we can look at sets of explanatory variables—some numerical variables like Ed and some dummy variables like Female—simultaneously. This is what **multiple regression** is all about. (See Figure 12.23.)

$$SSD_x = \sum\left(X-\bar{X}\right)^2 \qquad S_x^2 = \frac{SSD_X}{n-1}$$

$$SSD_Y = \sum\left(Y-\bar{Y}\right)^2 \qquad S_y^2 = \frac{SSD_Y}{n-1}$$

$$SCD_{XY} = \sum\left(X-\bar{X}\right)\left(Y-\bar{Y}\right) \qquad S_{XY} = \frac{SCD_{XY}}{n-1}$$

Regression coefficients

$$b = \frac{SCD_{XY}}{SSD_X} \qquad a = \bar{Y} - b\bar{X}$$

Analysis of variance

	Sum of Squares	df	Mean Square	F_c
Regression	SSR	2 − 1	SSR/df_n	$\frac{SSR/df_n}{SSE/df_d}$
Error	SSE	$n-2$	SSE/df_d	
Total	SST	$n-1$		

where,

Total variation	=	Variation explained	+	Remaining variation
SST		SSR		SSE
$\sum\left(Y-\bar{Y}\right)^2$	=	$\sum\left(\hat{Y}-\bar{Y}\right)^2$	+	$\sum\left(Y-\hat{Y}\right)^2$

Standard Errors

$$S_e = \sqrt{\frac{SSE}{n-2}} = \sqrt{MSE}$$

$$S_b = \frac{S_e}{\sqrt{SSD_X}} \qquad S_{\bar{Y}|X} = S_e\sqrt{0+\frac{1}{n}+\frac{\left(X^*-\bar{X}\right)^2}{SSD_X}}$$

$$S_r = \sqrt{\frac{1-r^2}{n-2}} \qquad S_{\hat{Y}|X} = S_e\sqrt{1+\frac{1}{n}+\frac{\left(X^*-\bar{X}\right)^2}{SSD_X}}$$

Coefficients of correlation and determination

$$r = \frac{SCD_{XY}}{\sqrt{SSD_X SSD_Y}} \qquad R^2 = r^2 = 1-\frac{SSE}{SST}=\frac{SSR}{SST}$$

Figure 12.23 Simple regression: Summary.

12.10 The Need for Multiple Regression

In the preceding examples, we found that we could conclude, with a very small probability of being wrong, salary depended on education. We could not conclude that salary depended on sex. An end of chapter exercise will ask you to see whether it depends on experience. It is important to understand that we cannot generally explore multiple explanatory variables one at a time. In the regression using education, we were not controlling for sex or experience; in the regression using sex, we were not controlling for education and experience. If these explanatory variables are correlated, we may be crediting education with effects that are actually due to experience.

Likewise, the lack of an effect of sex may be because we were not comparing men and women of equal education and experience. This problem is called **omitted variable bias**, and it is why we must move in Chapter 13 to multiple regression—regression that includes multiple explanatory variables in the same regression.

12.11 Exercises

12.1 In the chapter, we used data on a sample of 50 employees to explore whether education or sex might help explain differences in employee salaries. Employees2.xls contains the data. Using these same data, explore whether salary depends on experience.

a. Create a scattergram of salary and experience.
b. Calculate the OLS sample regression line relating salary to experience.
c. Interpret your coefficients in words.
d. Calculate the sum of squared errors, mean square error, and standard error of the estimate.
e. Create an ANOVA table. Can you conclude that salary depends on experience in the population? Use $\alpha = 0.05$. What is the p-value of this test?
f. How much of the variation in salary can be explained by experience?
g. Can you conclude that the true population slope coefficient is different from zero? Use $\alpha = 0.05$. What is the p-value of this test?
h. Find a 95% confidence interval estimate for the mean salary of employees with 10 years of experience.
i. Find a 95% confidence interval estimate for the salary of an individual employee with 10 years of experience.

12.2 Refer back to the NLSY1.xls data file, that contains height, weight, age, and sex for a sample of 281 young adults.

a. Create a scattergram of weight and height.
b. Calculate the OLS sample regression line relating weight to height.
c. Interpret your coefficients in words.
d. Calculate the sum of squared errors, mean square error, and standard error of the estimate.
e. Create an ANOVA table. Can you conclude that weight depends on height in the population? Use $\alpha = 0.05$. What is the p-value of this test?
f. How much of the variation in weight can be explained by height?
g. Can you conclude that the true population slope coefficient is different from zero? Use $\alpha = 0.05$. What is the p-value of this test?
h. Find a 95% confidence interval estimate for the mean weight of young adults who are 70 inches tall.
i. Find a 95% confidence interval estimate for the weight of an individual young adult who is 70 inches tall.

12.3 Continue with these same data.
a. Create a scattergram of weight and age.
b. Calculate the OLS sample regression line relating weight to age.
c. Interpret your coefficients in words.
d. Calculate the sum of squared errors, mean square error, and standard error of the estimate.
e. Create an ANOVA table. Can you conclude that weight depends on age in the population? Use $\alpha = 0.05$. What is the p-value of this test?
f. How much of the variation in weight can be explained by age?
g. Can you conclude that the true population slope coefficient is different from zero? Use $\alpha = 0.05$. What is the p-value of this test?
h. Find a 95% confidence interval estimate for the mean weight of young adults who are 25 years old.
i. Find a 95% confidence interval estimate for the weight of an individual young adult who is 25 years old.

12.4 Continue with these same data.
a. Create a scattergram of weight and sex.
b. Calculate the OLS sample regression line relating weight to sex.
c. Interpret your coefficients in words.
d. Calculate the sum of squared errors, mean square error, and standard error of the estimate.
e. Create an ANOVA table. Can you conclude that weight depends on sex in the population? Use $\alpha = 0.05$. What is the p-value of this test?
f. How much of the variation in weight can be explained by sex?
g. Can you conclude that the true population slope coefficient is different from zero? Use $\alpha = 0.05$. What is the p-value of this test?

12.5 Continue with these same data.
a. Create a scattergram of height and sex.
b. Calculate the OLS sample regression line relating height to sex.
c. Interpret your coefficients in words.
d. Calculate the sum of squared errors, mean square error, and standard error of the estimate.
e. Create an ANOVA table. Can you conclude that height depends on sex in the population? Use $\alpha = 0.05$. What is the p-value of this test?
f. How much of the variation in height can be explained by sex?
g. Can you conclude that the true population slope coefficient is different from zero? Use $\alpha = 0.05$. What is the p-value of this test?

12.6 Continue with these same data. How do your results in the previous question relate to your results in questions 9.3 and 11.2?

12.7 Refer back to the Students2.xls data file, that contains a variety of information on a sample of 100 students at a large university. Suppose you expect that students who did better in high school receive more financial aid.

a. Create a scattergram of financial aid and high-school GPA.
b. Calculate the OLS sample regression line relating financial aid to high-school GPA.
c. Interpret your coefficients in words.
d. Calculate the sum of squared errors, mean square error, and standard error of the estimate.
e. Create an ANOVA table. Can you conclude that financial aid depends on high-school GPA in the population? Use $\alpha = 0.05$. What is the p-value of this test?
f. How much of the variation in financial aid can be explained by high-school GPA?
g. Can you conclude that the true population slope coefficient is different from zero? Use $\alpha = 0.05$. What is the p-value of this test?
h. Find a 95% confidence interval estimate for the mean financial aid of students whose high-school GPAs were 4.0.
i. Find a 95% confidence interval estimate for the financial aid of an individual student whose high-school GPA was 4.0.

12.8 Continue with these same data. Suppose you expect that students who did better in high school—because they are smarter or have better study habits—do better in college as well.

a. Create a scattergram of college GPA and high-school GPA.
b. Calculate the OLS sample regression line relating college GPA to high-school GPA.
c. Interpret your coefficients in words.
d. Calculate the sum of squared errors, mean square error, and standard error of the estimate.
e. Create an ANOVA table. Can you conclude that college GPA depends on high-school GPA in the population? Use $\alpha = 0.05$. What is the p-value of this test?
f. How much of the variation in college GPA can be explained by high-school GPA?
g. Can you conclude that the true population slope coefficient is different from zero? Use $\alpha = 0.05$. What is the p-value of this test?
h. Find a 95% confidence interval estimate for the mean college GPA of students whose high-school GPAs were 4.0.
i. Find a 95% confidence interval estimate for the college GPA of an individual student whose high-school GPA was 4.0.

12.9 Continue with these same data. Suppose you expect that students who study more do better in college as well.

a. Create a scattergram of college GPA and study time.
b. Calculate the OLS sample regression line relating college GPA to study time.
c. Interpret your coefficients in words.
d. Calculate the sum of squared errors, mean square error, and standard error of the estimate.

e. Create an ANOVA table. Can you conclude that college GPA depends on study time in the population? Use $\alpha = 0.05$. What is the p-value of this test?
f. How much of the variation in college GPA can be explained by study time?
g. Can you conclude that the true population slope coefficient is different from zero? Use $\alpha = 0.05$. What is the p-value of this test?
h. Find a 95% confidence interval estimate for the mean college GPA of students who study 30 hours per week.
i. Find a 95% confidence interval estimate for the college GPA of an individual student who studies 30 hours per week.

13

Multiple Regression

13.1 Extensions of Regression Analysis

In the last chapter, we saw how we could use ordinary least squares (OLS) regression to estimate a linear relationship between two numeric variables. In this chapter, we generalize the procedure to allow for two complications. First, the relationship if it exists may involve more than two variables. And second, the relationship if it exists may be nonlinear. The actual algebra gets a great deal messier, and the algebra of "simple" regression was messy enough already. The concepts, though, are fairly straightforward extensions of Chapter 12. We will let the computer do the algebra and concentrate on these concepts.

13.1.1 When There Is More Than One Cause

We ended Chapter 12 with an allusion to the issue of multiple causation. We found, in the Employees2.xls data, evidence that salary depended on education and (in Exercise 12.1) job experience. We found no such evidence for sex. However, we were looking at these explanatory variables just one at a time. If education and job experience are correlated and we exclude one, the included one will pick up some of the effect of the other. The included one will seem more or less important than it really is. Indeed, the same goes for sex. If women have more or less education and experience, and these variables are both excluded, sex will pick up some of their effect. (In this case, the finding that sex has no effect would actually be troubling.) The problem is called **omitted variable bias** and, to avoid it, we need to include all our explanatory variables in the same regression. In such a regression, each individual slope coefficient represents the effect of its variable, *ceteris paribus*—that is, holding constant the other variables in the regression.

13.1.2 When the Line Is Not Straight

All our previous problems also assumed that the relationship was a straight line. In our salary example this meant that each additional year of education or experience had the same effect on salary. But clearly this need not be the case. For example, additional experience might be subject to diminishing returns. This would imply that the true slope with respect to experience flattens out. It is not just a constant "β" and we would not want to estimate it as such. We will look at a couple of simple ways in which we can allow for curvature.

13.2 The Population Regression Line

As before, when we estimate a sample regression line among variables, it is because we have reason to think that there really is a relationship among these variables in the population. With just two variables we were able to graph this relationship, as we did in Figure 12.1 (page 270). With additional variables, we need more dimensions and graphs lose much of their appeal. Still, if you compare Figure 13.1 with Figure 12.2, you should see that what we are proposing is a fairly obvious extension. Instead of just one variable X, with an effect of β on the dependent variable, there are $k - 1$ variables, $X_2, X_3,..., X_k$, with effects of $\beta_2, \beta_3,..., \beta_k$ on the dependent variable. As in Figure 12.2, Figure 13.1 shows this assumed relationship in a couple of ways. Version 1 emphasizes that the *expected*, or *average*, value of Y depends on the Xs. Version 2 emphasizes that the *actual* value of Y—even in the population—is the sum of a deterministic part that depends on the Xs and a random part that does not.

A couple of asides on notation and terminology. First, I have called the intercept α and the slopes βs to agree with the notation of the last chapter. But I have also included α in my numbering of the coefficients, as if it were β_1. Indeed, there is nothing wrong with calling it β_1, if that makes more sense to you. The important point is that there are k coefficients, of which $k - 1$ are slopes. You may well come across other authors who count things differently, α, or β_0, might be the intercept and $\beta_1 \ldots \beta_k$ the slopes. In that case, there would be $k + 1$ coefficients, of which k are slopes. All degrees of freedom would seem to be off by one. So, as you move beyond this book, always make sure you know whether k is the number of *coefficients* or the number of *slopes*. In this book, it is always the number of *coefficients*.

Secondly, I will continue to call the equation a "line." Technically, the relationship becomes a "plane," and then a "hyperplane," as we go into additional dimensions. But, as a practical matter, we will (almost) always be able to think of ourselves as looking at the slope of the line relating Y and X_2, *ceteris paribus*, the line relating Y and X_3, *ceteris paribus*, and so on.

The logic, then, parallels what we have already been through with simple regression. We do not have access to the whole population; hence we will be taking a sample and estimating the line from just this sample.

1. $$E\left(Y_i \middle| \vec{X}_i\right) = \alpha_1 + \beta_2 X_{2i} + \beta_3 X_{3i} + \ldots + \beta_k X_{ki}$$

or

2. $$Y_i \middle| \vec{X}_i = \alpha_1 + \beta_2 X_{2i} + \beta_3 X_{3i} + \ldots + \beta_k X_{ki} + \varepsilon_i$$

where ε_i is a normally distributed random variable with zero mean and constant variance, which is uncorrelated with the Xs or itself.

Figure 13.1 The population regression line.

But we want this sample line to be a good estimate of the unseen population line. We want to be able to use our estimate to make good predictions about average or individual values of Y for new values of the Xs. Moreover, we could be wrong about there being relationships among these variables in the population at all. Hence, we also want to test hypotheses about these relationships.

Again, if it were not for the random variation around the population regression line, none of this would present a problem. Our sample points would all be on the population regression line; we would get an exact estimate of the population regression line. Our predictions would be exactly right. And, as long as the slopes were not zero, we would conclude that we had been right to think the variables were related.

Again, though, because of the random variation around the population regression line, our sample points will almost certainly not be on the population regression line. Again, there are whole distributions of sample slopes (and intercepts) that we could get from this population. There is a whole distribution of predictions about average or individual values of Y for new values of the Xs. We will need confidence intervals again.

Moreover, suppose some of the population regression slopes were actually zero; that the value for Y did not really depend on some of the Xs at all. There are still whole distributions of sample slopes we could get for those Xs, most of which are not zero. Hence, we will need to test hypotheses about slopes again. Are our sample slopes so different from zero that we can conclude the true population slopes are not zero, with an acceptable probability of being wrong?

13.3 The Sample Regression Line

13.3.1 The Best Sample Line: Ordinary Least Squares

Our criterion for the best sample regression is still OLS. Our sample regression line is of the form $\hat{Y} = a_1 + b_2\ X_2 + b_3\ X_3 + \ldots + b_k\ X_k$. As before, $\hat{Y}$ is the value of Y predicted by the equation, as opposed to the actual data value, Y. As before, we want the prediction errors—the $(Y - \hat{Y})$s—to be small. As before, we square these errors, to keep positives and negatives from canceling out, and minimize the sum of the squared errors, $\text{SSE} = \Sigma(Y_i - \hat{Y}_i)^2$. Conceptually, imagine shifting the line (by changing a_1) and rotating it in all dimensions (by changing the bs). As you do, the line will get closer to some points and further from others. Some of the squared errors will get smaller and some of the squared errors will get larger. According to the OLS criterion, our best line is the one with the combination of intercept and slopes that minimizes the sum of squared errors.

In Chapter 12, all this led to two equations in two unknowns—a and b—and the algebra was quite doable. Here, we have k equations in k unknowns—$a_1\ b_2,\ b_3, \ldots,\ b_k$—and, as a practical matter, the solution requires the techniques of linear algebra. This is why we are going to turn the algebra over to the computer. Conceptually, though, we are doing the same thing as in Chapter 12.

13.3.2 Finding the Intercept and Slopes: A Job for the Computer

Spreadsheets like Excel, Lotus, and Quattro Pro will calculate the regression for you if you have no other statistical program available. Typically, though, multiple regression is done using a dedicated statistical program. Some of the better-known programs are SAS, SPSS, Shazam, Stata, and EViews, but there are many others. All allow you to transform your data easily (which will turn out to be useful). All allow you to specify a dependent variable and a list of explanatory variables. All present the results in a manner such as in Figure 13.2. And all support more advanced techniques, which will be valuable if you take additional courses in statistics or econometrics. It is worth learning to use such a package.

13.3.3 Interpreting the Intercept and Slopes

The output in Figure 13.2 uses the Employee2.xls data that we used in Chapter 12 and have already referred back to in this chapter. The regression relates salary to education, experience, and sex. The constant, $a_1 = -1.6521$, is our estimate for the salary (in $ thousands) of a male employee with no education and no experience. It is of limited interest since we have no employees with no education. The other coefficients are the interesting ones. We estimate that an additional year of education increases salary by 2.5659 ($ thousand), *experience and sex held constant*; an additional year of experience increases salary by 0.5010 ($ thousand), *education and sex held constant*; and being female instead of male decreases salary by 2.2872 ($ thousand), *education and experience held constant.* Find these numbers in Figure 13.2.

Note the difference in the interpretation of the coefficients for numerical and dummy variables. The coefficients of the numerical variables

Multiple Regression

Dependent Variable: Salary

R Square: 0.9820	**Adjusted R Square:** 0.9808	**Standard Error:** 1.3283

ANOVA	**Sum of Squares**	**df**	**Mean Square**	**F_c**	**2-Tailed p-Value**
Regression	4429.8539	3	1476.6180	836.9373	0.0000
Error	81.1583	46	1.7643		
Total	4511.0122	49			

Variable	**Coefficient**	**Standard Error**	**t_c**	**2-Tailed p-Value**
Constant	−1.6521	1.1096	−1.4888	0.1434
Ed	2.5659	0.0797	32.2134	0.0000
Exp	0.5010	0.0144	34.7279	0.0000
Female	−2.2872	0.3979	−5.7478	0.0000

Figure 13.2 A typical multiple regression output.

are estimated effects on salary of one more unit. The coefficient of the dummy variable is an estimated effect of being in one category instead of the other. Still, all of the coefficients are estimated effects of a change in their variable, *the others in the regression being held constant.* And note that these results are indeed quite different from the results that we got in our three separate simple regressions in Chapter 12.

13.4 Evaluating the Sample Regression Line

13.4.1 The Sum of Squared Errors

Most of the measures we used to evaluate sample regressions in Chapter 12 carry over in much the way one might expect. Moreover, any decent statistical program will provide them automatically. The most basic is the sum of squared errors. We have minimized it but it is not zero. How big is it? The ANOVA table in Figure 13.2 reports that the original squared variation around the mean was SST = 4511.0122 (the same as in all the simple regressions we ran with these data). With this set of three explanatory variables, we are able to explain or account for SSR = 4429.8539 leaving SSE = 81.1583 unexplained. Find these numbers in the printout.

13.4.2 The Mean Square Error and Standard Error of the Estimate

The formulas for the mean square error, s_e^2 and standard error of the estimate, s_e adjust as shown in Figure 13.3. We lose an additional degree of freedom with each additional coefficient so, instead of dividing through by $n-2$, we divide through by $n-k$. Note that this is really just a generalization of the formula. In Chapter 12, k was always 2; we had just an intercept and a slope. Now it can vary. Note too that a standard regression output like that in Figure 13.2 does the algebra. The mean square error, 81.1583/46 = 1.7643, and the standard error or the estimate, $\sqrt{1.7643} = 1.3283$, are both explicitly reported. Find them.

13.4.3 R^2 and $\bar{R}^2$: The Coefficients of Determination

Figure 13.4 shows two versions of the coefficient of determination. The R^2 is calculated as before; it is the proportion of the original variation that the regression can explain, or one minus the proportion of the original

$$s_e^2 = \frac{\sum \left(Y_i - \hat{Y}_i\right)^2}{n-k} = \frac{\text{SSE}}{n-k} = \text{MSE}$$

$$s_e = \sqrt{\frac{\sum \left(Y_i - \hat{Y}_i\right)^2}{n-k}} = \sqrt{\frac{\text{SSE}}{n-k}} = \sqrt{\text{MSE}}$$

Figure 13.3 The mean square error and standard error of the estimate.

$$R^2 = \frac{\text{SSR}}{\text{SST}} = 1 - \frac{\text{SSE}}{\text{SST}} \quad \bar{R}^2 = 1 - \frac{\text{SSE}/n-k}{\text{SST}/n-1} = 1 - \frac{s_e^2}{s_r^2} \quad F_c = \frac{\text{SSR}/k-1}{\text{SSE}/n-k}$$

Figure 13.4 R^2 and $\bar{R}^2$: The coefficients of determination and the F_c statistic.

variation that the regression cannot explain. However there is a problem with this measure as we begin to consider regressions of differing lengths. Adding an additional explanatory variable can never lead to the regression explaining less. Indeed, the additional variable, no matter how irrelevant—the phase of the moon or the price of tea in China—is likely to reduce the SSE by some small, random amount. Hence, R^2 always increases with the addition of another variable. It is biased in favor of longer regressions.

A better measure is the **adjusted R^2**, often written as $\bar{R}^2$ ("R-bar squared"). $\bar{R}^2$ takes degrees of freedom into account. The addition of another variable affects both SSE and $n - k$. $\bar{R}^2$ will go up with the addition of another variable only if the reduction in SSE is great enough to compensate for the loss of a degree of freedom. $\bar{R}^2$ is not strictly between zero and one, as R^2 is. A regression that uses many explanatory variables, yet explains almost nothing, can have a negative $\bar{R}^2$. Still, it is generally interpreted in the same manner as R^2—as the proportion of the original variation that can be explained, or accounted for, by the regression. Again, a standard regression output such as that in Figure 13.2 will report both R^2 and $\bar{R}^2$ explicitly. With $R^2 = 0.9820$ and $\bar{R}^2 = 0.9808$ in our example (find them), education, experience, and sex can explain a little more than 98% of the variation in Salary.

13.4.4 Testing the Sample Regression Line

Finally, we can use the ANOVA table to test whether Salary is actually independent of this set of explanatory variables; that there really is no relationship. Formally,

1. H_o: Salary is independent of Education, Experience, and Sex — H_o: $\beta_2 = \beta_3 = \beta_4 = 0$

   H_a: Salary is not independent of Education, Experience, and Sex — H_a: Slopes do not all = 0

As always, F_c is the ratio of two variances. As with simple regression, the variance in the denominator is just the mean square error, s_e^2, our estimator of σ_e^2, the constant variance around the population regression line. Again, as with simple regression, the numerator is also an estimator of σ_e^2. Under the null hypothesis, there is no real relationship between these explanatory variables and salary. Still, we would expect essentially any three variables to have some small, random correlation with salary. Hence, the SSR will not be zero. Rather, the SSR/df$_n$—where the degrees of freedom equal the number of coefficients minus one—will have an expected value of σ_e^2. The result is that, under the null hypothesis, the F_c

statistic has an expected value of one. Of course, if there *is* a relationship between our explanatory variables and salary, we would expect to explain more than just a random amount of variation in salary. The SSR/df_n will be larger; hence, F_c will be larger. We are looking for a value of F_c so large that we can conclude (with an acceptable probability of being wrong) that there is, in fact, a relationship.

2. Reject H_o if $F_c > 2.812$ ($df_n = 3$, $df_d = 46$; $\alpha = 0.05$).
3. Figure 13.2 includes all the information we need; indeed it presents both F_c and its associated *p*-value explicitly. Find them. $F_c = 836.9373$
4. $\therefore$ Reject H_o

The *p*-value is 0.0000 to four decimals. There is essentially no chance of getting a relationship this strong just by chance.

13.5 Evaluating the Sample Regression Slopes

We know that, taken together, the regression is able to explain a significant amount of the variation in salary. But which of the variables actually contribute? It is perfectly possible that only one or two of the explanatory variables are actually relevant. Indeed, recall that when we ran three separate simple regressions in Chapter 12, we found that education and experience appeared to matter but sex did not. To see which variables in a multiple regression really matter, we need to test their individual coefficients. To conclude that a variable really matters, we must be able to reject the hypothesis that its individual slope coefficient is zero.

1. H_o: $\beta_i = 0$
   H_a: $\beta_i \neq 0$.

I have written the hypotheses as a two-tailed test though, following the reasoning of the last chapter, we might argue for one-tailed alternatives—positive for education and experience and negative for female.

Our test statistic follows the *t* distribution. The degrees of freedom equal *n* minus the number of coefficients estimated, $n - k$. Since 46 degrees of freedom is not in the table, we would use 45. And, because this is a two-tailed test, we need to remember to divide α in half to look it up. Alternatively, we can use the special spreadsheet function = TINV(0.05,46) = 2.013.

2. Reject H_o if $|t_c| > 2.013$ ($df = 46$; $\alpha = 0.05$).

Our test statistic is calculated just as in simple regression:

$$t_{ic} = \frac{b_i - \beta_i}{s_{b_i}} = \frac{b_i - 0}{s_{b_i}} = \frac{b_i}{s_{b_i}}.$$

The calculation of the s_{b_i} values is a good deal messier, but the standard computer output includes them. Refer back to Figure 13.2 again. Next to each coefficient is its standard error, its *t*-value for this test and even its *p*-value. Find them. The *p*-values for each of the slopes is 0.0000 to four decimals. There is essentially no chance that any of these coefficients is just randomly different from zero. All three explanatory variables appear to contribute to the variation in salary.

Note that this result is different from the result of our individual simple regressions that found no separate effect of sex. Apparently there is a separate effect of sex when we control for education and experience.

It has been a while since you began this book; turn back and reread pages 1 and 2. What we have done here is address the third scenario. It is this regression that allows you to report back to your boss something like the following from page 2.

> Based on this sample, I estimate that we are rewarding employees an average of about $2500 for each additional year of education, and an average of about $500 for each additional year of experience. However, we appear to have a problem with gender equity. I estimate that females are being paid nearly $2300 less than equally qualified males.

Make sure you understand how this statement follows from the results in Figure 13.2.

13.6 Predictions Using the Regression Line

13.6.1 Using the Regression Line to Predict Y Given the X_i

Since our regression seems a good fit—it can explain about 98% of the variation in salary and all the explanatory variables have significant coefficients—we may want to use it to make predictions. We need to choose values for each of our explanatory variables. Suppose we want to predict the salary of a male with 10 years of education and 20 years of experience. Ed = 10, Exp = 20, and Female = 0. Predicted Salary is

$$\widehat{\text{Salary}} = -1.6521 + 2.5659 \times 10 + 0.5010 \times 20 - 2.2872 \times 0 = 34.0269 \text{ (\$ thousand)}.$$

13.6.2 Confidence Intervals for Y Given the X_i

As before, we can think of $\hat{Y}$ ($\widehat{\text{Salary}}$) as an estimate of the mean or expected value of Y (Salary), given these values for education, experience, and sex. Or we can think of it as a prediction for an individual employee given these values. The predictions are the same—about $34,027—but the confidence intervals around them are different.

Both confidence intervals are centered on $\hat{Y}$ and are $t_{(n-k)}$ standard errors around $\hat{Y}$. (See Figure 13.5.) The only difference is in the standard errors. The standard error for the mean is smaller because there is only one source of error—we are using a sample regression line, which is bound to be somewhat wrong. The standard error for an individual case is larger because—in addition to this source of error—the individual points are scattered around the line even in the population.

1. The mean Y given X_i	2. a. An individual Y given the X_i
$\hat{Y} \pm t_{(n-k)} s_{\bar{Y}\|X_i}$	$\hat{Y} \pm t_{(n-k)} s_{\hat{Y}\|X_i}$
	b. An approximation
	$\hat{Y} \pm t_{(n-k)} s_e$

Figure 13.5 Confidence intervals for Y given the X_i.

Unfortunately, since the standard errors vary with the values of the X_i, computer programs cannot simply include them in their standard outputs. Some programs allow you to request standard errors for particular combinations of the X_i. For these values of the X_i—Ed = 10, Exp = 20, Female = 0—the standard errors are

$s_{\bar{Y}|X}$ = 0.3562 ($ thousand)

The 95% CI for the mean Salary of male employees with 10 years of education and 20 years of experience:

95% CI: $34,027 ± 2.013 × $356.2
95% CI: $34,027 ± $717

$s_{\hat{Y}|X}$ = 1.3752 ($ thousand)

The 95% CI for the Salary of an individual male employee with 10 years of education and 20 years of experience:

95% CI: $34,027 ± 2.013 × $1,375.2
95% CI: $34,027 ± $2,768.

If your program does not give you these standard errors, s_e—always included in the standard output—can sometimes be used as an approximation for $s_{\hat{Y}|X}$.

To understand this approximation, and its limitations, refer back to the simple regression case in Chapter 12 (Section 12.6.2) For simple regression, $s_{\hat{Y}|X} = s_e\sqrt{1+(1/n)+((X^* - \bar{X})^2/SSD_X)}$. So, $s_{\hat{Y}|X} = s_e$ times a correction of $\sqrt{1+}$. How much greater than 1 this correction is depends on sample size (the second term) and closeness to the mean of the data (the third term). It is always at least a little greater than 1 because $1/n$ is never really zero and we are seldom exactly at the mean of the data. So the true $s_{\hat{Y}|X}$ is always at least a little greater than s_e. Still, $1/n$ is never very large. So, if we are close to the mean of the data, the approximation, $s_{\hat{Y}|X} \approx s_e$, is reasonably good.

The formulas for multiple regression are messier, but it is still true that $s_{\hat{Y}|X} = s_e$ times a correction of $\sqrt{1+}$. How much greater than 1 the correction is still depends on sample size and closeness to the mean of the data. If we are close to the mean of the data, the approximation, $s_{\hat{Y}|X} \approx s_e$, is still reasonably good.

For our predicted salary of an individual male with 10 years of education and 20 years of experience, an approximate 95% confidence interval would be

95% CI: $34,027 ± 2.013 × $1,328.3

95% CI: $34,027 ± $2,674.

Comparing it to the actual confidence interval previously, it is a little too narrow. But it is probably close enough for many purposes. Unfortunately, there is no similar approximation for $s_{\overline{Y}|X}$.

13.7 Categorical Variables

So far our explanatory variables have been numeric, like education and experience, or dummy variables—categorical variables with just two categories. Notice that a single dummy variable distinguishes between two categories. Female is enough to distinguish between females and males. We did not need a separate variable Male. Indeed, had we included variables for both Female and Male, the computer would have kicked one out for being redundant. We already know who the males are—those who are not female.

The use of Female instead of Male, though, was completely arbitrary. What if we had included Male instead of Female? The coefficients would have changed in such a way as to mean exactly the same thing. Refer back to Figure 13.2 again. The coefficient for Female was –2.2872, meaning that women earned 2.2872 *less* than men with the same education and experience. If we had included Male instead, the coefficient for Male would have been + 2.2872, meaning that men earned 2.2872 *more* than women with the same education and experience. The meaning is exactly the same.

The constant would also have changed by the amount of the male/female coefficient. In our regression, male was the "omitted case," so you can think of the regression constant as the y intercept for men. Men with zero education and experience are predicted to earn –1.6521 ($ thousand). And you can think of the constant plus the coefficient for Female as the y intercept for women. Women with zero education and experience are predicted to earn $-1.6521 - 2.2872 = -3.9393$ ($ thousand). If we had included Male instead of Female, female would have been the omitted case, so the constant would have been the y intercept for women, and it would have equaled –3.9393 ($ thousand). And the constant plus the coefficient for Male, $-3.9393 + 2.2872 = -1.6521$, would have been the y intercept for males.

In short, as long as you always think of the regression constant as the y intercept for the omitted case, and the coefficients for the dummy variables as *shifts relative* to the omitted case, you get the same interpretation regardless of which variable you include and which you make the missing case.

Now, suppose that we thought an employee's job type should be included. Recall that Type was defined as line = 1, office = 2, and management = 3. This is a categorical variable with more than two categories. To use this variable we would need to be able to assume that the intervals from 1 to 2 and 2 to 3 were somehow equal. And that is very unlikely.

Instead, we need to create a set of dummy variables. With three categories we need two dummies. Suppose we decide on Office (yes = 1, no = 0) and Manage (yes = 1, no = 0). This would make line workers the omitted case. The regression constant would be the y intercept for line workers; the coefficient for Office would be the shift for office workers relative to line workers; the coefficient for Manage would be the shift for managers relative to line workers.

The means of actually creating Office and Manage will vary considerably with your statistical program. All have menus or commands—transform, compute, generate, replace, etc.—designed for the purpose of creating new variables from the existing data. For dummies, typically, you do it in two steps. First you create a variable, setting its value to 0 for all cases; second you change the value to 1 for the appropriate cases. For example:

compute Office = 0 compute Manage = 0
replace Office = 1 if Type = 2 replace Manage = 1 if Type = 3.

If you skip the first step, typically, the new variables will have missing values rather than zeros for all the cases not equal to 1.

Figure 13.6 shows the results with Office and Manage included. Office employees earn an estimated 0.3141 ($ thousand) less than line employees, other things equal; management employees earn an estimated 0.1451 ($ thousand) more than line employees, other things equal.

Evaluating the results, we would ordinarily look first at the signs and sizes of these coefficients. Are they plausible? These differences seem small. However, it is important to remember that we are controlling for the effects of education, experience, and sex. So perhaps they are not unreasonable.

If so, we would want to know if they are significantly different from zero. That is, can we reject the hypothesis that the true βs are zero.

1. H_o: $\beta_i = 0$
   H_a: $\beta_i \neq 0$.

Multiple Regression

Dependent Variable: Salary

***R* Square:** 0.9823		**Adjusted *R* Square:** 0.9803		**Standard Error:** 1.3477	
ANOVA	**Sum of Squares**	**df**	**Mean Square**	F_c	**2-Tailed *p*-Value**
Regression	4431.1001	5	886.2200	487.9574	0.0000
Error	79.9121	44	1.8162		
Total	4511.0122	49			

Variable	Coefficient	Standard Error	t_c	2-Tailed *p*-Value
Constant	−1.2660	1.2229	−1.0352	0.3062
Ed	2.5425	0.0866	29.3662	0.0000
Exp	0.4995	0.0153	32.7116	0.0000
Female	−2.2169	0.4220	−5.2535	0.0000
Office	−0.3141	0.4645	−0.6761	0.5025
Manage	0.1451	0.5486	0.2645	0.7927

Figure 13.6 Results with the addition of Office and Manage.

Our test statistic follows the t distribution, now with 44 degrees of freedom. For an α of 0.05, our criterion is 2.015.

2a. Reject H_o if $|t_c| > 2.015$ (df = 44; $\alpha = 0.05$).

Since the computer output includes p-values, we can recast our criterion in terms of these and avoid the need for t tables or spreadsheet functions. Recall that the p-values are the probabilities of getting a test statistic this extreme just by chance. Hence, they are the αs, the probabilities of rejecting wrongly, we need to be willing to accept in order to reject H_0. For an α of 0.05, then our criterion is just:

2b. Reject H_o if p-value < 0.05.

With p-values of 0.5025 and 0.7927, these coefficients could easily have arisen just randomly. We cannot reject the hypothesis that the true slopes are zero. We cannot conclude that salary depends on being an office versus line employee or a management versus line employee, other things equal.

This example raises several related issues. First, the t test is a test of individual coefficients. We have tested individually whether the effect of office versus line is zero and whether the effect of management versus line is zero. This is not quite the same thing as testing whether the three-way effect is zero. For that we would need to test the pair of coefficients together. There are such tests and they would have given the same answer in this case. This pair of coefficients could easily be just randomly different from zero. Such tests are not difficult but they are beyond the scope of this course.

Second, how do we respond to the fact that these new coefficients are not statistically significant? Do we drop Office and Manage? There is no simple answer. Other things equal, simpler is better and more efficient. This would argue for dropping them. On the other hand, we have not shown that job type does not matter; we have only failed to show that it does. If it really does and we leave it out, we risk omitted-variable bias. The coefficients on the other variables will pick up the effects of the omitted variables and be distorted. The decision should probably rest on the strength of the original case for including the variable. For example, economists speak of "the *law* of demand," which says that the quantity of a good demanded varies inversely with the price. So if we are estimating a demand curve, Price really *must* be an explanatory variable, whether it is statistically significant or not. On the other hand, the case is not nearly so strong here for including job type. There is no law that says that job type must be a determinant of salary, education, experience, and sex held constant. And both coefficients are very far from being statistically significant. So we would probably drop these variables.

Third, note the changes in the other coefficients. Every time we add or drop a variable, all the coefficients are reestimated and change. This is because of correlation among the explanatory variables—a condition known as **multicollinearity**. If multicollinearity is strong, the addition or deletion of a variable can change the results for the other variables a great deal. And this can make the decision about adding or deleting a variable both important and difficult. In the current example, there is actually very little change in the original coefficients when we add Office and Manage. The coefficient for Ed is still a little more than 2.5 ($ thousand); that for Exp is still around 0.5 ($ thousand); and that for Female is still around –2.2 ($ thousand). The estimates are robust. And this makes the decision to delete the insignificant job type variables much easier.

Finally, look at the overall regression information. The two job type variables do reduce SSE slightly; hence, R^2 rises slightly. However, all the other overall regression statistics get worse. The s_e rises and $\bar{R}^2$ falls because the reduction in SSE is not enough to compensate for the two degrees of freedom lost. These changes are consistent with all the preceding discussion. Office and Manage seem to add little or nothing.

13.8 Estimating Curved Lines

Years ago, a student in my class who came from a farm decided to try a regression using annual rainfall to predict the corn crop yield. He had several years worth of data from his farm. Much to his surprise, he did not find a significant relationship. He came to see me because he was sure that he must have done something wrong. When we plotted the data we got something like Figure 13.7. Clearly, there was a relationship; it was just not a straight line. There could be too much or too little rain. We needed to estimate a curve.

There are all sorts of functional forms that plot as curves. We will stick with two that are simple to estimate and interpret: the quadratic function and the Cobb–Douglas function.

13.8.1 The Quadratic Function

Figure 13.8 gives the quadratic function; it graphs as a parabola, either rising and then falling or falling and then rising. The curve in Figure 13.7 is a quadratic. Figure 13.9 shows several others.

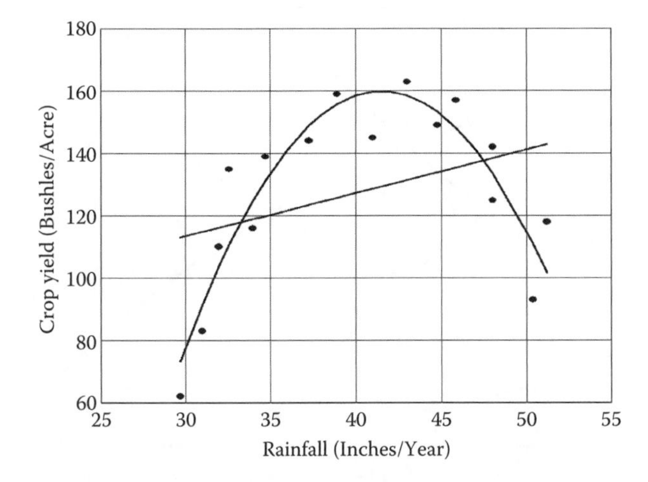

Figure 13.7 Rainfall and crop yield.

$$Y = a_1 + b_2X + b_3X^2$$

Figure 13.8 The Quadratic Function.

The signs on b_2 and b_3, the coefficients for X and X^2, are the keys to the shape. Since X^2 grows faster than X, the final part of the equation eventually dominates; hence the sign on b_3 determines whether the eventual slope is positive or negative. In Figure 13.9, then, since b_3 is negative for all of the curves through point A, they all eventually turn down. And since b_3 is positive for all of the curves through point B, they all eventually turn up.

The sign of b_2 relative to b_3 determines the turning point. If the signs disagree, the curve turns at a positive value for X, so we see the turning point in quadrant 1. If they agree, the curve turns at a negative value for X, so it has already turned before entering quadrant 1. And if $b_2 = 0$, the curve turns right at the Y axis, where $X = 0$.

Several additional points should be made. First, we will be creating new variables, X_i^2s , and estimating the coefficients for equations like that in Figure 13.8. If the b_3s—the coefficients for the X_i^2s—are significantly different from zero, we will conclude that we have a curve rather than a straight line, regardless of whether or not the b_2s are significant. Recall that two of the six curves in Figure 13.9 have b_2s of zero. In assessing whether a relationship is nonlinear rather than linear, it is the coefficients on the squared terms that matter.

Second, the interpretation of the coefficients must change somewhat. We can no longer think of them as slopes. There is no single slope. And we cannot talk of changing just X (or X^2), other things held constant. We cannot change X holding X^2 constant.

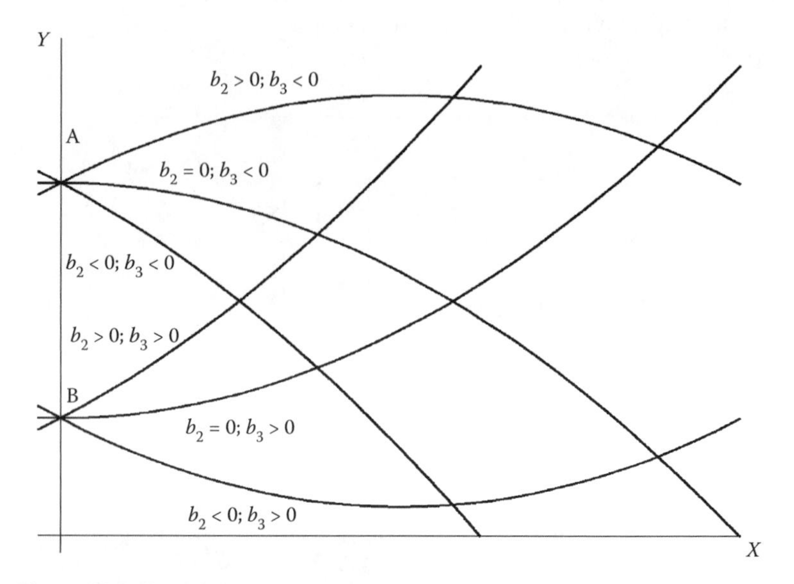

Figure 13.9 Quadratics of various shapes.

The slope at a particular point is $b_2 + 2b_3 X$. If b_2 is zero or agrees in sign with b_3, we can talk in general terms about increases in X having a positive or negative and increasing effect on Y. Even if the coefficients differ in sign, it may be that the data are all actually on the upward- or downward-sloping side of the turning point. The turning point in terms of X can be calculated as $X = -b_2/(2b_3)$. Thus, if our equation looks like the top one in Figure 13.9, but all our data are on the upward-sloping side, we can say that increases in X have a positive but declining effect on Y. Finally we may find, as in Figure 13.7, that our data straddle the turning point. In this case, we would say that the increases in rainfall have a positive but declining effect on crop yield up until about 41½ inches and a negative and increasing effect beyond that.

Third, a multivariate relationship can be nonlinear in more than one dimension. It can also be linear in some dimensions but nonlinear in others. In our salary example, we can test to see whether the effects of education and experience have nonlinear effects on salary. Perhaps one or both is subject to "diminishing returns."

Finally, we cannot use squares of the dummy variables. Intuitively, it takes more than two points to differentiate a curve from a straight line. On a practical level $D^2 = D$; the squares of zero and one are zero and one. It is similar to including both Male and Female. If we tried to include both, the computer would kick one out for being redundant.

Let us return to the salary example to see if the effects of education and experience are nonlinear. As in the previous example we need to create new variables, and the exact means of doing so will vary with your statistical program. Use the menu or command—transform, compute, generate, etc.—to create new variables Ed2 and Exp2. For example,

compute Ed2 = Ed^2   compute Exp2 = Exp^2

Most programs limit special characters in the variable names; some use "**" instead of "^" for exponentiation.

Figure 13.10 presents the results. Actually, I first tried including Office and Manage too. Due to multicollinearity, the addition of new variables can change the significance of the variables already in the equation. Hence, we do not want to give up on variables too easily. In this case, though, Office and Manage performed no better, and Figure 13.10 shows the results with them removed.

Because we already know that education, experience, and sex matter, our main concern is with the coefficients on Ed2 and Exp2. Our alternative hypotheses might be two-tailed or, if we are expecting diminishing returns, one-tailed negative.

1. H_o: $\beta_i = 0$ H_o: $\beta_i = 0$
   H_a: $\beta_i \neq 0$ H_a: $\beta_i < 0$.

Our test statistic follows the t distribution, with 44 degrees of freedom. For an α of 0.05, our two-tailed criterion is 2.015; our one-tailed criterion is –1.680.

2a. Reject H_o if $|t_c| > 2.015$ (df = 44; $\alpha = 0.05$)   Reject H_o if $t_c < -1.680$ (df = 44; $\alpha = 0.05$).

Multiple Regression					
Dependent Variable: Salary					
R Square: 0.9822		**Adjusted R Square:** 0.9802		**Standard Error:** 1.3508	
ANOVA	**Sum of Squares**	**df**	**Mean Square**	**F_c**	**2-Tailed p-Value**
Regression	4430.7330	5	886.1466	485.6853	0.0000
Error	80.2792	44	1.8245		
Total	4511.0122	49			
Variable	**Coefficient**	**Standard Error**	**t_c**	**2-Tailed p-Value**	
Constant	1.1957	7.1960	0.1662	0.8688	
Ed	2.1617	1.0568	2.0455	0.0468	
Ed^2	0.0153	0.0388	0.3952	0.6946	
Exp	0.4627	0.0589	7.8604	0.0000	
Exp^2	0.0009	0.0014	0.6611	0.5120	
Female	–2.3590	0.4178	–5.6464	0.0000	

Figure 13.10 Results with the addition of squares for education and experience.

Again, since the computer output includes p-values, it makes sense to recast these criteria in terms of these. Recall that the p-values are the probabilities of getting a test statistic this extreme just by chance. They are also two-tailed, meaning that there is only half this probability in each tail. With the one-tail alternative, the reported p-value can be twice as high since we are using just one of the tails. For an α of 0.05, then, our criteria are just:

2b. Reject H_0 if p-value < 0.05 Reject H_0 if $b_i < 0$ and p-value < 0.10.

With a one-tailed alternative, we fail to reject immediately since we were looking for negative coefficients and got positives.

With a two-tailed alternative we do little better. With p-values of 0.6946 and 0.5120, these coefficients could easily have arisen just randomly. We cannot reject the hypothesis that the true coefficients for Ed2 and Exp2 are zero. We cannot conclude that the effects of education or experience are nonlinear. If there were very strong reasons to believe the relationship should be nonlinear and our coefficients made sense, we could decide to adopt this regression even though our results do not support our expectations. After all, we have not shown that the effects are linear; we have just failed to show that they are not. If we exclude the squares, we could be making a type II error. In this case, though, any theory behind expecting a curve would probably be that of "diminishing returns," and our new coefficients have the wrong signs for that. Between the two, we would prefer the simpler, linear form.

The Cobb–Douglas function itself:

$$Y = A_1 X_2^{b_2} X_3^{b_3}$$

The log–log transformation:

$$\ln(Y) = a_1 + b_2 \ln(X_2) + b_3 \ln(X_3)$$

Figure 13.11 The Cobb–Douglas function.

13.8.2 The Cobb–Douglas (Log–Log) Function

Figure 13.11 gives the **Cobb–Douglas** function, named for those who introduced it as a useful functional form in statistical work. The A_1 is just a constant. The explanatory variables are multiplied; their coefficients are exponents. The function is useful because it has some desirable properties and is easy to estimate. For instance, if you have had basic economics, you know that the responsiveness of one variable to another is often measured by elasticity rather than slope, where elasticity is the percent change of one variable given a one percent change in another; and it can be shown that the exponents in the Cobb–Douglas function are elasticities. That is, b_2 is the elasticity of Y with respect to X_2, the percent change in Y with a 1% change in X_2.

You can also think of it as just a differently shaped curve that might fit better than the quadratic in some cases. Figure 13.12 shows several examples. If b_2 is negative, the Cobb–Douglas is downward sloping; the more negative b_2 is, the sharper the bend. If b_2 is positive, it is upward sloping. If b_2 is positive but less than one, it flattens out; if b_2 is greater than one, it gets steeper.

The secret to estimating the Cobb–Douglas is that it is linear in the logs. That is, if we take the logs of both sides of the equation, we end up with a linear function that we can estimate in the usual way. Indeed, this transformed version is often called "log-linear." This term is somewhat

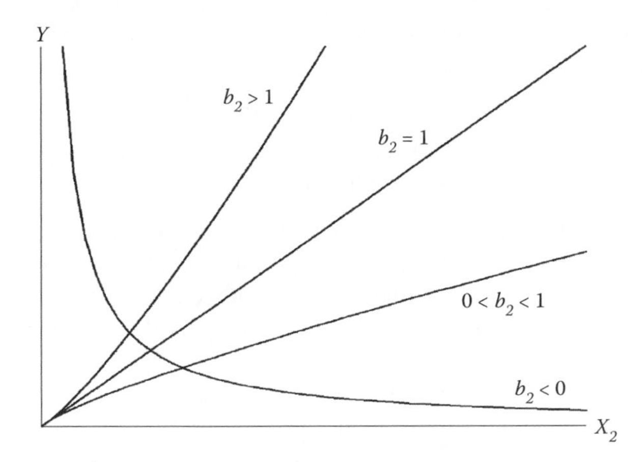

Figure 13.12 Cobb–Douglas functions of various shapes.

ambiguous, though, and should probably be avoided. Less ambiguous terms are **log–log** or **double-log**, to indicate that we are taking logs of both left- and right-hand side variables.

Ln(X) is the natural (base e) logarithm of X. Any base can be used as long as you are consistent, but natural logs are most common. The "anti-log" for ln(X) is e^X, often written exp(X). Many calculators, as well as all spreadsheets and statistical packages have built in ln(.) and exp(.) functions so the procedure is not tedious. First create new variables that are the logs of your original variables. Then, estimate the regression using the logs instead of the original variables. The b_is are estimates of the exponents in the original Cobb–Douglas function—the percent change in Y with a 1% change in X_i. The constant, a_1, is an estimate of Ln(A_1) in the original. If we want an estimate of the original A_1 we can just calculate exp(a_1). Of course we have seldom been terribly interested in the constant.

The interpretation of results is different in some respects from that of quadratics. First, with the quadratic, the curvature is caused by the squared terms; if their coefficients are significant we have evidence of curvature; if not we do not. With the Cobb–Douglas, the entire functional form is different; hence we must look elsewhere to decide whether we have evidence for curvature.

Second, unlike the quadratic, the Cobb–Douglas does not completely reverse direction. Hence, there is no ambiguity about whether a relationship is positive or negative.

Finally, we cannot take the logs of the dummy variables; the log of zero is undefined. We can still use the dummy variable itself in our regression. However, the coefficient for the dummy is not an elasticity. Indeed, we cannot even think of changing a dummy variable by 1%; we can only change it from zero to one. Instead, the coefficient $b = \ln(1 + p)$ where p is the proportional change in Y as the dummy goes from zero to one. Thus $p = \exp(b) - 1$. And multiplying p by 100, gives us the percentage change in Y as the dummy goes from zero to one.

Let us return one final time to the Salary example, to see how a log–log function fits these data. As in the previous examples we need to create new variables, and the exact means of doing so will vary with your statistical program. For example,

compute lnSal = ln(Salary)
compute lnEd = ln(Ed)
compute lnExp = ln(Exp).

Recall that, since Female is a dummy variable, we do not calculate its log; we do, though, include it in our regression. Figure 13.13 presents the results. (Again, I actually tried including Office and Manage too; again, they performed no better and Figure 13.13 shows the results with them removed.)

Unlike the last two examples, in which we were concerned primarily with additions to the regression, in this case we are really starting from scratch.

We estimate that an increase of 1% in education increases salary by about 0.75%, other things equal; an increase of 1% in experience increases

Multiple Regression					
Dependent Variable: lnSal					
R Square: 0.9147		Adjusted R Square: 0.9092		Standard Error: 0.0717	
ANOVA	**Sum of Squares**	**df**	**Mean Square**	F_c	**2-Tailed p-Value**
Regression	2.5364	3	0.8455	164.4860	0.0000
Error	0.2364	46	0.0051		
Total	2.7728	49			
Variable	**Coefficient**	**Standard Error**	t_c	**2-Tailed p-Value**	
Constant	1.4383	0.1482	9.7030	0.0000	
lnEd	0.7516	0.0580	12.9533	0.0000	
lnExp	0.1382	0.0091	15.1255	0.0000	
Female	−0.0616	0.0217	−2.8430	0.0066	

Figure 13.13 Results in log–log form.

salary by about 0.14%, other things equal; and being female decreases salary by about (exp(−0.0616) − 1) × 100 = −5.98%, other things equal.

We again want to see if these effects are significantly different from zero. Are they large enough that we can conclude (with an acceptable probability of being wrong) that the true effects are not zero?

1. H_o: $\beta_i = 0$
   H_o: $\beta_i \neq 0$.

I have written the hypotheses as two-tailed though, again, we might argue for one-tailed alternatives—positive for education and experience and negative for female.

Our two-tailed test statistic follows the t distribution with 46 degrees of freedom.

2a. Reject H_o if $|t_c| > 2.013$ (df = 46; $\alpha = 0.05$) or simply
2b. Reject H_o if p-value < 0.05.

All the coefficients have the expected signs (if we are using one-tailed tests) and p-values of much less than 0.05. There is little chance that any of these coefficients is just randomly different from zero.

Looking beyond the individual coefficients it is important to remember that our dependent variable is lnSal, not Salary. The LnSal values are much smaller and vary less than Salary values. This is the reason that the sums of squares and related measures are all so much smaller than they were in the linear form (Section 13.3.2). And $\bar{R}^2 = 0.9092$ means we have explained about 91% of the variation in lnSal, not Salary.

How do we decide between this form and the original, linear form? The coefficients for each have sensible signs and magnitudes and are strongly significant. If one had been much stronger by these criteria, these would probably decide the issue. In this case they probably do not.

Each explains more than 90% of the variation in its dependent variable, so each is a pretty good overall fit. The $\bar{R}^2$s are not strictly comparable since the first measures the variation in Salary explained while the second measures the variation in lnSal explained. It is possible to calculate an alternative $\bar{R}^2$ for the second one in terms of Salary instead of lnSal, and choose the model that explains the most variation in Salary. We will not do so. As a practical matter, there is seldom a great difference in a model's ability to explain ln$Y$ and Y. So if there is a substantial difference in $\bar{R}^2$s, this might decide the issue. In this case, the linear form (Section 13.3.2), with an $\bar{R}^2$ of 0.9808 clearly has a bit of an edge.

In other cases, we might decide on grounds of theory. In economics, for example, production is assumed to be subject to "diminishing returns." Therefore, the "production function" relating inputs to output should *not* be linear. In this case, we would have a preference for the Cobb–Douglas, even if it were not stronger statistically. Indeed, production functions were the original application of this form.

Finally, we might find the interpretation of coefficients easier as slopes or as elasticities. If we want to be able to talk about the slope, it helps if the estimated "slope" is constant; if we want to be able to talk about "the elasticity," it helps if the estimated elasticity is constant.

13.9 Additional Examples

In Exercises 12.2 through 12.4, you calculated simple regressions relating a young adult's weight to his or her height, age, and sex. We now have the tools with which to (1) relate weight to all three variables simultaneously; and (2) examine whether the effects of height and age are nonlinear:

a. Return to the NLSY1.xls data file, and estimate the simplest regression you can to predict a young adult's weight based on his or her height, age, and sex.
b. Interpret your coefficients in words.
c. Does weight really depend on height? age? sex? Test the hypotheses that it does not. Use $\alpha = 0.05$.
d. Does a nonlinear form fit the data better? For each regression you try, explain why you tried it. For each, interpret your results.
e. How well does your best regression explain the variation in the weight of young adults?
f. Predict the weight of an individual 68 inch tall, 25-year-old man.
g. Predict the weight of an individual 68 inch tall, 25-year-old woman.
h. Give (approximate) 95% confidence intervals around these predictions.

a. The simplest regression relating a young adult's weight to his or her height, age, and sex would be just a linear one. Figure 13.14 presents the results.
b. –132.9995 is our estimated weight in pounds for a male who is zero inches tall and zero years old. Of course there are no such people, so the fact that it is negative is of no concern.

Multiple Regression					
Dependent Variable: Weight					
R Square: 0.4423		Adjusted *R* Square: 0.4362		Standard Error: 22.9340	
ANOVA	**Sum of Squares**	**df**	**Mean Square**	F_c	**2-Tailed *p*-Value**
Regression	115532.0066	3	38510.6689	73.2184	0.0000
Error	145693.5877	277	525.9696		
Total	261225.5943	280			
Variable	**Coefficient**	**Standard Error**	t_c	**2-Tailed *p*-Value**	
Constant	−132.9995	38.7310	−3.4339	0.0007	
Height	3.6233	0.5065	7.1537	0.0000	
Age	1.9313	0.6157	3.1368	0.0019	
Female	−13.9578	3.9907	−3.4976	0.0005	

Figure 13.14 Results of Weight regression: Linear form.

3.6233 is our estimated increase in weight in pounds with each additional inch in height, holding age, and sex constant.

1.9313 is our estimated increase in weight in pounds with each additional year in age, holding height, and sex constant.

–13.9578 is our estimated decrease in weight in pounds if the individual is female instead of male, holding height, and age constant.

It is worth noting that the coefficients for Height and Female are both substantially smaller than those we found in Chapter 12 looking at each individually. We have multicollinearity between height and sex; hence, those earlier estimates suffered from omitted-variable bias. Without Female in the equation, Height picked up the effect of sex; without Height in the equation, Female picked up the effect of height.

c. 1. H_o: $\beta_i = 0$
H_a: $\beta_i \neq 0$.

Again I have written the hypotheses as two-tailed, though we might argue for one-tailed alternatives for Height (positive) and Female (negative).

2. Reject H_o if p-value < 0.05 for the two-tailed test

Reject H_o if sign correct and p-value < 0.10 for the one-tailed tests.

All the coefficients have the expected signs (if we are using one-tailed tests) and p-values of much less than 0.05. There is little chance that any of these coefficients is just randomly different from zero. Height, age, and sex all help explain a young adult's weight.

d. To test for curvature we can try a quadratic for height and age by adding squares. Recall that we cannot use the square of the dummy variable. Figure 13.15 presents the results.

We might be troubled at first by the negative coefficient for Height (and even Age). However, recall that a quadratic with a negative coefficient for X and a positive coefficient for $X2$ is U-shaped, and has its minimum at $X = -b_2/(2b_3)$. For Height and Age these values are Height = – (–12.0666)/(2 × 0.1159) ≈ 52 and Age = – (–11.6493)/(2 × 0.2758) ≈ 21, and all the young adults in our sample are at least 52 inches tall and at least 21 years old. Hence the downward-sloping half of the U is irrelevant; in our sample Weight rises with Height and Age.

Since we already know that height, age, and sex matter, our main concern here is with the coefficients of the squares. Those coefficients have p-values of 0.1102 and 0.3695—greater than our 0.05 rejection criterion—so we fail to reject H_0. These coefficients could too easily have arisen just randomly.

Notice too that the coefficients for Height and Age are no longer significant. What we have here is strong multicollinearity between Height and Height2, and between Age and Age2. And when either Height or Height2 can explain variation in Weight, neither one gets credit for doing so.

We might decide that the squared terms add little, and that the quadratic is no improvement over a straight line. Notice, though,

Multiple Regression

Dependent Variable: Weight

R Square: 0.4490		**Adjusted R Square: 0.4390**		**Standard Error: 22.8784**	
ANOVA	**Sum of Squares**	**df**	**Mean Square**	**F_c**	**2-Tailed p-Value**
Regression	117284.7146	5	23456.9429	44.8146	0.0000
Error	143940.8797	275	523.4214		
Total	261225.5943	280			

Variable	**Coefficient**	**Standard Error**	**t_c**	**2-Tailed p-Value**
Constant	561.9013	381.7083	1.4721	0.1421
Height	−12.0666	9.8039	−1.2308	0.2195
Height2	0.1159	0.0723	1.6027	0.1102
Age	−11.6493	15.0426	−0.7744	0.4393
Age2	0.2758	0.3068	0.8989	0.3695
Female	−14.0065	3.9818	−3.5177	0.0005

Figure 13.15 Results of Weight regression: Quadratic form (1).

that the least significant variable is not Age2, but Age. Instead of removing Age2 again, we might remove Age. If we do (not shown) Age2 becomes highly significant. Moreover, now, the least significant variable is not Height2, but Height. Instead of removing Height2 again, we might remove Height.

Figure 13.16 presents the results. All the coefficients have the expected signs (if we are using one-tailed tests) and p-values of much less than 0.05. There is little chance that any of these coefficients is just randomly different from zero. Height, age, and sex all help explain a young adult's weight, and the effects of height and age are slightly nonlinear.

The improvement is minimal. $\bar{R}^2$ rises from 0.4362 to 0.4388 only so we are still explaining between 43 and 44% of the variation in weights. And there is a cost in terms ease of interpretation; we can no longer speak of the effect of being an inch taller or a year older. In this case the improvement may not be worth the cost.

We can also try a Cobb–Douglas to test for a different shape curve. For this we need the natural logs of Weight, Height, and Age. Recall that we cannot take the log of the dummy variable. Figure 13.17 presents the results.

−2.7179 is our estimate of ln(A) in the Cobb–Douglas and is not a very interesting number.

1.6258 is our estimated percent increase in weight with each additional 1% increase in height, holding age, and sex constant.

0.2891 is our estimated percent increase in weight with each additional 1% increase in age, holding height, and sex constant.

(exp(−0.0946) − 1) × 100 = −9.03% is our estimated percent decrease in weight if the individual is female instead of male, holding height, and age constant.

Multiple Regression

Dependent Variable: Weight

R Square:	**Adjusted R Square:**	**Standard Error:**
0.4448	**0.4388**	**22.8822**

ANOVA	**Sum of Squares**	**df**	**Mean Square**	**F_c**	**2-Tailed p-Value**
Regression	116190.2425	3	38730.0808	73.9698	0.0000
Error	145035.3518	277	523.5933		
Total	261225.5943	280			

Variable	**Coefficient**	**Standard Error**	**t_c**	**2-Tailed p-Value**
Constant	11.4510	20.0344	0.5716	0.5681
Height2	0.0270	0.0037	7.2430	0.0000
Age2	0.0395	0.0125	3.1509	0.0018
Female	−13.7930	3.9767	−3.4685	0.0006

Figure 13.16 Result of Weight regression: Quadratic form (3).

Multiple Regression					
Dependent Variable: lnWgt					
R Square: 0.4675		Adjusted R Square: 0.4617		Standard Error: 0.1454	
ANOVA	Sum of Squares	df	Mean Square	F_c	2-Tailed p-Value
Regression	5.1398	3	1.7133	81.0577	0.0000
Error	5.8548	277	0.0211		
Total	10.9946	280			
Variable	Coefficient	Standard Error	t_c	2-Tailed p-Value	
Constant	−2.7179	0.9698	−2.8025	0.0054	
lnHgt	1.6258	0.2163	7.5173	0.0000	
lnAge	0.2891	0.0950	3.0417	0.0026	
Female	−0.0946	0.0253	−3.7442	0.0002	

Figure 13.17 Result of Weight regression: Log–log form.

Again, all the coefficients have the expected signs (if we are using one-tailed tests) and p-values of much less than 0.05. There is little chance that any of these coefficients is just randomly different from zero. Height, age, and sex all help explain a young adult's weight, and the effects of height and age are slightly nonlinear.

Again, the improvement is modest. $\bar{R}^2$ rises to 0.4617 but, remember, this measures how well we have explained the variation in lnWgt, not Weight, so it is not directly comparable to the others. With this small a difference all we should probably say is that this form probably explains a little more of the variation in Weight than the straight line or quadratic.

So does a nonlinear form fit the data better? Yes and the log–log form probably fits best. But the improvement is so small that it may not be worth the cost in terms of interpretation. In this case pounds, inches, and years are probably more natural than percentages.

e. The linear regression explains 43.62% of the variation in weight. The log–log form explains 46.17% of the variation in lnWgt; we would expect it to explain roughly that much of the variation in Weight as well.

f–h. If we use the linear form, the predicted weight for a 68 inch tall, 25-year-old man is

$$\widehat{\text{Weight}} = -132.9995 + 3.6233 \times 68$$
$$+1.9313 \times 25 - 13.9578 \times 0 = 161.6674.$$

The standard error of this individual prediction is $s_{\hat{Y}/X} = 23.0556$. However, this is not something we would calculate ourselves and few computer programs give it easily. For an approximate 95% confidence interval, we can use $s_{\hat{Y}/X} \approx s_e = 22.9340$. = TINV(0.05,277) = 1.969, so

$$95\% \text{ CI: } 161.6674 \pm 1.969 \times 22.9340$$

$$95\% \text{ CI: } 161.6674 \pm 45.1570$$

$$95\% \text{ CI: } 116.5104 < \text{Weight} < 206.8244.$$

For an individual woman of the same height and age, the predicted weight is just 13.9578 less.

$$\widehat{\text{Weight}} = -132.9995 + 3.6233 \times 68 + 1.9313 \times 25 - 13.9578 \times 1$$
$$= 147.7096.$$

The standard error of this individual prediction is $s_{\hat{Y}/X} = 23.0767$. Again, though, this is not something we are likely to know. For an approximation, we can use $s_{\hat{Y}/X} \approx s_e = 22.9340$.

$$95\% \text{ CI: } 147.7096 \pm 1.969 \times 22.9340$$

$$95\% \text{ CI: } 147.7096 \pm 45.1570$$

$$95\% \text{ CI: } 102.5526 < \text{Weight} < 192.8666.$$

If we use the log–log form, the calculations are a bit more complicated. We need to enter the natural logs of 68 and 25; and we need to take the antilog of our answer. For a 68-inch tall, 25-year-old man

$$\widehat{\text{lnWgt}} = -2.7179 + 1.6258 \times \ln(68) + .2891 \times \ln(25) - 0.0946 \times 0$$
$$= -2.7179 + 1.6258 \times 4.2195 + .2891 \times 3.2189 - 0.0946 \times 0$$
$$= 5.0728$$

$$\widehat{\text{Weight}} = \exp(5.0728) = 159.6131.$$

Again, we are unlikely to know $s_{\hat{Y}/X}$, and so would use $s_{\hat{Y}/X} \approx s_e = .1454$. Moreover, we need to remember that s_e is for lnWgt, not Weight. So we need to find the approximate confidence interval around lnWgt first; then find the antilogs of the lower and upper bounds.

$$95\% \text{ CI: } 5.0728 \pm 1.969 \times 0.1454$$

$$95\% \text{ CI: } 5.0728 \pm 0.2863$$

$$95\% \text{ CI: } 4.7865 < \text{lnWgt} < 5.3590$$

$$95\% \text{ CI: } 119.8763 < \text{Weight} < 212.5219.$$

For an individual woman of the same height and age:

$$\widehat{\text{lnWgt}} = -2.7179 + 1.6258 \times \ln(68) + .2891 \times \ln(25) - 0.0946 \times 1$$
$$= -2.7179 + 1.6258 \times 4.2195 + .2891 \times 3.2189 - 0.0946 \times 1$$
$$= 4.9782$$

$$\widehat{\text{Weight}} = \exp(4.9782) = 145.2059.$$

95% CI: $4.9782 \pm 1.969 \times 0.1454$

95% CI: 4.9782 ± 0.2863

95% CI: $4.6919 < \text{lnWgt} < 5.2644$

95% CI: $109.0558 < \text{Weight} < 193.3390.$

Consider the following example:

Economists theorize that the demand for money depends positively on people's income and negatively on the interest rate. Money1.xls contains U.S. economic data on real money, in $ billions (RM2), real disposable personal income in (RDPI) $ billions, and the six month treasury bill rate of interest as a percent (TBR) from 1959 through 2002. (Source: Economic Report of the President, 2004, Tables B-31, B-60, B-69, and B-73. RM2 is calculated by the author from M2 and the consumer price index.)

a. Estimate the simplest regression you can to predict money demand based on disposable income and the interest rate.
b. Interpret your coefficients in words.
c. Does real money demand really depend on real disposable income? The interest rate? Test the hypotheses that it does not. Use $\alpha = 0.05$.
d. Does a nonlinear form fit the data better? For each regression you try, explain why you tried it. For each, interpret your results.
e. How well does your best regression explain the variation in RM2?
f. Predict RM2 for a year in which RDPI = $6000 (billion) and TBR = 5%.
g. Give an (approximate) 95% confidence interval around this prediction.

a. The simplest regression relating RM2 to RDPI and TRB would be just a linear one. Figure 13.18 presents the results.
b. 1170.4054 is our estimated real money demand (in $ billions) in a year in which RDPI (in $ billions) and the TBR (%) were both zero. There are no such years, so it is of no concern.

   0.5330 is our estimate of the increase in money demand (in $ billions) with an increase of one ($ billion) in disposable personal income.

   26.9478 is our estimate of the increase in money demand (in $ billions) with an increase of one percent in the interest rate.

Multiple Regression					
Dependent Variable: RM2					
R Square: 0.9354		Adjusted *R* Square: 0.9323		Standard Error: 242.9170	
ANOVA	Sum of Squares	df	Mean Square	F_c	2-Tailed *p*-Value
Regression	35047041.3334	2	17523520.6667	296.9652	0.0000
Error	2419355.1257	41	59008.6616		
Total	37466396.4591	43			
Variable	Coefficient	Standard Error	t_c	2-Tailed *p*-Value	
Constant	1170.4054	130.4567	8.9716	0.0000	
RDPI	0.5330	0.0220	24.2734	0.0000	
TBR	26.9478	14.6213	1.8430	0.0726	

Figure 13.18 The demand for money: Linear form.

c. We have explicit one-tailed tests—positive for the effect of RDPI and negative for the effect of TBR.

1. $H_o{:}\beta_2 = 0$ $\quad$ $H_o{:}\beta_3 = 0$
   $H_a{:}\beta_2 > 0$ $\quad$ $H_a{:}\beta_3 < 0.$

   Recall that the reported *p*-values include both tails. Since we are using only one tail, we can reject with a reported *p*-value twice 0.05.

2. Reject H_o if $b_2 > 0$ and *p*-value < 0.10 $\quad$ Reject H_o if $b_3 < 0$ and *p*-value < 0.10

   The coefficient for RDPI has the expected sign according to economic theory and a very low *p*-value. There is little chance that this coefficient is just randomly different from zero. Things are not so good with the coefficient for TBR. We were looking for a negative effect and got a positive one. We cannot reject the null hypothesis regardless of the *p*-value.

d. To test for curvature, we can try a quadratic for both RDPI and TBR. Since we have already established that there is a relationship between RDPI and RM2, our main concern there is whether a curve fits better than a straight line. And for that the coefficients of the squared term is the key. For TBR and RM2, we are still looking for any evidence of a negative relationship.

With the quadratic we no longer have straightforward one tail tests since the slope changes from positive to negative, or negative to positive, depending on the coefficients of both X and X^2. Therefore, we use a two-tailed alternative.

1. $H_o{:}\beta_2 = 0$
   $H_a{:}\beta_2 \neq 0$
2. Reject H_o if *p*-value < 0.05.

Multiple Regression					
Dependent Variable: RM2					
***R* Square:** 0.9517		**Adjusted *R* Square:** 0.9468		**Standard Error:** 215.3127	
ANOVA	**Sum of Squares**	**df**	**Mean Square**	F_c	**2-Tailed *p*-Value**
Regression	35658373.2060	4	8914593.3015	192.2924	0.0000
Error	1808023.2531	39	46359.5706		
Total	37466396.4591	43			
Variable	**Coefficient**	**Standard Error**	t_c	**2-Tailed *p*-Value**	
Constant	384.8924	245.2918	1.5691	0.1247	
RDPI	0.9509	0.1687	5.6385	0.0000	
RDPI2	−4.7182e-05	1.8787e-05	−2.5114	0.0163	
TBR	57.9432	69.8607	0.8294	0.4119	
TBR2	−4.5665	4.2389	−1.0773	0.2880	

Figure 13.19 The demand for money: Quadratic form (1).

Figure 13.19 presents the results. They are mixed but hopeful. The coefficients for RDPI and RDPI^2 are positive and negative, respectively, and both have very low *p*-values; there is very little chance that they are just randomly different from zero. The curve first rises and then falls. This agrees with economic theory as long as we are on the upward-sloping side of the curve, so that RDPI has a positive but diminishing effect. We can find the turning point: RDPI $= -b_2/(2b_3) = -(0.9509)/(2 \times -0.000047182) \approx 10077$. This is well beyond our largest value of RDPI, so we are on the upward-sloping side. The RDPI has the predicted positive effect.

Neither TBR nor TBR^2 has a significant effect on RM2. On the other hand, the coefficient for TBR^2 is the more significant of the two, with a *p*-value of 0.2880 and has the predicted negative sign. This is hopeful. It may be that, if we remove TBR, TBR^2 will remain negative and become more significant.

Figure 13.20 presents the results. As a practical matter, they are little changed. The coefficients for RDPI and RDPI^2 are little changed and still highly significant. You can confirm that the turning point is still well above the actual values for RDPI. We are still on the upward-sloping side of the curve; the effect of RDPI is still positive as predicted. The coefficient for TBR^2 remains negative, as hoped, but does not become much more significant. There is a 28.25% chance of a coefficient this large arising just randomly when there is no real effect. So we cannot reject the null hypothesis that the true value is zero.

If we had no strong reason for including it, we could try removing it. But remember, we have not shown that the TBR has no effect; we have just failed to confirm that it does. With the correct sign and theory clearly on the side of including it, we would generally do so.

Multiple Regression					
Dependent Variable: RM2					
R Square: 0.9509		Adjusted R Square: 0.9472		Standard Error: 214.4712	
ANOVA	Sum of Squares	df	Mean Square	F_c	2-Tailed p-Value
Regression	35626481.4023	3	11875493.8008	258.1748	0.0000
Error	1839915.0568	40	45997.8764		
Total	37466396.4591	43			
Variable	**Coefficient**	**Standard Error**	t_c	**2-Tailed p-Value**	
Constant	420.3773	240.5878	1.7473	0.0883	
RDPI	1.0378	0.1317	7.8787	0.0000	
RDPI2	−5.6840e-05	1.4687e-05	−3.8702	0.0004	
TBR2	−1.1654	1.0698	−1.0894	0.2825	

Figure 13.20 The demand for money: Quadratic form (2).

We can also try a Cobb–Douglas to test for a different shape curve. The results (not shown) again have the wrong sign for lnTBR, so it is clearly not an improvement.

Finally, you should know that a number of issues have been sort of swept under the rug in this example. First real world data are messy. There are several different measures of money supply, income, price deflators, and interest rates, all of which would have given somewhat different results.

Second, I have assumed that the quantity of money demanded depends on the interest rate. But perhaps the interest rate depends on the quantity of money. Or perhaps they are jointly determined.

Third, I have assumed that the quantity of money demanded depends on disposable income and the interest rate in the same year. But perhaps the effect takes time. Perhaps demand in the current year depends on the other variables with a lag. The demand for money might depend on income and/or interest rates in a previous year.

And finally, I have assumed that the relationship is a constant one over more than 40 years. But 40 years ago there was no internet banking, there were no ATMs, there was no branch banking in many states. A checking account did not pay interest, you could not write a check on a savings account, and few people had credit cards. These have all changed. And in doing so, they have also probably changed the relationship among the demand for money, income, and the interest rate.

If you take a follow-up course in applied statistics or econometrics, you will learn how to deal with some of these complications. You now have a solid basis for doing so. Multiple regression is the primary tool you will use. It is the primary tool used in applied business and economic research.

13.10 Exercises

13.1 Refer back to the Students2.xls data file that contains a variety of information on a sample of 100 students at a large university. Suppose you expect that a student's weight depends on his or her height and sex.

a. Estimate the simplest regression you can to predict a student's weight based on his or her height and sex.
b. Interpret your regression coefficients in words.
c. Does a student's weight really depend on his or her height? Sex? Test the hypotheses that it does not. Use $\alpha = 0.05$. Use $\alpha = 0.01$. What is the p-value for each estimated slope?
d. Does a nonlinear form fit the data better? For each regression you try, explain why you tried it. For each, interpret your results. Decide which you prefer and explain why.
e. How well does your preferred regression explain the variation in students' weights?
f. Predict the weight of a 68-inch tall, male student.
g. Predict the weight of a 68-inch tall, female student.
h. Give (approximate) 95% confidence intervals around these predictions.

13.2 Continue with these same data. In Exercise 12.7, you found that a student's financial aid depends on his or her high school GPA. Suppose you wonder if it depends also on participation in a varsity sport and/or music ensemble. You also wonder if there is discrimination based on sex.

a. Estimate the simplest regression you can to predict a student's financial aid based on his or her high school GPA, participation in a varsity sport, participation in a music ensemble, and sex.
b. Interpret your regression coefficients in words.
c. Does a student's financial aid really depend on his or her high school GPA? Participation in a varsity sport? Participation in a music ensemble? Sex? Test the hypotheses that it does not. Use $\alpha = 0.05$. Use $\alpha = 0.01$. What is the p-value for each estimated slope?
d. Does a nonlinear form fit the data better? Note: *Aid* has cases with zero values; if you try the log-log form, these cases will be dropped. For each regression you try, explain why you tried it. For each, interpret your results. Decide which you prefer and explain why.
e. How well does your preferred regression explain the variation in financial aid?
f. Predict the financial aid of an individual student with a 4.000 high school GPA who does not participate in either a varsity sport or music ensemble.
g. Predict the financial aid of this student if he or she does participate in a varsity sport, a music ensemble, both.
h. Give (approximate) 95% confidence intervals around these predictions.

13.3 Continue with these same data. In Exercises 12.8 and 12.9, you found that a student's college GPA depends separately on his or her high school GPA and study time. But in neither regression were you holding the other constant.

a. Estimate the simplest regression you can to predict a student's college GPA based on both his or her high school GPA and study time.
b. Interpret your regression coefficient in words.
c. Does a student's college GPA really depend on his or her high school GPA? Study time? Test the hypotheses that it does not. Use $\alpha = 0.05$. Use $\alpha = 0.01$. What is the *p*-value for each estimated slope?
d. Are high school GPA and study time subject to diminishing returns? For each regression you try, explain why you tried it. For each, interpret your results. Decide which you prefer and explain why.
e. How well does your preferred regression explain the variation in a college GPA?
f. Predict the college GPA of an individual student with a 4.000 high school GPA who studies 30 hours per week.
g. Give an (approximate) 95% confidence interval around this prediction.

13.4 Continue with these same data. Suppose someone argues that other student attributes can affect college GPA as well. Possibilities in the data include sex, year in school, major, holding a job, participation in a varsity sport, participation in a music ensemble, and belonging to a fraternity or sorority.

a. Add these to the full quadratic model in Exercise 13.3.d. Note: year in school and major are neither numeric or dummy variables. (A sophomore is not twice a freshman; an economics major is not twice a mathematics major.) You will need to create a set of dummy variables for each of these categorical variables and include all but one from each set in the regression.
b. Remove variables with insignificant effects one by one starting with the least significant until you are left with just those that have significant effects, or that you believe are logically necessary even if their effects are insignificant.
c. Interpret your regression coefficients in words.
d. How well does your preferred regression explain the variation in a college GPA?

13.5 Continue with these same data. Now someone else argues that some of the other student attributes above have their effects, if any, by competing for time with study time.

a. Estimate the simplest regression you can to predict a students' study time based on his or her sex, year in school, major, holding a job, participation in a varsity sport, participation in a music ensemble, and belonging to a fraternity or sorority.
b. Remove variables with insignificant effects one by one starting with the least significant until you are left with just those that have significant effects, or that you believe are logically necessary even if their effects are insignificant.
c. Interpret your regression coefficients in words.

d. How well does your best regression explain the variation in students' study times?

13.6 Continue with these same data.

a. Create a model to explain entertainment expenditure based on other variables in the data. For each explanatory variable, justify its inclusion. That is, why might it affect students' entertainment spending?
b. Estimate the simplest regression you can to represent your model.
c. Interpret your regression coefficients in words.
d. Does a student's entertainment spending really depend on each of the variables in your model? Test the hypotheses that it does not. Use $\alpha = 0.05$. Use $\alpha = 0.01$. What is the p-value for each estimated slope?
e. Try to improve on your results. This might involve trying nonlinear forms (if your model includes numeric variables). It might involve removing insignificant variables one by one starting with the least significant until you are left with those that are significant, or that you believe are logically necessary even if insignificant.
f. How well does your best regression explain the variation in students' entertainment spending?

14

Time-Series Analysis

14.1 Exploiting Patterns over Time

In the last five chapters, we have been exploring hypothesized relationships among variables—relationships that we could think of as causal in some sense. Choice of major really could depend on sex. A GPA really could depend on major. Salary really could depend on education, experience, and sex. The weight of young adults really could depend on their height, age, and sex.

Now, suppose you want to predict next quarter's sales by your firm. You could attempt to estimate a causal relationship between your sales and the various things that determine them. You would select a sample of past quarters and record for those quarters both the value of your sales and the values of all the things that determined them. You would then use these data to estimate a regression relating your sales to all those determinants.

It might work. Three problems are likely to arise, though. First, in many cases, the causes of the variation in your past sales would be quite numerous. Assuming you could actually measure them all, your multiple regression would be extremely long and complex. Second, you would often have no way of measuring many of them; your regression results then would be unreliable. And finally, predicting next quarter would probably require plugging in values of all those explanatory variables for next quarter—values that you would not know until next quarter. There are ways of dealing with some of these problems. But in some cases, it is possible to do better using a different approach.

The alternative is, essentially, to look for patterns in your sales over time. If you can identify stable patterns over a period of many previous quarters, you can probably assume these patterns will continue for at least a little longer. And since you know where next quarter fits into these long-term patterns, you can combine these into a prediction of next quarter's sales.

It is important to recognize the difference between this approach and our previous one. Suppose you find a stable, upward trend in your sales over time. Time is not the cause of improving sales; sales are improving for all sorts of underlying business and economic reasons. But if those underlying reasons do not change dramatically, time can serve as a proxy for the whole set of them. And, of course, if these underlying reasons do change dramatically, your prediction will be completely wrong.

Time-series analysis has a long tradition in business and economics; it is uncommon in most other disciplines. This is because business and economics have the luxury of recurring data.

Business data are recorded on a daily, weekly, monthly, quarterly, and annual basis, just in the natural course of business. Likewise, the government reports data on the economy on a regular basis. The Consumer Price Index is reported monthly, the Gross Domestic Product is reported quarterly, and so on. Most other disciplines rely primarily on experimental or survey data collected at a particular moment.

This chapter introduces classical time-series analysis. There is a great deal more that is beyond the scope of this course.

14.2 The Basic Components of a Time Series

Figure 14.1 presents the traditional four types of variation over time. **Secular variation** refers to the **long-term trend**. It presumably reflects underlying forces that transcend season or business cycle. For example, the aging of the U.S. population is causing medical expenditures to rise. Yes, these expenditures may go up and down with the season or with the health of the economy; but taking the long view, the aging of the population means that the trend is up.

Cyclical variation refers to business cycle variation. Some businesses and markets are affected a great deal by the state of the economy. Consumer durables like automobiles, for example, tend to have strong cyclical variation. Sales are very good during booms, when consumers feel confident; they are very poor during recessions, when wary consumers make do with their old models.

Seasonal variation refers to variation for seasonal reasons. Some of these reasons may be weather related; ice cream sells best in the summer. Some may be cultural; in the United States fourth quarter sales of toys tend to be strongest because of Christmas gift giving.

Finally, **random variation** refers to variation that follows no pattern. It is unpredictable; we can only hope that it is small.

Figure 14.2 presents sales for a hypothetical company over the last 10 years or 40 quarters. The file Sales1.xls contains the raw sales data. Starting with seasonal variation, there is a clear four-quarter pattern. The first quarters (1, 5, 9, 13, etc.) tend to be higher than the quarters before and after. The second quarters (2, 6, 10, 14, etc.) tend to be much lower. The pattern is not perfect. The amount of the decline is not uniform; indeed, there is no decline at all from 13 to 14. But the pattern is strong enough that it surely pays to predict a higher value for next quarter if it is a first quarter than if it is a second quarter.

1. Secular (long-term trend) variation
2. Cyclical (business cycle) variation
3. Seasonal variation
4. Random variation

Figure 14.1 The four types of variation over time.

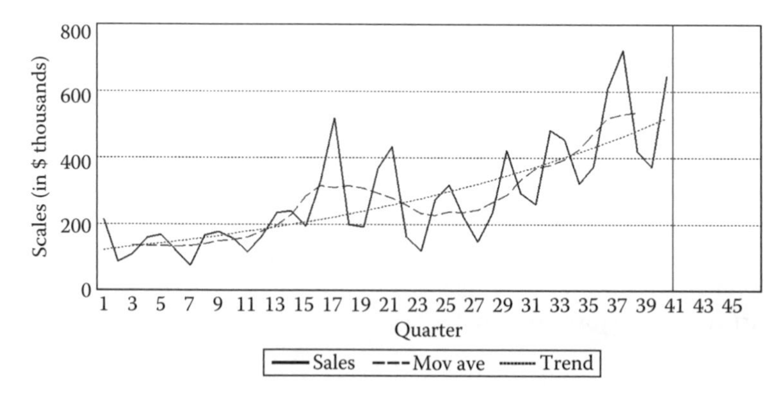

Figure 14.2 Company sales: A time-series decomposition.

The dashed line in Figure 14.2 is a **moving average**. We will cover the calculation of moving averages in the next section. For now, it is just an estimate of "deseasonalized" sales. That is, it is an estimate of what sales would have been if there had been no seasonal variation. Hence, it represents an estimate of the combined long-term trend, business cycle, and random variation. The dotted line is an estimate of the long-term trend. Clearly, there is an upward trend. Surely we would want to take this upward trend into account in our prediction for next quarter.

Finally, there is also substantial variation of the moving average—the dashed line around the trend—the dotted line. This movement represents an estimate of the combined business cycle and random variation. It is somewhat irregular due to the random component. But, clearly, years four and five (quarters 13–20) were a relatively strong period—above trend. And years six and seven (quarters 21–28) were relatively weak—below trend. Notice that the dashed line ends above the trend; apparently we are currently in something of a boom period. Surely we would want to take this into account in our prediction for next quarter as well.

What we have done is "decompose" the variation over time into its components. To make a prediction for next quarter requires reassembling these components. We would start with the trend; the trend line is just a regression estimate with quarter (1 to 40) as the explanatory variable. Simply plugging in 41 gives us our trend estimate for the next quarter. Next, we would want to take into account that we appear to be in a boom period. This is the diciest part; there is no mathematical formula for the business cycle. Still, if there is a clear business cycle in the data, we would want to take it into account. Finally, we would apply an index of seasonal variation, computed along the way, to take into account that the next quarter is a first quarter, and first quarters are strong. We would have our estimate for next quarter's sales.

Finally, a reminder of how all this is different from what we have been doing up until now. In the last few chapters, we have been looking for what we could think of as causal relationships. Presumably we were looking for the relationships that we thought were there in the population.

In that sense, not finding a significant relationship was something of a disappointment. Now, when we use regression to estimate the trend line, for example, it is not because we think the passage of time causes a change in sales. We are just assuming that the true causes continue to have a reasonably consistent effect that can be represented by time. And it is perfectly alright to find no trend; likewise, there could be no cyclical or seasonal variation. If your sales have no trend, cyclical, or seasonal variation, your job of predicting sales is very easy—just predict the previous average for sales. However, things are unlikely to be that simple. Chances are there is some systematic variation in your sales. And the techniques of time-series analysis are intended to help you exploit this systematic variation to improve your predictions.

14.3 Moving Averages

Moving averages are just averages of several periods in a row, where each successive average drops the earliest period and adds in the next. They are used to smooth out period-to-period fluctuations that are thought to be largely random in the hopes of uncovering more fundamental patterns. If this is all they are intended to do, an odd number of periods is usually chosen and they are centered.

Consider a three-period moving average. The notion is that period two is affected by fundamental, longer-term forces but also by random shocks. The fundamental, longer-term forces will tend to be the same as those affecting periods one and three while the random ones will not be. Hence, by averaging period two with the periods just before and after, we should tend to average away the random shocks making it easier to identify the fundamentals.

Figure 14.3 illustrates the calculations; Figure 14.4 shows them graphed. MA3 is a three-period moving average. The first number in E3 is =AVERAGE(D2:D4). And, of course, by using relative references the

	A	B	C	D	E	F	G
1	Period	Year	Quarter	Sales	MA3		MA9
2	1	1	Q1	214.40	–		–
3	2	1	Q2	86.31	136.84	← =AVERAGE(D2:D4)	–
4	3	1	Q3	109.80	117.95		–
5	4	1	Q4	157.73	145.16		–
6	5	2	Q1	167.95	149.59	=AVERAGE(D2:D10) →	142.07
7	6	2	Q2	123.08	122.39		135.70
:	:	:	:	:	:		:
:	:	:	:	:	:		:
36	35	9	Q3	373.28	435.25		446.56
37	36	9	Q4	608.95	568.49		489.38
38	37	10	Q1	723.25	583.84		–
39	38	10	Q2	419.32	505.01		–
40	39	10	Q3	372.45	479.09		–
41	40	10	Q4	645.50	–		–

Figure 14.3 Calculating moving averages.

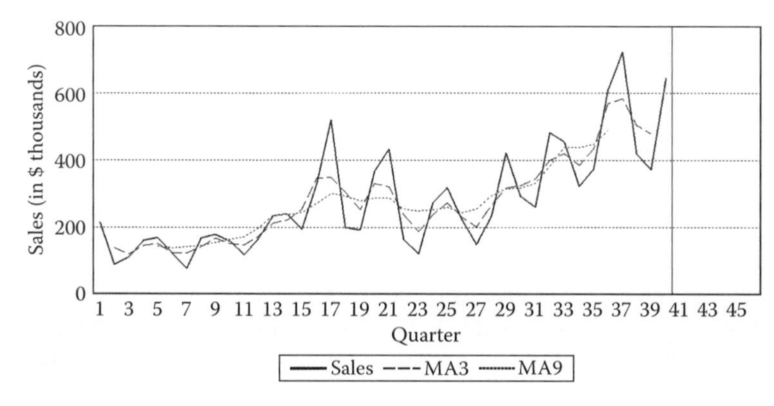

Figure 14.4 Graph of sales, MA3, and MA9.

rest can be found with **COPY**. MA9 is a nine-period moving average. The first number in G6 is =AVERAGE(D2:D10). And, again, by using relative references the rest can be found with **COPY**.

Comparing MA3 and MA9 two things should be apparent. First, MA9 uses up a lot more data. Period 5 is the first and period 36 the last, for which we have four periods before and after. Indeed, you need to be careful not to copy the formula down into the cells at the bottom with "–" in them, since it will give you invalid answers. And, of course, this lack of observation for the last four periods means that it tells us less that is relevant to the present.

On the other hand, MA9 is a lot smoother. By comparison, MA3 still shows many of the short-term movements in the original data. The moving average that balances best between smoothing and saying something relevant to the present will vary with the data and the purpose. Fortunately, with spreadsheets, it is easy to try several.

14.4 Seasonal Variation

As already mentioned, when data are "sub-annual" there may well be a seasonal pattern. Ice cream sells best in the summer; toys sell best at Christmas. Knowing the seasonal pattern to your business can be valuable for at least two reasons. First, it may help you make better business decisions, such as when inventories should be high and when they should be low. If you run a toy store, you do not want to be stuck with low stock in December or high stock in January. An **index of seasonal variation** can tell you what percentage of "average" your sales tend to be for each month or quarter.

Second, you do not want to confuse seasonal variation with longer-term variation. If your toy sales rise 50% from quarter 3 to quarter 4, you should probably not assume that this as a long-term trend and that quarter 1 to follow will show a similar rise. An index of seasonal variation can be used for this as well—to "deseasonalize" data—to translate your raw data into estimates for an average quarter. Changes in the deseasonalized numbers represent long-term, cyclical, and/or random changes.

14.4.1 Using a Moving Average to Create a Seasonal Index

14.4.1.1 Finding a Centered, Seasonally Balanced Moving Average

We can use a moving average computed in a particular way as the basis for an index of seasonal variation. Computed properly, it represents an estimate of sales stripped of their seasonal variation. Ratios of actual sales to moving average values give us period-by-period estimates of the amount by which actual sales exceed or fall short of this moving average. We can then average all the first quarter ratios, all the second quarter ratios, and so on to find an overall index for each quarter. This is where we are headed.

In introducing moving averages, I said that, if all we are trying to do is smooth out period-to-period fluctuations that are thought to be largely random, an odd number of periods is usually chosen and the average is centered. This would make sense for annual data, for example.

However, our sales data are quarterly; hence the period-to-period fluctuations are seasonal as well as random. We should calculate our moving average in a manner that takes this into account. By taking an average of four successive quarters, we would give each quarter or season an equal influence on the value of the moving average. Thus, the moving average would have no net seasonal effect. It would represent our estimate for the long-term trend and cyclical components stripped of the seasonal effects.

However, with an even number of quarters in a year, we cannot align our moving average with the middle quarter; there is no middle quarter. We need a way of dealing with four quarters in such a way that there is also a middle quarter. Figure 14.5 shows the trick for doing so. The first number in E4 is =AVERAGE(D2:D5,D3:D6). This is the

	A	B	C	D	E	F	G
1	Period	Year	Quarter	Sales	Mov Ave		SSI
2	1	1	Q1	214.40	–		–
3	2	1	Q2	86.31	–		–
4	3	1	Q3	109.80	136.25	← =AVERAGE(D2:D5,D3:D6)	80.58
5	4	1	Q4	157.73	135.04		116.80
6	5	2	Q1	167.95	135.43		124.01
7	6	2	Q2	123.08	132.24	=(D7/E7)×100 →	93.07
:	:	:	:	:	:		:
:	:	:	:	:	:		:
36	35	9	Q3	373.28	473.72		78.80
37	36	9	Q4	608.95	519.23		117.28
38	37	10	Q1	723.25	531.10		136.18
39	38	10	Q2	419.32	535.56		78.30
40	39	10	Q3	372.45	–		–
41	40	10	Q4	645.50	–		–

Figure 14.5 Calculating a centered, seasonally balanced moving average.

same as the average of =AVERAGE(D2:D5) and =AVERAGE(D3:D6). The =AVERAGE(D2:D5) would be centered between quarters 2 and 3; =AVERAGE(D3:D6) would be centered between quarters 3 and 4. The average of the two is centered at quarter 3.

We can think of the value 136.25 as this average of averages or simply the average of eight quarters. Either way, it includes quarter 1 (D2) once, quarter 2 (D3) twice, quarter 3 (D4) twice, quarter 4 (D5) twice, and quarter 1 (D6) once. We have two of each quarter or season; hence, there should be no net seasonal effect. And yet there is a middle quarter, quarter 3.

Again, by using relative references, the rest of the values can be found with **COPY.** Again, we lose cases at the beginning and the end. Quarter 3 of the first year is the first quarter that can be at the center of the moving average; quarter 2 of the last year is the last quarter that can be at the center. Again, you need to be careful not to copy the formula down into the cells at the bottom with "–" in them, since it will give you invalid answers.

These moving average values are the values plotted in Figure 14.2. If you compare this moving average with those in Figure 14.4 this one is much smoother. This is because those in Figure 14.4 assume that the period-to-period variation is just random, whereas there is actually systematic seasonal variation. Thus MA3, for example, includes just three quarters leaving out the fourth; MA9 includes three quarters twice and one quarter three times. In both cases, the imbalance in the treatment of the four quarters means that a net seasonal effect remains. When seasonal variation is present, you should always include all seasons equally (twice) in your average, calculated so that there is a middle period.

Finally, I have presented all this with quarterly data. What if we have monthly data instead? Everything would be the same except we would have 12 "seasons" instead of four. Our moving average should include all 12 months equally. Again, we would have the problem that there is no middle month. An average of January through December would be centered between June and July; an average of February through January would be centered between July and August. The average of the two would include each month equally (twice) and be centered on July.

14.4.1.2 Finding Specific Seasonal Indexes

For all but the first two and last two quarters we now have a figure for our sales, and an estimate of what our sales would have been stripped of the seasonal effect. By comparing the two, we can create a **specific seasonal index (SSI)** for each quarter. Each is just SSI = (Sales/MA) × 100. The final column of Figure 14.5 illustrates. Those quarters for which sales exceed the moving average have a value greater than 100; those quarters for which sales fall short of the moving average have a value less than 100.

14.4.1.3 Finding an Overall Seasonal Index

This procedure gives us nine different first quarter values, nine different second quarter values, and so on; we want just one of each. The obvious answer is to average them. There are various conventions concerning

	A	B	C	D	E	F	G	H	I	J
1	Period	Year	Quarter	Sales	Mov Ave		SSI			Index
2	1	1	Q1	214.40	–		–	Q1:	135.15	135.47
3	2	1	Q2	86.31	–		–	Q2:	84.74	84.94
4	3	1	Q3	109.80	136.25		80.58	Q3:	66.90	67.06
5	4	1	Q4	157.73	135.04		116.80	Q4:	112.27	112.53
6	5	2	Q1	167.95	135.43		124.01		399.06	400.00
7	6	2	Q2	123.08	132.24		93.07			
:	:	:	:		:		:			
:	:	:	:		:		:			
36	35	9	Q3	373.28	473.72		78.80			
37	36	9	Q4	608.95	519.23		117.28			
38	37	10	Q1	723.25	531.10		136.18			
39	38	10	Q2	419.32	535.56		78.30			
40	39	10	Q3	372.45	–		–			
41	40	10	Q4	645.50	–		–			

Figure 14.6 Calculating the overall seasonal index.

how far back to go. We would not generally go all the way back to the beginning of a very long time series. For current purposes, though, we will use all of them. In Figure 14.6 the formula in I2 is just =AVERAGE(G6,G10,G14,G18,G22,G26,G30,G34,G38) = 135.15, the average of the nine first quarter specific seasonal indexes. Be careful in copying the formula down; the formula for quarter 3 starts over with G4 at the top.

One minor problem remains; we want 100 to represent "average;" that is no positive or negative seasonal effect. This means that the four quarterly values should sum to 400 and generally they will not. A final step is to re-weight each by 400 over their sum. In Figure 14.6, then, the formula in J2 is just =I2*400/$I$6 = 135.47.

Again, if we have monthly instead of quarterly data little would change except that, for 100 to represent "average", the 12 monthly values should sum to 1200. The final step, then, would be to re-weight each by 1200 over their sum.

14.4.2 Using a Seasonal Index

Now that we have a seasonal index, we can use it to "seasonalize" forecasts. That is, if we expect $2000 (thousand) in annual sales, we should not expect $500 per quarter; according to our index numbers above, we should expect more than $500 in quarters 1 and 4 and less than $500 in quarters 2 and 3.

And we can use it to "deseasonalize" raw data. That is, if sales go from $600 in quarter 1 to $400 in quarter 2, should we be concerned? After all, according to our index numbers above, sales are expected to be lower in quarter 2. We need to translate these two numbers to their equivalents for an "average" quarter in order to compare them. This deseasonalizing of raw data is called **seasonal adjustment.** It allows comparison of data without regard to season.

Figure 14.7 gives the formulas. Of course, both express the same relationship. The first is just solved for (seasonal) sales in terms of seasonally

1. $\text{Sales} = \text{Sales}_{SA} \times \left(\frac{\text{Index}}{100}\right)$

2. $\text{Sales}_{SA} = \text{Sales} \times \left(\frac{100}{\text{Index}}\right)$

Figure 14.7 The relationship between sales and seasonally adjusted sales.

adjusted sales and the index, while the second is solved for seasonally adjusted sales in terms of (seasonal) sales and the index.

Consider the following example:

Use the seasonal index numbers in Figure 14.6.

a. Suppose you expect annual sales of $2000 (thousands) for the coming year with no trend or cyclical variation. Predict sales for each quarter.
b. Suppose actual first quarter sales turn out to be $600 (thousands) and you expect them to continue at this level for the rest of the year. Predict annual sales.
c. Suppose actual second quarter sales then turn out to be $400 (thousand). Have seasonally adjusted sales risen or fallen?

	//	H	I	J	K	L	M
1				Index			
2		Q1:	135.15	135.47	500	677.332	←=K2*(J2/100)
3		Q2:	84.74	84.94	500	424.715	
4		Q3:	66.90	67.06	500	335.300	
5		Q4:	112.27	112.53	500	562.653	
6			399.06	400.00	2000	2000.000	
7							
8					600	442.914	←=K8*(100/J2)
9						1771.656	←=L8*4
10							
11					600	442.914	
12					400	470.904	

Figure 14.8 Using a seasonal index.

a. Predicted annual sales of $2000 would mean $500 in an average quarter—a quarter with an index of exactly 100. Since our quarters have indexes different from 100, we need to adjust our predictions up (Q1, Q4) and down (Q2, Q3) accordingly. Figure 14.8 illustrates. The first quarter estimate is just =K2*(J2/100) =500 × (135.47/100) = 677.332 and we can just copy down for the rest. Notice that the new numbers still sum to $2000.
b. Once we know actual first quarter sales, we can deseasonalize it to get its average or seasonally adjusted equivalent. That is, =K8*(100/J2) = 600 × (100/135.47) = 442.914. And since four seasonally adjusted quarters make up a year, predicted annual sales is just four times as much. That is, =L8*4 = 442.914 × 4 = 1771.656.

c. We already know the seasonally adjusted value for the first quarter; we just need to find the same for the second quarter. That is:

for Q1: =K11*(100/J2) = 600 × (100/135.47) = 442.914

for Q2: =K12*(100/J3) = 400 × (100/84.94) = 470.904.

Sales went down less than expected; on a seasonally adjusted basis they went up.

14.5 The Long-Term Trend

In the last example (part a) you were asked to predict sales by quarter, assuming *no trend or cyclical variation.* This condition was necessary because season need not be the only cause of variation from quarter to quarter. If there is an upward trend, for example, fourth quarter numbers will tend to be higher than first quarter numbers simply because they occur later in the upward trend. This is trend, not seasonal variation and, through the use of our moving-average approach to measuring seasonal variation, we have avoided confusing the two. To see, refer back to Figure 14.2. The moving average has smoothed out the variation in sales, but it still includes a clear upward trend, as well as cyclical variation around that trend. Our next step is to measure that trend.

Using sub-annual data, it is normal to find the centered, seasonally-balanced moving average first; then measure the trend in that moving average. If our data are annual instead, there would be no seasonal variation and we would measure the trend in the data themselves. In the first example to follow we will use annual data; then we will come back to these sales data and find the trend in their moving average.

To find the trend, we run a simple regression. The variable for which we want the trend is the dependent variable; the pattern in this variable is what we are trying to explain.

Time is the explanatory variable. The time periods can be represented with any set of consecutive numbers; the choice just determines the zero time period and thus the Y intercept. If you number the periods 1, 2, 3,…, period zero is the period before the beginning of your data; the Y intercept will be your estimate of Y in that period. If you have annual data and use actual year numbers 1995, 1996, 1997,…, period zero is over 2000 years ago and, since it is way outside your sample, the Y intercept will be pretty meaningless. While not logically wrong, you should generally avoid the latter since you may introduce unnecessary rounding errors.

It really makes sense to speak of a trend only if we observe a fairly constant rate of change. Hence we will look at just two functional forms, linear and exponential. The linear form grows at a constant amount per period; the exponential grows by a constant percentage per period.

14.5.1 A Linear Trend

Back in Chapter 2 (Section 2.3), we created time-series graphs of annual data on nominal GDP, Consumption, and Services. The data

$$\hat{GDP} = a + b\text{Time}$$

Figure 14.9 A straight-line trend in GDP.

are in Services1.xls. Since these are annual data, there are no seasonal effects. Figure 14.9 gives the formula for a straight-line trend in GDP.

If we have not updated the data beyond 1999, we can use the two-digit Year variable (70, 71, 72, etc.) for time. Indeed, this would have been common before the turn of the century; the numbers are not huge (like 1970, 1971, 1972, etc.) but their meaning is clear. With the turn of the century, though, year 2000 becomes year 100 and this is not so intuitive. Suppose, instead, we decide just to make Time = 0 in 1970 and number from there. This will mean the *Y*-intercept is our estimate for 1970. You may want to try numbering the periods differently; it should affect only your *Y* intercept. Figure 14.10 gives the regression results with Time = 0 in 1970.

With an $\bar{R}^2$ of 0.9847, there is clearly a very strong trend. The amount $452.2530 (billion) is our estimate of GDP in 1970, and $257.9527 (billion) is our estimate of the increase per year in GDP.

Despite the good fit, there are a couple of reasons for concern. First, our estimate for 1970 (year zero) is quite far from the mark The actual value is $1035.6 (billion). When we get large errors at the extremes, it may mean that a curved line would fit better.

And at the level of interpretation we do not generally think of economic and business variables as varying by some average amount; more often, we think of them as varying by some proportion or percentage. In this example annual increases of $257 (billion) would have been huge increases in the early years—increases of nearly 25%. Annual increases of

Multiple Regression

Dependent Variable: GDP

***R* Square:**	**Adjusted *R* Square:**	**Standard Error:**
0.9853	**0.9847**	**245.6191**

ANOVA	**Sum of Squares**	**df**	**Mean Square**	F_c	**2-Tailed *p*-Value**
Regression	97314146.0743	1	97314146.0743	1613.0648	0.0000
Error	1447889.4673	24	60328.7278		
Total	98762035.5416	25			

Variable	**Coefficient**	**Standard Error**	t_c	**2-Tailed *p*-Value**
Constant	452.2530	93.6254	4.8305	0.0001
Time	257.9527	6.4226	40.1630	0.0000

Figure 14.10 Linear trend in GDP.

\$257 (billion) would have been very modest increases in the latter years—increases of less than 4%. This does not seem plausible.

14.5.2 An Exponential (Semilog) Trend

The answer to this second concern is the exponential trend. It grows by a constant percentage instead of a constant amount. Figure 14.11 illustrates. You may be familiar with this function as the compound interest formula. Suppose the interest rate is 5%. Then, $(1 + r) = 1.05$. In time period zero $(1 + r)^0 = 1$, so $Y = A(1 + r)^0 = A(1.05)^0 = A(1) = A$. Thus, A is the starting amount. Then, as the value of Time increases to 1, 2, 3, and so on, A is multiplied by $(1.05)^1$, $(1.05)^2$, $(1.05)^3$, and so on. In each period Y is 1.05 times as great as it was in the previous period. In other words, in each time period, Y is 5% greater.

The secret to estimating this percentage growth rate is that, like the Cobb–Douglas in Chapter 13, the exponential can be made linear by taking logs. If we take logs of both sides, we get $\ln(Y) = \ln(A) + \ln(1 + r)^{\text{Time}} = \ln(A) + \text{Time} \ln(1 + r)$. Since A is a constant, $\ln(A)$ is just another constant; call it a. And since, for an exponential r *is* constant $(1 + r)$ is also a constant. Hence $\ln(1 + r)$ is just another constant; call it b. Our formula, then, is just $\ln(Y) = a + b$ Time.

Notice that the result of taking logs is different here from the result we got with the Cobb–Douglas. With the Cobb–Douglas, we ended up with a **log-log** form—that is, with logs on both sides of the equation. With the exponential, we end up with a **semilog** form—that is, with logs only on the left-hand side. We do not take the log of time. And the interpretation of the coefficients is different as well. With the Cobb–Douglas they were elasticities, the percentage change in Y with a 1% change in X. With the exponential, $b = \ln(1 + r)$. Thus, $r = \exp(b) - 1$. And multiplying by 100 gives us the percentage change in Y per time period.

Actually, this procedure may look a little familiar. When we entered dummy variables in the Cobb–Douglas, we did not (could not) take their logs. Hence we were really entering them in semilog form, and our interpretation was the same—the percentage change in Y as the dummy increased by one (from zero to one). The only difference here is that we are estimating the average percentage change in Y over a number of periods.

Figures 14.12 displays two exponential curves on an arithmetic graph. There are several possible objections to this graph. Dating back to when

The exponential function itself:

$$Y = A(1+r)^{\text{Time}}$$

The semilog transformation:

$$\ln(Y) = a + b\text{Time} \quad \text{where} \quad b = \ln(1+r), \quad \text{so} \quad r = \exp(b) - 1$$

Figure 14.11 The exponential function.

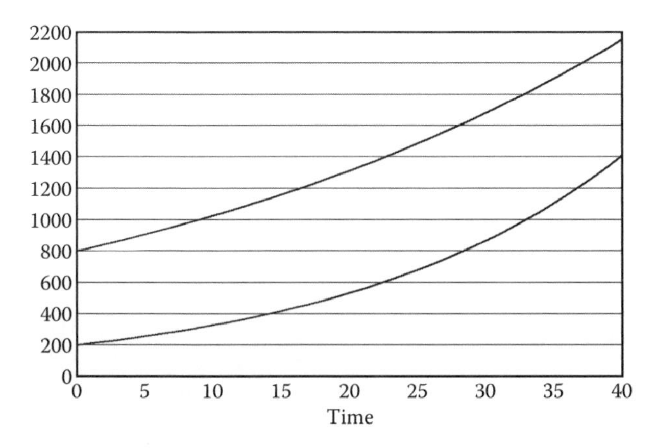

Figure 14.12 Exponential curves on an arithmetic graph.

such graphs were done by hand, the first is that drawing accurate curves is a lot more difficult than drawing accurate straight lines, which require just two points and a ruler. With computers, though, this objection is really no longer relevant.

Second, the lines tend to give the false impression that growth is accelerating when, in fact, each line shows growth at a constant percentage rate. And finally it is hard to look at these curves and be sure what is growing faster.

As we first saw in Chapter 2 (Section 2.3), there is a fix. Figure 14.13 shows the same information on a semilog graph. This is the graphical equivalent of the semilog transformation in Figure 14.11. Movements in the y-direction are *proportional* moves. The distance between \$100 and \$200, \$200 and \$400, \$400 and \$800, and so forth, are all increases of 100% and all are equal distance on the graph. It is now apparent that the growth rate of each line is constant and that the growth rate of the lower line is greater.

Figure 14.14 gives the semilog regression result. With an $\bar{R}^2$ of 0.9821, we again find a strong trend. 7.0451 is our estimate for $\ln(A)$, so

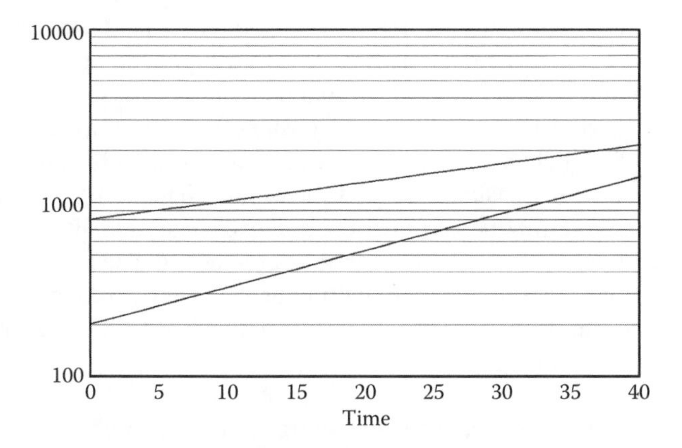

Figure 14.13 Exponential curves on a semilog graph.

Multiple Regression					
Dependent Variable: lnGDP					
R Square: 0.9828		Adjusted *R* Square: 0.9821		Standard Error: 0.0826	
ANOVA	**Sum of Squares**	**df**	**Mean Square**	**F_c**	**2-Tailed *p*-Value**
Regression	9.3398	1	9.3398	1369.6303	0.0000
Error	.1637	24	0.0068		
Total	9.5035	25			
Variable	**Coefficient**	**Standard Error**	**t_c**	**2-Tailed *p*-Value**	
Constant	7.0451	0.0315	223.8137	0.0000	
Time	0.0799	0.0022	37.0085	0.0000	

Figure 14.14 Semilog trend in GDP.

exp(7.0451) = 1147.2 (billion) is our estimate for GDP in year zero. 0.0799 is our estimate for $\ln(1 + r)$, so exp(0.0799) = 1.0832 is our estimate for $1 + r$. We estimate that GDP averaged 8.32% annual growth in this period.

How do we decide between this result and the straight line? First, remember that we have explained 98.21% of the variation in lnGDP here, not GDP, so the $\bar{R}^2$s are not strictly comparable. Still, if one of the two had been much larger than the other, this might well have decided the issue. With this small a difference—0.9847 versus 0.9821—it probably does not.

We did much better here in our estimate for year zero—1147.2 (billion) versus 452.2530 (billion) when the actual value was 1035.6 (billion)—but this should probably not decide the issue either. After all, our goal is to predict the future so year zero is not an especially interesting year.

Given roughly comparable fits, we would probably decide based on the way we want to interpret or use our results. And, generally, these considerations will favor the exponential. We do usually think about growth rates in percentage rather than dollar terms. Moreover, we often want to compare growth rates and the scale or units may not be comparable. In the end of chapter exercises, I ask you to compare the growth rate in GDP with the growth rates in Consumption and Services. The Services are just a part of Consumption and Consumption is just a part of GDP. Thus, the *dollar* change in Services will almost certainly be less than the *dollar* change in Consumption, and the *dollar* change in Consumption will almost certainly be less than the *dollar* change in GDP. But this is almost certainly not what we care about. Services could be growing faster or slower than Consumption (or GDP) in percentage terms, meaning that they are becoming more or less important in the economy. To see this we would use the semilog form.

The units could even be different. Suppose you find straight-line trends in the sales of hamburger and automobiles. You find that hamburger sales are increasing by X pounds per year, while automobile sales are increasing

	A	B	C	D	E	//	N	O
1	Period	Year	Quarter	Sales	Mov Ave		lnMA	
2	1	1	Q1	214.40				
3	2	1	Q2	86.31				
4	3	1	Q3	109.80	136.25		4.9145	← =LN(E4)
5	4	1	Q4	157.73	135.04		4.9056	
6	5	2	Q1	167.95	135.43		4.9085	
7	6	2	Q2	123.08	132.24		4.8846	
:	:	:	:	:	:		:	
:	:	:	:	:	:		:	
36	35	9	Q3	373.28	473.72		6.1606	
37	36	9	Q4	608.95	519.23		6.2523	
38	37	10	Q1	723.25	531.10		6.2749	
39	38	10	Q2	419.32	535.56		6.2833	
40	39	10	Q3	372.45				
41	40	10	Q4	645.50				

Figure 14.15 Mov Ave and lnMA for sales.

by *Y* automobiles per year. A comparison of *X* and *Y* is meaningless; the units are completely different. If, on the other hand, you find semilog trends, you will get comparable, unit-less percentage growth rates.

Return now to the earlier sales example:

> Using the moving average for Sales estimated earlier, find linear and exponential trends.

Figure 14.15 repeats the relevant portion of Figures 14.5 and 14.6. Mov Ave is the dependent variable for the linear trend; lnMA is the dependent variable for the semilog trend. Period is the explanatory variable for each. We have 36 cases, from Period = 3 through 38. Figures 14.16 and 14.17 present the results.

The linear regression has an $\bar{R}^2$ of 0.7777. Using this regression we estimate the increase in Mov Ave to be \$9.8826 (thousand) per period. Since Mov Ave and Sales share the same trend, this is also our estimate for the increase in Sales. Notice that this is the increase per quarter. Since it is a constant amount, we can translate it into the increase per year by just multiplying by four. The annual trend growth rate in Sales is $\$9.8826 \times 4 = \$39.5304$ (thousand) per year.

The semilog regression has an $\bar{R}^2$ of 0.8228. Using this regression, we estimate the increase in lnMA to be 0.0369. The rate of increase in Mov Ave, then, is exp(0.0369) – 1 = 0.0376, or 3.76%. Again, since Mov Ave and Sales share the same trend, this is also our estimate for the rate of increase in Sales. Again, this is an increase per quarter. And, since the line is not linear, we need to be careful in translating it into an annual rate. We need to multiply the slope coefficient by four *before* taking the antilog. Our estimate for the annual rate of trend growth is

$$\exp(0.0369 \times 4) - 1 = \exp(0.1476) - 1 = 0.1590, \text{ or } 15.90\%.$$

How do we decide which trend to use? Remember that the second regression has explained 82.28% of the variation in lnMA, not Mov Ave.

Multiple Regression Dependent Variable: Mov Ave					
R Square: 0.7840		**Adjusted R Square:** 0.7777		**Standard Error:** 55.4419	
ANOVA	**Sum of Squares**	**df**	**Mean Square**	**F_c**	**2-Tailed p-Value**
Regression	379428.5321	1	379428.5321	123.4395	0.0000
Error	104509.2441	34	3073.8013		
Total	483937.7762	35			
Variable	**Coefficient**	**Standard Error**	**t_c**	**2-Tailed p-Value**	
Constant	73.2642	20.4422	3.5840	0.0010	
Period	9.8826	0.8895	11.1103	0.0000	

Figure 14.16 Linear trend in sales.

Still, this is enough better than 77.77% that we might well choose the semilog regression on this basis. And, as before, constant percentage growth rates are generally more plausible than constant dollar growth rates. Thus, we would have a preference for the semilog form.

Once we have decided on the semilog form, we can compute the trend values for each period. Figure 14.18 illustrates. The first value is =exp(4.7 753 + 0.0369*A2). The rest of the values can be found with **COPY**.

14.5.3 The Limitations of "Curve-Fitting"

There are, of course, other functional forms that might fit the existing data better. For example, if the data rose for the first 20 periods and fell for the last 20 periods, a quadratic function would probably fit the data much

Multiple Regression Dependent Variable: lnMA					
R Square: 0.8278		**Adjusted R Square:** 0.8228		**Standard Error:** 0.1800	
ANOVA	**Sum of Squares**	**df**	**Mean Square**	**F_c**	**2-Tailed p-Value**
Regression	5.2968	1	5.2968	163.4626	0.0000
Error	1.1017	34	0.0324		
Total	6.3985	35			
Variable	**Coefficient**	**Standard Error**	**t_c**	**2-Tailed p-Value**	
Constant	4.7753	0.0664	71.9469	0.0000	
Period	0.0369	0.0029	12.7853	0.0000	

Figure 14.17 Semilog trend in sales.

	A	B	C	D	E	//	N	O	P
1	Period	Year	Quarter	Sales	Mov Ave		lnMA		Trend
2	1	1	Q1	214.40			=EXP(4.7753+0.0369*A2) →		123.00
3	2	1	Q2	86.31					127.63
4	3	1	Q3	109.80	136.25		4.9145		132.42
5	4	1	Q4	157.73	135.04		4.9056		137.40
6	5	2	Q1	167.95	135.43		4.9085		142.57
7	6	2	Q2	123.08	132.24		4.8846		147.92
:	:	:	:	:	:		:		:
:	:	:	:	:	:		:		:
36	35	9	Q3	373.28	473.72		6.1606		431.30
37	36	9	Q4	608.95	519.23		6.2523		447.51
38	37	10	Q1	723.25	531.10		6.2749		464.33
39	38	10	Q2	419.32	535.56		6.2833		481.79
40	39	10	Q3	372.45					499.90
41	40	10	Q4	645.50					518.69

Figure 14.18 Computing the trend.

better than either the linear or semilog. However, it is important to keep our purpose and implicit assumptions in mind. Our purpose is to use the past pattern to predict future values. This works only if time is a good proxy for the true causes of variation, so we are really assuming that the true causes of variation have a consistent pattern. And if the data first rose and then fell, this would seem to suggest that there has been a change in the pattern in the true causes. Something fundamental has changed. In such a case, even a very good fit to the *past* data will likely tell us very little about the *future*. In such a case, we should probably not be doing time-series analysis.

14.6 The Business Cycle

Comparing the moving average with the trend, the difference represents cyclical and random variation. More or less by definition, there is little that we can say about the random component. Unfortunately, there is not much more that we can say about the cyclical component either. It is systematic; looking back at Figure 14.2 there are clear broad swings of the moving average around the long-term trend. We should expect these to continue. But clearly, past swings have varied quite a bit in their length and severity. These give only modest guidance as to how long and severe the future swings will be.

14.7 Putting It All Together: Forecasting

14.7.1 Recapping Our Decomposition

We have covered a great deal in this chapter; a recap is probably in order. Our goal is to use the patterns in past data to predict future values. To do so, we needed first to identify these patterns in the past data. This is what we have done so far.

14.7.1.1 Seasonal Variation

Starting with seasonal variation, we created a seasonally balanced moving average. This moving average represented our estimate of past sales stripped of their seasonal variation. Thus by comparing actual sales, which had seasonal variation with the moving average, which did not we were able to create specific seasonal indexes. From these, we created an overall index of seasonal variation.

14.7.1.2 Long-Term Trend

Moving on to trend, the moving average includes trend as well as cyclical and random variation. To separate out the trend we used this moving average (or its log) in a regression with time as the explanatory variable. The predictions from this regression represented our estimate of trend.

14.7.1.3 Cyclical and Random Variation

The remaining variation around the trend is the cyclical and random variation. We saw that there were long but irregular swings around the trend.

We are ready to project these patterns into the future. We should be modest in doing so. Recall once more that we are not looking at causal relationships here. We are assuming that time is a reasonable proxy for the true causes of variation. For this to be true, these true causes must be varying over time in reasonably consistent ways. The further out we project the less likely this is to be true. But hopefully we can safely project a few periods into the future.

14.7.2 Projecting the Trend

The first step is to project the trend into the future. All we need to do is add values 41–46 for time and copy down the regression formula. Figures 14.19 and 14.20 show the spreadsheet and graph with trend projected six quarters into the future.

14.7.3 Projecting the Business Cycle

Projecting the business cycle variation around the trend is the most subjective part. We did not develop any quantitative measure of this variation. But we know two things. First, the moving average, which represents trend, cyclical, and random variation, ended above trend. So it is reasonable to assume that the next quarter or two, at least, will still be above trend. Second, we cannot stay above or below trend forever.

One approach is to start from where the moving average ended and try to project a continuation of the pattern so far. The moving average was above the trend when it ended in period 38. But it was pretty flat. While it would probably remain above trend for a couple more quarters, it is not clear that it would remain so for long. Figures 14.21 and 14.22 show one guess for how the pattern might be projected around the trend. Notice in Figure 14.21 that the projections are round numbers. Since they are, at best, educated guesses, there is no reason to pretend any great precision.

	A	B	C	D	E	//	G	H	I	//	P
1	Period	Year	Quarter	Sales	Mov Ave				Index		Trend
2	1	1	Q1	214.40			Q1:	135.15	135.47		123.00
3	2	1	Q2	86.31			Q2:	84.74	84.94		127.63
4	3	1	Q3	109.80	136.25		Q3:	66.90	67.06		132.42
5	4	1	Q4	157.73	135.04		Q4:	112.27	112.53		137.40
6	5	2	Q1	167.95	135.43			399.06	400.00		142.57
7	6	2	Q2	123.08	132.24						147.92
:	:	:	:	:	:						:
:	:	:	:	:	:						:
36	35	9	Q3	373.28	473.72						431.30
37	36	9	Q4	608.95	519.23						447.51
38	37	10	Q1	723.25	531.10						464.33
39	38	10	Q2	419.32	535.56						481.79
40	39	10	Q3	372.45							499.90
41	40	10	Q4	645.50							518.69
42	41	11	Q1			=EXP(4.7753 + 0.0369*A42) →					538.18
43	42	11	Q2								558.41
44	43	11	Q3								579.40
45	44	11	Q4								601.18
46	45	12	Q1								623.78
47	46	12	Q2								647.23

Figure 14.19 Projecting the trend (spread sheet).

An alternative approach would have been to project a gradual return to trend and then no further business cycle. This is not going to be right. But if we really do not know which side of trend is more likely, it is a reasonable assumption.

Finally, a real-world option, not available to us in this hypothetical example, is to factor in other economic information. That is, are economists predicting strong economic activity for the next year or so? If so, raise your projections a bit.

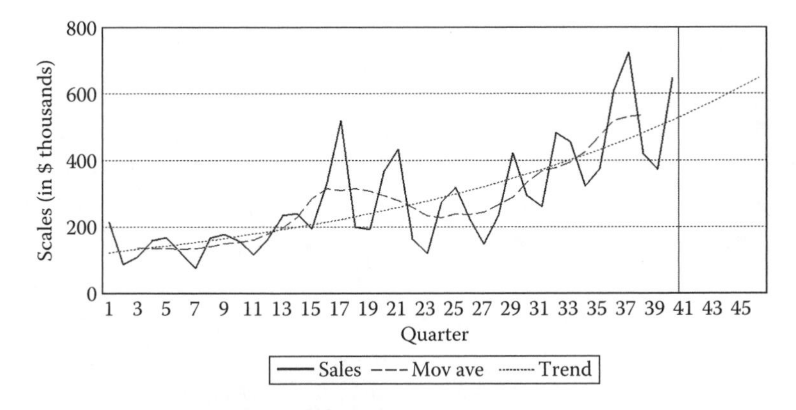

Figure 14.20 Projecting the trend (graph).

	A	B	C	D	E	//	G	H	I	//	P
1	Period	Year	Quarter	Sales	Mov Ave				Index		Trend
2	1	1	Q1	214.40			Q1:	135.15	135.47		123.00
3	2	1	Q2	86.31			Q2:	84.74	84.94		127.63
4	3	1	Q3	109.80	136.25		Q3:	66.90	67.06		132.42
5	4	1	Q4	157.73	135.04		Q4:	112.27	112.53		137.40
6	5	2	Q1	167.95	135.43			399.06	400.00		142.57
7	6	2	Q2	123.08	132.24						147.92
:	:	:	:	:	:						:
:	:	:	:	:	:						:
36	35	9	Q3	373.28	473.72						431.30
37	36	9	Q4	608.95	519.23						447.51
38	37	10	Q1	723.25	531.10						464.33
39	38	10	Q2	419.32	535.56						481.79
40	39	10	Q3	372.45	540.00						499.90
41	40	10	Q4	645.50	538.00						518.69
42	41	11	Q1		535.00						538.18
43	42	11	Q2		533.00						558.41
44	43	11	Q3		530.00						579.40
45	44	11	Q4		535.00						601.18
46	45	12	Q1		543.00						623.78
47	46	12	Q2		565.00						647.23

Figure 14.21 Projecting the business cycle around the trend (spreadsheet).

14.7.4 Projecting Seasonal Variation

Finally we need to factor in seasonal variation. This is what our overall seasonal index is for. Figures 14.23 and 14.24 illustrate. Our trend projection for period 41 is 538.18 (cell P42). The projection for the business cycle is to fall a little below the trend, to 535 (cell E42). But period 41 is a first quarter and our first quarters are our strongest. So, factoring in the seasonal effect, we project period 41 sales (cell D42) to be:

$$\text{Sales} = \text{D42} = \text{E42*(J2/100)} = 535 \times (135.47/100) = 724.75.$$

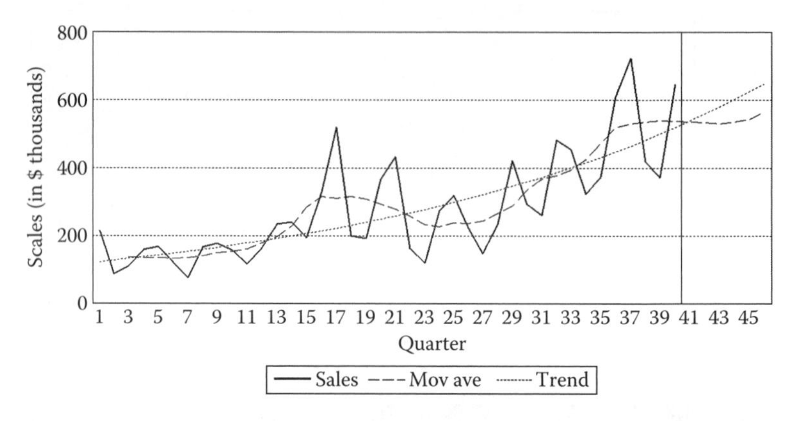

Figure 14.22 Projecting the business cycle around the trend (graph).

	A	B	C	D	E	//	H	I	J	//	P
1	Period	Year	Quarter	Sales	Mov Ave				Index		Trend
2	1	1	Q1	214.40			Q1:	135.15	135.47		123.00
3	2	1	Q2	86.31			Q2:	84.74	84.94		127.63
4	3	1	Q3	109.80	136.25		Q3:	66.90	67.06		132.42
5	4	1	Q4	157.73	135.04		Q4:	112.27	112.53		137.40
6	5	2	Q1	167.95	135.43			399.06	400.00		142.57
7	6	2	Q2	123.08	132.24						147.92
:	:	:	:	:	:						:
:	:	:	:	:	:						:
36	35	9	Q3	373.28	473.72						431.30
37	36	9	Q4	608.95	519.23						447.51
38	37	10	Q1	723.25	531.10						464.33
39	38	10	Q2	419.32	535.56						481.79
40	39	10	Q3	372.45	540.00						499.90
41	40	10	Q4	645.50	538.00						518.69
42	41	11	Q1	724.75	← =E42*(J2/100)						538.18
43	42	11	Q2	452.75	533.00						558.41
44	43	11	Q3	355.42	530.00						579.40
45	44	11	Q4	602.04	535.00						601.18
46	45	12	Q1	735.58	543.00						623.78
47	46	12	Q2	479.93	565.00						647.23

Figure 14.23 Factoring in the seasonal variation. (spreadsheet).

We can use **COPY** to get periods 42 through 44 (quarters 2 through 4); we need to start over with period 45, the first quarter of year 12.

We now have projections for the next six quarterly sales values that reflect and extend past trend, cyclical, and seasonal variation.

14.8 Another Example

The Excel file Constructionl.xls contains monthly data on construction employment in Illinois, in thousands of workers, from 1990 through 2003. (Source: US Department of Labor, Bureau of Labor Statistics.)

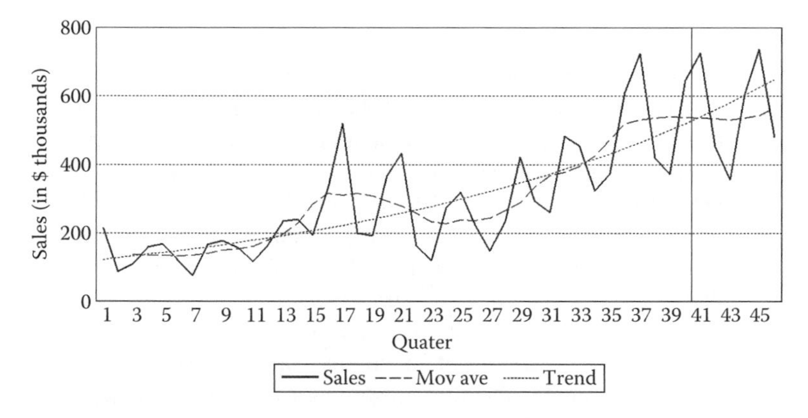

Figure 14.24 Factoring in the seasonal variation (graph).

a. Create a graph of construction employment in Illinois from 1990 through 2003. What types of systematic variation do you see?
b. Find a centered, seasonally balanced moving average.
c. Using the moving average in part b, find specific seasonal indexes. Using these, find an overall seasonal index for construction employment in Illinois. In what months is construction employment highest? Lowest?
d. Using the moving average in part b, find straight-line and exponential trends. Interpret each slope coefficient in words. Decide what trend to use and briefly explain your decision.
e. Add the trend to your spreadsheet. Add the moving average and trend to your graph.
f. Extend the trend through 2004. Add these values to your graph.
g. We now need to add the cyclical variation around the trend. Describe where we seem to be in the business cycle. Continue the business cycle around the trend through 2004. Add these values to your graph.
h. We now need to add the seasonal variation around the business cycle. Using the overall seasonal index from part b, factor in the seasonal variation. These are your final projections for 2004. Add them to your graph.

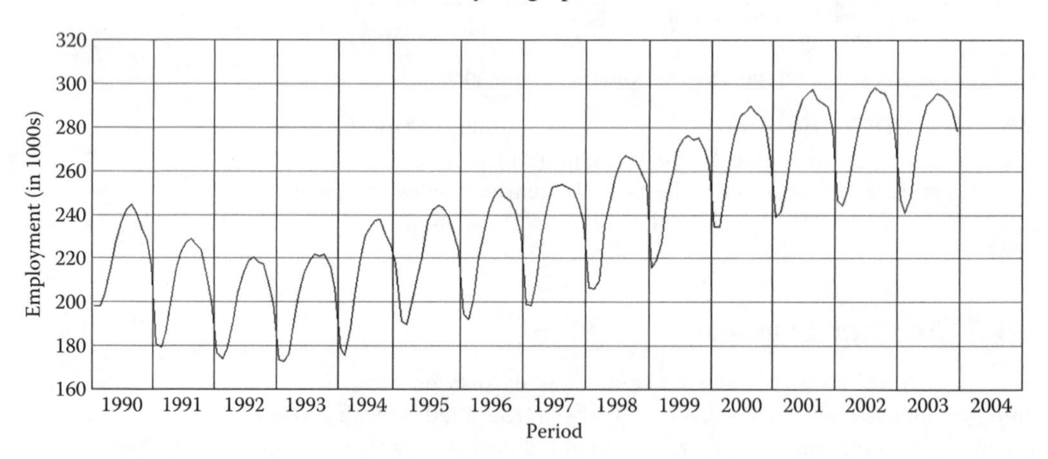

Figure 14.25 Construction employment in Illinois (a).

a. Figure 14.25 graphs the raw data. The seasonal variation is obvious, with employment lowest around February and highest in late summer or fall. There is also a definite upward trend over the period. And there appears to be some cyclical variation around that trend, with a trough around 1992–1993 and a peak around 2001–2002.

   Be sure to notice that the Y axis goes down just to 160. Thus, it tends to exaggerate the extent of the variation.
b. Figure 14.26 shows the first and last 15 cases of the data. Since these are monthly data, the centered, seasonally balanced moving average is an average of 24 months, constructed so that there is a middle month. The first month that can be at the middle is July 1990. The formula for this month (cell E8) is:

MA =AVERAGE(D2:D13,D3:D14) = 223.15.

That is, we average January 1990–December 1990 plus February 1990–January 1991. We have two of each type of month so there should be no net seasonal effect. But there is a middle month. There are 11 months ahead of July–January 1990 once and February–June 1990 twice. And there are 11 months after July–August–December 1990 twice and January 1991 once.

Again, we can use **COPY** for the rest. Again, we need to be careful not to go beyond June of 2003—the last month that can be at the center.

c. For all but the first six and last six months, we now have a figure for actual employment, and an estimate of what employment would have been stripped of the seasonal effect. By comparing the two, we can create a SSI for each month. The formula for July 1990 (cell F8) is

	A	B	C	D	E	F	G	H	I	J
1	Period	Year	Month	Employ	MA	SSI				Index
2	1	1990	January	197.9				January:	86.67	86.67
3	2	1990	February	197.9				February:	86.02	86.02
4	3	1990	March	203.8				March:	89.44	89.44
5	4	1990	April	215.9				April:	96.84	96.84
6	5	1990	May	227.6				May:	101.75	101.75
7	6	1990	June	236.2				June:	105.88	105.88
8	7	1990	July	242.2	223.15	108.54		July:	107.76	107.76
9	8	1990	August	244.6	221.66	110.35		August:	108.61	108.61
10	9	1990	September	240.7	220.12	109.35		September:	107.55	107.55
11	10	1990	October	233.9	218.76	106.92		October:	106.46	106.46
12	11	1990	November	229.1	217.63	105.27		November:	103.74	103.74
13	12	1990	December	216.4	216.53	99.94		December:	99.28	99.28
14	13	1991	January	181.0	215.34	84.05			1200.00	1200.00
15	14	1991	February	179.1	214.06	83.67				
16	15	1991	March	185.7	212.80	87.27				
:	:	:	:	:	:	:				
:	:	:	:	:	:	:				
155	154	2002	October	295.4	277.09	106.61				
156	155	2002	November	289.9	277.22	104.58				
157	156	2002	December	276.9	277.34	99.84				
158	157	2003	January	246.8	277.28	89.01				
159	158	2003	February	240.8	277.06	86.91				
160	159	2003	March	248.2	276.88	89.64				
161	160	2003	April	269.2	276.68	97.30				
162	161	2003	May	281.1	276.46	101.68				
163	162	2003	June	290.3	276.43	105.02				
164	163	2003	July	292.8						
165	164	2003	August	295.6						
166	165	2003	September	294.7						
167	166	2003	October	292.2						
168	167	2003	November	288.0						
169	168	2003	December	278.0						

Figure 14.26 Construction employment in Illinois (b, c).

SSI = (Employ/MA) × 100 =(D8/E8)*100 = 108.54.

We average all the January SSIs, all the February SSIs, and so on to get an overall index. For January, the formula is

I2 =AVERAGE(F14,F26,F38,F50,F62,F74,F86,
F98,F110,F122,F134,F146,F158) = 86.67.

In this case, the 12 monthly index numbers add up to 1200 to two decimals; still it is not exact. To make it so, we weight each by 1200 over their sum. For January (cell J2), the formula is

Index =I2*(1200/$I$14).

Construction employment is highest in August, with July and September close behind. It is lowest in February, followed by January and then March.

d. Figures 14.27 and 14.28 show the regression results for linear and semi-log trends.

According to the linear regression, employment is growing by about 0.5739 (thousand) jobs per month. Annually, that is $0.5739 \times 12 = 6.887$ (thousand) jobs per year. According to the semilog regression, employment is growing by about $\exp(0.0024) - 1 = 0.0024$ or 0.24% per month. Annually, that is $\exp(0.0024 \times 12) - 1 = 0.0292$ or 2.92% per year.

Both $\bar{R}^2$s are about 0.91; while they are not strictly comparable, they are so close that we would probably not decide on this basis anyway. Probably we would choose based on the interpretation that we preferred. We might want to compare the pattern in this industry in Illinois with the same industry in other states or nationally. We might also want to compare it with other industries in Illinois. If such comparisons are of interest, percentage growth rates are definitely preferable. We would assume exponential growth and use the semilog regression.

Multiple Regression

Dependent Variable: Mov Ave

***R* Square:**	**Adjusted *R* Square:**	**Standard Error:**
0.9117	**0.9111**	**8.0946**

ANOVA	Sum of Squares	df	Mean Square	F_c	2-Tailed *p*-Value
Regression	104208.6303	1	104208.6303	1590.4132	0.0000
Error	10090.5404	154	65.5230		
Total	114299.1707	155			

Variable	Coefficient	Std Error	t_c	2-Tailed *p*-Value
Constant	189.2125	1.3780	137.3090	0.0000
Period	0.5739	0.0144	39.8800	0.0000

Figure 14.27 Construction employment in Illinois—linear trend (d).

Multiple Regression					
Dependent Variable: lnMA					
R Square: 0.9110		Adjusted *R* Square: 0.9104		Standard Error: 0.0340	
ANOVA	Sum of Squares	df	Mean Square	F_c	2-Tailed *p*-Value
Regression	1.8204	1	1.8204	1576.3563	0.0000
Error	0.1778	154	0.0012		
Total	1.9982	155			
Variable	Coefficient	Standard Error	t_c	2-Tailed p-Value	
Constant	5.2619	0.0058	909.5692	0.0000	
Period	0.0024	0.0001	39.7034	0.0000	

Figure 14.28 Construction employment in Illinois—semilog trend (d).

e. Figure 14.30 adds the exponential trend to the spreadsheet. The formula for January 1990 (cell L2) is

Trend =EXP(5.2619 + 0.0024*A2).

We can use **COPY** for the rest.

Figure 14.29 adds the moving average and trend to the graph.

f. Figures 14.29 and 14.30 also include the trend extended through 2004. We can get these by just extending the Period column by 12, and using **COPY** to extend the trend.

g. It is clear from Figure 14.29 that construction employment was below trend in 2003. While we can only guess at how fast it will recover, it is unreasonable to think it will happen instantaneously. Figures 14.30 and 14.31 offer a more plausible path. Again, I have used round numbers, since these are no more than educated guesses.

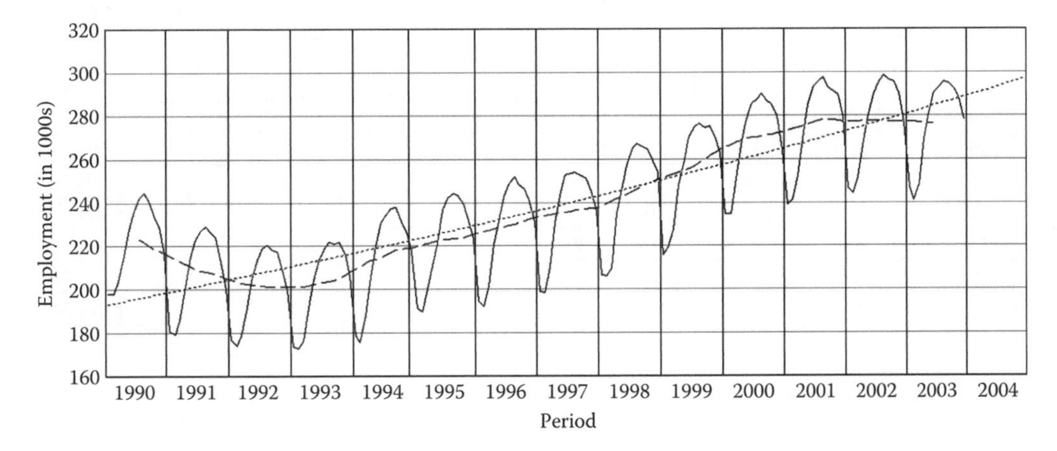

Figure 14.29 Construction employment in Illinois (e, f).

	A	B	C	D	E	//	H	I	J	K	L
1	Period	Year	Month	Employ	MA				Index		Trend
2	1	1990	January	197.9			January:	86.67	86.67	e.	193.31
3	2	1990	February	197.9			February:	86.02	86.02	↓	193.78
4	3	1990	March	203.8			March:	89.44	89.44		194.24
5	4	1990	April	215.9			April:	96.84	96.84		194.71
6	5	1990	May	227.6			May:	101.75	101.75		195.18
7	6	1990	June	236.2			June:	105.88	105.88		195.64
8	7	1990	July	242.2	223.15		July:	107.76	107.76		196.11
9	8	1990	August	244.6	221.66		August:	108.61	108.61		196.59
10	9	1990	September	240.7	220.12		September:	107.55	107.55		197.06
11	10	1990	October	233.9	218.76		October:	106.46	106.46		197.53
12	11	1990	November	229.1	217.62		November:	103.74	103.74		198.01
13	12	1990	December	216.4	216.53		December:	99.28	99.28		198.48
14	13	1991	January	181.0	215.34			1200.00	1200.0		198.96
15	14	1991	February	179.1	214.06						199.44
16	15	1991	March	185.7	212.80						199.92
:	:	:	:	:	:						:
:	:	:	:	:	:						:
155	154	2002	October	295.4	277.09						279.08
156	155	2002	November	289.9	277.22						279.75
157	156	2002	December	276.9	277.34						280.42
158	157	2003	January	246.8	277.28						281.10
159	158	2003	February	240.8	277.06						281.77
160	159	2003	March	248.2	276.88						282.45
161	160	2003	April	269.2	276.68						283.13
162	161	2003	May	281.1	276.46						283.81
163	162	2003	June	290.3	276.43						284.49
164	163	2003	July	292.8	276.00	g.					285.17
165	164	2003	August	295.6	276.00	↓					285.86
166	165	2003	September	294.7	276.00						286.55
167	166	2003	October	292.2	276.00						287.23
168	167	2003	November	288.0	277.00						287.92
169	168	2003	December	278.0	277.00						288.62
170	169	2004	January	240.1	← h.					f.	289.31
171	170	2004	February	238.3	277.00					↓	290.01
172	171	2004	March	248.7	278.00						290.70
173	172	2004	April	269.2	278.00						291.40
174	173	2004	May	282.9	278.00						292.10
175	174	2004	June	295.4	279.00						292.80
176	175	2004	July	300.7	279.00						293.51
177	176	2004	August	304.1	280.00						294.21
178	177	2004	September	301.1	280.00						294.92
179	178	2004	October	299.1	281.00						295.63
180	179	2004	November	291.5	281.00						296.34
181	180	2004	December	280.0	282.00						297.05

Figure 14.30 Construction employment in Illinois (e–h).

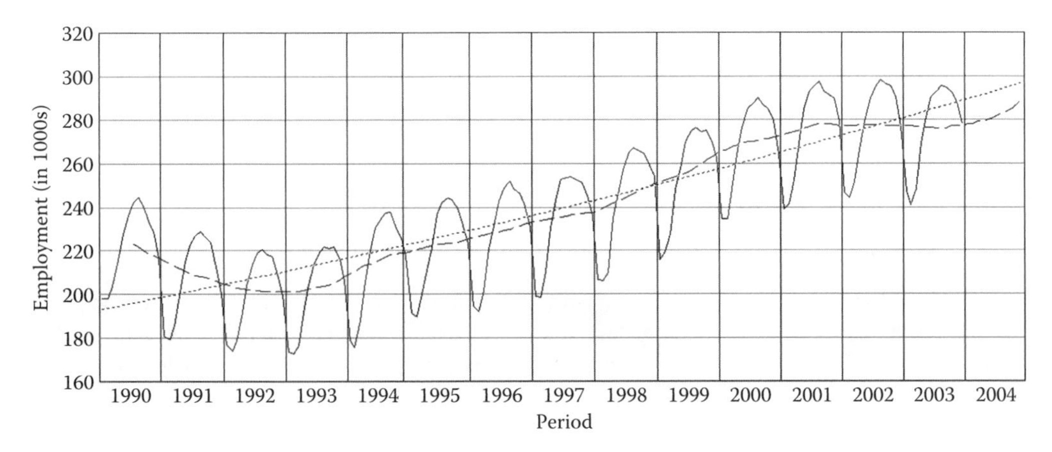

Figure 14.31 Construction employment in Illinois (g).

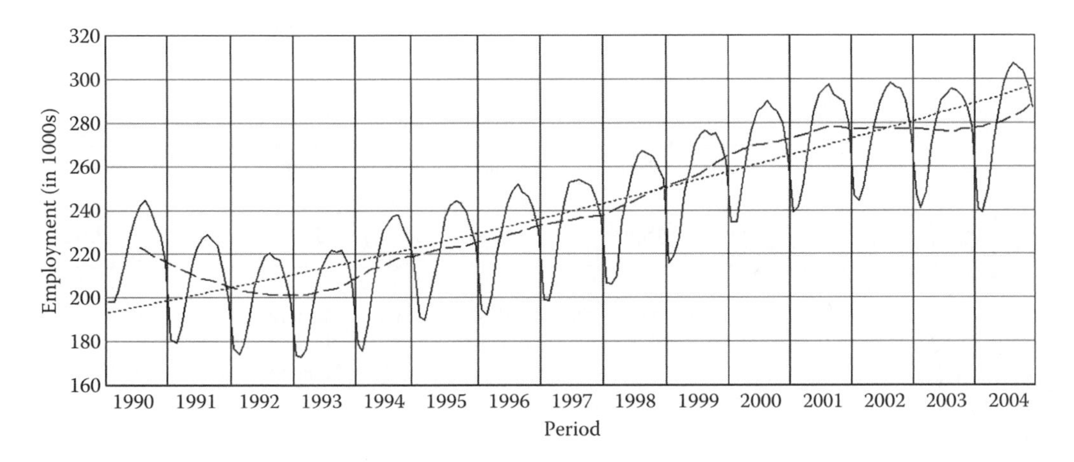

Figure 14.32 Construction employment in Illinois (h).

h. Finally, we need to factor in the seasonal variation. Figures 14.30 and 14.32 do. The formula for January 2004 (cell D170) is

Employ =E170*(J2/100) = 277.00 × (86.67/100) = 240.1.

And we can use **COPY** for the rest.
We now have projections for the next 12 monthly construction employment values that reflect and extend past (f) trend, (g) business cycle, and (h) seasonal variation.

14.9 Exercises

14.1 Your actual quarterly ice cream sales for last year are given below along with the seasonal index for each quarter. Does there seem to be any longer term (trend or cyclical) variation or does the variation seem to be just seasonal and random? Explain.

Quarter	Q1	Q2	Q3	Q4
Index	67.8	90.5	148.1	93.6
Sales	$10,520	$16,742	$33,322	$26,226

14.2 Continue with the seasonal indexes above.
   a. Suppose you expect annual sales of $100,000 for next year with no trend or cyclical variation. Predict sales for each quarter.
   b. Suppose actual first quarter sales turn out to be $20,000 and you expect them to continue at this level with no trend or cycle for the rest of the year. Predict sales for the remaining quarters.
   c. Suppose actual second quarter sales then turn out to be $30,000. Have seasonally adjusted sales risen or fallen?

14.3 Your sales for the last six months are given below, along with the seasonal index for each month. Does there seem to be any longer term—trend or cyclical—variation or does the variation seem to be just seasonal and random? Explain.

Month	January	February	March	April	May	June
Index	84.5	89.8	92.3	102.6	113.0	125.4
Sales	$50,900	$50,750	$48,880	$50,830	$52,000	$53,720

14.4 The Excel file Sales2.xls contains quarterly data on sales (in $ thousands), uncorrected for seasonal variation.
   a. Find a centered, seasonally balanced moving average for these sales.
   b. Using the moving average above, construct an overall seasonal index by quarters.
   c. If you expect $60 thousand in sales for the year with no trend or cycle, how much should you expect in each quarter?
   d. If quarter 1 sales are actually $15 thousand and you expect this level to continue with no trend or cycle, how much should you expect for the year?

14.5 Refer back to the Servicesl.xls data file that contains annual data on GDP, Consumption, and Services for 26 years.
   a. Find the trend for each. Interpret your results. Which is growing fastest? Slowest?
   b. Which form, linear or semilog, did you use? Why?

14.6 Refer back to the Pricesl.xls data file that contains annual data on the Consumer Price Index (CPI) data for all items and five components for 37 years.
   a. Find the trend for each. Interpret your results. Which is growing fastest? Which is growing slowest? Which is the least consistent?
   b. Which form, linear or semilog, did you use? Why?

14.7 The Excel file Government1.xls contains monthly data on government employment in Illinois, on thousands of workers from 1990 through 2003. (Source: U.S. Department of Labor, Bureau of Labor Statistics.)

a. Create a graph of government employment in Illinois from 1990 through 2003. Which types of systematic variation do you see? How are they different from the Construction1.xls data in the chapter?
b. Find a centered, seasonally balanced moving average.
c. Using the moving average in part b, find specific seasonal indexes. Using these, find an overall seasonal index for government employment in Illinois. In which months is government employment highest? Lowest?
d. Using the moving average in part b, find straight-line and exponential trends. Interpret each slope coefficient in words. Decide what trend to use and briefly explain your decision.
e. Add the trend to your spreadsheet. Add the moving average and trend to your graph.
f. Extend the trend through 2004. Add these values to your graph.
g. We now need to add the cyclical variation around the trend. Describe where we seem to be in the business cycle. Continue the business cycle around the trend through 2004. Add these values to your graph.
h. We now need to add the seasonal variation around the business cycle. Using the overall seasonal index from part b, factor in the seasonal variation. These are your final projections for 2004. Add them to your graph.

14.8 The Excel file Toysrus1.xls contains quarterly data on net sales and net earnings for Toys "R" Us, Inc. from 2000 through 2004. (Source: Toys "R" Us Annual Reports, 2001–2004. The company's fiscal year runs from February through January; hence Q1 sales are for February through April, etc.)

a. Create a graph of either net sales or net earnings from 2000 through 2004. What types of systematic variation do you see?
b. Find a centered, seasonally balanced moving average.
c. Using the moving average in part b, find specific seasonal indexes. Using these, find an overall seasonal index for net sales or net earnings. In which quarter is net sales or net earnings highest? Lowest?
d. Using the moving average in part b, find straight-line and exponential trends. Interpret each slope coefficient in words. Decide what trend to use and briefly explain your decision.
e. Add the trend to your spreadsheet. Add the moving average and trend to your graph.
f. Extend the trend through 2005. Add these values to your graph.
g. We now need to add the cyclical variation around the trend. Describe where we seem to be in the business cycle.

Continue the business cycle around the trend through 2005. Add these values to your graph.

h. We now need to add the seasonal variation around the business cycle. Using the overall seasonal index from part b, factor in the seasonal variation. These are your final projections for 2005. Add them to your graph.

Appendix A

Examples, Exercises, and Data Files

A number of data files are used repeatedly though the text in the hope that doing so will help you understand how the various techniques of the course are connected. The files below are used for in-chapter examples (I) and end-of-chapter exercises (E), as listed. Keeping your work for the early chapter examples and exercises will help in doing the later ones.

Data File	**Ch Ol**	**Ch 02**	**Ch 03**	**Ch 04**	**Ch 05**	**Ch 06**	**Ch 07**
Construction1.xls							
Diet1.xls							
Employees1.xls	I	E	E				E
Employees2.xls							
Governmentl.xls							
Moneyl.xls							
Nickelsl.xls							E
NLSYl.xls		I E	I E				E
Pricesl.xls		E					
Regressionl.xls			I				
Salesl.xls							
Sales2.xls							
Servicesl.xls		I E					
Studentsl.xls		I E	I E				E
Students2.xls		E	E				E
Toysrusl.xls							

Data File	**Ch 08**	**Ch 09**	**Ch 10**	**Ch 11**	**Ch 12**	**Ch 13**	**Ch 14**
Construction1.xls							I
Diet1.xls		E					
Employees1.xls							
Employees2.xls	E	E	E	E	I E	I	
Governmentl.xls							E
Moneyl.xls						I	
Nickelsl.xls	E	E	E	E	E	E	
NLSYl.xls	E	E		E	E	I	
Pricesl.xls							E
Regressionl.xls					I		
Salesl.xls							I
Sales2.xls							E
Servicesl.xls							I E
Studentsl.xls	E	I	I	I	E	E	
Students2.xls	E	E	E	E	E	E	
Toysrusl.xls							E

Constructionl.xls: Monthly construction employment in Illinois for a sample of 14 years. Taken from the St. Louis Federal Reserve web site, FRED II, 2004. Original source, the U.S. Department of Labor, Bureau of Labor Statistics.

Period: Period (1–168)
Year: Year (1990–2002)
Month: Month (Jan, Feb, …, Dec)
Employ: Employment (in thousands)

Dietl.xls: Before and after weights for a sample of 25 dieters (hypothetical).

ID: Individual identification number
Before: Weight before dieting (in pounds)
After: Weight after dieting (in pounds)

Employeesl.xls: Personal information for a sample of 50 employees (hypothetical).

ID: Employee identification number
Ed: Education (in years)
Exp: Job experience (in years)
Female: Female (1 = yes; 0 = no)
Salary: Salary (in $ thousands)

Employees2.xls: Personal information for a sample of 50 employees (hypothetical). It is the same as Employeel.xls but with additional information on job type.

ID: Employee identification number
Ed: Education (in years)
Exp: Job experience (in years)
Type: Job type (1 = line, 2 = office, 3 = management)
Female: Female (1 = yes, 0 = no)
Salary: Salary (in $ thousands)

Governmentl.xls: Monthly government employment in Illinois for a sample of 14 years. Taken from the St. Louis Federal Reserve web site, FRED II, 2004. Original source, U.S. Department of Labor, Bureau of Labor Statistics.

Period: Period (1–168)
Year: Year (1990–2002)
Month: Month (Jan, Feb, …, Dec)
Employ: Employment (in thousands)

Moneyl.xls: Annual figures for the real quantity of money, real disposable personal income, and the six-month treasury bill rate, for a sample of 44 years. Taken from the Economic Report of the President, 2004, Tables B-31, B-60, B-69, B-75. Real M2 is calculated by the author from M2 and the consumer price index. Original sources, U.S. Department of Commerce, Bureau of Economic Analysis; Department of Labor, Bureau of Labor Statistics; Board of Governors, Federal Reserve System; and Department of the Treasury.

Date: Year (1959–2002)
RM2: Real M2 money supply (in $ billions)
TBR: Six-month treasury bill rate (in %)
RDPI: Real personal disposable income (in $ billions)

Nickelsl.xls: Personal and market information for a sample of 50 consumers (hypothetical).

Customer: Nickels customer (1 = yes, 0 = no)
Female: Female (1 = yes, 0 = no)
Age: Age (in years)
Income: Income (in $ thousands)
Source: Primary source of market information (1 = newspaper, 2 = radio, 3 = other)

NLSY1.xls: Personal information for a sample of 281 young adults. Taken from the National Longitudinal Survey of Youth.

ID: Individual identification number
Female: Female (1 = yes, 0 = no)
Age: Age (in years)
Height: Height (in inches)
Weight: Weight (in pounds)

Pricesl.xls: Annual Consumer Price Index (CPI) data for a sample of 37 years (1982–1984 = 100). Taken from the Economic Report of the President, 2004, Table B-60. Original source, the U.S. Department of Labor, Bureau of Labor Statistics.

Year: Year (1967–2003)
All items: All-item price index
Apparel: Apparel price index
Energy: Energy price index
Food: Food and beverages price index
Medical: Medical care price index

Regressionl.xls: A sample of 10 (X,Y) points.

Case: Case identification number
X: Explanatory variable
Y: Dependent variable

Salesl.xls: Quarterly sales for a sample of 10 years (hypothetical).

Period: Period (1–40)
Year: Year (1–10)

Qtr: Quarter (Q1, Q2, Q3, Q4)
Sales: Sales (in $ thousands)

Sales2.xls: Quarterly sales for a sample of eight years (hypothetical).
Year: Year (1997–2004)
Qtr: Quarter (Q1, Q2, Q3, Q4)
Sales: Sales (in $ thousands)

Services1.xls: Macroeconomic data for a sample of 26 years. Taken from the Economic Report of the President, 1997, Table B-l. Original source, the U.S. Department of Commerce, Bureau of Economic Analysis.
Year: Year (1970–1995)
GDP: Gross domestic product (in $ billions)
Consump: Consumption expenditures (in $ billions)
Services: Spending on services (in $ billions)

Studentsl.xls: Personal information for a sample of 50 students (hypothetical).
ID: Student identification number
Female: Female (1 = yes, 0 = no)
Major: Major (1 = Natural Science, 2 = Social Science, 3 = Humanities, 4 = Fine Arts, 5 = Business, 6 = Nursing)
GPA : Grade point average (4-point scale)

Students2.xls: Personal information for a sample of 100 students (hypothetical).
ID: Student identification number
Female: Female (1 = yes, 0 = no)
Height: Height (in inches)
Weight: Weight (in pounds)
Year: Year in school (1–5)
Major: Academic major (1 = Mathematics, 2 = Economics, 3 = Biology, 4 = Psychology, 5 = English, 6 = Other)
Aid: Financial aid (in dollars per year)
Job: Holds a job (1 = yes, 0 = no)
Earnings: Earnings from job (in dollars per month)
Sports: Participates in a varsity sport (1 = yes, 0 = no)
Music: Participates in a music ensemble (1 = yes, 0 = no)
Greek: Belongs to a fraternity or sorority (1 = yes, 0 = no)
Entertain: Spending on entertainment (in dollars per week)

Study:	Time spent studying (in hours per week)
HS_GPA:	High school grade point average (5-point scale)
Col_GPA:	College grade point average (4-point scale)

Toysrus1.xls: Quarterly sales and earnings data for Toys "R" Us, Inc., for a sample of five years. Taken from Toys "R" Us Annual Reports, 2001–2004. (The company's fiscal year runs from February through January; hence, Q1 Sales are for February through April, etc.)

Year:	Year (2000–2004)
Qtr:	Quarter (Q1, Q2, Q3, Q4)
Sales:	Net sales (in $ millions)
Earnings:	Net earnings (in $ millions)

Appendix B: Answers to Odd-Numbered Exercises

Chapter 1: Introduction to Statistics

No exercises.

Chapter 2: Describing Data: Tables and Graphs

2.1 a.

		Bin	Frequency	Cumulative Frequency	Relative Frequency	Cumulative Relative
1.5–2.0	2	2	4	4	8.0%	8.0%
2.0–2.5	2.5	2.5	9	13	18.0%	26.0%
2.5–3.0	3	3	14	27	28.0%	54.0%
3.0–3.5	3.5	3.5	14	41	28.0%	82.0%
3.5–4.0	4	4	9	50	18.0%	100.0%
		More	0			
			50			

b. The pie chart would not make sense for the cumulative relative frequency since each slice would include all the previous slices. The whole pie should represent 100%, yet just one of the slices—the last one—would represent 100%.

c. i

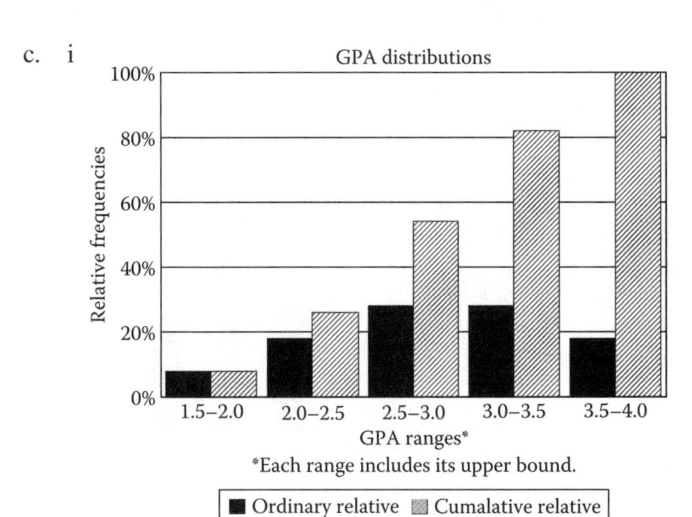

c. ii

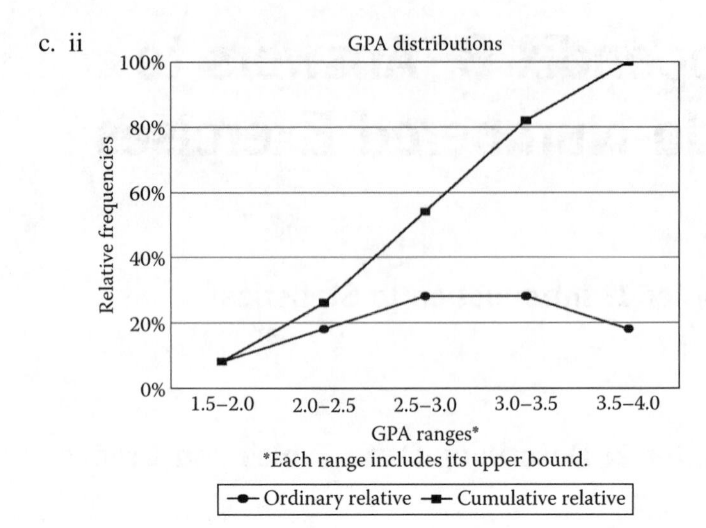

2.3 a.

Height Distribution by Sex

Height (inches)	Males	Females
56–60	0%	3%
60–64	1%	46%
64–68	25%	45%
68–72	53%	6%
72–76	20%	0%
76–80	0%	0%
80–84	1%	0%
	100%	100%

Each range includes its upper bound.

b. Relative frequencies. To compare groups of different sizes.

c. i

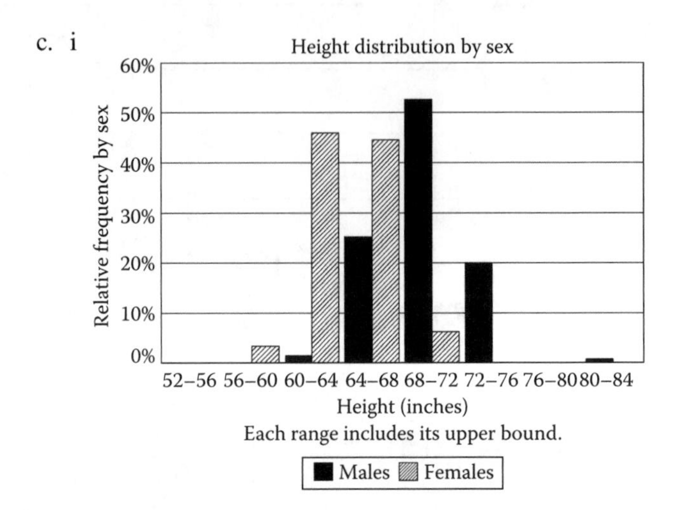

c. ii

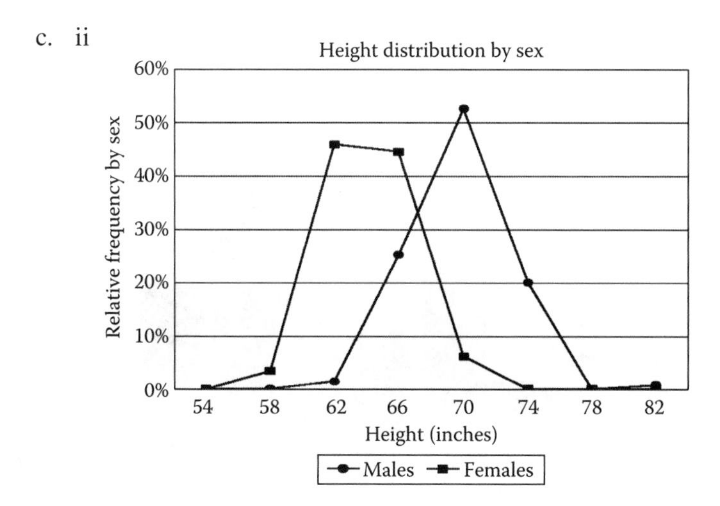

d.

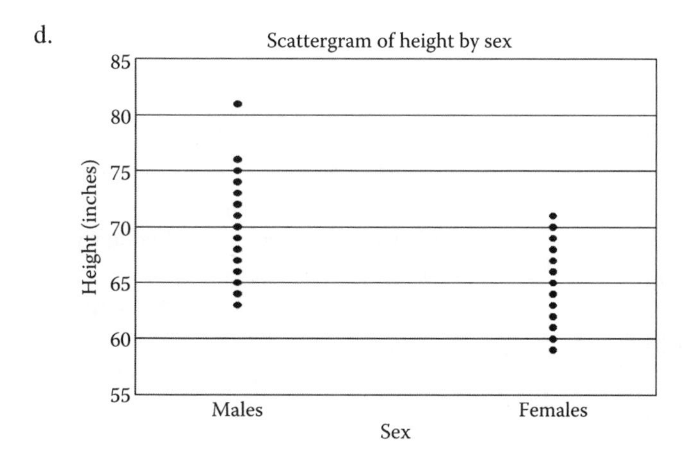

2.5 a.

		Bin	Frequency	Relative Frequency
20–25	25	25	1	2.0%
25–30	30	30	4	8.0%
30–35	35	35	10	20.0%
35–40	40	40	6	12.0%
40–45	45	45	4	8.0%
45–50	50	50	15	30.0%
50–55	55	55	8	16.0%
55–60	60	60	1	2.0%
60–65	65	65	1	2.0%
		More	0	
			50	100.0%

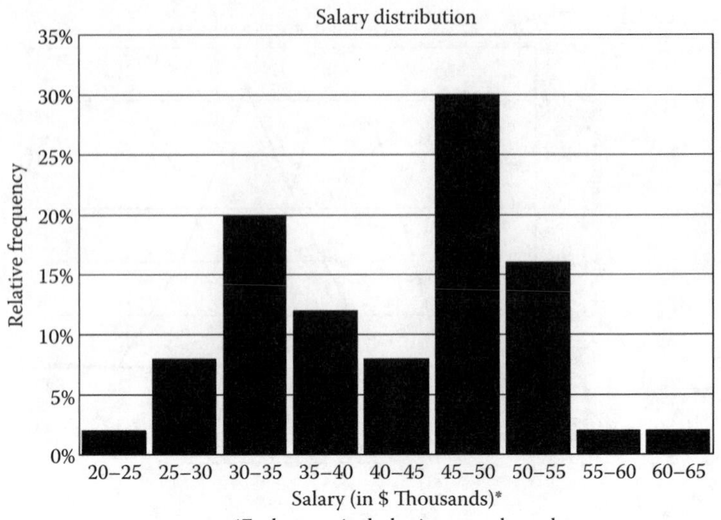

b.

			Males		Females	
		Bin	Frequency	Relative Frequency	Frequency	Relative Frequency
20–25	25	25	0	0.0%	1	5.6%
25–30	30	30	1	3.1%	3	16.7%
30–35	35	35	5	15.6%	5	27.8%
35–40	40	40	5	15.6%	1	5.6%
40–45	45	45	4	12.5%	0	0.0%
45–50	50	50	11	34.4%	4	22.2%
50–55	55	55	5	15.6%	3	16.7%
55–60	60	60	0	0.0%	1	5.6%
60–65	65	65	1	3.1%	0	0.0%
		More	0		0	
			32	100.0%	18	100.0%

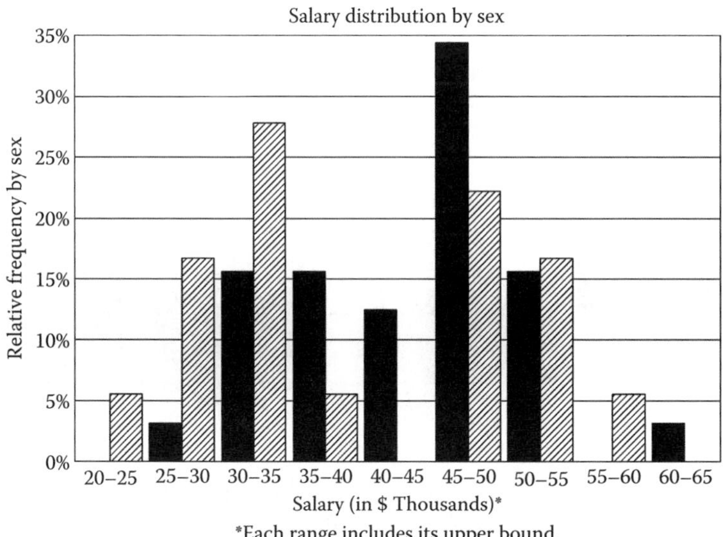
Salary distribution by sex
35%
30%
25%
20%
15%
10%
5%
0%
Relative frequency by sex
20–25
25–30
30–35
35–40
40–45
45–50
50–55
55–60
60–65
Salary (in $ Thousands)*
*Each range includes its upper bound.

Males
Females

c. i

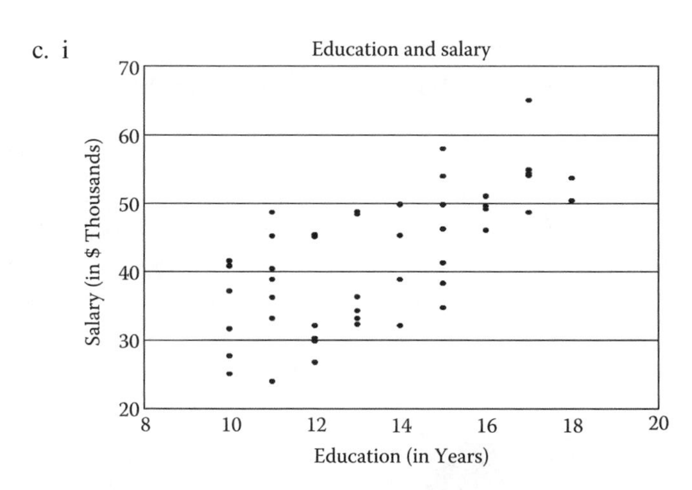
Education and salary
70
60
50
40
30
20
Salary (in $ Thousands)
8
10
12
14
16
18
20
Education (in Years)

c. ii

Experience and salary
70
60
50
40
30
20
Salary (in $ Thousands)
0
5
10
15
20
25
30
35
40
45
Experience (in Years)

c. iii

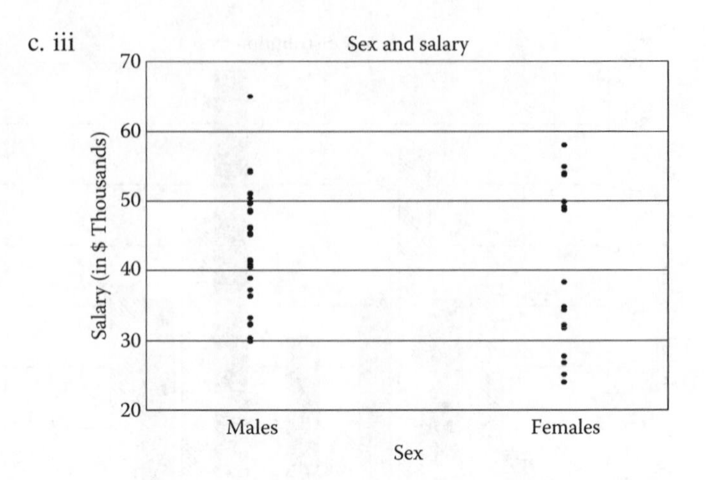

There appears to be a tendency for salaries to be higher for those with more experience, those with more education, and those who are male.

2.7 a.

		Bin	Athletes	Non–athletes	Athletes	Non–athletes
Male	0	0	15	32	62.5%	42.1%
Female	1	1	9	44	37.5%	57.9%
		More	0	0	0.0%	0.0%
			24	76	100.0%	100.0%

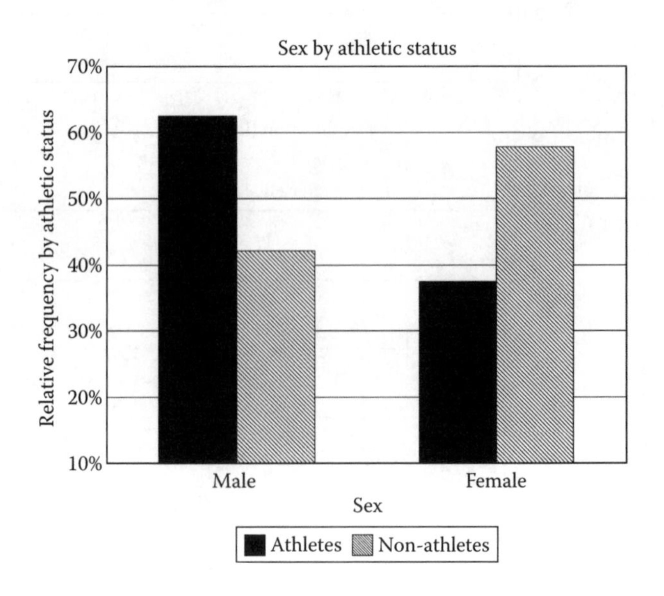

There may be a tendency for fewer females to participate in varsity sports. The difference seems large to be just random.

b.

		Bin	Athletes	Non–athletes	Athletes	Non–athletes
0.0–1.5	1,500	1500	1	11	4.2%	14.5%
1.5–3.0	3,000	3000	3	10	12.5%	13.2%
3.0–4.5	4,500	4500	3	15	12.5%	19.7%
4.5–6.0	6,000	6000	6	14	25.0%	18.4%
6.0–7.5	7,500	7500	5	15	20.8%	19.7%
7.5–9.0	9,000	9000	5	8	20.8%	10.5%
9.0–10.5	10,500	10500	1	3	4.2%	3.9%
		More	0	0	0.0%	0.0%
			24	76	100.0%	100.0%

b. i

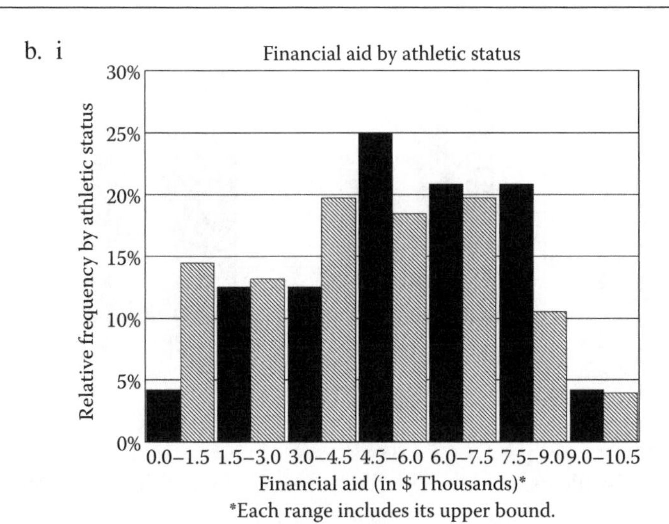

b. ii

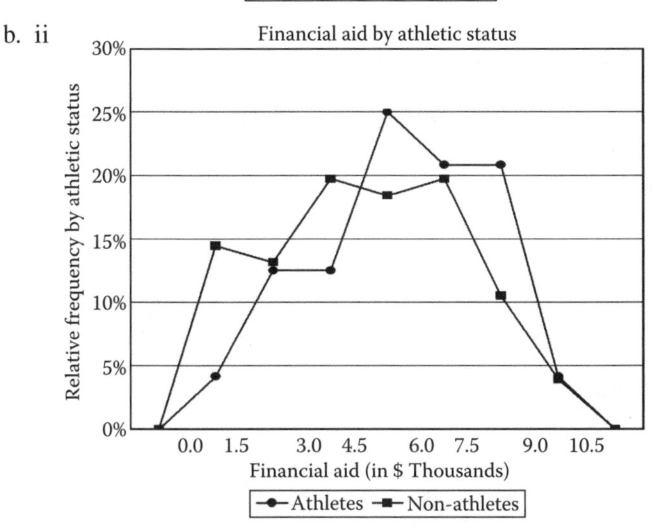

Athletes appear to get slightly more financial aid. It is hard to know if the difference is too large to be just random.

c.

		Bin	Athletes	Non–athletes	Athletes	Non–athletes
No Job	0	0	13	31	54.2%	40.8%
Job	1	1	11	45	45.8%	59.2%
		More	0	0	0.0%	0.0%
			24	76	100.0%	100.0%

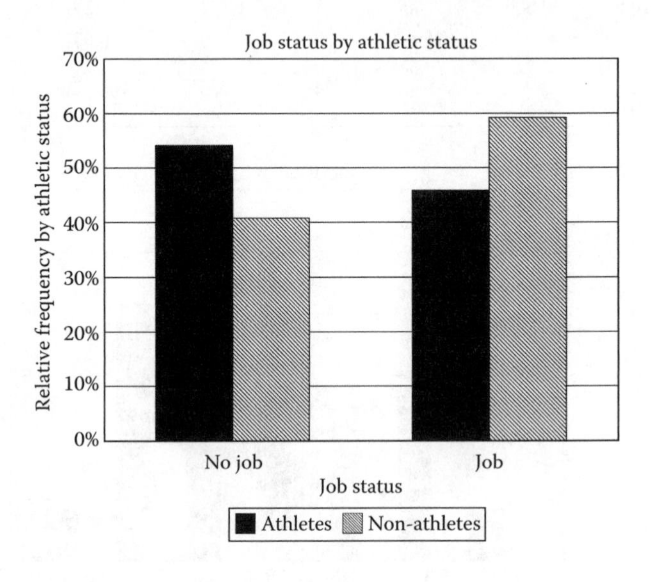

Athletes appear less likely to hold a job. The difference seems rather large to be just random.

d.

		Bin	Athletes	Non–athletes	Athletes	Non–athletes
Non–member	0	0	22	59	91.7%	77.6%
Member	1	1	2	17	8.3%	22.4%
		More	0	0	0.0%	0.0%
			24	76	100.0%	100.0%

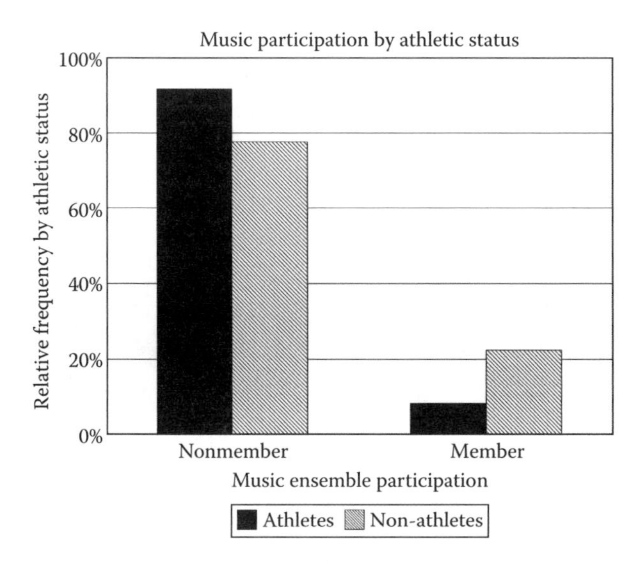

Athletes appear less likely to participate in music ensembles. The difference seems rather large to be just random.

e.

		Bin	Athletes	Non–athletes	Athletes	Non–athletes
Non-member	0	0	1	57	12.5%	75.0%
Member	1	1	7	19	87.5%	25.0%
		More	0	0	0.0%	0.0%
			8	76	100.0%	100.0%

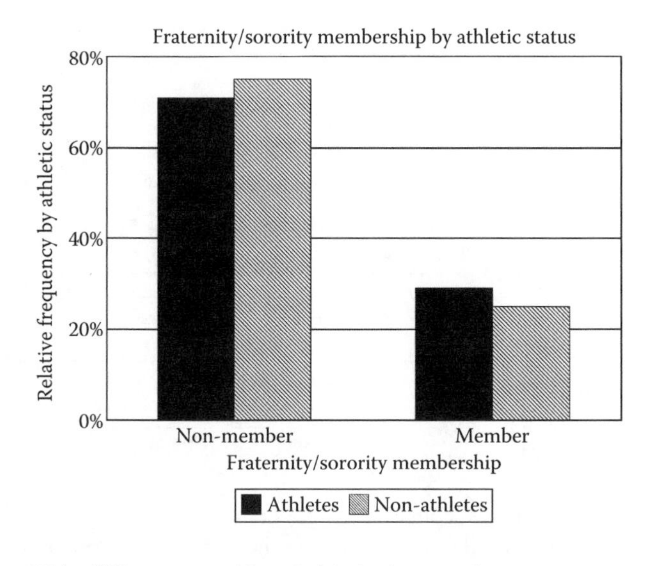

This difference could probably be just random.

f.

		Bin	Athletes	Non–athletes	Athletes	Non–athletes
0–10	10	10	1	9	4.2%	11.8%
10–20	20	20	7	11	29.2%	14.5%
20–30	30	30	5	14	20.8%	18.4%
30–40	40	40	2	24	8.3%	31.6%
40–50	50	50	3	12	12.5%	15.8%
50–60	60	60	6	6	25.0%	7.9%
		More	0	0	0.0%	0.0%
			24	76	100.0%	100.0%

f. i

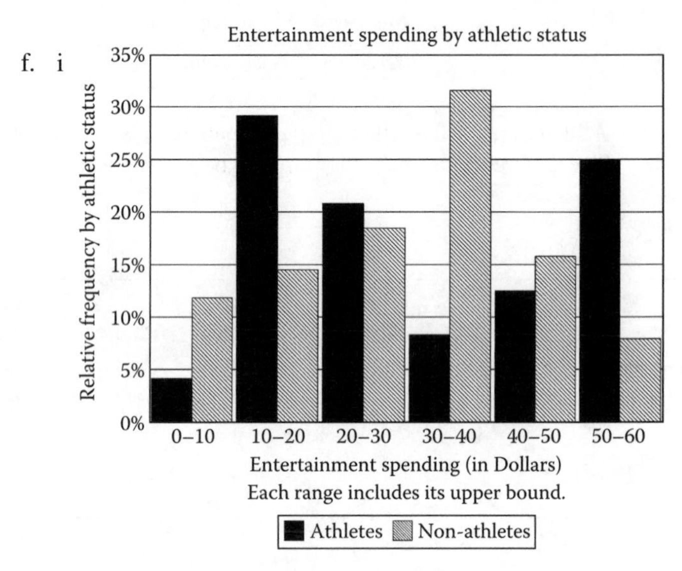

f. ii

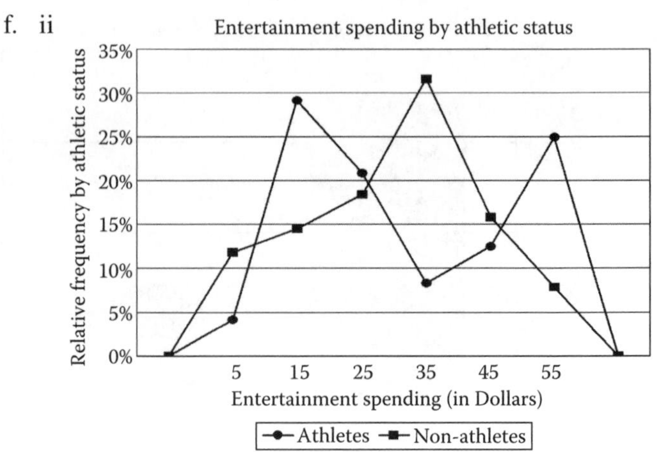

Athletes seem to spend either more or less than nonathletes. It is hard to know whether the difference is too large to be just random.

g.

		Bin	Athletes	Non–athletes	Athletes	Non–athletes
5–10	10	10	0	1	0.0%	1.3%
10–15	15	15	3	4	12.5%	5.3%
15–20	20	20	8	12	33.3%	15.8%
20–25	25	25	7	25	29.2%	32.9%
25–30	30	30	4	21	16.7%	27.6%
30–35	35	35	2	10	8.3%	13.2%
35–40	40	40	0	3	0.0%	3.9%
		More	0	0	0.0%	0.0%
			24	76	100.0%	100.0%

g. i

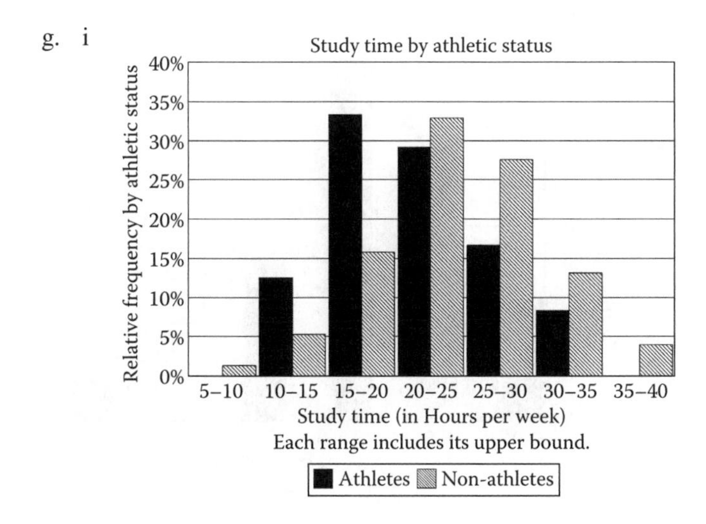

g. ii

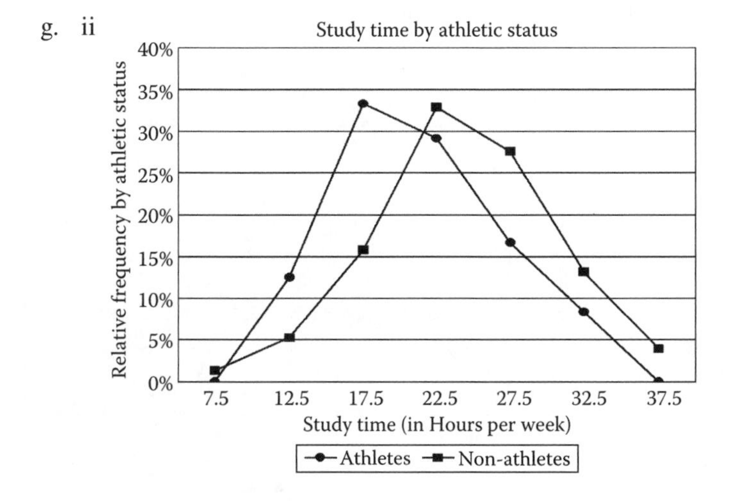

Athletes seem to study somewhat less. It is hard to know if the difference is too large to be just random.

h.

		Bin	Athletes	Non–athletes	Athletes	Non–athletes
2.0–2.5	2.5	2.5	3	6	12.5%	7.9%
2.5–3.0	3.0	3.0	7	19	29.2%	25.0%
3.0–3.5	3.5	3.5	9	27	37.5%	35.5%
3.5–4.0	4.0	4.0	4	14	16.7%	18.4%
4.0–4.5	4.5	4.5	0	8	0.0%	10.5%
4.5–5.0	5.0	5.0	1	2	4.2%	2.6%
		More	0	0	0.0%	0.0%
			24	76	100.0%	100.0%

Each range includes its upper bound.

h. i

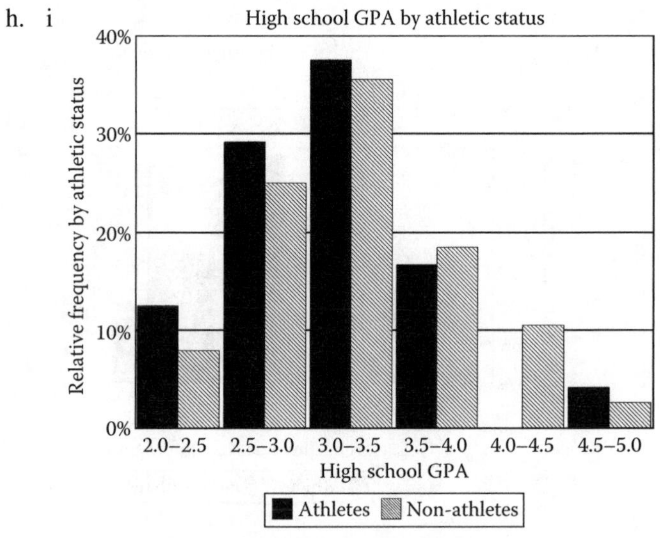

h. ii

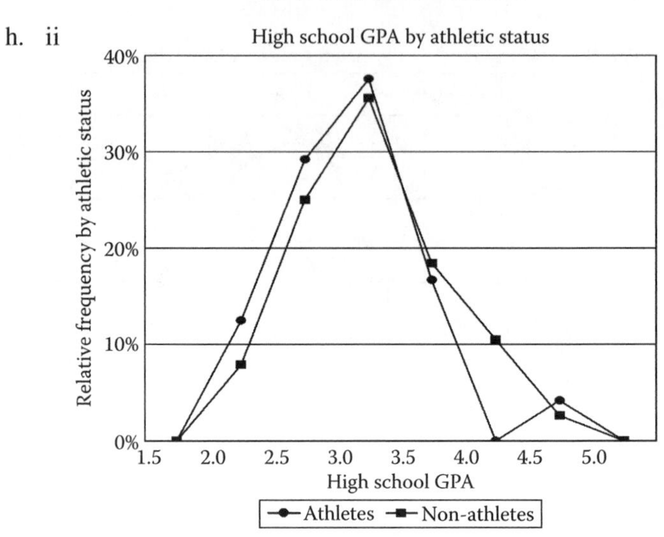

Differences this small could probably be just random.

i.

		Bin	Athletes	Non–athletes	Athletes	Non–athletes
1.5–2.0	2.0	2.0	2	1	8.3%	1.3%
2.0–2.5	2.5	2.5	7	13	29.2%	17.1%
2.5–3.0	3.0	3.0	8	24	33.3%	31.6%
3.0–3.5	3.5	3.5	5	20	20.8%	26.3%
3.5–4.0	4.0	4.0	2	18	8.3%	23.7%
		More	0	0	0.0%	0.0%
			24	76	100.0%	100.0%

Each range includes its upper bound.

i. i

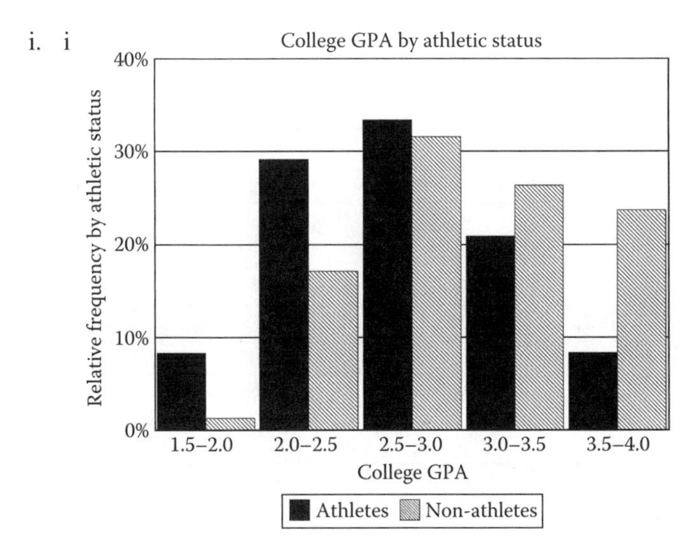

i. ii

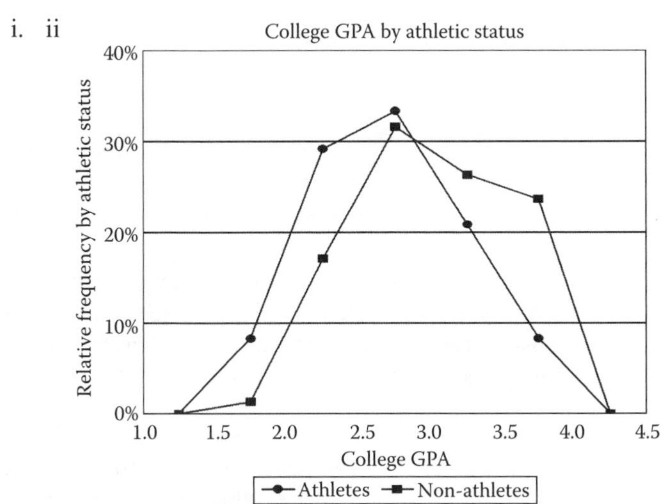

Athletes seem to have somewhat lower GPAs. It is hard to know if the difference is too large to be just random.

2.9 a. i

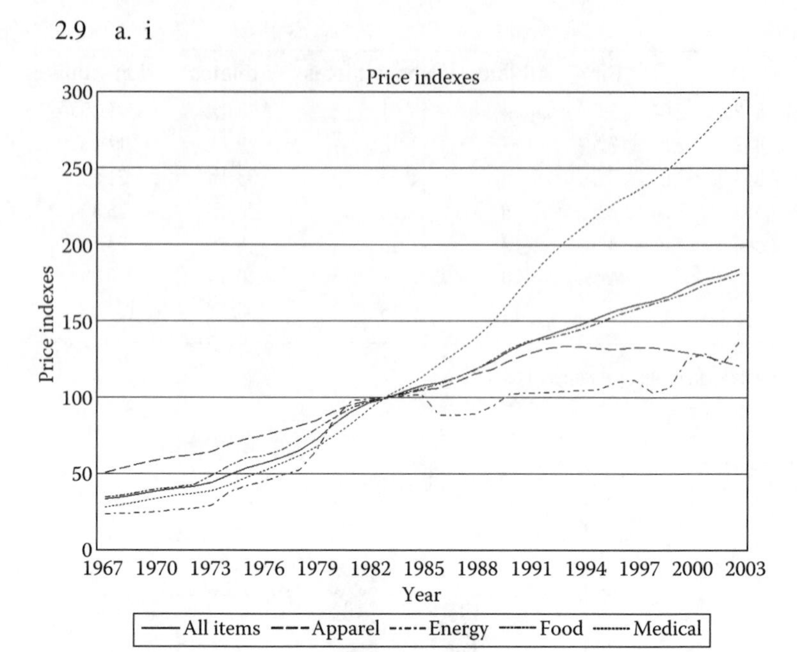

a. ii

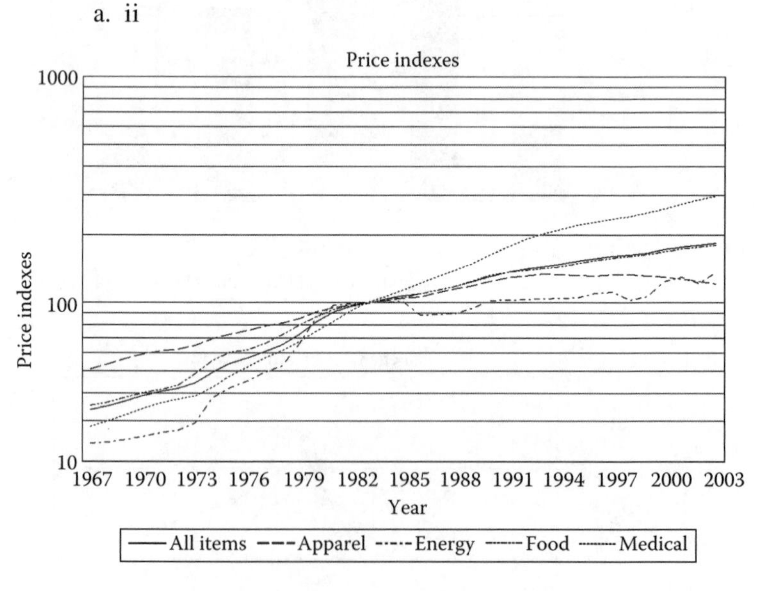

b. Medical care prices are growing most rapidly. Until recently, energy prices had grown least rapidly; now apparel prices have, actually falling in recent years. Energy prices appear to be the most volatile.

c. Both graphs are technically correct. However, the arithmetic graph probably gives an exaggerated impression of what is happening with medical care. While medical care prices are certainly rising most rapidly, they are not *accelerating* as one might be led to believe based on the arithmetic graph.

Chapter 3: Describing Data: Summary Statistics

3.1 a.

$$\overline{\text{GPA}} = 146.497 / 50 = 2.930$$

$$\text{Median}_{\text{GPA}} = (2.890 + 2.981) / 2 = 2.936$$

$$\text{Mode}_{\text{GPA}} = \text{none (no numbers repeated)}$$

$$\text{Range}_{\text{GPA}} = 3.978 - 1.750 = 2.228$$

$$S^2_{\text{GPA}} = 16.065 / 49 = 0.328$$

$$S_{\text{GPA}} = \sqrt{0.328} = 0.573$$

$$CV_{\text{GPA}} = 0.573 / 2.930 \times 100 = 19.5\%$$

b. Same as part a.

c.

	"X"	F	F "X"	("X"-Mean)	("X"-Mean)^2	F ("X"-Mean)^2
1.5–2.0	1.75	4	7.00	−1.15	1.3225	5.2900
2.0–2.5	2.25	9	20.25	−0.65	0.4225	3.8025
2.5–3.0	2.75	14	38.50	−0.15	0.0225	0.3150
3.0–3.5	3.25	14	45.50	0.35	0.1225	1.7150
3.5–4.0	3.75	9	33.75	0.85	0.7225	6.5025
		50	145.00			17.6250
						49

Mean = 2.900

Variance = 0.3597
Standard deviation = 0.5997

3.3 a. Males appear to have the higher mean and also the higher standard deviation.

b.

Males	Females
$\overline{\text{GPA}} = 59.268 / 20 = 2.963$	$\overline{\text{GPA}} = 87.229 / 30 = 2.908$
$\text{Median}_{\text{GPA}} = 2.938$	$\text{Median}_{\text{GPA}} = 2.926$
$\text{Mode}_{\text{GPA}} = \text{none}$	$\text{Mode}_{\text{GPA}} = \text{none}$
$\text{Range}_{\text{GPA}} = 2.228$	$\text{Range}_{\text{GPA}} = 1.821$
$S^2_{\text{GPA}} = 7.953 / 19 = 0.4186$	$S^2_{\text{GPA}} = 8.075 / 29 = 0.2785$
$S_{\text{GPA}} = \sqrt{0.4186} = 0.6470$	$S_{\text{GPA}} = \sqrt{0.2785} = 0.5277$
$CV_{\text{GPA}} = 21.83\%$	$CV_{\text{GPA}} = 18.15\%$

c. Same as part b.

d.

Males

	"X"	F	F "X"	("X"-Mean)	("X"-Mean)^2	F ("X"-Mean)^2
1.5–2.0	1.75	2	3.50	−1.15	1.3225	2.6450
2.0–2.5	2.25	3	6.75	−0.65	0.4225	1.2675
2.5–3.0	2.75	6	16.50	−0.15	0.0225	0.1350
3.0–3.5	3.25	5	16.25	0.35	0.1225	0.6125
3.5–4.0	3.75	4	15.00	0.85	0.7225	2.8900
		20	58.00			7.5500
						19

Mean = 2.900

Variance = 0.3974
Standard deviation = 0.6304

Females

	"X"	F	F "X"	("X"-Mean)	("X"-Mean)^2	F ("X"-Mean)^2
1.5–2.0	1.75	2	3.50	−1.15	1.3225	2.6450
2.0–2.5	2.25	6	13.50	−0.65	0.4225	2.5350
2.5–3.0	2.75	8	22.00	−0.15	0.0225	0.1800
3.0–3.5	3.25	9	29.25	0.35	0.1225	1.1025
3.5–4.0	3.75	5	18.75	0.85	0.7225	3.6125
		30	87.00			10.0750
						29

Mean = 2.900

Variance = 0.3474
Standard deviation = 0.5894

e. The results agree with our expectations, though the difference, especially in means, is very small. Indeed, the approximations in part d are the same.

3.5 a.

Total	Males	Females
$\overline{\text{HGT}} = 67.44$	$\overline{\text{HGT}} = 70.41$	$\overline{\text{HGT}} = 64.68$
$S_{\text{HGT}} = 3.94$	$S_{\text{HGT}} = 2.77$	$S_{\text{HGT}} = 2.66$

b. Yes, knowing that the employee is Male/Female raises/lowers his/her expected height and reduces the variation in height.

3.7 a.

Total	Males	Females
$\overline{SAL} = 42.27$	$\overline{SAL} = 43.41$	$\overline{SAL} = 40.24$
$S_{SAL} = 9.59$	$S_{SAL} = 8.19$	$S_{SAL} = 11.68$

b. Yes, knowing that the employee is Male/Female raises/lowers his/her expected salary and reduces the variation in salary for men (though not for women).

3.9 a.

$$\overline{\text{Height}} = 6761/100 = 67.61$$

$$\text{Median}_{\text{Height}} = (68+68)/2 = 68$$

$$\text{Mode}_{\text{Height}} = 70$$

$$\text{Range}_{\text{Height}} = 79-58 = 20$$

$$S^2_{\text{Height}} = 2223.79/99 = 22.46$$

$$S_{\text{Height}} = \sqrt{22.46} = 4.74$$

$$CV_{\text{Height}} = 67.61/4.74 = 7.01\%$$

b. Same as part a.

c.

	Bin	Frequency	
58.1311	58.1311	1	
77.0889	77.0889	97	97%
	More	2	
		100	

3.11 a. The mean height looks greater for males; the standard deviations look quite similar.

b.

Females

count = 53	variance = 14.44	maximum = 74
sum = 3446	standard deviation = 3.80	median = 65
mean = 65.02	cv = 5.84%	minimum = 59
		range = 15

Males

count = 47	variance = 15.56	maximum = 78
sum = 3315	standard deviation = 3.94	median = 71
mean = 70.53	CV = 5.59%	minimum = 58
		range = 20

c.

Table for "Females"

		"X"	*F*	*F"X"*	*"X"-Mean*	*("X"-Mean)²*	*F("X"-Mean)²*
56–60	60	58	7	406	–6.49	42.13	294.89
60–64	64	62	19	1178	–2.49	6.20	117.86
64–68	68	66	16	1056	1.51	2.28	36.45
68–72	72	70	9	630	5.51	30.35	273.18
72–76	76	74	2	148	9.51	90.43	180.86
76–80	80	78	0	0	13.51	182.50	0.00
			53	3418			903.25
							52

Mean = 64.49

Variance = 17.37
Standard deviation = 4.17

Table for "Males"

		"X"	*F*	*F"X"*	*"X"*-Mean	(*"X"*-Mean)²	*F*(*"X"*-Mean)²
56–60	60	58	1	58	–12.00	144.00	144.00
60–64	64	62	1	62	–8.00	64.00	64.00
64–68	68	66	13	858	–4.00	16.00	208.00
68–72	72	70	16	1120	0.00	0.00	0.00
72–76	76	74	14	1036	4.00	16.00	224.00
76–80	80	78	2	156	8.00	64.00	128.00
			47	3290			768.00
							46

Mean = 70.00

Variance = 16.70
Standard deviation = 4.09

d. The results agree with our expectations.

3.13 a. The mean aid looks somewhat greater for athletes; the standard deviations look quite similar.

b.

Athletes

count = 24
sum = 139000
mean = 5791.67
variance = 5759058
standard deviation = 2399.80
CV = 41.44%
maximum = 9500
median = 6000
minimum = 0
range = 9500

Non-athletes

count = 76
sum = 372500
mean = 4901.32
variance = 6913465
standard deviation = 2629.35
CV = 53.65%
maximum = 10000
median = 5000
minimum = 0
range = 10000

c.

Table for "Athletes"

		"X"	F	F"X"	"X"-Mean	("X"-Mean)²	F("X"-Mean)²
0.0–1.5	1500	750	1	750	–4875	23765625	23765625
1.5–3.0	3000	2250	3	6750	–3375	11390625	34171875
3.0–4.5	4500	3750	3	11250	–1875	3515625	10546875
4.5–6.0	6000	5250	6	31500	–375	140625	843750
6.0–7.5	7500	6750	5	33750	1125	1265625	6328125
7.5–9.0	9000	8250	5	41250	2625	6890625	34453125
9.0–10.5	10000	9750	1	9750	4125	17015625	17015625
			24	135000			127125000
							23

Mean = 5625.0 Variance = 5527173.9
Standard deviation = 2351.0

Table for "Non-athletes"

		"X"	F	F"X"	"X"-Mean	("X"-Mean)²	F("X"-Mean)²
0.0–1.5	1500	750	11	8250	–3947	15581717	171398892
1.5–3.0	3000	2250	10	22500	–2447	5989612	59896122
3.0–4.5	4500	3750	15	56250	–947	897507	13462604
4.5–6.0	6000	5250	14	73500	553	305402	4275623
6.0–7.5	7500	6750	15	101250	2053	4213296	63199446
7.5–9.0	9000	8250	8	66000	3553	12621191	100969529
9.0–10.5	10000	9750	3	29250	5053	25529086	76587258
			76	357000			489789474
							75

Mean = 4697.4 Variance = 6530526.3
Standard deviation = 2555.5

d. The results agree with our expectations.

Chapter 4: Basic Probability

4.1 $P(B) = .18 + .07 = .25$

4.3 a. $P(W = 3) = 0.20 \times 0.20 \times 0.20 = 0.20^3 = .008$
$P(W = 1) = (0.20 \times 0.80 \times 0.80 + (0.80 \times 0.20 \times 0.80$

b. $+(0.80 \times 0.80 \times 0.20) = 3(0.20)^1(0.80)^2 = 0.384$

c. $P(W \geq 1) = 1 - P(W = 0) = 1 - 0.80^3 = 1 - 0.512 = 0.488$

4.5

Condition	P(C)	P(G|C)	P(G ∩ C)	P(C|G)
Running properly	0.80	0.90	0.72	0.973 = Answer
Not	0.20	0.10	0.02	0.027
	1.00		P(G) = 0.74	1.000

$$P(\text{running properly}|\text{good}) = \frac{P(\text{good} \cap \text{running properly})}{P(\text{good})} = \frac{0.72}{0.74} = 0.973$$

4.7 a. $P(S=3) = 0.92 \times 0.88 \times 0.85 = 0.68816$

b. $P(S=0) = 0.08 \times 0.12 \times 0.15 = 0.00144$

c. $P(S=1) = (0.92 \times 0.12 \times 0.15) + (0.08 \times 0.88 \times 0.15)$

$+ (0.08 \times 0.12 \times 0.85)$

$= 0.01656 + 0.01056 + 0.00816 = 0.03528$

d. $P(S \geq 1) = 1 - P(S=0) = 1 - 0.00144 = 0.99856$

4.9 a. $P(D=2) = (2/3)1 \times 4(0.05)^2 + (1/3)(0.20)^2 = 0.0017 + 0.0133$

$= 0.0150$

b.

Line	P(L)	P(2D|L)	P(2D ∩ L)	P(L|2D)
A or B	2/3	0.05^2	0.0017	0.1111
C	1/3	0.20^2	0.0133	0.8889 = Answer
	1.00		P(2D) = 0.0150	1.0000

$$\text{P(line 3}|\text{2 defects)} = \frac{P(\text{2 defects} \cap \text{line 3})}{P(\text{2 defects})} = \frac{0.0133}{0.0150} = 0.8889$$

4.11

Bin	P(B)	P(G|B)	P(G ∩ B)	P(B|G)
2G	1/3	1	1/3	2/3 = Answer
1G	1/3	1/2	1/6	1/3
0G	1/3	0	0	0
	1		P(G) = 1/2	1

$$P(\text{1st bin}|\text{1st part good}) = \frac{(\text{1st part good} \cap \text{1st bin})}{P(\text{1st part good})} = \frac{1/3}{1/2} = 2/3$$

Chapter 5: Probability Distributions

5.1 a. {3,6; 4,5; 5,4; 6,3} $P(9) = 4/36 = 1/9$

$P(\text{1st, last} = 9; \text{2nd, 3rd, 4th} \neq 9) = 1/9 \times 8/9 \times 8/9 \times 8/9 \times 1/9$

$= (1/9)^2(8/9)^3 = 0.00867$

b. $P(s = 2|n = 5, \pi = 1/9) = C_2^5(1/9)^2(8/9)^3$

$$= \frac{{}^{10}\not{5}!}{{}_1\not{2}! \times \not{3}!_1}(1/9)^2(8/9)^3 = 10 \times 0.00867 = 0.086'$$

5.3 $P(r = 3|n = 5, \pi = 0.25) = C_3^5(0.25)^3(0.75)^2 = 0.0879$

$P(r = 4|n = 5, \pi = 0.25) = C_4^5(0.25)^4(0.75)^1 = 0.0146$

$P(r = 5|n = 5, \pi = 0.25) = C_5^5(0.25)^5(0.75)^0 = \underline{0.0010}$

0.1035

5.5 a.

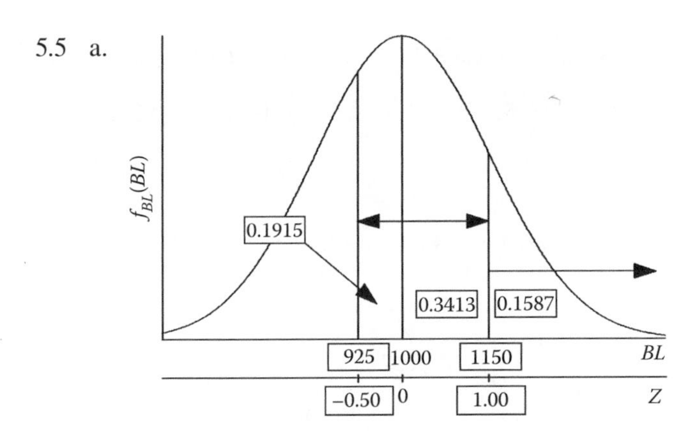

$$Z_a = \frac{BL_a - \mu_{BL}}{\sigma_{BL}} = \frac{1150 - 1000}{150} = 1.00$$

$P(BL > 1150) = 0.5 - 0.3413 = 0.1587$

b. $$Z_b = \frac{BL_b - \mu_{BL}}{\sigma_{BL}} = \frac{925 - 1000}{150} = -0.50$$

$P(925 < BL < 1000) = 0.1915$

$P(1000 < BL < 1150) = 0.3413$

$P(925 < BL < 1150) = 0.1915 + 0.3413 = 0.5328$

c.

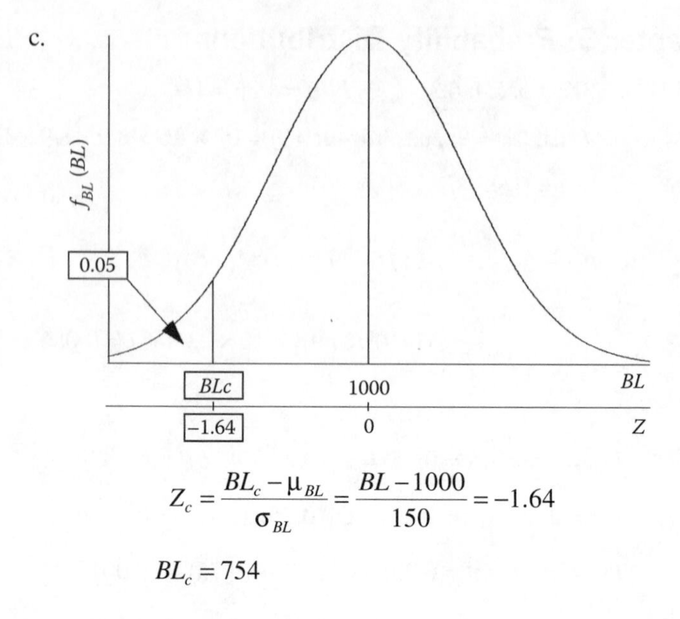

$$Z_c = \frac{BL_c - \mu_{BL}}{\sigma_{BL}} = \frac{BL - 1000}{150} = -1.64$$

$$BL_c = 754$$

5.7 $P(\text{SSS} \geq 2 | n = 5, \pi = 0.33) = 1 - \{P(\text{SSS} = 0) + P(\text{SSS} = 1)\}$

$$P(\text{SSS} = 0 | n = 5, \pi = 0.33) = C_0^5 (0.33)^0 (0.67)^5 = 0.1350$$

$$P(\text{SSS} = 1 | n = 5, \pi = 0.33) = C_1^5 (0.33)^1 (0.67)^4 = \underline{0.3325}$$

$$0.4675$$

$$P(\text{SSS} \geq 2 | n = 5, \pi = 0.33) = 1 - 0.4675 = 0.5325$$

5.9

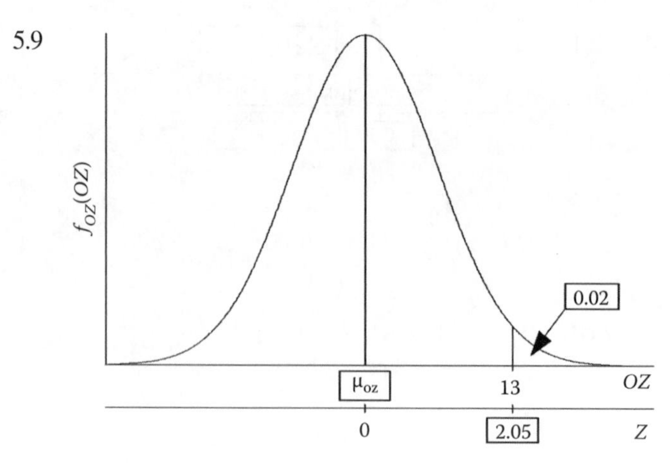

$$Z = \frac{OZ - \mu_{oz}}{\sigma_{oz}} = \frac{13 - \mu_{oz}}{0.5} = 2.05$$

$$\mu_{oz} = 11.975$$

5.11

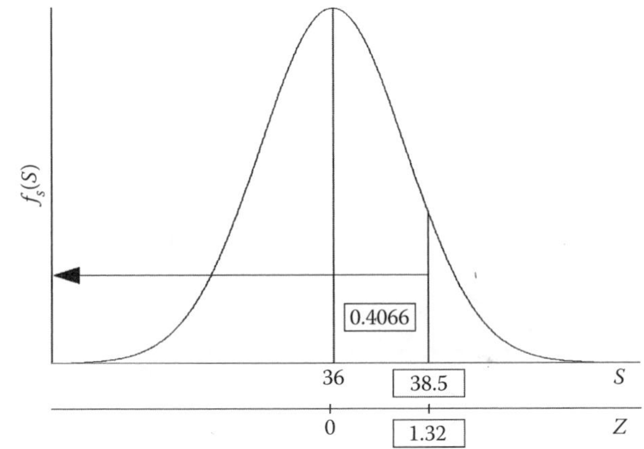

S = "Shows," $\pi_s = P(S) = 0.90$. We want $S < 38.5$.

We could have used NS = "No Shows," $\pi_{NS} = P(NS) = 0.10$. $NS > 1.5$.

$$\mu_S = n \times \pi_S = 40 \times 0.90 = 36$$

$$\sigma_S = \sqrt{n \times \pi_S \times (1 - \pi_S)} = \sqrt{40 \times 0.90 \times 0.10} = \sqrt{3.6} = 1.897$$

$$Z = \frac{38.5 - 36}{1.897} = \frac{2.5}{1.897} = 1.32$$

$$P(S < 38.5) = 0.4066 + 0.5000 = 0.9066$$

Chapter 6: Sampling and Sampling Distributions

6.1 a. This is a binomial and, with $n = 10$, the normal approximation would not be good.

$$\begin{aligned} P(d \geq 8\%) &= P(d \geq 0.8) = 1 - P(d = 0) \\ &= 1 - P(d = 0 | n = 10, \pi_d = 0.05) \\ &= 1 - C_0^{10}(0.05)^0(0.95)^{10} = 1 - 0.5987 = 0.4013 \end{aligned}$$

b. This is a binomial and, with $n = 100$, the normal approximation should be fairly good.

As a binomial

$$P(d \geq 8\%) = P(d \geq 8) = 1 - P(d < 8) = 1 - 0.8720 = 0.1280$$

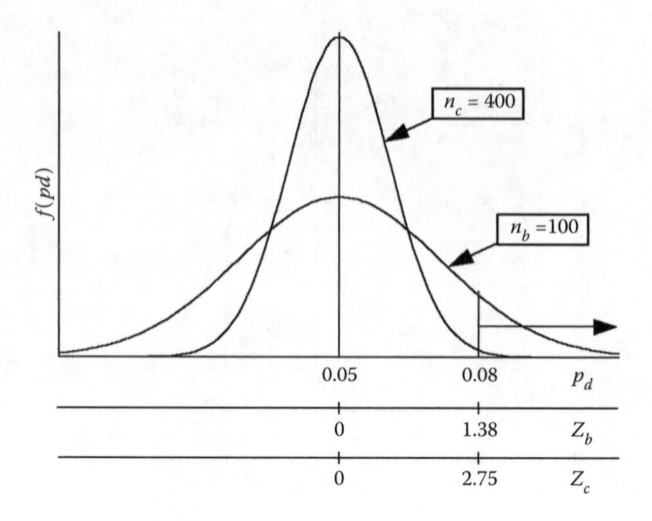

As a normal

$$\mu_p = \pi_d = 0.05$$

$$\sigma_p = \sqrt{\pi_d \times (1 - \pi_d) / n} = 0.0218$$

$$Z = \frac{0.08 - 0.05}{0.0218} = 1.38$$

$$P(d \geq 8\%) = 0.5000 - 0.4162 = 0.0838$$

c. This is a binomial and, with $n = 400$, the normal approximation should be quite good.

As a binomial

$$P(d \geq 8\%) = P(d \geq 32) = 1 - P(d < 32)$$
$$= 1 - 0.9933 = 0.0067$$

As a normal

$$\mu_p = \pi_d = 0.05$$

$$\sigma_d = \sqrt{\pi_d \times (1 - \pi_d) / n} = 0.0109$$

$$Z = \frac{0.08 - 0.05}{0.0109} = 2.75$$

$$P(d \geq 8\%) = 0.5000 - 0.4970 = 0.0030$$

6.3 a.

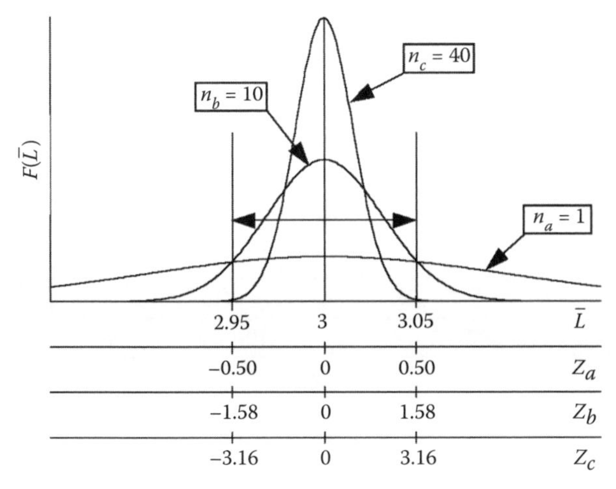

$$Z_a = \frac{3.05 - 3.00}{.10/\sqrt{1}} = \frac{0.05}{0.10} = 0.50$$

$$P_a(3.00 < \bar{L} < 3.05) = 0.1915$$

$$P_a(2.95 < \bar{L} < 3.05) = 0.1915 \times 2$$

$$= 0.3830$$

b. $$Z_a = \frac{3.05 - 3.00}{0.10/\sqrt{10}} = \frac{0.05}{0.0316} = 1.58$$

$$P_a(3.00 < \bar{L} < 3.05) = 0.4429$$

$$P_a(2.95 < \bar{L} < 3.05) = 0.4429 \times 2$$

$$= 0.8858$$

c. $$Z_c = \frac{3.05 - 3.00}{0.10/\sqrt{40}} = \frac{0.05}{0.0158} = 3.16$$

$$P_c(3.00 < \bar{L} < 3.05) = 0.4992$$

$$P_c(2.95 < \bar{L} < 3.05) = 0.4992 \times 2$$

$$= 0.9984$$

6.5 a.

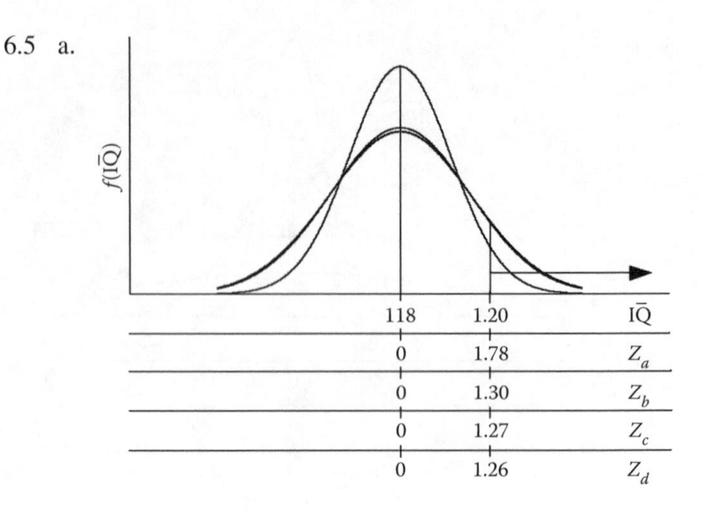

$$\sigma_{\overline{\text{IQ}}} = \frac{10}{\sqrt{40}}\sqrt{\frac{80-40}{80-1}} = 1.581 \times 0.712 = 1.125$$

$$Z_a = \frac{120-118}{1.125} = 1.78$$

$$P_a(\overline{\text{IQ}} > 120) = 0.5000 - 0.4625 = 0.0375$$

b.

$$\sigma_{\overline{\text{IQ}}} = \frac{10}{\sqrt{40}}\sqrt{\frac{800-40}{800-1}} = 1.581 \times 0.975 = 1.542$$

$$Z_b = \frac{120-118}{1.542} = 1.30$$

$$P_b(\overline{\text{IQ}} > 120) = 0.5000 - 0.4032 = 0.0968$$

c.

$$\sigma_{\overline{\text{IQ}}} = \frac{10}{\sqrt{40}}\sqrt{\frac{8000-40}{8000-1}} = 1.581 \times 0.998 = 1.577$$

$$Z_c = \frac{120-118}{1.577} = 1.27$$

$$P_c(\overline{\text{IQ}} > 120) = 0.5000 - 0.3980 = 0.1020$$

d.

$$\sigma_{\overline{\text{IQ}}} = \frac{10}{\sqrt{40}} = 1.581$$

$$Z_d = \frac{120-118}{1.581} = 1.26$$

$$P_d(\overline{\text{IQ}} > 120) = 0.5000 - 0.3962 = 0.1038$$

6.7 a.

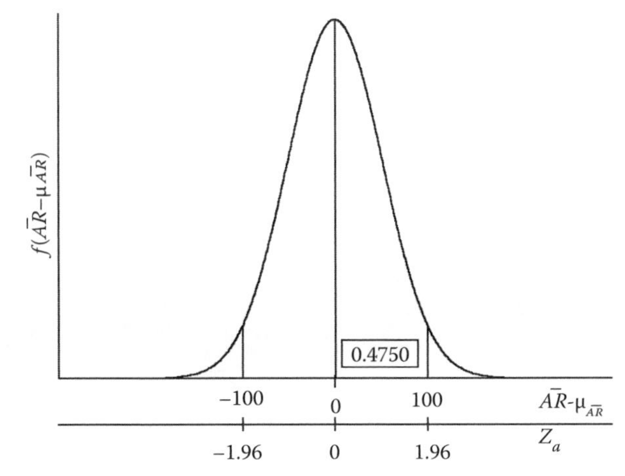

$$1.96\sigma_{\overline{AR}} \leq \$100$$

$$\sigma_{\overline{AR}} \leq \$100/1.96 = 51.02$$

$$\sigma_{\overline{AR}} \leq \$1000/\sqrt{n} = 51.02$$

$$\sigma^2_{\overline{AR}} \leq \$1000^2/n = 51.02^2$$

$$n = \$1000^2/51.02^2 = 384$$

b. 95

c. 95

6.9 a.

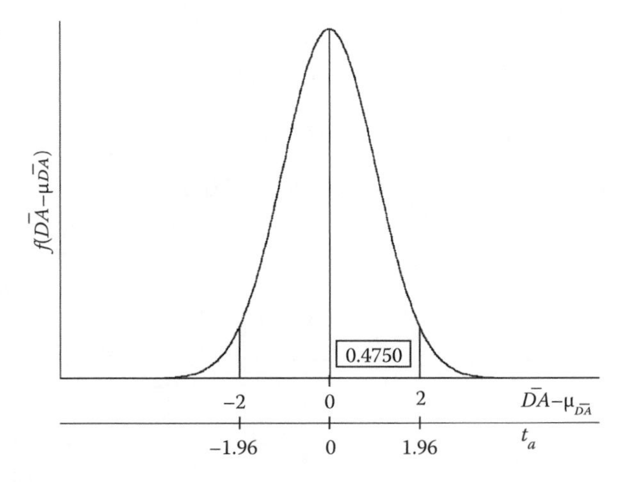

Guess that n will be large, take t from the bottom of the t table, 1.960.

$$1.96\sigma_{\overline{DA}} \leq \$2$$

$$\sigma_{\overline{DA}} \leq \$2/1.96 = 1.020$$

$$\sigma_{\overline{DA}} = \$20 / \sqrt{n} = 1.020$$

$$\sigma^2_{\overline{DA}} = \$20^2 / n = 1.020^2$$

$$n = \$20^2 / 1.020^2 = 384$$

b. 95

c. 95

Chapter 7: Estimation and Confidence Intervals

All z- and t-values are taken from the tables. For t, when the table doesn't list the exact degrees of freedom, the closest degrees of freedom is used. However, if the Excel value with the exact degrees of freedom differs, it is also reported [in brackets] in the margin.

7.1 $P_{\text{def}} = 5 / 100 = 0.05$

$$s_p = \sqrt{0.05 \times 0.95 / 100} = \sqrt{.000475} = 0.0218$$

a. 75% CI: $0.05 \pm 1.15 \times 0.0218$

or 75% CI: 0.05 ± 0.0251

or 75% CI: $0.0249 < p_{\text{def}} < 0.0751$

b. 95% CI: $0.05 \pm 1.96 \times 0.0218$

or 95% CI: 0.05 ± 0.0427

or 95% CI: $0.0073 < p_{\text{def}} < 0.0927$

c. $1.96 \times s_p \leq 0.02$

$$s_p \leq 0.02 / 1.96 = 0.0102$$

$$s_p = \sqrt{p \times (1 - p) / n} = 0.0102$$

$$n = p \times (1 - p) / 0.0102^2$$

Using $p = 0.05$ $\quad n = 0.05 \times 0.95 / 0.0102^2 = 456$

increase $= 456 - 100 = 356$

Using $p = 0.50$ $\quad n = 0.50 \times 0.50 / 0.0102^2 = 2401$

increase $= 2401 - 100 = 2301$

7.3 $1.96 \times s_p \leq 0.02$

$$s_p \leq 0.02 / 1.96 = 0.0102$$

$$s_p = \sqrt{p \times (1-p) / n} = 0.0102$$

$$n = p \times (1-p) / 0.0102^2$$

Using $p = 0.50$ $\quad n = 0.50 \times 0.50 / 0.0102^2 = 2401$

7.5 a. 85% CI: $p_f \pm z \times s_p$

85% CI: $0.60 \pm 1.44 \times 0.0693$

or 85% CI: 0.60 ± 0.0998

or 85% CI: $0.5002 < p_f < 0.6998$

b. 95% CI: $p_f \pm z \times s_p$

or 95% CI: $0.60 \pm 1.96 \times 0.0693$

or 95% CI: 0.60 ± 0.1358

or 95% CI: $0.4642 < p_f < 0.7358$

c. $1.96 \times s_p \leq 0.05$

$$s_p \leq 0.05 / 1.96 = 0.0255$$

$$s_p = \sqrt{p \times (1-p) / n} = 0.0255$$

$$n = p \times (1-p) / 0.0255^2$$

Using $p = 0.60$ $\quad n = 0.60 \times 0.40 / 0.0255^2 = 369$

increase $= 369 - 50 = 319$

Using $p = 0.50$ $\quad n = 0.50 \times 0.50 / 0.0255^2 = 384$

increase $= 384 - 50 = 334$

7.7 From Exercise 3.4, the interval that includes roughly 95% of the data is

$67.4377 \pm 2 \times 3.9430$

or 67.4377 ± 7.8859

or $59.5518 \Leftrightarrow 75.3237$ a range of over 15 inches.

From Exercise 7.6, the 95% confidence interval estimate for $\overline{\text{HGT}}$ is

$$95\%\text{ CI: } 67.4377 \pm 1.960 \times 0.2352 \quad [1.968]$$

$$\text{or } 95\%\text{ CI: } 67.4377 \pm 0.4610$$

$$\text{or } 95\%\text{ CI: } 66.9767 < \overline{\text{HGT}} < 67.8987$$ a range of less than an inch

The first assumes that the *sample data* follow roughly a normal distribution. It is based on the *standard deviation of the sample* and tells us roughly *how scattered the data* are.
The second assumes that *all possible sample means* for samples of size n follow a normal distribution. It is based on the *standard error of the mean* and tells us *how precise our estimate of the mean* is.
The standard error $(s_{\bar{X}})$ = the standard deviation (s_X) divided by $\sqrt{n}$.

7.9 From Exercise 3.6, the interval that includes roughly 95% of the data is

$$42.2660 \pm 2 \times 9.5949$$

$$\text{or } 42.2660 \pm 19.1897$$

$$\text{or } 23.0763 \Leftrightarrow 61.4557$$ a range of over $38,000.

From Exercise 7.6, the 95% confidence interval estimate for $\overline{\text{SAL}}$ is:

$$95\%\text{ CI: } 42.2660 \pm 2.009 \times 1.3569 \quad [2.010]$$

$$\text{or } 95\%\text{ CI: } 42.2660 \pm 2.7261$$

$$\text{or } 95\%\text{ CI: } 39.5399 < \overline{\text{SAL}} < 44.9921$$ a range of less than $6,000.

The first assumes that the *sample data* follow roughly a normal distribution. It is based on the *standard deviation of the sample* and tells us roughly *how scattered the data* are.
The second assumes that *all Possible sample means* for samples of size n follow a normal distribution. It is based on the *standard error of the mean* and tells us *how precise our estimate of the mean* is.
The standard error $(s_{\bar{X}})$ = the standard deviation (s_X) divided by $\sqrt{n}$.

7.11 $p_{\text{female}} = 18 / 50 = 0.36$

$$s_p = \sqrt{0.36 \times 0.64 / 50} = \sqrt{0.0046} = 0.0679$$

a. 95% CI: $0.36 \pm 1.96 \times 0.0679$

or 95% CI: 0.36 ± 0.1330

or 95% CI: $0.2270 < p_{\text{female}} < 0.4930$

b. $1.96 s_p \leq 0.05$

$s_p \leq 0.05 / 1.96 = 0.0255$

$s_p = \sqrt{p \times (1-p) / n} = 0.0255$

$n = p \times (1-p) / 0.0255^2$

Using $p = 0.36$ $\quad n = 0.36 \times 0.64 / 0.0255^2 = 354$

increase $= 354 - 50 = 304$

Using $p = 0.50$ $\quad n = 0.500 \times 0.500 / 0.0255^2 = 384$

increase $= 384 - 50 = 334$

7.13 $p_{\text{female}} = 53 / 100 = 0.53$

$s_p = \sqrt{0.53 \times 0.47 / 100} = \sqrt{0.00249} = 0.0499$

a. 95% CI: $0.53 \pm 1.96 \times 0.0499$

or 95% CI: 0.53 ± 0.0978

or 95% CI: $0.4322 < p_{\text{female}} < 0.6278$

b. $1.96 s_p \leq 0.05$

$s_p \leq 0.05 / 1.96 = 0.0255$

$s_p = \sqrt{p \times (1-p) / n} = 0.0255$

$n = p \times (1-p) / 0.0255^2$

Using $p = 0.53$ $\quad n = 0.53 \times 0.47 / 0.0255^2 = 383$

increase $= 383 - 100 = 283$

Using $p = 0.50$ $\quad n = 0.500 \times 0.500 / 0.0255^2 = 384$

increase $= 384 - 100 = 284$

7.15 a. For proportion female:

$$95\%\ \text{CI: } 0.550 \pm 1.96 \times 0.111$$

or $95\%\ \text{CI: } 0.550 \pm 0.218$

or $95\%\ \text{CI: } 0.332 < p_{\text{female}} < 0.768$

b. For mean age:

$$95\%\ \text{CI: } 43.250 \pm 2.093 \times 2.524$$

or $95\%\ \text{CI: } 43.250 \pm 5.284$

or $95\%\ \text{CI: } 37.966 < \overline{\text{AGE}} < 48.534$

c. For mean income:

$$95\%\ \text{CI: } 55.270 \pm 2.093 \times 4.131$$

or $95\%\ \text{CI: } 55.270 \pm 8.647$

or $95\%\ \text{CI: } 46.623 < \overline{\text{INC}} < 63.917$

d. For proportion newspaper:

$$95\%\ \text{CI: } 0.500 \pm 1.96 \times 0.112$$

or $95\%\ \text{CI: } 0.500 \pm 0.219$

or $95\%\ \text{CI: } 0.281 < p_{\text{newspaper}} < 0.719$

Chapter 8: Tests of Hypotheses: One-Sample Tests

All rejection criteria are taken from the tables. For *t,* when the table doesn't list the exact degrees of freedom, the closest degrees of freedom is used. However, if the Excel value with the exact degrees of freedom differs, it is also reported [in brackets].

8.1 H_o: $\pi_{\text{on time}} \geq 0.95$ (Claim is true.)
H_a: $\pi_{\text{on time}} < 0.95$ (Claim is not true.)

a. Reject H_0 if $Z_c < -2.33$ ($\alpha = 0.01$)

b. Reject H_o if $Z_c < -3.09$ ($\alpha = 0.001$)

$$Z_c = \frac{0.91 - 0.95}{0.0154} = -2.5955$$

∴ Reject H_o – It is not true.
∴ Fail to Reject H_o – It could be true.

c. The *p*-value = 0.5000–0.4953 = 0.0047

8.3 H_o: $\mu_{\text{strength}} \geq 25$ lbs (Thread is strong enough.)
H_a: $\mu_{\text{strength}} < 25$ lbs (Thread is not strong enough.)

a. Reject H_o if $t_c < -1.833$ (df = 9; $\alpha = 0.05$)

b. Reject H_o if $t_c < -2.821$ (df = 9; $\alpha = 0.01$)

$$t_c = \frac{23-25}{1.044} = -1.9165$$

∴ Reject H_o—It is not strong enough.
∴ Fail to Reject H_o—It could be strong enough.

c. The p-value is between 0.0500 and 0.0250.
=TDIST(1.9165,9,1) = 0.0438.

8.5 H_o: $\pi_{\text{defective}} \leq 0.05$ (Nothing is wrong.)
H_o: $\pi_{\text{defective}} > 0.05$ (Something is wrong.)
Reject H_o if $Z_c > 1.65$ ($\alpha = 0.05$)

$$Z_c = \frac{x/50 - 0.05}{\sqrt{0.05 \times 0.95/50}} = \frac{x/50 - 0.05}{0.0308} = 1.65$$

$x/50 = 1.65 \times 0.0308 + 0.05 = 0.1009$
$x = 0.1009 \times 50 = 5.0428$ More than five defectives.

8.7 a. H_o: $\mu_{\text{height}} \leq 67$ inches (Mean height for young men is 67 inches or less.)
H_a: $\mu_{\text{height}} > 67$ inches (Mean height for young men is more than 67 inches.)
Reject H_o if $t_c > 1.658$ [1.656] (df = 134; $\alpha = 0.05$)

$$t_c = \frac{70.4148 - 67}{0.2382} = 14.3334$$

∴ Reject H_o—The mean height for young men is more than 67 inches.
The p-value is less than 0.0005.
=TDIST(14.3334,134,1) = 0.0000.

b. H_o: $\mu_{\text{height}} \geq 67$ inches (Mean height for young women is 67 inches or more.)
H_a: $\mu_{\text{height}} < 67$ inches (Mean height for young women is less than 67 inches.)
Reject H_o if $t_c < -1.655$ (df = 145; $\alpha = 0.05$)

$$t_c = \frac{64.6849 - 67}{0.2199} = -10.5288$$

∴ Reject H_o—The mean height for young women is less than 67 inches.
The p-value is less than 0.0005.
=TDIST(10.5288,145,1) = 0.0000.

8.9 a. H_o: $\mu_{\text{height}} = 67$ inches (Mean height for all such students is 67 inches.)
H_a: $\mu_{\text{height}} \neq 67$ inches (Mean height for all such students is not 67 inches.)
Reject H_o if $|t_c| > 1.984$ (df = 99; $\alpha = 0.05$)

$$t_c = \frac{67.61 - 67}{0.4739} = 1.2871$$

∴ Fail to Reject H_o—The mean height for all such students could be 67 inches.
The *p*-value is slightly greater than 0.2000.
=TDIST(1.2871,99,2) = 0.2011.

b. H_o: $\mu_{height} = 150$ lbs (Mean weight for all such students is 150 pounds.)
H_a: $\mu_{height} \neq 150$ lbs (Mean weigh for all such students is not 150 pounds.)
Reject H_o if $|t_c| > 1.984$ (df = 99; $\alpha = 0.05$)

$$t_c = \frac{142.69 - 150}{2.9027} = -2.5184$$

∴ Reject H_o—The mean weight for all such students is not 150.
The *p*-value is between 0.01 and 0.02.
=TDIST(2.5184,99,2) = 0.0134.

c. H_o: $\mu_{spending} = \$25$ (Mean entertainment spending for all such students is \$25.)
H_a: $\mu_{spending} \neq \$25$ (Mean entertainment spending for all such students is not \$25.)
Reject H_o if $|t_c| > 1.984$ (df = 99; $\alpha = 0.05$)

$$t_c = \frac{31.34 - 25}{1.5060} = 4.2099$$

∴ Reject H_o—The mean entertainment spending for all such students is not \$25.
The *p*-value is less than 0.0010.
=TDIST(4.2099,99,2) = 0.0001.

d. H_o: $\mu_{study} = 20$ hours (Mean study time for all such students is 20.)
H_a: $\mu_{study} \neq 20$ hours (Mean study time for all such students is not 20.)
Reject H_o if $|t_c| > 1.984$ (df = 99; $\alpha = 0.05$)

$$t_c = \frac{24.06 - 20}{0.6090} = 6.6667$$

∴ Reject H_o – The mean study time for all such students is not 20 hours.
The *p*-value is less than 0.0010.
=TDIST(6.6667,99,2) = 0.0000.

e. H_o: $\mu_{GPA} = 3.000$ (Mean GPA for all such students is 3.000.)
H_a: $\mu_{GPA} \neq 3.000$ (Mean GPA for all such students is not 3.000.)
Reject H_o if $|t_c| > 1.984$ (df = 99; $\alpha = 0.05$)

$$t_c = \frac{2.9332 - 3.000}{0.0567} = -1.1775$$

∴ Fail to Reject H_o—The mean GPA for all such students could be 3.000.
The p-value is between 0.2000 and 0.5000.
=TDIST(1.1775,99,2) = 0.2418.

8.11 H_o: $\pi_{market} \geq 0.50$ (Nickels has at least half the market.)
H_a: $\pi_{market} < 0.50$ (Nickels has less than half the market.)
Reject H_o if $Z_c < -1.65$ ($\alpha = 0.05$)

$$Z_c = \frac{.40 - .50}{0.0693} = -1.4434$$

∴ Fail to Reject H_o—Nickels could have at least half the market.
The p-value is 0.5000 – 0.4251 = 0.0749.

8.13 a. H_o: $\pi_{female} \geq 0.50$ (At least half of non-Nickels customers are female.)
H_a: $\pi_{female} < 0.50$ (Less than half of non-Nickels customers are female.)
Reject H_o if $Z_c < -1.65$ ($\alpha = 0.05$)

$$Z_c = \frac{.4333 - .50}{0.0905} = -0.7369$$

∴ Fail to Reject H_o—At least half could be female.
The p-value is 0.5000 – 0.4251 = 0.0749.

b. H_o: $\mu_{age} \geq 30$ (Mean age of non-Nickels customers is at least 30.)
H_a: $\mu_{age} < 30$ (Mean age of non-Nickels customers is less than 30.)
Reject H_o if $t_c < -1.699$ (df = 29; $\alpha = 0.05$)

$$t_c = \frac{25.50 - 30}{1.5416} = -2.9191$$

∴ Reject H_o—Mean age is less than 30.
The p-value is between 0.0025 and 0.0050.
=TDIST(2.9191,29,1) = 0.0034.

c. H_o: $\mu_{income} \geq \$40,000$ (Mean income of non-Nickels customers is \$40,000 or more.)
H_a: $\mu_{income} < \$40,000$ (Mean income of non-Nickels customers is less than \$40,000.)
Reject H_o if $t_c < -1.699$ (df = 29; $\alpha = 0.05$)

$$t_c = \frac{28.60637 - 40}{2.4802} = -4.5937$$

∴ Reject H_o—Mean income is less than $40,000.

The p-value is less than 0.0005.
=TDIST(4.5937,29,1) = 0.0000.

d. H_o: $\pi_{paper} \geq 1/3$ (At least 1/3 of non-Nickels customers rely on newspaper.)
H_a: $\pi_{paper} < 1/3$ (Less than 1/3 of non-Nickels customers rely on newspaper.)
Reject H_o if $Z_c < -1.65$ ($\alpha = 0.05$)

$$Z_c = \frac{0.1667 - 1/3}{0.0680} = -2.4495$$

∴ Reject H_o—Proportion who rely on newspaper is less than 1/3.
The p-value is 0.5000 – 0.4929 = 0.0071.

Chapter 9: Tests of Hypotheses: Two-Sample Tests

All rejection criteria are taken from the tables. For t, when the table doesn't list the exact degrees of freedom, the closest degrees of freedom is used. However, if the Excel value with the exact degrees of freedom differs, it is also reported [in brackets].

9.1 H_o: $\pi_{SC\text{–}Hourly} \geq \pi_{SC\text{–}Other}$ (Expectation is wrong.)
H_a: $\pi_{SC\text{–}Hourly} < \pi_{SC\text{–}Other}$ (Expectation is right.)
Reject H_o if $Z_c < -1.65$ ($\alpha = 0.05$) Reject H_o if $Z_c < -2.33$ ($\alpha = 0.01$)

Hourly-Wage	At Least Some College		
Worker	**Yes**	**No**	**Total**
Yes	102	244	346
No	216	158	374
Total	318	402	720

$p_{SC\text{–}Hourly} = 102/346 = 0.2948$
$p_{SC\text{–}Other} = 216/374 = 0.5775$
$p_{SC\text{–}Pooled} = 318/720 = 0.4417$

$$Z_c = \frac{0.2948 - 0.5775}{0.370} = -7.63$$

∴ Reject H_o ∴ Reject H_o
The p-value is virtually zero.

9.3 a. H_o: $\mu_{Female} = \mu_{Male}$ (Mean heights are equal.)
H_a: $\mu_{Female} \neq \mu_{Male}$ (Mean heights differ.)
Reject H_o if $|t_c| > 1.960$ [1.969] (df = 279; $\alpha = 0.05$) Reject H_o if $|t_c| > 2.576$ [2.594] (df = 279; $\alpha = .01$)

	Female	Male
Mean	64.6849	70.4148
Standard Deviation	2.6568	2.7681
Count	146	135

$$t_c = \frac{64.6849 - 70.4148}{0.3237} = -17.702$$

∴ Reject H_o ∴ Reject H_o

The p-value is virtually zero.

b. Yes, the sample standard deviations seem very similar.

c. Without this assumption, the degrees of freedom and standard error would change as follows. (The changes are in bold.)

Reject H_o if $|t_c| > 1.960$ [1.969] (df = **275;** $\alpha = 0.05$) Reject H_o if $|t_c| > 2.576$ [2.594] (df = **275**; $\alpha = 0.01$)

$$t_c = \frac{64.6849 - 70.4148}{\mathbf{0.3242}} = \mathbf{-17.674}$$

∴ Reject H_o ∴ Reject H_o

The p-value is still virtually zero.

9.5 a. H_o: $\pi_{\text{Cust}} = \pi_{\text{Noncust}}$ (Proportions female are equal.)
H_a: $\pi_{\text{Cust}} \neq \pi_{\text{Noncust}}$ (Proportions female differ.)

Reject H_o if $|Z_c| > 1.96$ ($\alpha = 0.05$) Reject H_o if $|Z_c| > 2.58$ ($\alpha = 0.01$)

	Customer	**Noncustomer**	**Total**
Male	9	17	26
Female	11	13	24
Total	20	30	50

$P_{\text{Cust-}F}$	$P_{\text{Noncust.-}F}$	$P_{p\text{-}F}$
0.5500	0.4333	0.4800

$$Z_c = \frac{0.5500 - 0.4333}{0.1442} = 0.8089$$

∴ Fail to Reject H_o ∴ Fail to Reject H_o

The p-value is $(0.5000 - 0.2910) \times 2 = 0.4180$.

b. H_o: $\mu_{\text{Cust}} = \mu_{\text{Noncust}}$ (Mean ages are equal.)
H_a: $\mu_{\text{Cust}} \neq \mu_{\text{Noncust}}$ (Mean ages differ.)

Reject H_o if $|t_c| > 2.009$ [2.011] (df = 48; $\alpha = 0.05$) Reject H_o if $|t_c| > 2.678$ [2.682] (df = 48; $\alpha = 0.01$)

	Customer	**Noncustomer**
Mean	43.2500	25.5000
Standard Deviation	11.2898	8.4435
Count	20	30

$$t_c = \frac{43.2500 - 25.5000}{2.7917} = 6.3580$$

∴ Reject H_o ∴ Reject H_o

The p-value is virtually zero.

c. H_o: $\mu_{Cust} = \mu_{Noncust}$ (Mean Incomes are equal.)
H_a: $\mu_{Cust} \neq \mu_{Noncust}$ (Mean Incomes differ.)

Reject H_o if $|t_c| > 2.009$ [2.011] (df = 48; $\alpha = 0.05$)   Reject H_o if $|t_c| > 2.678$ [2.682] (df = 48; $\alpha = 0.01$)

	Customer	Noncustomer
Mean	55.2700	28.6067
Standard Deviation	18.4755	13.5845
Count	20	30

$$t_c = \frac{55.2700 - 28.6067}{4.5333} = 5.8817$$

∴ Reject H_o ∴ Reject H_o

The p-value is virtually zero.

d. H_o: $\pi_{Cust} = \pi_{Noncust}$ (Proportions relying on newspaper are equal.)
H_a: $\pi_{Cust} = \pi_{Noncust}$ (Proportions relying on newspaper differ.)

Reject H_o if $|Z_c| > 1.96$ ($\alpha = 0.05$)   Reject H_o if $|Z_c| > 2.58$ ($\alpha = 0.01$)

	Customer	Noncustomer	Total
Newspaper	10	5	15
Radio	6	14	20
Other	4	11	15
Total	20	30	50
	$P_{\text{Cust-News}}$	$P_{\text{Noncust-News}}$	$P_{\text{p-News}}$
	0.5000	0.1667	0.3000

$$Z_c = \frac{0.5000 - 0.1667}{0.1323} = 2.5198$$

∴ Reject H_o ∴ Fail to Reject H_o

The p-value is $(0.5000 - 0.4941) \times 2 = 0.0118$.

9.7 a. H_o: $\mu_{Female} = \mu_{Male}$ (Mean educations are equal.)
H_a: $\mu_{Female} \neq \mu_{Male}$ (Mean educations differ.)

Reject H_o if $|t_c| > 1.676$ [1.677] (df = 48; α = 0. 10)

Reject H_o if $|t_c| > 2.009$ [2.011] (df = 48; α = 0.05)

	Female	**Male**
Mean	13.8333	13.3750
Standard Deviation	2.4793	2.3793
Count	18	32

$$t_c = \frac{13.8333 - 13.3750}{0.7116} = 0.644$$

$\therefore$ Fail to Reject H_o $\qquad$ $\therefore$ Fail to Reject H_o

The p-value is greater than $0.2500 \times 2 = 0.5000$. =TDIST(0.644,48,2) = 0.5226.

b. H_o: $\mu_{Female} = \mu_{Male}$ (Mean years of experience are equal.)
H_a: $\mu_{Female} \neq \mu_{Male}$ (Mean years of experience differ.)

Reject H_o if $|t_c| > 1.676$ [1.677] (df = 48; α = 0.10)

Reject H_o if $|t_c| > 2.009$ [2.011] (df = 48; α = 0.05)

	Female	**Male**
Mean	17.3333	21.4375
Standard Deviation	14.7089	12.5182
Count	18	32

$$t_c = \frac{17.3333 - 21.4375}{3.9290} = -1.045$$

$\therefore$ Fail to Reject H_o $\qquad$ $\therefore$ Fail to Reject H_o

The p-value is between $0.2500 \times 2 = 0.5000$ and $0.1000 \times 2 = 0.2000$. =TDIST(1.045,48,2) = 0.3014.

c. H_o: $\pi_{Female} = \pi_{Male}$ (Proportions in management are equal.)
H_a: $\pi_{Female} \neq \pi_{Male}$ (Proportions in management differ.)

Reject H_o if $|Z_c| > 1.65$ (α = 0.10)

Reject H_o if $|Z_c| > 1.96$ (α = 0.05)

	Female	**Male**	**Total**
Line	6	19	25
Office	8	7	15
Management	4	6	10
Total	18	32	50
	P_{Female}	P_{male}	P_{pooled}
	0.2222	0.1875	0.2000

$$Z_c = \frac{0.2222 - 0.1875}{0.1179} = 0.29$$

∴ Fail to Reject H_o ∴ Fail to Reject H_o

The p-value is $(0.5000 - 0.1141) \times 2 = 0.7718$.

d. H_o: $\mu_{Female} = \mu_{Male}$ (Mean salaries are equal.)
H_a: $\mu_{Female} = \mu_{Male}$ (Mean salaries differ.)

Reject H_o if $|t_c| > 1.676$ [1.677] (df = 48; $\alpha = 0.10$) Reject H_o if $|t_c| > 2.009$ [2.011] (df = 48; $\alpha = 0.05$)

	Female	Male
Mean	40.2389	43.4063
Standard Deviation	11.6771	8.1862
Count	18	32

$$t_c = \frac{40.2389 - 43.4063}{2.8194} = -1.1234$$

∴ Fail to Reject H_o ∴ Fail to Reject H_o

The p-value is between $0.2500 \times 2 = 0.5000$ and $0.1000 \times 2 = 0.2000$. =TDIST(1.1234,48,2) = 0.2668.

9.9 a. H_o: $\mu_{Female} = \mu_{Male}$ (Mean heights are equal.)
H_a: $\mu_{Female} \neq \mu_{Male}$ (Mean heights differ.)

Reject H_o if $|t_c| > 1.984$ (df = 98; $\alpha = 0.05$) Reject H_o if $|t_c| > 2.626$ [2.627] (df = 98; $\alpha = 0.01$)

	Female	Male
Mean	65.0189	70.5319
Standard Deviation	3.8003	3.9445
Count	53	47

$$t_c = \frac{65.0189 - 70.5319}{0.7751} = -7.1125$$

∴ Reject H_o ∴ Reject H_o

The p-value is virtually zero.

b. H_o: $\mu_{Female} = \mu_{Male}$ (Mean weights are equal.)
H_a: $\mu_{Female} \neq \mu_{Male}$ (Mean weights differ.)

Reject H_o if $|t_c| > 1.984$ (df = 98; $\alpha = 0.05$) Reject H_o if $|t_c| > 2.626$ [2.627] (df = 98; $\alpha = 0.01$)

	Female	Male
Mean	119.8868	168.4043
Standard Deviation	14.0379	17.7723
Count	53	47

$$t_c = \frac{119.8868 - 168.4043}{3.1858} = -15.2292$$

$\therefore$ Reject H_o $\therefore$ Reject H_o

The *p*-value is virtually zero.

c. H_o: $\pi_{Female} = \pi_{Male}$ (Proportions majoring in economics are equal.)
H_a: $\pi_{Female} \neq \pi_{Male}$ (Proportions majoring in economics differ.)
Reject H_o if $|Z_c| > 1.96$ ($\alpha = 0.05$) Reject H_o if $|Z_c| > 2.58$ ($\alpha = 0.01$)

	Female	Male	Total
Economics	7	9	16
Other	46	38	84
Total	53	47	100

$p_{F\text{-}Econ}$	$p_{M\text{-}Econ}$	$p_{p\text{-}Econ}$
0.1321	0.1915	0.1600

$$Z_c = \frac{0.1321 - 0.1915}{0.0735} = -0.8089$$

$\therefore$ Fail to Reject H_o $\therefore$ Fail to Reject H_o

The *p*-value is $(0.5000 - 0.2910) \times 2 = 0.4180$.

d. H_o: $\mu_{Female} = \mu_{Male}$ (Mean aid is equal.)
H_a: $\mu_{Female} \neq \mu_{Male}$ (Mean aid differs.)
Reject H_o if $|t_c| > 1.984$ ($df = 98$; $\alpha = 0.05$) Reject H_o if $|t_c| > 2.626$ [2.627] ($df = 98$; $\alpha = 0.01$)

	Female	Male
Mean	5047.17	5191.49
Standard Deviation	2607.80	2601.32
Count	53	47

$$t_c = \frac{5047.17 - 5191.49}{521.8925} = -0.2765$$

$\therefore$ Fail to Reject H_o $\therefore$ Fail to Reject H_o

The *p*-value is 0.7827.

e. H_o: $\pi_{Female} = \pi_{Male}$ (Proportions with jobs are equal.)
H_a: $\pi_{Female} \neq \pi_{Male}$ (Proportions with jobs differ.)
Reject H_o if $|Z_c| > 1.96$ ($\alpha = 0.05$) Reject H_o if $|Z_c| > 2.58$ ($\alpha = 0.01$)

	Female	**Male**	**Total**
Nonjob	22	22	44
Job	31	25	56
Total	53	47	100

$p_{F\text{-}Job}$	$p_{M\text{-}Job}$	$p_{p\text{-}Job}$
0.5849	0.5319	0.5600

$$Z_c = \frac{.5849 - .5319}{.0995} = .5328$$

∴ Fail to Reject H_o ∴ Fail to Reject H_o

The p-value is $(0.5000 - 0.2019) \times 2 = 0.5962$.

f. H_o: $\pi_{Female} = \pi_{Male}$ (Proportions in sports are equal.)
H_a: $\pi_{Female} \neq \pi_{Male}$ (Proportions in sports differ.)
Reject H_o if $|Z_c| > 1.96$ ($\alpha = 0.05$) Reject H_o if $|Z_c| > 2.58$ ($\alpha = 0.01$)

	Female	**Male**	**Total**
Nonsports	44	32	76
Sports	9	15	24
Total	53	47	100

$p_{F\text{-}Sports}$	$p_{M\text{-}Sports}$	$p_{p\text{-}Sports}$
0.1698	0.3191	0.2400

$$Z_c = \frac{0.1698 - 0.3191}{0.0856} = -1.7452$$

∴ Fail to Reject H_o ∴ Fail to Reject H_o

The p-value is $(0.5000 - 0.4599) \times 2 = 0.0802$.

g. H_o: $\pi_{Female} = \pi_{Male}$ (Proportions in music are equal.)
H_a: $\pi_{Female} \neq \pi_{Male}$ (Proportions in music differ.)
Reject H_o if $|Z_c| > 1.96$ ($\alpha = 0.05$) Reject H_o if $|Z_c| > 2.58$ ($\alpha = 0.01$)

	Female	**Male**	**Total**
Nonmusic	42	39	81
Music	11	8	19
Total	53	47	100

p_F-Music	p_M-Music	p_p-Music
0.2075	0.1702	0.1900

$$Z_c = \frac{0.2075 - 0.1702}{0.0786} = 0.4750$$

∴ Fail to Reject H_o ∴ Fail to Reject H_o

The p-value is $(0.5000 - 0.1808) \times 2 = 0.6384$.

h. H_o: $\pi_{Female} = \pi_{Male}$ (Proportions in fraternities/sororities are equal.)

H_a: $\pi_{Female} \neq \pi_{Male}$ (Proportions in fraternities/sororities differ.)

Reject H_o if $|Z_c| > 1.96$ ($\alpha = 0.05$) Reject H_o if $|Z_c| > 2.58$ ($\alpha = 0.01$)

	Female	Male	Total
Nonfraternities/sororities	39	35	74
Fraternities/sororities	14	12	26
Total	53	47	100
	$p_{F\text{-}F/S}$	$P_{M\text{-}F/S}$	$p_{p\text{-}F/S}$
	0.2642	0.2553	0.2600

$$Z_c = \frac{0.2642 - 0.2553}{0.0879} = 0.1005$$

∴ Fail to Reject H_o ∴ Fail to Reject H_o

The p-value is $(0.5000 - 0.0398) \times 2 = 0.9204$.

i. H_o: $\mu_{Female} = \mu_{Male}$ (Mean entertainment spending is equal.)

H_a: $\mu_{Female} \neq \mu_{Male}$ (Mean entertainment spending differs.)

Reject H_o if $|t_c| > 1.984$ (df = 98; $\alpha = 0.05$) Reject H_o if $|t_c| > 2.626$ [2.627] (df = 98; $\alpha = 0.01$)

	Female	Male
Mean	27.7736	35.3617
Standard Deviation	14.8384	14.4228
Count	53	47

$$t_c = \frac{27.7736 - 35.3617}{2.9343} = -2.5860$$

∴ Reject H_o ∴ Fail to Reject H_o

The p-value is .0112.

j. H_o: $\mu_{Female} = \mu_{Male}$ (Mean study times are equal.)

H_a: $\mu_{Female} \neq \mu_{Male}$ (Mean Study times differ.)

Reject H_o if $|t_c| > 1.984$ (df = 98; $\alpha = 0.05$) Reject H_o if $|t_c| > 2.626$ [2.627] (df = 98; $\alpha = 0.01$)

	Female	Male
Mean	25.0377	22.9574
Standard Deviation	5.4946	6.5838
Count	53	47

$$t_c = \frac{25.0377 - 22.9574}{1.2083} = 1.7217$$

$\therefore$ Fail to Reject H_o $\therefore$ Fail to Reject H_o

The p-value is .0883.

k. H_o: $\mu_{Female} = \mu_{Male}$ (Mean GPAs are equal.)
   H_a: $\mu_{Female} \neq \mu_{Male}$ (Mean GPAs differ.)
   Reject H_o if $|t_c| > 1.984$ (df = 98; $\alpha = 0.05$) Reject H_o if $|t_c| > 2.626$ [2.627] (df = 98; $\alpha = 0.01$)

	Female	Male
Mean	3.0230	2.8319
Standard Deviation	0.5166	0.6096
Count	53	47

$$t_c = \frac{3.0230 - 2.8319}{0.1126} = 1.6967$$

$\therefore$ Fail to Reject H_o $\therefore$ Fail to Reject H_o

The p-value is 0.0929.

9.11 a. H_o: $\pi_{Sports} = \pi_{Nonsports}$ (Proportions female are equal.)
   H_a: $\pi_{sports} \neq \pi_{Nonsports}$ (Proportions female differ.)
   Reject H_o if $|Z_C| > 1.96$ ($\alpha = 0.05$) Reject H_o if $|Z_C| > 2.58$ ($\alpha = 0.01$)

	Sports	Nonsports	Total
Male	15	32	47
Female	9	44	53
Total	24	76	100

$P_{S\text{-Female}}$	$P_{NS\text{-Female}}$	$P_{P\text{-Female}}$
0.3750	.0.5789	0.5300

$$Z_c = \frac{0.3750 - 0.5789}{0.1169} = -1.7452$$

$\therefore$ Fail to Reject H_o $\therefore$ Fail to Reject H_o

The p-value is $(0.5000 - 0.4599) \times 2 = 0.0802$.
Note that this is mathematically equivalent to 9.9.f.

b. H_o: $\mu_{Sports} = \mu_{Nonsports}$ (Mean heights are equal.)
   H_o: $\mu_{Sports} \neq \mu_{Nonsports}$ (Mean heights differ.)

Reject H_o if $|t_c| > 1.984$ (df = 98; α = 0.05) Reject H_o if $|t_c| > 2.626$ [2.627] (df = 98; α = 0.01)

	Sports	Nonsports
Mean	68.4167	67.3553
Standard Deviation	4.9775	4.6668
Count	24	46

$$t_c = \frac{68.4167 - 67.3553}{1.1102} = 0.9560$$

$\therefore$ Fail to Reject H_o $\therefore$ Fail to Reject H_o

The p-value is 0.3414.

c. H_o: $\mu_{Sports} = \mu_{Nonsports}$ (Mean weights are equal.)
H_o: $\mu_{Sports} \neq \mu_{Nonsports}$ (Mean weights differ.)

Reject H_o if $|t_c| > 1.984$ (df = 98; α = 0.05) Reject H_o if $|t_c| > 2.626$ [2.627] (df = 98; α = 0.01)

	Sports	Nonsports
Mean	147.5000	141.1711
Standard Deviation	31.5388	28.2382
Count	24	46

$$t_c = \frac{147.5000 - 141.1711}{6.8011} = 0.9306$$

$\therefore$ Fail to Reject H_o $\therefore$ Fail to Reject H_o
The p-value is .3544.

d. H_o: $\pi_{Sports} = \pi_{Nonsports}$ (Proportions majoring in economics are equal.)
H_a: $\pi_{Sports} \neq \pi_{Nonsports}$ (Proportions majoring in economics differ.)

Reject H_o if $|Z_c| > 1.96$ (α = 0.05) Reject H_o if $|Z_c| > 2.58$ (α = 0.01)

	Sports	Nonsports	Total
Economics	4	12	16
Other	20	64	84
Total	24	76	100

$P_{S\text{-}Econ}$	$P_{NS\text{-}Econ}$	$P_{P\text{-}Econ}$
0.1667	0.1279	0.1600

$$Z_c = \frac{0.1667 - 0.1579}{0.0858} = 0.1022$$

$\therefore$ Fail to Reject H_o $\therefore$ Fail to Reject H_o

The p-value is $(0.5000 - 0.0398) \times 2 = 0.9204$.

e. H_o: $\mu_{Sports} = \mu_{Nonsports}$ (Mean aid is equal.)
H_a: $\mu_{Sports} \neq \mu_{Nonsports}$ (Mean aid differs.)

Reject H_o if $|t_c| > 1.984$ (df = 98; $\alpha = 0.05$) Reject H_o if $|t_c| > 2.626$ [2.627] (df = 98; $\alpha = 0.01$)

	Sports	Nonsports
Mean	5791.67	4901.32
Standard Deviation	2399.80	2629.35
Count	24	76

$$t_c = \frac{5791.67 - 4901.32}{603.4683} = 1.4754$$

$\therefore$ Fail to Reject H_o $\therefore$ Fail to Reject H_o

The p-value is 0.1433.

f. H_o: $\pi_{Sports} = \pi_{Nonsports}$ (Proportions with jobs are equal.)
H_a: $\pi_{Sports} \neq \pi_{Nonsports}$ (Proportions with jobs differ.)

Reject H_o if $|Z_c| > 1.96$ ($\alpha = 0.05$) Reject H_o if $|Z_c| > 2.58$ ($\alpha = 0.01$)

	Sports	Nonsports	Total
Nonjob	13	31	44
Job	11	45	56
Total	24	76	100

$P_{S\text{-}Job}$	$P_{NS\text{-}Job}$	$P_{P\text{-}Job}$
0.4583	0.5921	0.5600

$$Z_c = \frac{0.4583 - 0.5921}{0.1162} = -1.1510$$

$\therefore$ Fail to Reject H_o $\therefore$ Fail to Reject H_o

The p-value is $(0.5000 - 0.3749) \times 2 = 0.2502$.

g. H_o: $\pi_{Sports} = \pi_{Nonsports}$ (Proportions in music are equal.)
H_a: $\pi_{Sports} \neq \pi_{Nonsports}$ (Proportions in music differ.)

Reject H_o if $|Z_c| > 1.96$ ($\alpha = 0.05$) Reject H_o if $|Z_c| > 2.58$ ($\alpha = 0.01$)

	Sports	Nonsports	Total
Nonmusic	22	59	81
Music	2	17	19
Total	24	76	100

$P_{S\text{-}Music}$	$P_{NS\text{-}Music}$	$P_{P\text{-}Music}$
0.0833	0.2237	0.1900

$$Z_c = \frac{0.0833 - 0.2237}{0.0919} = 1.5279$$

$\therefore$ Fail to Reject H_o $\therefore$ Fail to Reject H_o

The p-value is $(0.5000 - 0.4370) \times 2 = 0.1260$.

h. H_o: $\pi_{Sports} = \pi_{Nonsports}$ (Proportions in fraternities/sororities are equal.)
H_a: $\pi_{Sports} \neq \pi_{Nonsports}$ (Proportions in fraternities/sororities differ.)
Reject H_o if $|Z_c| > 1.96$ ($\alpha = 0.05$) Reject H_o if $|Z_c| > 2.58$ ($\alpha = 0.01$)

	Sports	Nonsports	Total
Nonfraternities/sororities	17	57	74
Fraternities/sororities	7	19	26
Total	24	76	100

$P_{S\text{-}F/S}$	$P_{NS\text{-}F/S}$	$P_{P\text{-}F/S}$
0.2917	0.2500	0.2600

$$Z_c = \frac{0.2917 - 0.2500}{0.1027} = 0.4057$$

$\therefore$ Fail to Reject H_o $\therefore$ Fail to Reject H_o

The p-value is $(0.5000 - 0.1591) \times 2 = 0.6818$.

i. H_o: $\mu_{Sports} = \mu_{Nonsports}$ (Mean entertainment spending is equal.)
H_a: $\mu_{Sports} \neq \mu_{Nonsports}$ (Mean entertainment spending differs.)
Reject H_o if $|t_c| > 1.984$ ($df = 98$; $\alpha = 0.05$) Reject H_o if $|t_c| > 2.626$ [2.627] ($df = 98$; $\alpha = 0.01$)

	Sports	Nonsports
Mean	32.7083	30.9079
Standard Deviation	17.4243	14.3343
Count	24	76

$$t_c = \frac{32.7083 - 30.9079}{3.5394} = 0.5087$$

$\therefore$ Fail to Reject H_o $\therefore$ Fail to Reject H_o

The p-value is 0.6121.

j. H_o: $\mu_{Sports} = \mu_{Nonsports}$ (Mean study times are equal.)
H_a: $\mu_{Sports} \neq \mu_{Nonsports}$ (Mean study times differ.)
Reject H_o if $|t_c| > 1.984$ ($df = 98$; $\alpha = 0.05$) Reject H_o if $|t_c| > 2.626$ [2.627] ($df = 98$; $\alpha = 0.01$)

	Sports	Nonsports
Mean	21.7917	24.7763
Standard Deviation	5.6413	6.0852
Count	24	76

$$t_c = \frac{21.7917 - 24.7763}{1.4011} = -2.1302$$

∴ Reject H_o ∴ Fail to Reject H_o

The p-value is 0.0357.

k. H_o: $\mu_{Sports} = \mu_{Nonsports}$ (Mean GPAs are equal.)
H_a: $\mu_{Sports} \neq \mu_{Nonsports}$ (Mean GPAs differ.)

Reject H_o if $|t_c| > 1.984$ (df = 98; $\alpha = 0.05$) Reject H_o if $|t_c| > 2.626$ [2.627] (df = 98; $\alpha = 0.01$)

	Sports	Nonsports
Mean	2.6734	3.0152
Standard Deviation	0.5003	0.5656
Count	24	76

$$t_c = \frac{2.6734 - 3.0152}{0.1290} = -2.6499$$

∴ Reject H_o ∴ Reject H_o

The p-value is 0.0094.

9.13 a. H_o: $\mu_{After\text{-}Before} \geq 0$ (No reduction in mean weights.)
H_a: $\mu_{After\text{-}Before} < 0$ (Reduction in mean weights.)

Reject H_o if $t_c < -1.711$ (df = 24; $\alpha = 0.05$) Reject H_o if $t_c < -2.492$ (df = 24; $\alpha = 0.01$)

	After	Before	Difference
Mean	148.24	150.04	–1.8000
Standard Deviation	31.8398	32.9197	3.3417
Count	25	25	25

$$t_c = \frac{-1.8000 - 0}{0.6683} = -2.693$$

∴ Reject H_o ∴ Reject H_o

The p-value is between 0.0050 and 0.0100.
=TDIST(2.693,24,1) = 0.0064.

Chapter 10: Tests of Hypotheses: Contingency and Goodness-of-Fit

10.1 H_o: The candidates are equally popular.
H_a: The candidates are not equally popular.

Reject H_o if $\chi_c^2 > 7.815$ (df = 3; α = 0.05) Reject H_o if $\chi_c^2 > 11.345$ (df = 3; α = 0.01)

	f_o	f_e			
Adams	33	30	3	9	0.3000
Baker	17	30	−13	169	5.6333
Clark	40	30	10	100	3.3333
Davis	30	30	0	0	0.0000
	120	120			9.2667

$$\chi_c^2 = 9.2667$$

∴ Reject H_o ∴ Fail to Reject H_o

The p-value of the test is between 0.0500 and 0.0250.
=CHIDIST(9.2667,3) = 0.0259.

10.3 a. H_o: Patronage is independent of sex.
H_a: Patronage is not independent of sex.

Reject H_o if $\chi_c^2 > 3.841$ (df = 1; α = 0.05) Reject H_o if $\chi_c^2 > 6.635$ (df = 1; α = 0.01)

f_o	Customer	Noncustomer	
Male	9	17	26
Female	11	13	24
	20	30	50
	0.4	0.6	

f_e	Customer	Noncustomer	
Male	10.4	15.6	26
Female	9.6	14.4	24
	20.0	30.0	50

f_o	f_e			
9	10.4	−1.4	1.96	0.1885
11	9.6	1.4	1.96	0.2042
17	15.6	1.4	1.96	0.1256
13	14.4	−1.4	1.96	0.1361
50	50.0			0.6544

$$\chi_c^2 = 0.6544$$

∴ Fail to Reject H_o ∴ Fail to Reject H_o

The p-value of the test is greater than 0.2500.
=CHIDIST(0.6544,1) = 0.4186.

b. Chi-square with one degree of freedom is the standard normal squared. That is, except for rounding, our χ^2 rejection criteria in the test above are the squares of our $|Z|$ rejection criteria in Chapter 9. And our calculated χ_c^2, is the square of our calculated $|Z_c|$ in Chapter 9. Hence the two tests are mathematically equivalent and always give the same conclusion.

c. H_o: Patronage is independent of source.
H_a: Patronage is not independent of source.

Reject H_o if $\chi_c^2 > 5.991$ (df = 2; $\alpha = 0.05$) Reject H_o if $\chi_c^2 > 9.210$ (df = 2; $\alpha = 0.01$)

f_o	Customer	Noncustomer	
News	10	5	15
Radio	6	14	20
Other	4	11	15
	20	30	50
	0.4	0.6	

f_e	Customer	Noncustomer	
News	6	9	15
Radio	8	12	20
Other	6	9	15
	20	30	50

f_o	f_e			
10	6	4	16	2.6667
6	8	–2	4	0.5000
4	6	–2	4	0.6667
5	9	–4	16	1.7778
14	12	2	4	0.3333
11	9	2	4	0.4444
50	50			6.3889

$$\chi_c^2 = 6.3889$$

∴ Reject H_o ∴ Fail to Reject H_o

The p-value of the test is between 0.0250 and 0.0500.
=CHIDIST(6.3889,2) = 0.0410.

d. The Z only works for two by two tables; we had to combine Radio and Other in Chapter 9. The χ_c^2 allows us to separate Radio and Other. Of course, this means the two tests are no longer mathematically equivalent, even though they give similar results in this case.

10.5 a. H_o: Majoring in economics is independent of sex.
H_a: Majoring in economics is not independent of sex.

Reject H_0 if $\chi_c^2 > 3.841$ (df = 1; $\alpha = 0.05$) Reject H_0 if $\chi_c^2 > 6.635$ (df = 1; $\alpha = 0.01$)

f_o	Female	Male	
Economics	7	9	16
Other	46	38	84
	53	47	100
	0.53	0.47	

f_e	Female	Male	
Economics	8.48	7.52	16
Other	44.52	39.48	84
	20.00	30.00	100

f_o	f_e			
7	8.48	−1.48	2.1904	0.2583
46	44.52	1.48	2.1904	0.0492
9	7.52	1.48	2.1904	0.2913
38	39.48	−1.48	2.1904	0.0555
100	100.00			0.6543

$$\chi_c^2 = 0.6543$$

∴ Fail to Reject H_o ∴ Fail to Reject H_o

The p-value of the test is greater than 0.2500.
=CHIDIST (0.6543,1) = 0.4186.

b. Chi-square with one degree of freedom is the standard normal squared. That is, except for rounding, our χ^2 rejection criteria in the test above are the squares of our $|Z|$ rejection criteria in Chapter 9. And our calculated χ_c^2 is the square of our calculated $|Z_c|$ in Chapter 9. Hence the two tests are mathematically equivalent and always give the same conclusion.

c. H_o: Choice of major is independent of sex.
H_a: Choice of major is not independent of sex.

Reject H_o if $\chi_c^2 > 11.070$ (df = 5; $\alpha = 0.05$) Reject H_o if $\chi_c^2 > 15.086$ (df = 5; $\alpha = 0.01$)

f_o	**Female**	**Male**	
Math	10	5	15
Economics	7	9	16
Biology	11	10	21
Psychology	6	10	16
English	9	7	16
Other	10	6	16
	53	47	100
	0.53	0.47	

f_e	**Female**	**Male**	
Math	7.95	7.05	15
Economics	8.48	7.52	16
Biology	11.13	9.87	21
Psychology	8.48	7.52	16
English	8.48	7.52	16
Other	8.48	7.52	16
	53.00	47.00	100

f_o	f_e			
10	7.95	2.05	4.2025	0.5286
7	8.48	−1.48	2.1904	0.2583
11	11.13	−0.13	0.0169	0.0015
6	8.48	−2.48	6.1504	0.7253
9	8.48	0.52	0.2704	0.0319
10	8.48	1.52	2.3104	0.2725
5	7.05	−2.05	4.2025	0.5961
9	7.52	1.48	2.1904	0.2913
10	9.87	0.13	0.0169	0.0017
10	7.52	2.48	6.1504	0.8179
7	7.52	−0.52	0.2704	0.0360
6	7.52	−1.52	2.3104	0.3072
100	100.00			3.8682

$$\chi_c^2 = 3.8682$$

∴ Fail to Reject H_o ∴ Fail to Reject H_o

The *p*-value of the test is greater than 0.2500.

=CHIDIST (3.8682,5) = 0.5685

d. The Z only works for two by two tables; we had to combine all other majors in Chapter 9. The χ_c^2 allows us to separate the "Others." Of course, this means the two tests are no longer mathematically equivalent, even though they give similar results in this case.

Examples 9.9.e through 9.9.h. can all be worked as 2 × 2 contingency tables. In each, χ^2 is just Z^2 and the p-values are the same (except for rounding).

e. H_o: Holding a job is independent of sex.
H_a: Holding a job is not independent of sex.

Reject H_o if $\chi_c^2 > 3.841$ (df = 1; $\alpha = 0.05$)

Reject H_o if $\chi_c^2 > 6.635$ (df = 1; $\alpha = 0.01$)

f_o	Female	Male	
Not	22	22	44
Job	31	25	56
	53	47	100
	0.53	0.47	

f_e	Female	Male	
Not	23.32	20.68	44
Job	29.68	26.32	56
	53.00	47.00	100

f_o	f_e			
22	23.32	−1.32	1.7424	0.0747
31	29.68	1.32	1.7424	0.0587
22	20.68	1.32	1.7424	0.0843
25	26.32	−1.32	1.7424	0.0662
100	100.00			0.2839

$$\chi_c^2 = 0.2839$$

∴ Fail to Reject H_o ∴ Fail to Reject H_o

The p-value of the test is greater than 0.2500.
=CHIDIST (0.2839,1) = 0.5942.

f. H_o: Participating in varsity sports is independent of sex.
H_a: Participating in varsity sports is not independent of sex.

Reject H_o if $\chi_c^2 > 3.841$ (df = 1; $\alpha = 0.05$)

Reject H_o if $\chi_c^2 > 6.635$ (df = 1; $\alpha = 0.01$)

f_o	Female	Male	
Not	44	32	76
Sports	9	15	24
	53	47	100
	0.53	0.47	

f_e	**Female**	**Male**	
Not	40.28	35.72	76
Sports	12.72	11.28	24
	53.00	47.00	100

f_o	f_e			
44	40.28	3.72	13.8384	0.3436
9	12.72	–3.72	13.8384	1.0879
32	35.72	–3.72	13.8384	0.3874
15	11.28	3.72	13.8384	1.2268
100	100.00			3.0457

$$\chi_c^2 = 3.0457$$

∴ Fail to Reject H_o ∴ Fail to Reject H_o

The p-value of the test is between 0.0500 and 0.1000. =CHIDIST(3.0457,1) = 0.0810.

g. H_o: Participating in a music ensemble is independent of sex.

H_a: Participating in a music ensemble is not independent of sex.

Reject H_o if $\chi_c^2 > 3.841$ (df = 1; $\alpha = 0.05$) Reject H_o if $\chi_c^2 > 6.635$ (df = 1; $\alpha = 0.01$)

f_o	**Female**	**Male**	
Not	42	39	81
Music	11	8	19
	53	47	100
	0.53	0.47	

f_e	**Female**	**Male**	
Not	42.93	38.07	81
Music	10.07	8.93	19
	53.00	47.00	100

f_o	f_e			
42	42.93	–0.93	0.8649	0.0201
11	10.07	0.93	0.8649	0.0859
39	38.07	0.93	0.8649	0.0227
8	8.93	–0.93	0.8649	0.0969
100	100.00			0.2256

$$\chi_c^2 = 0.2256$$

∴ Fail to Reject H_o. ∴ Fail to Reject H_o

The p-value of the test is greater than 0.2500.
=CHIDIST (0.2256,1) = 0.6348.

h. H_0: Fraternity/sorority membership is independent of sex.
H_a: Fraternity/sorority membership is not independent of sex.

Reject H_o if $\chi_c^2 > 3.841$ (df = 1; $\alpha = 0.05$) Reject H_o if $\chi_c^2 > 6.635$ (df = 1; $\alpha = 0.01$)

f_o	Female	Male	
Not	39	35	74
Fraternity/sorority	14	12	26
	53	47	100
	0.53	0.47	

f_e	Female	Male	
Not	39.22	34.78	74
Fraternity/sorority	13.78	12.22	26
	53.00	47.00	100

f_o	f_e			
39	39.22	−0.22	0.0484	0.0012
14	13.78	0.22	0.0484	0.0035
35	34.78	0.22	0.0484	0.0014
12	12.22	−0.22	0.0484	0.0040
100	100.00			0.0101

$$\chi_c^2 = 0.0101$$

∴ Fail to Reject H_o ∴ Fail to Reject H_o

The p-value of the test is greater than 0.2500.
=CHIDIST (0.0101,1) = 0.9200.

10.7 H_0: The majors are equally popular.
H_a: The majors are not equally popular.

Reject H_o if $\chi_c^2 > 11.070$ (df = 5; $\alpha = 0.05$) Reject H_o if $\chi_c^2 > 15.086$ (df = 5; $\alpha = 0.01$)

Major	f_o	f_e			
Math	15	16.6667	−1.6667	2.7778	0.1667
Economics	16	16.6667	−0.6667	0.4444	0.0267
Biology	21	16.6667	4.3333	18.7778	1.1267
Psychology	16	16.6667	−0.6667	0.4444	0.0267
English	16	16.6667	−0.6667	0.4444	0.0267
Other	16	16.6667	−0.6667	0.4444	0.0267
	100	100.0000			1.4000

$$\chi_c^2 = 1.4000$$

∴ Fail to Reject H_o ∴ Fail to Reject H_o

The p-value of the test is greater than 0.2500.
=CHIDIST (1.4000,5) = 0.9243.

10.9 a. H_0: The packages are equally popular.
H_a: The packages are not equally popular.

Reject H_o if $\chi_c^2 > 5.991$ (df = 2; $\alpha = 0.05$) Reject H_o if $\chi_c^2 > 9.210$ (df = 2; $\alpha = 0.01$)

Package	f_o	f_e			
A	32	28	4	16	0.5714
B	17	28	−11	121	4.3214
C	35	28	7	49	1.7500
	84	84			6.6428

$$\chi_c^2 = 6.6429$$

∴ Reject H_o ∴ Fail to Reject H_o

The p-value of the test is between 0.0500 and 0.0250.
=CHIDIST(6.6429,2) = 0.0361.

b. H_o: Popularity is independent of job type.
H_a: Popularity is not independent of job type.
Reject H_o if $\chi_c^2 > 5.991$ (df = 2; $\alpha = 0.05$) Reject H_o if $\chi_c^2 > 9.210$ (df = 2; $\alpha = 0.01$)

f_o	**Office**	**Line**	
A	32	48	80
B	17	13	30
C	35	55	90
	84	116	200
	0.42	0.58	

f_e	Office	Line	
A	33.6	46.4	80
B	12.6	17.4	30
C	37.8	52.2	90
	84.0	116.0	200

f_o	f_e			
32	33.6	−1.6	2.56	0.0762
17	12.6	4.4	19.36	1.5365
35	37.8	−2.8	7.84	0.2074
48	46.4	1.6	2.56	0.0552
13	17.4	−4.4	19.36	1.1126
55	52.2	2.8	7.84	0.1502
200	200.0			3.1381

$$\chi_c^2 = 3.1381$$

∴ Fail to Reject H_o ∴ Fail to Reject H_o

The p-value of the test is between 0.1000 and 0.2500.
=CHIDIST(3.1381,2) = 0.2082.

Chapter 11: Tests of Hypotheses: ANOVA and Tests of Variances

All rejection criteria are taken from the tables. For t and F, when the table doesn't list the exact degrees of freedom, the closest degrees of freedom is used. However, if the Excel value with the exact degrees of freedom differs, it is also reported [in brackets].

11.1 H_o: $\sigma_{ABC} = \sigma_{XYZ}$ (Standard deviations in returns are equal.)
H_a: $\sigma_{ABC} \neq \sigma_{XYZ}$ (Standard deviations in returns differ.)
Reject H_o if $F_c > 2.714$ Reject H_o if $F_c > 3.312$
($df_n = 9$, $df_d = 13$; $\alpha = 0.10$) ($df_n = 9$, $df_d = 13$; $\alpha = 0.05$)

$$F_c = \frac{3.90^2}{2.60^2} = 2.2500$$

∴ Fail to Reject H_o ∴ Fail to Reject H_o

The p-value is between 0.200 and 0.100.
=FDIST(2.2500,9,13) × 2 = 0.0894 × 2 = 0.1788.

11.3 a. H_o: Weights are independent of sex.
H_a: Weights are not independent of sex.
Reject H_o if $F_c > 3.841$ Reject H_o if $F_c > 6.635$
[3.875] [6.727]
($df_n = 1$, $df_d = 279$; $\alpha = 0.05$) ($df_n = 1$, $df_d = 279$; $\alpha = 0.01$)

	Female	Male	All	
Mean	134.521	169.015	151.093	
Standard Deviation	24.308	26.215		
Count	146	135		
Sample	$\overline{X}$	$\overline{X}-\overline{X}_p$	$\left(\overline{X}-\overline{X}_p\right)^2$	$n \times = \left(\overline{X}-\overline{X}_p\right)^2$
Female	134.521	−16.572	274.631	40096.05
Male	169.015	17.922	321.208	43363.14
			BSS =	83459.19
				85678.44
				92087.97
			WSS =	177766.41

$$F_c = \frac{\frac{83459.19}{1}}{\frac{177766.41}{279}} = \frac{83459.19}{637.156} = 130.9871$$

∴ Reject H_o ∴ Reject H_o

The p-value is less than 0.005. = FDIST(130.9871,1,279) = 0.0000.

$$\text{BSS} = 83459.19$$
$$\text{WSS} = \underline{177766.41}$$
$$\text{TSS} = 261225.59$$

31.95% of the variation in weight (BSS/TSS) can be explained by sex.

b. F with df = {1,v} is t with df = {v} squared. That is, except for rounding, our F rejection criteria above are the squares of our $|t|$ rejection criteria in Chapter 9. And our calculated F_c is the square of our calculated $|t_c|$ in Chapter 9. Hence the two tests are mathematically equivalent and always give the same conclusion.

c. H_o: $\sigma_f = \sigma_m$ (Standard deviations in weights are equal.)
H_a: $\sigma_f \neq \sigma_m$ (Standard deviations in weights differ.)

Reject H_o if $F_c > 1.271$ [1.394] ($df_n = 134$, $df_d = 145$; $\alpha = 0.10$)

Reject H_o if $F_c > 1.374$ [1.549] ($df_n = 134$, $df_d = 145$; $\alpha = 0.05$)

$$F_c = \frac{26.215^2}{24.308^2} = 1.1630$$

∴ Fail to Reject H_o ∴ Fail to Reject H_o

The p-value is greater than 0.100.
= FDIST(1.1630,134,145) × 2 = 0.3722

11.5 a. H_0: Ages are independent of source of market information.
H_a: Ages are not independent of source of market information.

Reject H_0 if $F_c > 3.204$ [3.195] ($df_n = 2$, $df_d = 47$; $\alpha = 0.05$)

Reject H_0 if $F_c > 5.110$ [5.087] ($df_n = 2$, $df_d = 47$; $\alpha = 0.01$)

	Newspaper	Radio	Other	All
Mean	35.933	31.000	31.400	32.600
Standard Deviation	12.708	13.650	12.620	
Count	15	20	15	
Sample	$\overline{X}$	$\overline{X} - \overline{X}_p$	$\left(\overline{X} - \overline{X}_p\right)^2$	$n \times = \left(\overline{X} - \overline{X}_p\right)^2$
Newspaper	35.933	3.333	11.111	166.667
Radio	31.000	−1.600	2.560	51.200
Other	31.400	−1.200	1.440	21.600
			BSS =	239.467
				2260.933
				3540.000
				2229.600
			WSS =	8030.533

$$F_c = \frac{\frac{239.467}{2}}{\frac{8030.533}{47}} = \frac{119.733}{170.862} = 0.7008$$

∴ Fail to Reject H_0 ∴ Fail to Reject H_0

The p-value is greater than 0.100.
=FDIST(0.7008,2,47) = 0.5013.

$$\text{BSS} = 239.467$$
$$\text{WSS} = \underline{8030.533}$$
$$\text{TSS} = 8270.000$$

2.90% of the variation in age (BSS/TSS) can be explained by source.

b. H_0: Incomes are independent of source of market information.
H_a: Incomes are not independent of source of market information.

Reject H_0 if $F_c > 3.204$ [3.195] ($df_n = 2$, $df_d = 47$; $\alpha = 0.05$)

Reject H_0 if $F_c > 5.110$ [5.087] ($df_n = 2$, $df_d = 47$; $\alpha = 0.01$)

	Newspaper	Radio	Other	All
Mean	46.947	35.520	36.600	39.272
Standard Deviation	19.562	21.799	18.343	
Count	15	20	15	
Sample	$\overline{X}$	$\overline{X}-\overline{X}_p$	$\left(\overline{X}-\overline{X}_p\right)^2$	$n \times = \left(\overline{X}-\overline{X}_p\right)^2$
Newspaper	46.947	7.675	58.901	883.508
Radio	35.520	−3.752	14.078	281.550
Other	36.600	−2.672	7.140	107.094
			BSS =	1272.151
				5357.317
				9028.592
				4710.320
			WSS =	19096.229

$$F_c = \frac{\frac{1272.151}{2}}{\frac{19096.229}{47}} = \frac{636.076}{406.303} = 1.5655$$

∴ Fail to Reject H_o ∴ Fail to Reject H_o
The p-value is greater than 0.100.
=FDIST(1.5655,2,47) = 0.2197.

$$BSS = 1272.151$$

$$WSS = \underline{19096.229}$$

$$TSS = 20368.381$$

6.25% of the variation in income (BSS/TSS) can be explained by source.

11.7 a. H_o: Salaries are independent of job type.
H_a: Salaries are not independent of job type.
Reject H_o if $F_c > 2.425$ [2.419] ($df_n = 2$, $df_d = 47$; $\alpha = 0.10$)
Reject H_o if $F > 3.204$ [3.195] ($df_n = 2$, $df_d = 47$; $\alpha = 0.05$)

	Line	Office	Manage	All
Mean	41.384	38.533	50.070	42.266
Standard Deviation	8.733	9.313	8.359	
Count	25	15	10	
Sample	$\bar{X}$	$\bar{X} - \bar{X}_p$	$\left(\bar{X} - \bar{X}_p\right)^2$	$n \times = \left(\bar{X} - \bar{X}_p\right)^2$
Line	41.384	−0.882	0.778	19.448
Office	38.533	−3.733	13.933	208.992
Manage	50.070	7.804	60.902	609.024
			BSS =	837.464
				1830.494
				1214.213
				628.841
			WSS =	3673.548

$$F_c = \frac{\frac{837.464}{2}}{\frac{3673.548}{47}} = \frac{418.732}{78.161} = 5.3573$$

∴ Reject H_o ∴ Reject H_o

The p-value is between 0.010 and 0.005.

=FDIST(5.3573, 2,47) = 0.0080.

$$\text{BSS} = 837.464$$
$$\text{WSS} = \underline{3673.548}$$
$$\text{TSS} = 4511.012$$

18.56% of the variation in salary (BSS/TSS) can be explained by job type.

11.9 a. H_o: $\sigma_{\text{Female}} = \sigma_{\text{Male}}$ (Standard deviations in heights are equal.)

H_a: $\sigma_{\text{Female}} \neq \sigma_{\text{Male}}$ (Standard deviations in heights differ.)

Reject H_o if $F_c > 1.796$ [1.755] ($df_n = 46$, $df_d = 52$; $\alpha = 0.05$)

Reject H_o if $F_c > 2.164$ [2.101] ($df_n = 46$, $df_d = 52$; $\alpha = 0.01$)

$$F_c = \frac{3.944^2}{3.800^2} = 1.0773$$

∴ Fail to Reject H_o ∴ Fail to Reject H_o

The p-value is greater than $0.100 \times 2 = 0.200$.

=FDIST(1.0773,46,52) × 2 = 0.7911.

b. H_o: $\sigma_{\text{Female}} = \sigma_{\text{Male}}$ (Standard deviations in weights are equal.)

H_a: $\sigma_{Female} \neq \sigma_{Male}$ (Standard deviations in weights differ.)

Reject H_o if $F_c > 1.796$ [1.755] ($df_n = 46$, $df_d = 52$; $\alpha = 0.05$)	Reject H_o if $F_c > 2.164$ [2.101] ($df_n = 46$, $df_d = 52$; $\alpha = 0.01$)

$$F_c = \frac{17.772^2}{14.038^2} = 1.6028$$

∴ Fail to Reject H_o	∴ Fail to Reject H_o

The p-value is almost exactly 0.100.

=FDIST(1.6028,46,52) × 2 = 0.0998.

c. H_o: $\sigma_{Female} = \sigma_{Male}$ (Standard deviations in aid are equal.)

H_a: $\sigma_{Female} \neq \sigma_{Male}$ (Standard deviations in aid differ.)

Reject H_o if $F_c > 1.757$ [1.773] ($df_n = 52$, $df_d = 46$; $\alpha = 0.05$)	Reject H_o if $F_c > 2.109$ [2.132] ($df_n = 52$, $df_d = 46$; $\alpha = 0.01$)

$$F_c = \frac{2607.80^2}{2601.32^2} = 1.0050$$

∴ Fail to Reject H_o	∴ Fail to Reject H_o

The p-value is greater than 0.100 × 2 = 0.200.

=FDIST(1.0050,52,46) × 2 = 0.9909.

d. H_o: $\sigma_{Female} = \sigma_{Male}$ (Standard deviations in entertainment expenditures are equal.)

H_a: $\sigma_{Female} \neq \sigma_{Male}$ (Standard deviations in entertainment expenditures differ.)

Reject H_o if $F_c > 1.757$ [1.773] ($df_n = 52$, $df_d = 46$; $\alpha = 0.05$)	Reject H_o if $F_c > 2.109$ [2.132] ($df_n = 52$, $df_d = 46$; $\alpha = 0.01$)

$$F_c = \frac{14.838^2}{14.423^2} = 1.0585$$

∴ Fail to Reject H_o	∴ Fail to Reject H_o

The p-value is greater than 0.100 × 2 = 0.200.

=FDIST (1.0585,52,46) × 2 = 0.8482.

e. H_o: $\sigma_{Female} = \sigma_{Male}$ (Standard deviations in study time are equal.)

H_a: $\sigma_{Female} \neq \sigma_{Male}$ (Standard deviations in study time differ.)

Reject H_o if $F_c > 1.796$ [1.755] ($df_n = 46$, $df_d = 52$; $\alpha = 0.05$)	Reject H_o if $F_c > 2.164$ [2.101] ($df_n = 46$, $df_d = 52$; $\alpha = 0.01$)

$$F_c = \frac{6.5838^2}{5.4946^2} = 1.4357$$

∴ Fail to Reject H_o	∴ Fail to Reject H_o

The p-value is greater than $0.100 \times 2 = 0.200$.
=FDIST(1.4357,46,52) × 2 = 0.2062.

f. H_o: $\sigma_{Female} = \sigma_{Male}$ (Standard deviations in college GPAs are equal.)
H_a: $\sigma_{Female} \neq \sigma_{Male}$ (Standard deviations in college GPAs differ.)

Reject H_o if $F_c > 1.796$ [1.755]	Reject H_o if F > 2.164 [2.101]
($df_n = 46$, $df_d = 52$; $\alpha = 0.05$)	($df_n = 46$, $df_d = 52$; $\alpha = 0.01$)

$$F_c = \frac{0.6096^2}{0.5166^2} = 1.3925$$

∴ Fail to Reject H_o	∴ Fail to Reject H_o

The p-value is greater than $0.100 \times 2 = 0.200$.
=FDIST(1.3925,46,52) × 2 = 0.2469.

Chapter 12: Simple Regression and Correlation

All rejection criteria are taken from the tables. For t and F, when the table doesn't list the exact degrees of freedom, the closest degrees of freedom is used. However, if the Excel value with the exact degrees of freedom differs, it is also reported [in brackets].

12.1 a.

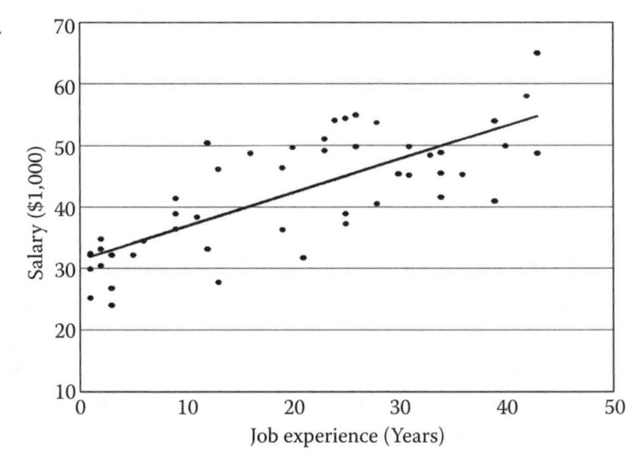

b. $\bar{X} = 19.960$ $\quad SSD_x = 8729.9200$ $\quad b = 0.5446$

$\bar{Y} = 42.266$ $\quad SSD_Y = 4511.0122$ $\quad a = 31.3967$

$n = 50$ $\quad SCD_{XY} = 4753.9320$

c. 31.3967 is our estimate for the salary of an employee with no job experience.
0.5446 is our estimate for the increase in salary with each additional year of job experience.

d. SSE = 1922.2291 $\quad$ MSE = 40.046 $\quad s_e = 6.3282$

e. H_o: Salaries are independent of job experience.
H_a: Salaries are not independent of job experience.
Reject H_o if $F_c > 4.034$ [4.043] ($df_n = 1$, $df_d = 48$; $\alpha = 0.05$).

ANOVA	Sum of Squares	df	Mean Square	F_c
Regression	2588.7831	1	2588.7831	64.6445
Error	1922.2291	48	40.0464	
Total	4511.0122	49		

$F_c = 64.6445$
$\therefore$ Reject H_o
The p-value of the test is less than 0.005. = FDIST(64.64 45,1,48) = 1.91E-10. There is almost no chance that an F_c this large would have arisen randomly.

f. $R^2 = 2588.7831/4511.0122 = 0.5739$.
57.30% of the variation in salary can be explained by experience.

g. Two-Tailed Test:
H_o: $\beta = 0$
H_a: $\beta \neq 0$
Reject H_o if $|t_c| > 2.009$ [2.011]
($df = 48$, $\alpha = 0.05$)

One-Tailed Test:
H_o: $\beta = 0$
H_a: $\beta > 0$
Reject H_o if $t_c > 1.676$ [1.677]
($df = 48$, $\alpha = 0.05$)

$$s_b = \frac{6.3282}{\sqrt{8729.920}} = 0.0677$$

$$t_c = \frac{0.5446 - 0}{0.0677} = 8.0402$$

$\therefore$ Reject H_o (two-tailed) $\therefore$ Reject H_o (one-tailed)

The p-value of the two-tailed test is less than $0.0005 \times 2 = 0.001$.
=TDIST(8.0402,48) = 1.91E-10, the same as that for the F test. That for the one-tailed test is half of that. The true population slope is almost certainly not zero.

h. and i. The predicted Salary for the mean or individual with 10 years of job experience is:

Salary = 31.3967 + 0.5446 (10) = 36.842.

The standard errors are:

$$s_{\bar{Y}|X} = 6.3282\sqrt{0 + \frac{1}{50} + \frac{(10 - 19.96)^2}{8729.92}} = 6.3282\sqrt{0.0314} = 1.121$$

$$s_{\hat{Y}}/_X = 6.3282\sqrt{1 + \frac{1}{50} + \frac{(10 - 19.96)^2}{8729.92}} = 6.3282\sqrt{1.0314} = 6.427$$

The 95% CI for the mean salary of employees with 10 years of job experience:
95% CI: $36.842 \pm 2.009 \times 1.121$
95% CI: 36.842 ± 2.252

The 95% CI for the salary of an individual employee with 10 years of job experience:
95% CI: $36.842 \pm 2.009 \times 6.427$
95% CI: 36.842 ± 12.911

As always, we are a lot more certain about the mean value along the regression line than we are about individual cases, which are scattered about the regression line even in the population.

12.3 a.

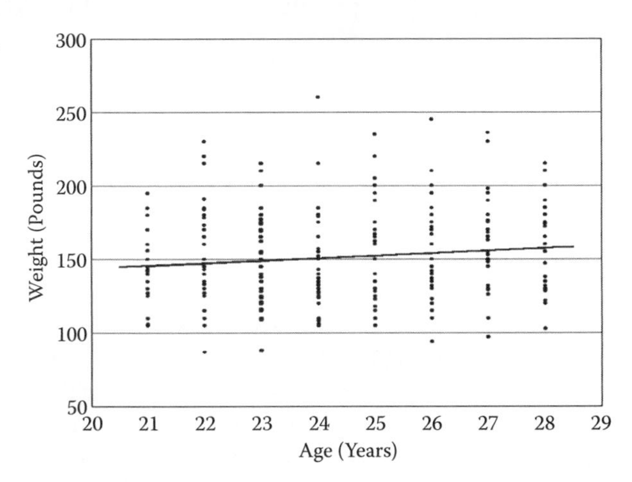

b.

$\bar{X} = 24.3345$ $\quad$ $\text{SSD}_x = 1388.5552$ $\quad$ $b = 1.7236$

$\bar{Y} = 151.0925$ $\quad$ $\text{SSD}_Y = 261225.5943$ $\quad$ $a = 109.1497$

$n = 281$ $\quad$ $\text{SCD}_{XY} = 2393.3025$

c. 109.1497 is our estimate for the weight of a newborn baby. This is a very poor estimate because it is based solely on adults, whose weights have (almost) stabilized.
1.7236 is our estimate for the increase in weight with each additional year of age.

d. SSE = 257100.5174 MSE = 921.5072 $s_e = 30.3563$

e. H_0: Weights of young adults are independent of age.
H_a: Weights of young adults are not independent of age.
Reject H_0 if $F_c > 3.841$ [3.875] ($\text{df}_n = 1$, $\text{df}_d = 279$; $\alpha = 0.05$).

ANOVA	Sum of Squares	df	Mean Square	F_c
Regression	4125.0769	1	4125.0769	4.4767
Error	257100.5174	279	921.5072	
Total	261225.5943	280		

$F_c = 4.4764$ $\therefore$ Reject H_0
The p-value of the test is between 0.050 and 0.025.
=FDIST(4.4764,1,279) = 0.0353.

f. $R^2 = 4125.0769/261225.5943 = 0.0158$.
Only 1.58% of the variation in the weights of young adults can be explained by age.

g. Two-Tailed Test:
H_o: $\beta = 0$
H_a: $\beta \neq 0$
Reject H_o if $|t_c| > 1.960$ [1.969]
(df = 279, $\alpha = 0.05$)

One-Tailed Test:
H_o: $\beta = 0$
H_a: $\beta > 0$
Reject H_o if $t_c > 1.645$ [1.650]
(df = 279, $\alpha = 0.05$)

$$s_b = \frac{30.3563}{\sqrt{1388.5552}} = 0.8146$$

$$t_c = \frac{1.7236 - 0}{0.8146} = 2.1158$$

$\therefore$ Reject H_o $\qquad$ $\therefore$ Reject H_o

The p-value of the two-tailed test is between $0.0250 \times 2 = 0.05$ and $0.0100 \times 2 = 0.02$.
=TDIST(2.1158,279,2) = 0.0353, the same as that for the F test. That for the one-tailed test is half of that. While age explains very little of the variation in the weights of young adults, they do seem to be gaining a little additional weight with age.

h. and i. The predicted Weight for the mean or individual who is 25 years old is

Weight = 109.1497 + 1.7236 (25) = 152.240.

The standard errors are

$$s_{\bar{Y}|X} = 30.36\sqrt{0 + \frac{1}{281} + \frac{(25 - 24.33)^2}{1388.56}} = 30.36\sqrt{0.0039} = 1.890$$

$$s_{\hat{Y}|X} = 30.36\sqrt{1 + \frac{1}{281} + \frac{(25 - 24.33)^2}{1388.56}} = 3036\sqrt{1.0039} = 30.415$$

The 95% CI for the mean weight of 25-year-old young adults:
95% CI: 152.240 ± 1.960 × 1.890
95% CI: 152.240 ± 3.705

The 95% CI for the weight of an individual 25-year-old young adult:
95% CI: 152.240 ± 1.960 × 30.415
95% CI: 152.240 ± 59.614

As always, we are a lot more certain about the mean value along the regression line than we are about individual cases, which are scattered about the regression line even in the population.

12.5 a.

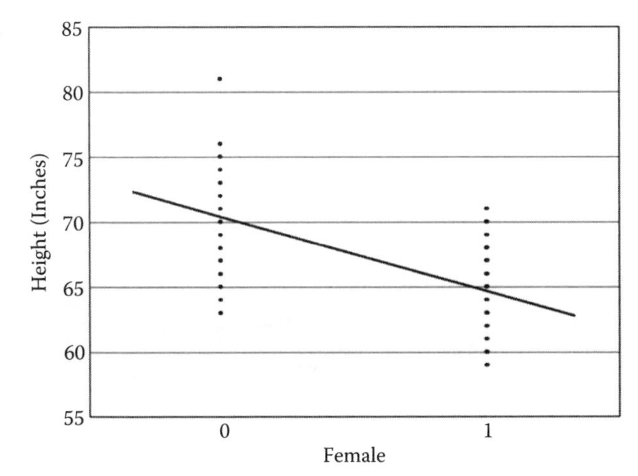

b.

$\bar{X} = 0.5196$ $SSD_x = 70.1423$ $b = -5.7299$

$\bar{Y} = 67.4377$ $SSD_Y = 4353.1601$ $a = 70.4148$

$n = 281$ $SCD_{XY} = -401.9075$

c. 70.4148 is our estimate for the height of a young adult male.
–5.7299 is our estimate for the difference in height of a young adult female; hence $70.4148 - 5.7299 \times 1 = 64.6849$ is our estimate for the height of a young adult female.

d. SSE = 2050.2772 MSE = 7.3487 $s_e = 2.7108$

e. H_o: Heights of young adults are independent of sex.
H_a: Heights of young adults are not independent of sex.
Reject H_o if $F_c > 3.841$ [3.875] ($df_n = 1$, $df_d = 279$; $\alpha = 0.05$).

ANOVA	Sum of Squares	df	Mean Square	F_c
Regression	2302.8829	1	2302.8829	313.3744
Error	2050.2772	279	7.3487	
Total	4353.1601	280		

$F_c = 313.3744$ ∴ Reject H_o
The p-value of the test is much less than 0.005.
=FDIST(313.3744,1,279) = 1.59E-47.

f. $R^2 = 2302.8829/4353.1601 = 0.5290$.
Almost 53% of the variation in the heights of young adults can be explained by sex.

g. Two-Tailed Test:
H_o: $\beta = 0$
H_a: $\beta \neq 0$
Reject Ho if $|t_c| > 1.960$
[1.969]
(df = 279, $\alpha = 0.05$)

One-Tailed Test:
H_o: $\beta = 0$
H_a: $\beta < 0$
Reject Ho if tc < -1.645
[−1.650]
(df = 279, $\alpha = 0.05$)

$$s_b = \frac{2.7108}{\sqrt{70.1423}} = 0.3237$$

$$t_c = \frac{-5.730 - 0}{0.3237} = -17.7024$$

$\therefore$ Reject H_o $\qquad\qquad$ $\therefore$ Reject H_o

The p-value of the two-tailed test is less than $0.0005 \times 2 = 0.001$.
=TDIST(17.7024,279) = 1.59E-47, the same as that for the F test. That for the one-tailed test is half of that. There is almost no chance that this difference in heights is just random.

12.7 a.

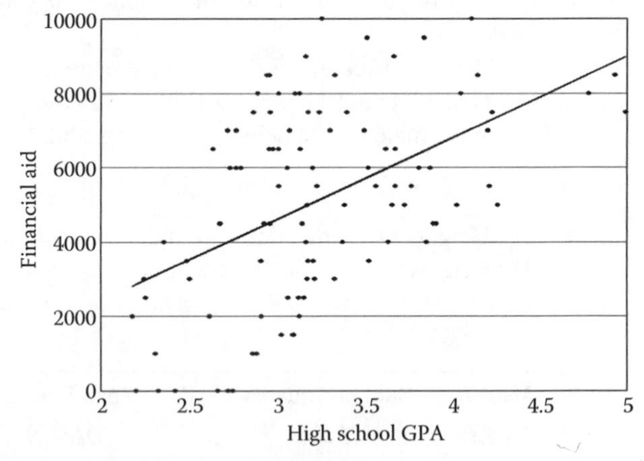

b.

$\bar{X} = 3.229$	$SSD_x = 32.926$	$b = 2189.73$
$\bar{Y} = 5115$	$SSD_Y = 665427500$	$a = -1955.89$
$n = 100$	$SCD_{XY} = 72100.24$	

c. −1955.89 is our estimate for the aid to a student with a zero HS_GPA. Of course, there are no such students.
2189.73 is our estimate for the increase in aid with an additional point of HS_GPA.

d. SSE = 507547244 $\qquad$ MSE = 5179054 $\qquad$ $s_e = 2275.75$

e. H_o: Financial aid is independent of HS_GPA.
H_a: Financial aid is not independent of HS_GPA.

Reject H_o if $F_c > 3.936$ [3.938] ($df_n = 1$, $df_d = 98$; $\alpha = 0.05$).

ANOVA	Sum of Squares	df	Mean Square	F_c
Regression	157880256	1	157880256	30.484
Error	507547224	98	5179054	
Total	665427500	99		

$F_c = 30.484$
$\therefore$ Reject H_o
The p-value of the test is much less than 0.005.
=FDIST (30.484,1,98) = 2.761E-07.

f. $R^2 = 157880256/665427500 = 0.2373$.
Almost 24% of the variation in the financial aid can be explained by HS_GPA.

g. Two-Tailed Test:
H_o: $\beta = 0$
H_a: $\beta \neq 0$
Reject H_o if $|t_c| > 1.984$ (df = 98, $\alpha = 0.05$)

One-Tailed Test:
H_o: $\beta = 0$
H_a: $\beta > 0$
Reject H_o if $t_c > 1.660$ [1.661] (df = 98, $\alpha = 0.05$)

$$s_b = \frac{2275.75}{\sqrt{32.926}} = 396.600$$

$$t_c = \frac{2189.73}{396.600} = 5.521$$

$\therefore$ Reject H_o (two-tailed) $\therefore$ Reject H_o (one-tailed)

The p-value of the two-tailed test is less than $0.0005 \times 2 = 0.001$.
=TDIST(5.521,98) = 2.761E-07, the same as that for the F test. That for the one-tailed test is half of that. The true population slope is almost certainly not zero.

h. and i. The predicted Aid for the mean or individual with a 4.000 is

Aid = −1955.89 + 2189.73 (4.000) = 6803.04.

The standard errors are

$$s_{\bar{Y}|X} = 2275.75\sqrt{0 + \frac{1}{100} + \frac{(4-3.229)^2}{32.926}} = 2275.75\sqrt{0.167} = 381.136$$

$$s_{\hat{Y}|X} = 2275.75\sqrt{1 + \frac{1}{100} + \frac{(4-3.229)^2}{32.926}} = 2275.75\sqrt{1.014} = 2307.448$$

The 95% CI for mean Aid for students with a 4.000 HS_GPA:

95% CI: $6803.04 \pm 1.984 \times 381.136$

95% CI: 6803.04 ± 756.35

The 95% CI for Aid for an individual student with a 4.000 HS_GPA:

95% CI: $6803.043 \pm 1.984 \times 2307$

95% CI: 6803.043 ± 4579.06

As always, we are a lot more certain about the mean value along the regression line than we are about individual cases, which are scattered about the regression line.

12.9 a.

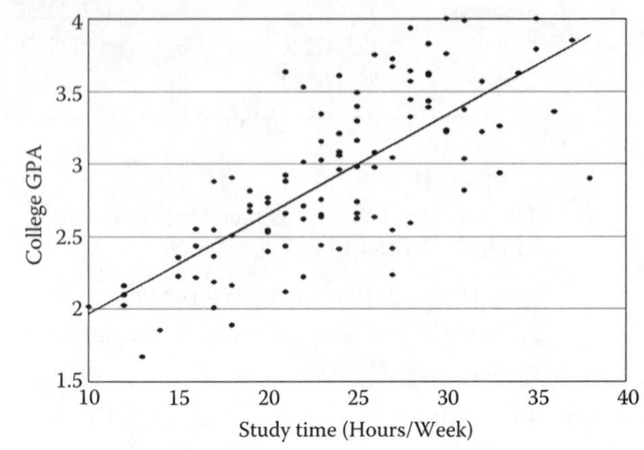

b.

$\bar{X} = 24.060$	$SSD_x = 3671.640$	$b = 0.0686$
$\bar{Y} = 2.933$	$SSD_Y = 31.879$	$a = 1.2838$
$n = 100$	$SCD_{XY} = 251.708$	

c. 1.2838 is our estimate for the Col_GPA of a student who studies zero hours.
0.0686 is our estimate for the increase in Col_GPA with an additional hour of study.

d. SSE = 14.623 MSE = 0.149 $s_e = 0.386$

e. H_0: College GPA is independent of hours of study.
H_a: College GPA is not independent of hours of study.
Reject H_0 if $F_c > 3.936$ [3.938] ($df_n = 1$, $df_d = 98$; $\alpha = 0.05$)

ANOVA	Sum of Squares	df	Mean Square	F_c
Regression	17.256	1	17.256	115.644
Error	14.623	98	0.149	
Total	31.879	99		

$F_c = 115.6442$
∴ Reject H_0
The p-value of the test is much less than 0.005.
=FDIST (115.644,1,98) = 2819E-18.

f. $R^2 = 17.256/31.879 = 0.5413$.
About 54% of the variation in college GPA can be explained by hours of study.

g. Two-Tailed Test:

H_o: $\beta = 0$

H_a: $\beta \neq 0$

Reject H_o if $|t_c| > 1.984$ (df = 98, $\alpha = 0.05$)

One-Tailed Test:

H_o: $\beta = 0$

H_a: $\beta > 0$

Reject H_o if $t_c > 1.660$ [1.661] (df = 98, $\alpha = 0.05$)

$$s_b = \frac{0.386}{\sqrt{3671.640}} = 0.0064$$

$$t_c = \frac{1.2838}{0.0064} = 10.7538$$

$\therefore$ Reject H_o $\qquad$ $\therefore$ Reject H_o

The p-value of the two-tailed test is less than $0.0005 \times 2 = 0.001$.

=TDIST(10.7538,98) = 2.819E-18, the same as that for the F test. That for the one-tailed test is half of that. The true population slope is almost certainly not zero.

h. and i. The predicted COL_GPA for the mean or individual who studies 30 hours is

COL_GPA = 1.2838 + 0.0686 (30) = 3.340

The standard errors are

$$s_{\bar{Y}|X} = 0.386\sqrt{0 + \frac{1}{100} + \frac{(30 - 24.060)^2}{3671.640}} = 0.386\sqrt{0.140} = 0.054$$

$$s_{\hat{Y}|X} = 0.386\sqrt{1 + \frac{1}{100} + \frac{(30 - 24.060)^2}{3671.640}} = 0.386\sqrt{1.010} = 0.390$$

The 95% CI for mean COL_GPA for students who studies 40 hours:

95% CI: $3.340 \pm 1.984 \times 0.054$

95% CI: 3.340 ± 0.107

The 95% CI for COL_GPA for an individual who studies 40 hours:

95% CI: $3.340 \pm 1.984 \times 0.390$

95% CI: 3.340 ± 0.774

As always, we are a lot more certain about the mean value along the regression line than we are about individual cases, which are scattered about the regression line even in the population.

Chapter 13: Multiple Regression

All rejection criteria are taken from the tables. For t and F, when the table doesn't list the exact degrees of freedom, the closest degrees of freedom is used. However, if the Excel value with the exact degrees of freedom differs, it is also reported [in brackets].

13.1 a.

Multiple Regression

Dependent Variable: Weight

***R* Square:** 0.8486			**Adjusted *R* Square:** 0.8455		**Standard Error:** 11.4103
ANOVA	**Sum of Squares**	**df**	**Mean Square**	F_c	**2-Tailed *p*-Value**
Regression	70784.4981	2	35392.2490	271.8408	0.0000
Error	12628.8919	97	130.1948		
Total	83413.3900	99			

Variable	Coefficient	Standard Error	t_c	2-Tailed *p*-Value
Constant	−34.5814	21.0801	−1.6405	0.1041
Height	2.8779	0.2979	9.6594	0.0000
Female	−32.6513	2.8151	−11.5988	0.0000

b. –34.5814 is the estimated weight of a zero inch tall male student.
2.8779 is the estimated increase in weight with each additional inch in height, sex held constant.
–32.6513 is the estimated decrease in weight if the student is female, height held constant.

c. For Height:

Two-Tailed Test:
H_o: $\beta = 0$
H_a: $\beta \neq 0$
Reject H_o if $|t_c| > 1.984$ [1.985]
(df = 97, $\alpha = 0.05$)
Reject H_o if $|t_c| > 2.626$ [2.627]
(df = 97, $\alpha = 0.01$)

One-Tailed Test:
H_o: $\beta = 0$
H_a: $\beta > 0$
Reject H_o if $t_c > 1.660$ [1.661]
(df = 97, $\alpha = 0.05$)
Reject H_o if $t_c > 2.364$ [2.365]
(df = 97, $\alpha = 0.01$)

$t_c = 9.6594$

∴ Reject H_o (either α)   ∴ Reject H_o (either α)

For Female:

Two-Tailed Test:
H_o: $\beta = 0$
H_a: $\beta \neq 0$
Reject H_o if $|t_c| > 1.984$ [1.985]
(df = 97, $\alpha = 0.05$)
Reject H_o if $|t_c| > 2.626$ [2.627]
(df = 97, $\alpha = 0.01$)

One-Tailed Test:
H_o: $\beta = 0$
H_a: $\beta < 0$
Reject H_o if $t_c > -1.660$ [1.661]
(df = 97, $\alpha = 0.05$)
Reject H_o if $t_c < -2.364$ [−2.365]
(df = 97, $\alpha = 0.01$)

$t_c = -11.5988$

∴ Reject H_0 (either α) ∴ Reject H_0 (either α)

p-values for both are zero to four decimals; there is almost no chance that these slopes are just randomly different from zero.

d. i

Multiple Regression

Dependent Variable: Weight

***R* Square:** 0.8528		**Adjusted *R* Square:** 0.8482		**Standard Error:** 11.3095	
ANOVA	**Sum of Squares**	**df**	**Mean Square**	F_c	**2-Tailed *p*-Value**
Regression	71134.6006	3	23711.5335	185.3853	0.0000
Error	12278.7894	96	127.9041		
Total	83413.3900	99			
Variable	**Coefficient**	**Standard Error**	t_c	**2-Tailed *p*-Value**	
Constant	309.8544	209.2325	1.4809	0.1419	
Height	−7.3669	6.1993	−1.1883	0.2376	
Height2	0.0758	0.0458	1.6545	0.1013	
Female	−32.4195	2.7937	−11.6045	0.0000	

Adding Height2 allows the effect of height to be nonlinear. (The effect of the dummy variable for Female cannot be nonlinear.)

The Height2 coefficient has a p-value of 0.1013, which is not significant. It *is* better than that for Height though; using Height2 instead of Height might be an improvement.

d. ii

Multiple Regression

Dependent Variable: Weight

***R* Square:** 0.8506		**Adjusted *R* Square:** 0.8476		**Standard Error:** 11.3335	
ANOVA	**Sum of Squares**	**df**	**Mean Square**	F_c	**2-Tailed *p*-Value**
Regression	70953.9795	2	35476.9897	276.1983	0.0000
Error	12459.4105	97	128.4475		
Total	83413.3900	99			
Variable	**Coefficient**	**Standard Error**	t_c	**2-Tailed *p*-Value**	
Constant	61.5589	11.0355	5.5783	0.0000	
Height2	0.0214	0.0022	9.7925	0.0000	
Female	−32.4937	2.7989	−11.6093	0.0000	

The Height2 coefficient has a slightly higher t_c statistic than the Height coefficient had in part a, and the $\bar{R}^2$ is slightly higher. It is not clear that this improvement is worth the added complexity, though, given that it is so small and given that there is no obvious reason to think the effect *should* be nonlinear.

The Cobb–Douglas (log–log) form offers another nonlinear form. (We cannot take the log of the dummy variable for Female.)

d. iii

Multiple Regression					
Dependent Variable: lnWeight					
R Square: 0.8474		Adjusted *R* Square: 0.8443		Standard Error: 0.0802	
ANOVA	**Sum of Squares**	**df**	**Mean Square**	F_c	**2-Tailed *p*-Value**
Regression	3.4636	2	1.7318	269.3733	0.0000
Error	0.6236	97	0.0064		
Total	4.0873	99			
Variable	**Coefficient**	**Standard Error**	t_c	**2-Tailed *p*-Value**	
Constant	–0.6036	0.5992	–1.0074	0.3162	
lnHeight	1.3454	0.1408	9.5548	0.0000	
Female	–0.2302	0.0197	–11.6616	0.0000	

The log–log form seems a slightly worse fit than the others (though remember, with different dependent variables, the $\bar{R}^2$ s are not strictly comparable).

In short, the third regression—with Height2 and Female—*is* the best fit by a very small amount, suggesting a very slight curvature. But with (i) such a small advantage in fit, and (ii) no obvious reason to think the effect of height *should* be nonlinear, we might well stick with the simpler to interpret, linear form.

e. If we stick with the linear form, we can explain 84.55% of the variation in Weight.

f. Weight = –34.5814 + 2.8779 × 68 – 32.6513 × 0 = 161.1158 or about 161 lbs.

g. Weight = – 34.5814 + 2.8779 × 68 – 32.6513 × 1 = 128.4645 or about 128 lbs.

h. For males:
95% CI: 161.1158 ± 1.984 × 11.4103 [1.985]
95% CI: 161.1158 ± 22.6463

For females:
95% CI: 128.4645 ± 1.984 × 11.4103 [1.985]
95% CI: 128.4645 ± 22.6463

13.3 a.

Multiple Regression					
Dependent Variable: Col_GPA					
***R* Square:** 0.6930		**Adjusted *R* Square:** 0.6867		**Standard Error:** 0.3176	
ANOVA	**Sum of Squares**	**df**	**Mean Square**	**F_c**	**2-Tailed *p*-Value**
Regression	22.0933	2	11.0466	109.5021	0.0000
Error	9.7854	97	0.1009		
Total	31.8787	99			
Variable	**Coefficient**	**Standard Error**	**t_c**	**2-Tailed *p*-Value**	
Constant	0.2865	0.1940	1.4766	0.1430	
HS_GPA	0.4085	0.0590	6.9248	0.0000	
Study	0.0552	0.0056	9.8770	0.0000	

b. 0.2865 is the estimated college GPA of a student with a zero high school GPA who does not study.
0.4085 is the estimated increase in college GPA with each additional point of high school GPA, study time held constant.
0.0552 is the estimated increase in college GPA with each additional hour of study, high school GPA held constant.

c. Two-Tailed Test:
H_o: $\beta = 0$
H_a: $\beta \neq 0$
Reject H_o if $|t_c| > 1.984$ [1.985]
(df = 97, $\alpha = 0.05$)
Reject H_o if $|t_c| > 2.626$ [2.627]
(df = 97, $\alpha = 0.01$)

One-Tailed Test:
H_o: $\beta = 0$
H_a: $\beta > 0$
Reject H_o if $t_c > 1.660$ [1.661]
(df = 97, $\alpha = 0.05$)
Reject H_o if $t_c > 2.364$ [2.365]
(df = 97, $\alpha = 0.01$)

For HS_GPA:

$$t_c = 6.9248$$

∴ Reject H_o (either α) ∴ Reject H_o (either α)

For Study:

$$t_c = 9.8770$$

∴ Reject H_o (either α) ∴ Reject H_o (either α)

p-values for both are zero to four decimals; there is almost no chance that these slopes are just randomly different from zero.

d. i

Multiple Regression
Dependent Variable: Col_GPA

***R* Square:**	**Adjusted *R* Square:**	**Standard Error:**
0.7011	**0.6885**	**0.3167**

ANOVA	**Sum of Squares**	**df**	**Mean Square**	**F_c**	**2-Tailed *p*-Value**
Regression	22.3508	4	5.5877	55.7129	0.0000
Error	9.5280	95	0.1003		
Total	31.8787	99			

Variable	**Coefficient**	**Standard Error**	**t_c**	**2-Tailed *p*-Value**
Constant	−0.3043	0.7982	−0.3813	0.7039
HS_GPA	0.4089	0.4697	0.8707	0.3861
HS_GPA2	−0.0017	0.0674	−0.0257	0.9796
Study	0.1090	0.0350	3.1154	0.0024
Study2	−0.0011	0.0007	−1.5635	0.1213

"Diminishing returns" means that the positive effects of the explanatory variables get smaller as the explanatory variables get larger. This implies positive coefficients for the unsquared variables and negative coefficients for the squared ones (see Figure 13.9 on page 312). The signs we get are right. Moreover, note that the reported *p*-values are for two-tailed tests; since we are predicting the signs we can divide those *p*-values in half. The *p*-value for HS_GPA2, even divided in half, is very high. This could too easily have arisen just randomly. But those for Study and Study2 are promising. Perhaps, by removing HS_GPA2, the results for the rest will improve.

d. ii

Multiple Regression
Dependent Variable: Col_GPA

***R* Square:**	**Adjusted *R* Square:**	**Standard Error:**
0.7011	**0.6918**	**0.3150**

ANOVA	**Sum of Squares**	**df**	**Mean Square**	**F_c**	**2-Tailed *p*-Value**
Regression	22.3507	3	7.4502	75.0651	0.0000
Error	9.5280	96	0.0993		
Total	31.8787	99			

Variable	**Coefficient**	**Standard Error**	**t_c**	**2-Tailed *p*-Value**
Constant	−0.2867	0.4046	−0.7085	0.4804
HS_GPA	0.3970	0.0589	6.7340	0.0000
Study	0.1091	0.0340	3.2133	0.0018
Study2	−0.0011	0.0007	−1.6104	0.1106

Again, the signs are right. The coefficient for Study2 has a one-tailed p-value of 0.1106/2 = 0.0553, so we still cannot quite reject the H_0 at the 0.05 level. Still, since it is very close and because it makes sense that studying is subject to diminishing returns, we might accept this as an improvement. The Cobb–Douglas (log–log) form offers another nonlinear form.

d. iii

Multiple Regression

Dependent Variable: lnCol_GPA

***R* Square:**	***Adjusted R* Square:**	**Standard Error:**
0.7048	0.6987	0.1098

ANOVA	Sum of Squares	df	Mean Square	F_c	2-Tailed p-Value
Regression	2.7917	2	1.3958	115.8142	0.0000
Error	1.1691	97	0.0121		
Total	3.9607	99			

Variable	Coefficient	Standard Error	t_c	2-Tailed p-Value
Constant	−0.8694	0.1301	−6.6812	0.0000
lnHS_GPA	0.4387	0.0680	6.4466	0.0000
lnStudy	0.4510	0.0433	10.4052	0.0000

The $\bar{R}^2$ is the highest so far (though remember, with different dependent variables, the $\bar{R}^2$ s are not strictly comparable). And with the coefficients between zero and one, the effects of both high school GPA and study time do show diminishing returns (see Figure 13.12 on page 315).

e. If we decide that the Cobb–Douglas is the best, we can explain 69.87% of the variation in lnCol_GPA.

f. lnCol_GPA = −0.8694 + 0.4387 × ln(4.000) + 0.4510 × (30) = 1.2727.
Col_GPA = exp(1.2727) =3.571.

g. 95% CI: 1.2727 ± 1.984 × 1.098 [1.985]
95% CI: 1.2727 ± .2179
95% CI: 1.0548 < lnCol_GPA < 1.4906
95% CI: 2.8714 < Col_GPA < 4.4399

13.5 a.

Multiple Regression					
Dependent Variable: Study					
R Square: 0.3198		Adjusted *R* Square: 0.2077		Standard Error: 5.4206	
ANOVA	**Sum of Squares**	**df**	**Mean Square**	F_c	**2-Tailed *p*-Value**
Regression	1174.0637	14	83.8617	2.8541	0.0015
Error	2497.5763	85	29.3833		
Total	3671.6400	99			
Variable	**Coefficient**	**Standard Error**	t_c	**2-Tailed *p*-Value**	
Constant	24.4994	2.3100	10.6057	0.0000	
Female	2.0667	1.1462	1.8032	0.0749	
Job	−3.0258	1.1908	−2.5410	0.0129	
Sports	−3.0942	1.3621	−2.2716	0.0256	
Music	−3.2064	1.5242	−2.1037	0.0384	
Greek	−1.3431	1.2685	−1.0588	0.2927	
Year2	−2.4211	1.6864	−1.4356	0.1548	
Year3	3.1292	1.8275	1.7123	0.0905	
Year4	0.1379	1.7801	0.0775	0.9384	
Year5	−0.8832	2.1612	−0.4087	0.6838	
Math	2.6682	1.9791	1.3482	0.1812	
Economics	4.1916	1.9450	2.1550	0.0340	
Biology	4.8352	1.9147	2.5253	0.0134	
Psychology	−0.0045	1.9873	−0.0023	0.9982	
English	−1.0880	2.0091	−0.5416	0.5895	

Note that Yearl is the omitted cases for year above; thus, all the year coefficients above are in comparison with Yearl. If a different year is chosen as the omitted one, all the year coefficients (and the constant) will change so as to maintain the same year-to-year differences. And Other is the omitted case for major above; thus all the major coefficients above are in comparison with Other. If a different major is chosen as the omitted one, all the major coefficients (and the constant) will change so as to maintain the same major-to-major differences.

b. Removing variables until all the remaining ones are significant at the .10 level or better for a two-tailed test gives the following result. It seems that having a job or participating in a sport or a music ensemble does take away from study time. But it also seems females study somewhat more, certain majors study more, juniors study more, and sophomores study less.

Multiple Regression

Dependent Variable: Study

R Square: 0.3041		Adjusted R Square: 0.2345		Standard Error: 5.3283	
ANOVA	**Sum of Squares**	**df**	**Mean Square**	F_c	**2-Tailed** p**-Value**
Regression	1116.4806	9	124.0534	4.3695	0.0001
Error	2555.1594	90	28.3907		
Total	3671.6400	99			
Variable	**Coefficient**	**Standard Error**	t_c	**2-Tailed p-Value**	
Constant	23.5115	1.2708	18.5007	0.0000	
Female	1.9869	1.1137	1.7841	0.0778	
Job	−2.9020	1.1348	−2.5572	0.0122	
Sports	−2.9657	1.3008	−2.2799	0.0250	
Music	−3.0314	1.4680	−2.0649	0.0418	
Year^2	−2.3121	1.3293	−1.7393	0.0854	
Year^3	3.4408	1.4425	2.3854	0.0192	
Math	3.0285	1.5947	1.8991	0.0608	
Economics	4.4315	1.6096	2.7531	0.0071	
Biology	5.3501	1.4498	3.6902	0.0004	

c. 23.5115 is the estimated study time of a male student without a job, who does not participate in a sport or music ensemble, who is not a sophomore or junior, and is not a mathematics, economics, or biology major.
1.9869 is the estimated increase in study time if the student is female, other things held constant.
–2.9020 is the estimated decrease in study time if the student holds a job, other things held constant.
–2.9657 is the estimated decrease in study time if the student participates in a sport, other things held constant.
–3.0314 is the estimated decrease in study time if the student participates in a music ensemble, other things held constant.
–2.3121 is the estimated decrease in study time if the student is a sophomore, other things held constant.
3.4408 is the estimated increase in study time if the student is a junior, other things held constant.
3.0285 is the estimated increase in study time if the student is a mathematics major, other things held constant.
4.4315 is the estimated increase in study time if the student is an economics major, other things held constant.

5.3501 is the estimated increase in study time if the student is a biology major, other things held constant.

d. We can explain about 23.45% of the variation in study time.

Chapter 14: Time-Series Analysis

14.1

Quarter	Q1	Q2	Q3	Q4	Total
Index	67.8	90.5	148.1	93.6	400.0
Seasonal	$10,520	$16,742	$33,322	$26,226	$ 86,810

Quarter	Q1	Q2	Q3	Q4	Total
Adjusted	$15,516	$18,499	$22,500	$28,019	$ 84,535

Seasonally adjusted sales show an upward pattern, which could be trend and/or cycle.

14.3

Month	January	February	March	April	May	June
Index	84.5	89.8	92.3	102.6	113	125.4
Seasonal	$50,900	$50,750	$48,880	$50,830	$52,000	$ 53,720

Month	January	February	March	April	May	June
Adjusted	$60,237	$56,514	$52,958	$49,542	$46,018	$ 42,839

Seasonally adjusted sales show a downward pattern, which could be trend and/or cycle.

14.5 a.

Multiple Regression
Dependent Variable: lnGDP

R Square:	Adjusted *R* Square:	Standard Error:
0.9828	0.9821	0.0826

ANOVA	Sum of Squares	df	Mean Square	F_c	2-Tailed *p*-Value
Regression	9.3398	1	9.3398	1369.6303	0.0000
Error	0.1637	24	0.0068		
Total	9.5035	25			

Variable	Coefficient	Standard Error	t_c	2-Tailed *p*-Value
Constant	7.0451	0.0315	223.8137	0.0000
Time	0.0799	0.0022	37.0085	0.0000

7.0451 is our estimate for lnGDP in time 0–1970.
1147.1903 is our estimate for GDP in time 0–1970.
0.0799 is our estimate for the increase in lnGDP per year.
8.32% is our estimate for the percentage growth in GDP per year.

Multiple Regression Dependent Variable: lnCons					
R Square: 0.9869		Adjusted R Square: 0.9864		Standard Error: 0.0755	
ANOVA	**Sum of Squares**	**df**	**Mean Square**	F_c	**2-Tailed p-Value**
Regression	10.3150	1	10.3150	1809.7216	0.0000
Error	0.1368	24	0.0057		
Total	10.4518	25			
Variable	**Coefficient**	**Standard Error**	t_c	**2-Tailed p-Value**	
Constant	6.5553	0.0288	227.7887	0.0000	
Time	0.0840	0.0020	42.5408	0.0000	

6.5553 is our estimate for lnCons in time 0–1970.
702.9606 is our estimate for Consumption in time 0–1970.
0.0840 is our estimate for the increase in lnCons per year.
8.76% is our estimate for the percentage growth in Consumption per year.

Multiple Regression Dependent Variable: lnServ					
R Square: 0.9901		Adjusted R Square: 0.9897		Standard Error: 0.0743	
ANOVA	**Sum of Squares**	**df**	**Mean Square**	F_c	**2-Tailed p-Value**
Regression	13.2847	1	13.2847	2403.5021	0.0000
Error	0.1327	24	0.0055		
Total	13.4176	25			
Variable	**Coefficient**	**Standard Error**	t_c	**2-Tailed p-Value**	
Constant	5.7328	0.0283	202.2901	0.0000	
Time	0.0953	0.0019	49.0255	0.0000	

5.7328 is our estimate for lnServ in time 0–1970.
308.8193 is our estimate for Services in time 0–1970.
0.0953 is our estimate for the increase in lnServ per year.
10.00% is our estimate for the percentage growth in GDP per year.

Spending on Services is growing fastest, at 10.00%; GDP, the slowest at 8.32%.

b. The semi-log form is preferable for comparisons like this, where the units or magnitudes are not comparable.

14.7 a.

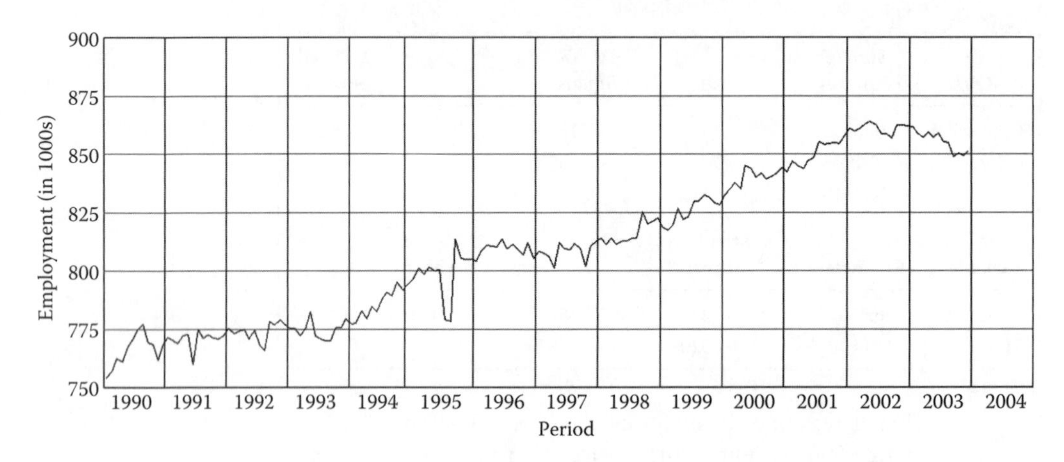

There appears to be an upward trend and something of a business cycle around that trend. But government employment is much less seasonal than construction.

c.

	Overall Seasonal Index	
January	100.056	100.041
February	100.042	100.027
March	100.032	100.018
April	100.024	100.010
May	100.204	100.189
June	99.946	99.931
July	99.875	99.860
August	99.814	99.799
September	100.175	100.160
October	100.003	99.988
November	99.971	99.957
December	100.033	100.019
	1200.175	1200.000

d.

Multiple Regression					
Dependent Variable: MA					
R Square: 0.9873		Adjusted *R* Square: 0.9748		Standard Error: 4.8961	
ANOVA	**Sum of Squares**	**df**	**Mean Square**	F_c	**2-Tailed *p*-Value**
Regression	143019.0895	1	143019.0895	5966.1733	0.0000
Error	3691.6359	154	23.9717		
Total	146710.7254	155			
Variable	**Coefficient**	**Standard Error**	t_c	**2-Tailed *p*-Value**	
Constant	752.9102	0.8335	903.3171	0.0000	
Time	0.6724	0.0087	77.2410	0.0000	

0.6724 is our estimate (in thousands) of the long-term-trend increase in government employment per month. That is 672 jobs per month.

Multiple Regression					
Dependent Variable: lnMA					
R Square: 0.9885		Adjusted *R* Square: 0.9771		Standard Error: 0.0057	
ANOVA	**Sum of Squares**	**df**	**Mean Square**	F_c	**2-Tailed *p*-Value**
Regression	0.2172	1	0.2172	6581.6962	0.0000
Error	0.0051	154	0.0000		
Total	0.2223	155			
Variable	**Coefficient**	**Standard Error**	t_c	**2-Tailed *p*-Value**	
Constant	6.6260	0.0010	6775.9593	0.0000	
Time	8.2853e-04	0.0000	81.1277	0.0000	

.00082853 is our estimate of the long-term-trend increase in lnMA per month.
.083% is our estimate of the percentage growth rate in government employment per month.
Percent change is generally preferable; it also fits slightly better in this case.

e.

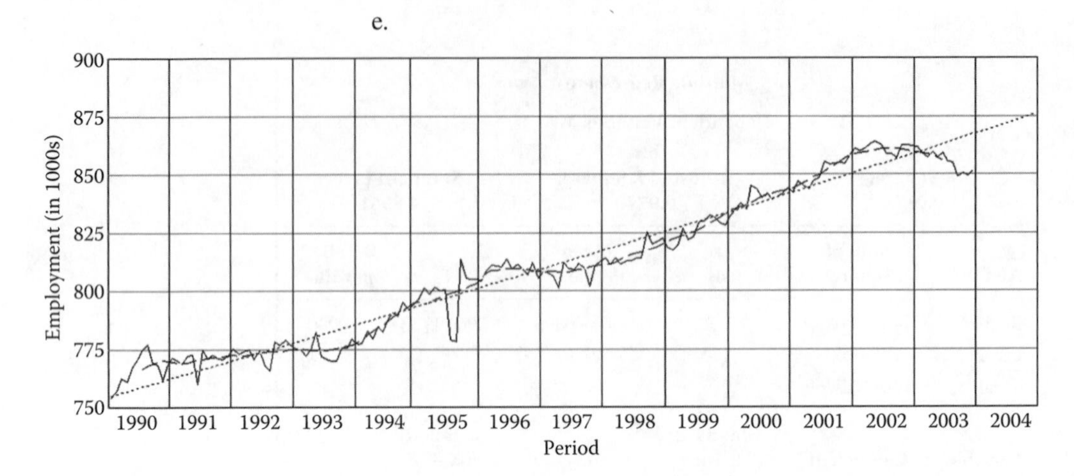

g. We end well below trend. Presumably it will take some time to get back up to trend. The projection below shows us gradually returning to the long-term trend.

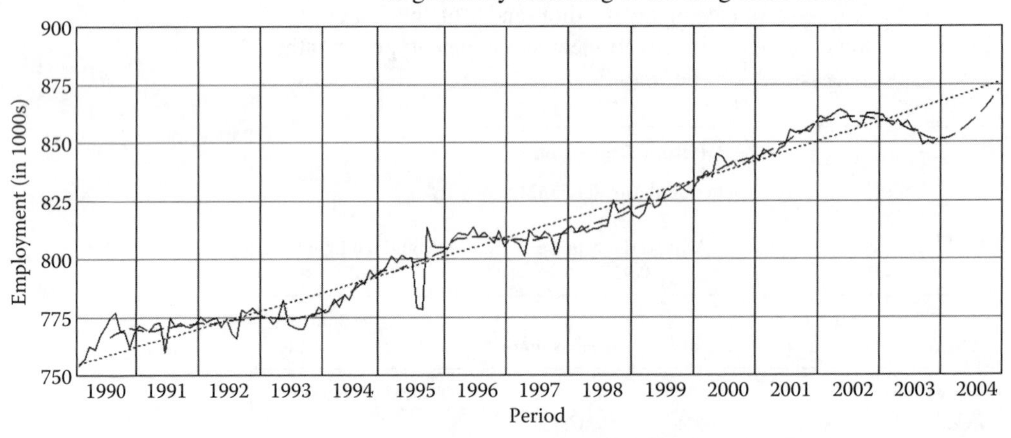

h. Using the seasonal index from part b, we add the seasonal variation back into our projections.

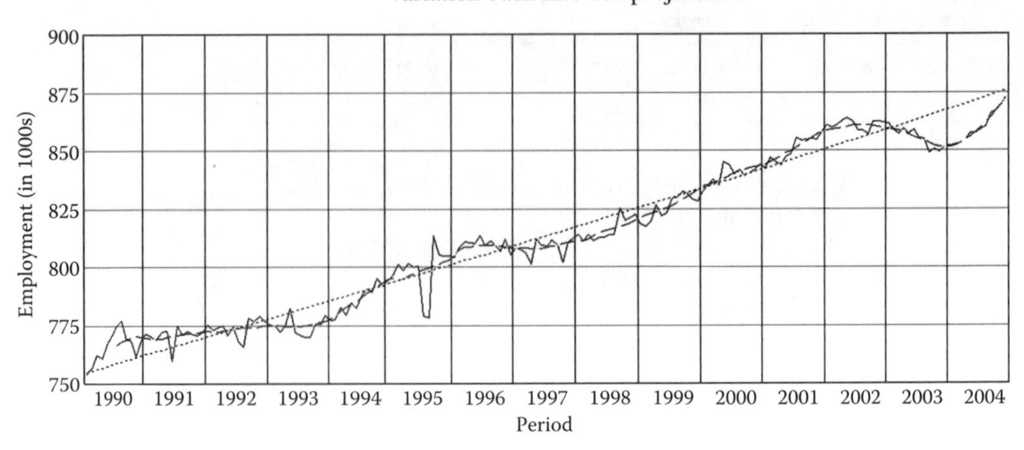

Appendix C

TABLE C.1 The Binomial Distribution

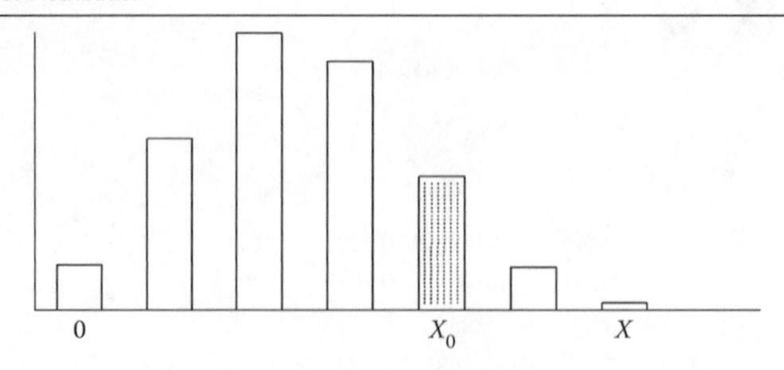

		Probability									
n	*x*	**0.05**	**0.10**	**0.15**	**0.20**	**0.25**	**0.30**	**0.35**	**0.40**	**0.45**	**0.50**
1	0	0.9500	0.9000	0.8500	0.8000	0.7500	0.7000	0.6500	0.6000	0.5500	0.5000
	1	0.0500	0.1000	0.1500	0.2000	0.2500	0.3000	0.3500	0.4000	0.4500	0.5000
2	0	0.9025	0.8100	0.7225	0.6400	0.5625	0.4900	0.4225	0.3600	0.3025	0.2500
	1	0.0950	0.1800	0.2550	0.3200	0.3750	0.4200	0.4550	0.4800	0.4950	0.5000
	2	0.0025	0.0100	0.0225	0.0400	0.0625	0.0900	0.1225	0.1600	0.2025	0.2500
3	0	0.8574	0.7290	0.6141	0.5120	0.4219	0.3430	0.2746	0.2160	0.1664	0.1250
	1	0.1354	0.2430	0.3251	0.3840	0.4219	0.4410	0.4436	0.4320	0.4084	0.3750
	2	0.0071	0.0270	0.0574	0.0960	0.1406	0.1890	0.2389	0.2880	0.3341	0.3750
	3	0.0001	0.0010	0.0034	0.0080	0.0156	0.0270	0.0429	0.0640	0.0911	0.1250
4	0	0.8145	0.6561	0.5220	0.4096	0.3164	0.2401	0.1785	0.1296	0.0915	0.0625
	1	0.1715	0.2916	0.3685	0.4096	0.4219	0.4116	0.3845	0.3456	0.2995	0.2500
	2	0.0135	0.0486	0.0975	0.1536	0.2109	0.2646	0.3105	0.3456	0.3675	0.3750
	3	0.0005	0.0036	0.0115	0.0256	0.0469	0.0756	0.1115	0.1536	0.2005	0.2500
	4	0.0000	0.0001	0.0005	0.0016	0.0039	0.0081	0.0150	0.0256	0.0410	0.0625
5	0	0.7738	0.5905	0.4437	0.3277	0.2373	0.1681	0.1160	0.0778	0.0503	0.0313
	1	0.2036	0.3281	0.3915	0.4096	0.3955	0.3602	0.3124	0.2592	0.2059	0.1563
	2	0.0214	0.0729	0.1382	0.2048	0.2637	0.3087	0.3364	0.3456	0.3369	0.3125
	3	0.0011	0.0081	0.0244	0.0512	0.0879	0.1323	0.1811	0.2304	0.2757	0.3125
	4	0.0000	0.0005	0.0022	0.0064	0.0146	0.0284	0.0488	0.0768	0.1128	0.1563
	5	0.0000	0.0000	0.0001	0.0003	0.0010	0.0024	0.0053	0.0102	0.0185	0.0313
6	0	0.7351	0.5314	0.3771	0.2621	0.1780	0.1176	0.0754	0.0467	0.0277	0.0156
	1	0.2321	0.3543	0.3993	0.3932	0.3560	0.3025	0.2437	0.1866	0.1359	0.0938
	2	0.0305	0.0984	0.1762	0.2458	0.2966	0.3241	0.3280	0.3110	0.2780	0.2344
	3	0.0021	0.0146	0.0415	0.0819	0.1318	0.1852	0.2355	0.2765	0.3032	0.3125
	4	0.0001	0.0012	0.0055	0.0154	0.0330	0.0595	0.0951	0.1382	0.1861	0.2344
	5	0.0000	0.0001	0.0004	0.0015	0.0044	0.0102	0.0205	0.0369	0.0609	0.0938
	6	0.0000	0.0000	0.0000	0.0001	0.0002	0.0007	0.0018	0.0041	0.0083	0.0156
7	0	0.6983	0.4783	0.3206	0.2097	0.1335	0.0824	0.0490	0.0280	0.0152	0.0078
	1	0.2573	0.3720	0.3960	0.3670	0.3115	0.2471	0.1848	0.1306	0.0872	0.0547
	2	0.0406	0.1240	0.2097	0.2753	0.3115	0.3177	0.2985	0.2613	0.2140	0.1641
	3	0.0036	0.0230	0.0617	0.1147	0.1730	0.2269	0.2679	0.2903	0.2918	0.2734
	4	0.0002	0.0026	0.0109	0.0287	0.0577	0.0972	0.1442	0.1935	0.2388	0.2734
	5	0.0000	0.0002	0.0012	0.0043	0.0115	0.0250	0.0466	0.0774	0.1172	0.1641

TABLE C.1 The Binomial Distribution (Continued)

		Probability									
n	*x*	**0.05**	**0.10**	**0.15**	**0.20**	**0.25**	**0.30**	**0.35**	**0.40**	**0.45**	**0.50**
	6	0.0000	0.0000	0.0001	0.0004	0.0013	0.0036	0.0084	0.0172	0.0320	0.0547
	7	0.0000	0.0000	0.0000	0.0000	0.0001	0.0002	0.0006	0.0016	0.0037	0.0078
8	0	0.6634	0.4305	0.2725	0.1678	0.1001	0.0576	0.0319	0.0168	0.0084	0.0039
	1	0.2793	0.3826	0.3847	0.3355	0.2670	0.1977	0.1373	0.0896	0.0548	0.0313
	2	0.0515	0.1488	0.2376	0.2936	0.3115	0.2965	0.2587	0.2090	0.1569	0.1094
	3	0.0054	0.0331	0.0839	0.1468	0.2076	0.2541	0.2786	0.2787	0.2568	0.2188
	4	0.0004	0.0046	0.0185	0.0459	0.0865	0.1361	0.1875	0.2322	0.2627	0.2734
	5	0.0000	0.0004	0.0026	0.0092	0.0231	0.0467	0.0808	0.1239	0.1719	0.2188
	6	0.0000	0.0000	0.0002	0.0011	0.0038	0.0100	0.0217	0.0413	0.0703	0.1094
	7	0.0000	0.0000	0.0000	0.0001	0.0004	0.0012	0.0033	0.0079	0.0164	0.0313
	8	0.0000	0.0000	0.0000	0.0000	0.0000	0.0001	0.0002	0.0007	0.0017	0.0039
9	0	0.6302	0.3874	0.2316	0.1342	0.0751	0.0404	0.0207	0.0101	0.0046	0.0020
	1	0.2985	0.3874	0.3679	0.3020	0.2253	0.1556	0.1004	0.0605	0.0339	0.0176
	2	0.0629	0.1722	0.2597	0.3020	0.3003	0.2668	0.2162	0.1612	0.1110	0.0703
	3	0.0077	0.0446	0.1069	0.1762	0.2336	0.2668	0.2716	0.2508	0.2119	0.1641
	4	0.0006	0.0074	0.0283	0.0661	0.1168	0.1715	0.2194	0.2508	0.2600	0.2461
	5	0.0000	0.0008	0.0050	0.0165	0.0389	0.0735	0.1181	0.1672	0.2128	0.2461
	6	0.0000	0.0001	0.0006	0.0028	0.0087	0.0210	0.0424	0.0743	0.1160	0.1641
	7	0.0000	0.0000	0.0000	0.0003	0.0012	0.0039	0.0098	0.0212	0.0407	0.0703
	8	0.0000	0.0000	0.0000	0.0000	0.0001	0.0004	0.0013	0.0035	0.0083	0.0176
	9	0.0000	0.0000	0.0000	0.0000	0.0000	0.0000	0.0001	0.0003	0.0008	0.0020
10	0	0.5987	0.3487	0.1969	0.1074	0.0563	0.0282	0.0135	0.0060	0.0025	0.0010
	1	0.3151	0.3874	0.3474	0.2684	0.1877	0.1211	0.0725	0.0403	0.0207	0.0098
	2	0.0746	0.1937	0.2759	0.3020	0.2816	0.2335	0.1757	0.1209	0.0763	0.0439
	3	0.0105	0.0574	0.1298	0.2013	0.2503	0.2668	0.2522	0.2150	0.1665	0.1172
	4	0.0010	0.0112	0.0401	0.0881	0.1460	0.2001	0.2377	0.2508	0.2384	0.2051
	5	0.0001	0.0015	0.0085	0.0264	0.0584	0.1029	0.1536	0.2007	0.2340	0.2461
	6	0.0000	0.0001	0.0012	0.0055	0.0162	0.0368	0.0689	0.1115	0.1596	0.2051
	7	0.0000	0.0000	0.0001	0.0008	0.0031	0.0090	0.0212	0.0425	0.0746	0.1172
	8	0.0000	0.0000	0.0000	0.0001	0.0004	0.0014	0.0043	0.0106	0.0229	0.0439
	9	0.0000	0.0000	0.0000	0.0000	0.0000	0.0001	0.0005	0.0016	0.0042	0.0098
	10	0.0000	0.0000	0.0000	0.0000	0.0000	0.0000	0.0000	0.0001	0.0003	0.0010
11	0	0.5688	0.3138	0.1673	0.0859	0.0422	0.0198	0.0088	0.0036	0.0014	0.0005
	1	0.3293	0.3835	0.3248	0.2362	0.1549	0.0932	0.0518	0.0266	0.0125	0.0054
	2	0.0867	0.2131	0.2866	0.2953	0.2581	0.1998	0.1395	0.0887	0.0513	0.0269
	3	0.0137	0.0710	0.1517	0.2215	0.2581	0.2568	0.2254	0.1774	0.1259	0.0806
	4	0.0014	0.0158	0.0536	0.1107	0.1721	0.2201	0.2428	0.2365	0.2060	0.1611
	5	0.0001	0.0025	0.0132	0.0388	0.0803	0.1321	0.1830	0.2207	0.2360	0.2256
	6	0.0000	0.0003	0.0023	0.0097	0.0268	0.0566	0.0985	0.1471	0.1931	0.2256
	7	0.0000	0.0000	0.0003	0.0017	0.0064	0.0173	0.0379	0.0701	0.1128	0.1611
	8	0.0000	0.0000	0.0000	0.0002	0.0011	0.0037	0.0102	0.0234	0.0462	0.0806

(*Continued*)

TABLE C.1 The Binomial Distribution (Continued)

		Probability									
n	*x*	**0.05**	**0.10**	**0.15**	**0.20**	**0.25**	**0.30**	**0.35**	**0.40**	**0.45**	**0.50**
	9	0.0000	0.0000	0.0000	0.0000	0.0001	0.0005	0.0018	0.0052	0.0126	0.0269
	10	0.0000	0.0000	0.0000	0.0000	0.0000	0.0000	0.0002	0.0007	0.0021	0.0054
	11	0.0000	0.0000	0.0000	0.0000	0.0000	0.0000	0.0000	0.0000	0.0002	0.0005
12	0	0.5404	0.2824	0.1422	0.0687	0.0317	0.0138	0.0057	0.0022	0.0008	0.0002
	1	0.3413	0.3766	0.3012	0.2062	0.1267	0.0712	0.0368	0.0174	0.0075	0.0029
	2	0.0988	0.2301	0.2924	0.2835	0.2323	0.1678	0.1088	0.0639	0.0339	0.0161
	3	0.0173	0.0852	0.1720	0.2362	0.2581	0.2397	0.1954	0.1419	0.0923	0.0537
	4	0.0021	0.0213	0.0683	0.1329	0.1936	0.2311	0.2367	0.2128	0.1700	0.1208
	5	0.0002	0.0038	0.0193	0.0532	0.1032	0.1585	0.2039	0.2270	0.2225	0.1934
	6	0.0000	0.0005	0.0040	0.0155	0.0401	0.0792	0.1281	0.1766	0.2124	0.2256
	7	0.0000	0.0000	0.0006	0.0033	0.0115	0.0291	0.0591	0.1009	0.1489	0.1934
	8	0.0000	0.0000	0.0001	0.0005	0.0024	0.0078	0.0199	0.0420	0.0762	0.1208
	9	0.0000	0.0000	0.0000	0.0001	0.0004	0.0015	0.0048	0.0125	0.0277	0.0537
	10	0.0000	0.0000	0.0000	0.0000	0.0000	0.0002	0.0008	0.0025	0.0068	0.0161
	11	0.0000	0.0000	0.0000	0.0000	0.0000	0.0000	0.0001	0.0003	0.0010	0.0029
	12	0.0000	0.0000	0.0000	0.0000	0.0000	0.0000	0.0000	0.0000	0.0001	0.0002
13	0	0.5133	0.2542	0.1209	0.0550	0.0238	0.0097	0.0037	0.0013	0.0004	0.0001
	1	0.3512	0.3672	0.2774	0.1787	0.1029	0.0540	0.0259	0.0113	0.0045	0.0016
	2	0.1109	0.2448	0.2937	0.2680	0.2059	0.1388	0.0836	0.0453	0.0220	0.0095
	3	0.0214	0.0997	0.1900	0.2457	0.2517	0.2181	0.1651	0.1107	0.0660	0.0349
	4	0.0028	0.0277	0.0838	0.1535	0.2097	0.2337	0.2222	0.1845	0.1350	0.0873
	5	0.0003	0.0055	0.0266	0.0691	0.1258	0.1803	0.2154	0.2214	0.1989	0.1571
	6	0.0000	0.0008	0.0063	0.0230	0.0559	0.1030	0.1546	0.1968	0.2169	0.2095
	7	0.0000	0.0001	0.0011	0.0058	0.0186	0.0442	0.0833	0.1312	0.1775	0.2095
	8	0.0000	0.0000	0.0001	0.0011	0.0047	0.0142	0.0336	0.0656	0.1089	0.1571
	9	0.0000	0.0000	0.0000	0.0001	0.0009	0.0034	0.0101	0.0243	0.0495	0.0873
	10	0.0000	0.0000	0.0000	0.0000	0.0001	0.0006	0.0022	0.0065	0.0162	0.0349
	11	0.0000	0.0000	0.0000	0.0000	0.0000	0.0001	0.0003	0.0012	0.0036	0.0095
	12	0.0000	0.0000	0.0000	0.0000	0.0000	0.0000	0.0000	0.0001	0.0005	0.0016
	13	0.0000	0.0000	0.0000	0.0000	0.0000	0.0000	0.0000	0.0000	0.0000	0.0001
14	0	0.4877	0.2288	0.1028	0.0440	0.0178	0.0068	0.0024	0.0008	0.0002	0.0001
	1	0.3593	0.3559	0.2539	0.1539	0.0832	0.0407	0.0181	0.0073	0.0027	0.0009
	2	0.1229	0.2570	0.2912	0.2501	0.1802	0.1134	0.0634	0.0317	0.0141	0.0056
	3	0.0259	0.1142	0.2056	0.2501	0.2402	0.1943	0.1366	0.0845	0.0462	0.0222
	4	0.0037	0.0349	0.0998	0.1720	0.2202	0.2290	0.2022	0.1549	0.1040	0.0611
	5	0.0004	0.0078	0.0352	0.0860	0.1468	0.1963	0.2178	0.2066	0.1701	0.1222
	6	0.0000	0.0013	0.0093	0.0322	0.0734	0.1262	0.1759	0.2066	0.2088	0.1833
	7	0.0000	0.0002	0.0019	0.0092	0.0280	0.0618	0.1082	0.1574	0.1952	0.2095
	8	0.0000	0.0000	0.0003	0.0020	0.0082	0.0232	0.0510	0.0918	0.1398	0.1833
	9	0.0000	0.0000	0.0000	0.0003	0.0018	0.0066	0.0183	0.0408	0.0762	0.1222
	10	0.0000	0.0000	0.0000	0.0000	0.0003	0.0014	0.0049	0.0136	0.0312	0.0611

TABLE C.1 The Binomial Distribution (Continued)

		Probability									
n	*x*	**0.05**	**0.10**	**0.15**	**0.20**	**0.25**	**0.30**	**0.35**	**0.40**	**0.45**	**0.50**
	11	0.0000	0.0000	0.0000	0.0000	0.0000	0.0002	0.0010	0.0033	0.0093	0.0222
	12	0.0000	0.0000	0.0000	0.0000	0.0000	0.0000	0.0001	0.0005	0.0019	0.0056
	13	0.0000	0.0000	0.0000	0.0000	0.0000	0.0000	0.0000	0.0001	0.0002	0.0009
	14	0.0000	0.0000	0.0000	0.0000	0.0000	0.0000	0.0000	0.0000	0.0000	0.0001
15	0	0.4633	0.2059	0.0874	0.0352	0.0134	0.0047	0.0016	0.0005	0.0001	0.0000
	1	0.3658	0.3432	0.2312	0.1319	0.0668	0.0305	0.0126	0.0047	0.0016	0.0005
	2	0.1348	0.2669	0.2856	0.2309	0.1559	0.0916	0.0476	0.0219	0.0090	0.0032
	3	0.0307	0.1285	0.2184	0.2501	0.2252	0.1700	0.1110	0.0634	0.0318	0.0139
	4	0.0049	0.0428	0.1156	0.1876	0.2252	0.2186	0.1792	0.1268	0.0780	0.0417
	5	0.0006	0.0105	0.0449	0.1032	0.1651	0.2061	0.2123	0.1859	0.1404	0.0916
	6	0.0000	0.0019	0.0132	0.0430	0.0917	0.1472	0.1906	0.2066	0.1914	0.1527
	7	0.0000	0.0003	0.0030	0.0138	0.0393	0.0811	0.1319	0.1771	0.2013	0.1964
	8	0.0000	0.0000	0.0005	0.0035	0.0131	0.0348	0.0710	0.1181	0.1647	0.1964
	9	0.0000	0.0000	0.0001	0.0007	0.0034	0.0116	0.0298	0.0612	0.1048	0.1527
	10	0.0000	0.0000	0.0000	0.0001	0.0007	0.0030	0.0096	0.0245	0.0515	0.0916
	11	0.0000	0.0000	0.0000	0.0000	0.0001	0.0006	0.0024	0.0074	0.0191	0.0417
	12	0.0000	0.0000	0.0000	0.0000	0.0000	0.0001	0.0004	0.0016	0.0052	0.0139
	13	0.0000	0.0000	0.0000	0.0000	0.0000	0.0000	0.0001	0.0003	0.0010	0.0032
	14	0.0000	0.0000	0.0000	0.0000	0.0000	0.0000	0.0000	0.0000	0.0001	0.0005
	15	0.0000	0.0000	0.0000	0.0000	0.0000	0.0000	0.0000	0.0000	0.0000	0.0000
16	0	0.4401	0.1853	0.0743	0.0281	0.0100	0.0033	0.0010	0.0003	0.0001	0.0000
	1	0.3706	0.3294	0.2097	0.1126	0.0535	0.0228	0.0087	0.0030	0.0009	0.0002
	2	0.1463	0.2745	0.2775	0.2111	0.1336	0.0732	0.0353	0.0150	0.0056	0.0018
	3	0.0359	0.1423	0.2285	0.2463	0.2079	0.1465	0.0888	0.0468	0.0215	0.0085
	4	0.0061	0.0514	0.1311	0.2001	0.2252	0.2040	0.1553	0.1014	0.0572	0.0278
	5	0.0008	0.0137	0.0555	0.1201	0.1802	0.2099	0.2008	0.1623	0.1123	0.0667
	6	0.0001	0.0028	0.0180	0.0550	0.1101	0.1649	0.1982	0.1983	0.1684	0.1222
	7	0.0000	0.0004	0.0045	0.0197	0.0524	0.1010	0.1524	0.1889	0.1969	0.1746
	8	0.0000	0.0001	0.0009	0.0055	0.0197	0.0487	0.0923	0.1417	0.1812	0.1964
	9	0.0000	0.0000	0.0001	0.0012	0.0058	0.0185	0.0442	0.0840	0.1318	0.1746
	10	0.0000	0.0000	0.0000	0.0002	0.0014	0.0056	0.0167	0.0392	0.0755	0.1222
	11	0.0000	0.0000	0.0000	0.0000	0.0002	0.0013	0.0049	0.0142	0.0337	0.0667
	12	0.0000	0.0000	0.0000	0.0000	0.0000	0.0002	0.0011	0.0040	0.0115	0.0278
	13	0.0000	0.0000	0.0000	0.0000	0.0000	0.0000	0.0002	0.0008	0.0029	0.0085
	14	0.0000	0.0000	0.0000	0.0000	0.0000	0.0000	0.0000	0.0001	0.0005	0.0018
	15	0.0000	0.0000	0.0000	0.0000	0.0000	0.0000	0.0000	0.0000	0.0001	0.0002
	16	0.0000	0.0000	0.0000	0.0000	0.0000	0.0000	0.0000	0.0000	0.0000	0.0000
17	0	0.4181	0.1668	0.0631	0.0225	0.0075	0.0023	0.0007	0.0002	0.0000	0.0000
	1	0.3741	0.3150	0.1893	0.0957	0.0426	0.0169	0.0060	0.0019	0.0005	0.0001
	2	0.1575	0.2800	0.2673	0.1914	0.1136	0.0581	0.0260	0.0102	0.0035	0.0010
	3	0.0415	0.1556	0.2359	0.2393	0.1893	0.1245	0.0701	0.0341	0.0144	0.0052

(*Continued*)

TABLE C.1 The Binomial Distribution (Continued)

		Probability									
n	*x*	**0.05**	**0.10**	**0.15**	**0.20**	**0.25**	**0.30**	**0.35**	**0.40**	**0.45**	**0.50**
	4	0.0076	0.0605	0.1457	0.2093	0.2209	0.1868	0.1320	0.0796	0.0411	0.0182
	5	0.0010	0.0175	0.0668	0.1361	0.1914	0.2081	0.1849	0.1379	0.0875	0.0472
	6	0.0001	0.0039	0.0236	0.0680	0.1276	0.1784	0.1991	0.1839	0.1432	0.0944
	7	0.0000	0.0007	0.0065	0.0267	0.0668	0.1201	0.1685	0.1927	0.1841	0.1484
	8	0.0000	0.0001	0.0014	0.0084	0.0279	0.0644	0.1134	0.1606	0.1883	0.1855
	9	0.0000	0.0000	0.0003	0.0021	0.0093	0.0276	0.0611	0.1070	0.1540	0.1855
	10	0.0000	0.0000	0.0000	0.0004	0.0025	0.0095	0.0263	0.0571	0.1008	0.1484
	11	0.0000	0.0000	0.0000	0.0001	0.0005	0.0026	0.0090	0.0242	0.0525	0.0944
	12	0.0000	0.0000	0.0000	0.0000	0.0001	0.0006	0.0024	0.0081	0.0215	0.0472
	13	0.0000	0.0000	0.0000	0.0000	0.0000	0.0001	0.0005	0.0021	0.0068	0.0182
	14	0.0000	0.0000	0.0000	0.0000	0.0000	0.0000	0.0001	0.0004	0.0016	0.0052
	15	0.0000	0.0000	0.0000	0.0000	0.0000	0.0000	0.0000	0.0001	0.0003	0.0010
	16	0.0000	0.0000	0.0000	0.0000	0.0000	0.0000	0.0000	0.0000	0.0000	0.0001
	17	0.0000	0.0000	0.0000	0.0000	0.0000	0.0000	0.0000	0.0000	0.0000	0.0000
18	0	0.3972	0.1501	0.0536	0.0180	0.0056	0.0016	0.0004	0.0001	0.0000	0.0000
	1	0.3763	0.3002	0.1704	0.0811	0.0338	0.0126	0.0042	0.0012	0.0003	0.0001
	2	0.1683	0.2835	0.2556	0.1723	0.0958	0.0458	0.0190	0.0069	0.0022	0.0006
	3	0.0473	0.1680	0.2406	0.2297	0.1704	0.1046	0.0547	0.0246	0.0095	0.0031
	4	0.0093	0.0700	0.1592	0.2153	0.2130	0.1681	0.1104	0.0614	0.0291	0.0117
	5	0.0014	0.0218	0.0787	0.1507	0.1988	0.2017	0.1664	0.1146	0.0666	0.0327
	6	0.0002	0.0052	0.0301	0.0816	0.1436	0.1873	0.1941	0.1655	0.1181	0.0708
	7	0.0000	0.0010	0.0091	0.0350	0.0820	0.1376	0.1792	0.1892	0.1657	0.1214
	8	0.0000	0.0002	0.0022	0.0120	0.0376	0.0811	0.1327	0.1734	0.1864	0.1669
	9	0.0000	0.0000	0.0004	0.0033	0.0139	0.0386	0.0794	0.1284	0.1694	0.1855
	10	0.0000	0.0000	0.0001	0.0008	0.0042	0.0149	0.0385	0.0771	0.1248	0.1669
	11	0.0000	0.0000	0.0000	0.0001	0.0010	0.0046	0.0151	0.0374	0.0742	0.1214
	12	0.0000	0.0000	0.0000	0.0000	0.0002	0.0012	0.0047	0.0145	0.0354	0.0708
	13	0.0000	0.0000	0.0000	0.0000	0.0000	0.0002	0.0012	0.0045	0.0134	0.0327
	14	0.0000	0.0000	0.0000	0.0000	0.0000	0.0000	0.0002	0.0011	0.0039	0.0117
	15	0.0000	0.0000	0.0000	0.0000	0.0000	0.0000	0.0000	0.0002	0.0009	0.0031
	16	0.0000	0.0000	0.0000	0.0000	0.0000	0.0000	0.0000	0.0000	0.0001	0.0006
	17	0.0000	0.0000	0.0000	0.0000	0.0000	0.0000	0.0000	0.0000	0.0000	0.0001
	18	0.0000	0.0000	0.0000	0.0000	0.0000	0.0000	0.0000	0.0000	0.0000	0.0000
19	0	0.3774	0.1351	0.0456	0.0144	0.0042	0.0011	0.0003	0.0001	0.0000	0.0000
	1	0.3774	0.2852	0.1529	0.0685	0.0268	0.0093	0.0029	0.0008	0.0002	0.0000
	2	0.1787	0.2852	0.2428	0.1540	0.0803	0.0358	0.0138	0.0046	0.0013	0.0003
	3	0.0533	0.1796	0.2428	0.2182	0.1517	0.0869	0.0422	0.0175	0.0062	0.0018
	4	0.0112	0.0798	0.1714	0.2182	0.2023	0.1491	0.0909	0.0467	0.0203	0.0074
	5	0.0018	0.0266	0.0907	0.1636	0.2023	0.1916	0.1468	0.0933	0.0497	0.0222
	6	0.0002	0.0069	0.0374	0.0955	0.1574	0.1916	0.1844	0.1451	0.0949	0.0518
	7	0.0000	0.0014	0.0122	0.0443	0.0974	0.1525	0.1844	0.1797	0.1443	0.0961
	8	0.0000	0.0002	0.0032	0.0166	0.0487	0.0981	0.1489	0.1797	0.1771	0.1442
	9	0.0000	0.0000	0.0007	0.0051	0.0198	0.0514	0.0980	0.1464	0.1771	0.1762

TABLE C.1 The Binomial Distribution (Continued)

		Probability									
n	*x*	**0.05**	**0.10**	**0.15**	**0.20**	**0.25**	**0.30**	**0.35**	**0.40**	**0.45**	**0.50**
	10	0.0000	0.0000	0.0001	0.0013	0.0066	0.0220	0.0528	0.0976	0.1449	0.1762
	11	0.0000	0.0000	0.0000	0.0003	0.0018	0.0077	0.0233	0.0532	0.0970	0.1442
	12	0.0000	0.0000	0.0000	0.0000	0.0004	0.0022	0.0083	0.0237	0.0529	0.0961
	13	0.0000	0.0000	0.0000	0.0000	0.0001	0.0005	0.0024	0.0085	0.0233	0.0518
	14	0.0000	0.0000	0.0000	0.0000	0.0000	0.0001	0.0006	0.0024	0.0082	0.0222
	15	0.0000	0.0000	0.0000	0.0000	0.0000	0.0000	0.0001	0.0005	0.0022	0.0074
	16	0.0000	0.0000	0.0000	0.0000	0.0000	0.0000	0.0000	0.0001	0.0005	0.0018
	17	0.0000	0.0000	0.0000	0.0000	0.0000	0.0000	0.0000	0.0000	0.0001	0.0003
	18	0.0000	0.0000	0.0000	0.0000	0.0000	0.0000	0.0000	0.0000	0.0000	0.0000
	19	0.0000	0.0000	0.0000	0.0000	0.0000	0.0000	0.0000	0.0000	0.0000	0.0000
20	0	0.3585	0.1216	0.0388	0.0115	0.0032	0.0008	0.0002	0.0000	0.0000	0.0000
	1	0.3774	0.2702	0.1368	0.0576	0.0211	0.0068	0.0020	0.0005	0.0001	0.0000
	2	0.1887	0.2852	0.2293	0.1369	0.0669	0.0278	0.0100	0.0031	0.0008	0.0002
	3	0.0596	0.1901	0.2428	0.2054	0.1339	0.0716	0.0323	0.0123	0.0040	0.0011
	4	0.0133	0.0898	0.1821	0.2182	0.1897	0.1304	0.0738	0.0350	0.0139	0.0046
	5	0.0022	0.0319	0.1028	0.1746	0.2023	0.1789	0.1272	0.0746	0.0365	0.0148
	6	0.0003	0.0089	0.0454	0.1091	0.1686	0.1916	0.1712	0.1244	0.0746	0.0370
	7	0.0000	0.0020	0.0160	0.0545	0.1124	0.1643	0.1844	0.1659	0.1221	0.0739
	8	0.0000	0.0004	0.0046	0.0222	0.0609	0.1144	0.1614	0.1797	0.1623	0.1201
	9	0.0000	0.0001	0.0011	0.0074	0.0271	0.0654	0.1158	0.1597	0.1771	0.1602
	10	0.0000	0.0000	0.0002	0.0020	0.0099	0.0308	0.0686	0.1171	0.1593	0.1762
	11	0.0000	0.0000	0.0000	0.0005	0.0030	0.0120	0.0336	0.0710	0.1185	0.1602
	12	0.0000	0.0000	0.0000	0.0001	0.0008	0.0039	0.0136	0.0355	0.0727	0.1201
	13	0.0000	0.0000	0.0000	0.0000	0.0002	0.0010	0.0045	0.0146	0.0366	0.0739
	14	0.0000	0.0000	0.0000	0.0000	0.0000	0.0002	0.0012	0.0049	0.0150	0.0370
	15	0.0000	0.0000	0.0000	0.0000	0.0000	0.0000	0.0003	0.0013	0.0049	0.0148
	16	0.0000	0.0000	0.0000	0.0000	0.0000	0.0000	0.0000	0.0003	0.0013	0.0046
	17	0.0000	0.0000	0.0000	0.0000	0.0000	0.0000	0.0000	0.0000	0.0002	0.0011
	18	0.0000	0.0000	0.0000	0.0000	0.0000	0.0000	0.0000	0.0000	0.0000	0.0002
	19	0.0000	0.0000	0.0000	0.0000	0.0000	0.0000	0.0000	0.0000	0.0000	0.0000
	20	0.0000	0.0000	0.0000	0.0000	0.0000	0.0000	0.0000	0.0000	0.0000	0.0000

TABLE C.2 The Standard Normal Distribution

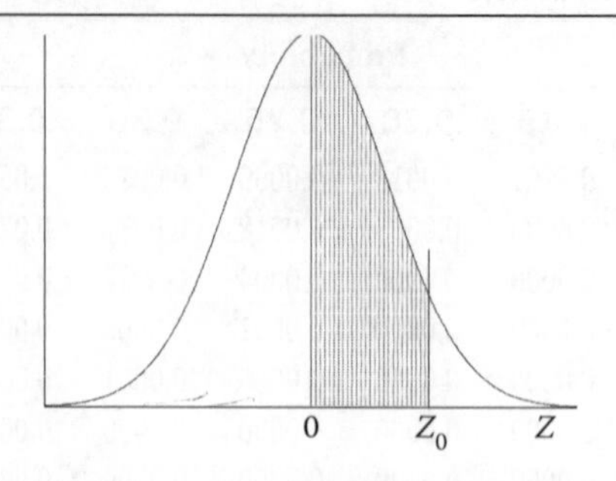

z	0.00	0.01	0.02	0.03	0.04	0.05	0.06	0.07	0.08	0.09
0.00	0.0000	0.0040	0.0080	0.0120	0.0160	0.0199	0.0239	0.0279	0.0319	0.0359
0.10	0.0398	0.0438	0.0478	0.0517	0.0557	0.0596	0.0636	0.0675	0.0714	0.0753
0.20	0.0793	0.0832	0.0871	0.0910	0.0948	0.0987	0.1026	0.1064	0.1103	0.1141
0.30	0.1179	0.1217	0.1255	0.1293	0.1331	0.1368	0.1406	0.1443	0.1480	0.1517
0.40	0.1554	0.1591	0.1628	0.1664	0.1700	0.1736	0.1772	0.1808	0.1844	0.1879
0.50	0.1915	0.1950	0.1985	0.2019	0.2054	0.2088	0.2123	0.2157	0.2190	0.2224
0.60	0.2257	0.2291	0.2324	0.2357	0.2389	0.2422	0.2454	0.2486	0.2517	0.2549
0.70	0.2580	0.2611	0.2642	0.2673	0.2704	0.2734	0.2764	0.2794	0.2823	0.2852
0.80	0.2881	0.2910	0.2939	0.2967	0.2995	0.3023	0.3051	0.3078	0.3106	0.3133
0.90	0.3159	0.3186	0.3212	0.3238	0.3264	0.3289	0.3315	0.3340	0.3365	0.3389
1.00	0.3413	0.3438	0.3461	0.3485	0.3508	0.3531	0.3554	0.3577	0.3599	0.3621
1.10	0.3643	0.3665	0.3686	0.3708	0.3729	0.3749	0.3770	0.3790	0.3810	0.3830
1.20	0.3849	0.3869	0.3888	0.3907	0.3925	0.3944	0.3962	0.3980	0.3997	0.4015
1.30	0.4032	0.4049	0.4066	0.4082	0.4099	0.4115	0.4131	0.4147	0.4162	0.4177
1.40	0.4192	0.4207	0.4222	0.4236	0.4251	0.4265	0.4279	0.4292	0.4306	0.4319
1.50	0.4332	0.4345	0.4357	0.4370	0.4382	0.4394	0.4406	0.4418	0.4429	0.4441
1.60	0.4452	0.4463	0.4474	0.4484	0.4495	0.4505	0.4515	0.4525	0.4535	0.4545
1.70	0.4554	0.4564	0.4573	0.4582	0.4591	0.4599	0.4608	0.4616	0.4625	0.4633
1.80	0.4641	0.4649	0.4656	0.4664	0.4671	0.4678	0.4686	0.4693	0.4699	0.4706
1.90	0.4713	0.4719	0.4726	0.4732	0.4738	0.4744	0.4750	0.4756	0.4761	0.4767
2.00	0.4772	0.4778	0.4783	0.4788	0.4793	0.4798	0.4803	0.4808	0.4812	0.4817
2.10	0.4821	0.4826	0.4830	0.4834	0.4838	0.4842	0.4846	0.4850	0.4854	0.4857
2.20	0.4861	0.4864	0.4868	0.4871	0.4875	0.4878	0.4881	0.4884	0.4887	0.4890
2.30	0.4893	0.4896	0.4898	0.4901	0.4904	0.4906	0.4909	0.4911	0.4913	0.4916
2.40	0.4918	0.4920	0.4922	0.4925	0.4927	0.4929	0.4931	0.4932	0.4934	0.4936
2.50	0.4938	0.4940	0.4941	0.4943	0.4945	0.4946	0.4948	0.4949	0.4951	0.4952
2.60	0.4953	0.4955	0.4956	0.4957	0.4959	0.4960	0.4961	0.4962	0.4963	0.4964
2.70	0.4965	0.4966	0.4967	0.4968	0.4969	0.4970	0.4971	0.4972	0.4973	0.4974
2.80	0.4974	0.4975	0.4976	0.4977	0.4977	0.4978	0.4979	0.4979	0.4980	0.4981
2.90	0.4981	0.4982	0.4982	0.4983	0.4984	0.4984	0.4985	0.4985	0.4986	0.4986
3.00	0.4987	0.4987	0.4987	0.4988	0.4988	0.4989	0.4989	0.4989	0.4990	0.4990
3.10	0.4990	0.4991	0.4991	0.4991	0.4992	0.4992	0.4992	0.4992	0.4993	0.4993
3.20	0.4993	0.4993	0.4994	0.4994	0.4994	0.4994	0.4994	0.4995	0.4995	0.4995
3.30	0.4995	0.4995	0.4995	0.4996	0.4996	0.4996	0.4996	0.4996	0.4996	0.4997
3.40	0.4997	0.4997	0.4997	0.4997	0.4997	0.4997	0.4997	0.4997	0.4997	0.4998
3.50	0.4998	0.4998	0.4998	0.4998	0.4998	0.4998	0.4998	0.4998	0.4998	0.4998

TABLE C.3 The *t* Distribution

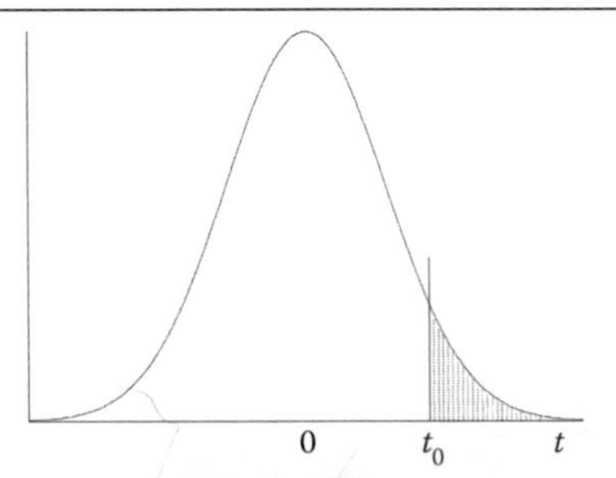

Probability

df	0.2500	1000	0500	0.0250	0.0100	0.0050	0.0025	0.0010	0.0005
1	1.000	3.078	6.314	12.706	31.821	63.657	127.321	318.309	636.619
2	0.816	1.886	2.920	4.303	6.965	9.925	14.089	22.327	31.599
3	0.765	1.638	2.353	3.182	4.541	5.841	7.453	10.215	12.924
4	0.741	1.533	2.132	2.776	3.747	4.604	5.598	7.173	8.610
5	0.727	1.476	2.015	2.571	3.365	4.032	4.773	5.893	6.869
6	0.718	1.440	1.943	2.447	3.143	3.707	4.317	5.208	5.959
7	0.711	1.415	1.895	2.365	2.998	3.499	4.029	4.785	5.408
8	0.706	1.397	1.860	2.306	2.896	3.355	3.833	4.501	5.041
9	0.703	1.383	1.833	2.262	2.821	3.250	3.690	4.297	4.781
10	0.700	1.372	1.812	2.228	2.764	3.169	3.581	4.144	4.587
11	0.697	1.363	1.796	2.201	2.718	3.106	3.497	4.025	4.437
12	0.695	1.356	1.782	2.179	2.681	3.055	3.428	3.930	4.318
13	0.694	1.350	1.771	2.160	2.650	3.012	3.372	3.852	4.221
14	0.692	1.345	1.761	2.145	2.624	2.977	3.326	3.787	4.140
15	0.691	1.341	1.753	2.131	2.602	2.947	3.286	3.733	4.073
16	0.690	1.337	1.746	2.120	2.583	2.921	3.252	3.686	4.015
17	0.689	1.333	1.740	2.110	2.567	2.898	3.222	3.646	3.965
18	0.688	1.330	1.734	2.101	2.552	2.878	3.197	3.610	3.922
19	0.688	1.328	1.729	2.093	2.539	2.861	3.174	3.579	3.883
20	0.687	1.325	1.725	2.086	2.528	2.845	3.153	3.552	3.850
21	0.686	1.323	1.721	2.080	2.518	2.831	3.135	3.527	3.819
22	0.686	1.321	1.717	2.074	2.508	2.819	3.119	3.505	3.792
23	0.685	1.319	1.714	2.069	2.500	2.807	3.104	3.485	3.768
24	0.685	1.318	1.711	2.064	2.492	2.797	3.091	3.467	3.745
25	0.684	1.316	1.708	2.060	2.485	2.787	3.078	3.450	3.725
26	0.684	1.315	1.706	2.056	2.479	2.779	3.067	3.435	3.707
27	0.684	1.314	1.703	2.052	2.473	2.771	3.057	3.421	3.690
28	0.683	1.313	1.701	2.048	2.467	2.763	3.047	3.408	3.674
29	0.683	1.311	1.699	2.045	2.462	2.756	3.038	3.396	3.659
30	0.683	1.310	1.697	2.042	2.457	2.750	3.030	3.385	3.646
31	0.682	1.309	1.696	2.040	2.453	2.744	3.022	3.375	3.633
32	0.682	1.309	1.694	2.037	2.449	2.738	3.015	3.365	3.622
33	0.682	1.308	1.692	2.035	2.445	2.733	3.008	3.356	3.611
34	0.682	1.307	1.691	2.032	2.441	2.728	3.002	3.348	3.601
35	0.682	1.306	1.690	2.030	2.438	2.724	2.996	3.340	3.591
40	0.681	1.303	1.684	2.021	2.423	2.704	2.971	3.307	3.551
45	0.680	1.301	1.679	2.014	2.412	2.690	2.952	3.281	3.520
50	0.679	1.299	1.676	2.009	2.403	2.678	2.937	3.261	3.496
55	0.679	1.297	1.673	2.004	2.396	2.668	2.925	3.245	3.476
60	0.679	1.296	1.671	2.000	2.390	2.660	2.915	3.232	3.460
70	0.678	1.294	1.667	1.994	2.381	2.648	2.899	3.211	3.435
80	0.678	1.292	1.664	1.990	2.374	2.639	2.887	3.195	3.416
100	0.677	1.290	1.660	1.984	2.364	2.626	2.871	3.174	3.390
120	0.677	1.289	1.658	1.980	2.358	2.617	2.860	3.160	3.373
150	0.676	1.287	1.655	1.976	2.351	2.609	2.849	3.145	3.357
∞	0.674	1.282	1.645	1.960	2.326	2.576	2.807	3.090	3.291

TABLE C.4 The *F* Distribution

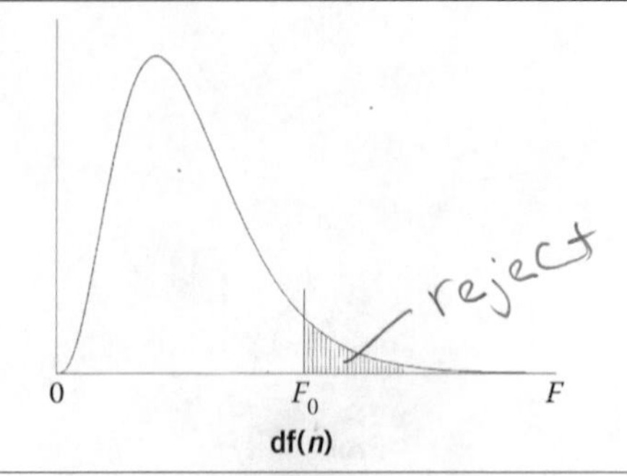

Prob	df(d)	1	2	3	4	5	6	7	8	9	10
0.100	1	39.863	49.500	53.593	55.833	57.240	58.204	58.906	59.439	59.858	60.195
0.050		161.448	199.500	215.707	224.583	230.162	233.986	236.768	238.883	240.543	241.882
0.025		647.789	799.500	864.163	899.583	921.848	937.111	948.217	956.656	963.285	968.627
0.010		4052.181	4999.500	5403.352	5624.583	5763.650	5858.986	5928.356	5981.070	6022.473	6055.847
0.005		16210.723	19999.500	21614.741	22499.583	23055.798	23437.111	23714.566	23925.406	24091.004	24224.487
0.100	2	8.526	9.000	9.162	9.243	9.293	9.326	9.349	9.367	9.381	9.392
0.050		18.513	19.000	19.164	19.247	19.296	19.330	19.353	19.371	19.385	19.396
0.025		38.506	39.000	39.165	39.248	39.298	39.331	39.355	39.373	39.387	39.398
0.010		98.503	99.000	99.166	99.249	99.299	99.333	99.356	99.374	99.388	99.399
0.005		198.501	199.000	199.166	199.250	199.300	199.333	199.357	199.375	199.388	199.400
0.100	3	5.538	5.462	5.391	5.343	5.309	5.285	5.266	5.252	5.240	5.230
0.050		10.128	9.552	9.277	9.117	9.013	8.941	8.887	8.845	8.812	8.786
0.025		17.443	16.044	15.439	15.101	14.885	14.735	14.624	14.540	14.473	14.419
0.010		34.116	30.817	29.457	28.710	28.237	27.911	27.672	27.489	27.345	27.229
0.005		55.552	49.799	47.467	46.195	45.392	44.838	44.434	44.126	43.882	43.686
0.100	4	4.545	4.325	4.191	4.107	4.051	4.010	3.979	3.955	3.936	3.920
0.050		7.709	6.944	6.591	6.388	6.256	6.163	6.094	6.041	5.999	5.964
0.025		12.218	10.649	9.979	9.605	9.364	9.197	9.074	8.980	8.905	8.844
0.010		21.198	18.000	16.694	15.977	15.522	15.207	14.976	14.799	14.659	14.546
0.005		31.333	26.284	24.259	23.155	22.456	21.975	21.622	21.352	21.139	20.967
0.100	5	4.060	3.780	3.619	3.520	3.453	3.405	3.368	3.339	3.316	3.297
0.050		6.608	5.786	5.409	5.192	5.050	4.950	4.876	4.818	4.772	4.735
0.025		10.007	8.434	7.764	7.388	7.146	6.978	6.853	6.757	6.681	6.619
0.010		16.258	13.274	12.060	11.392	10.967	10.672	10.456	10.289	10.158	10.051
0.005		22.785	18.314	16.530	15.556	14.940	14.513	14.200	13.961	13.772	13.618
0.100	6	3.776	3.463	3.289	3.181	3.108	3.055	3.014	2.983	2.958	2.937
0.050		5.987	5.143	4.757	4.534	4.387	4.284	4.207	4.147	4.099	4.060
0.025		8.813	7.260	6.599	6.227	5.988	5.820	5.695	5.600	5.523	5.461
0.010		13.745	10.925	9.780	9.148	8.746	8.466	8.260	8.102	7.976	7.874
0.005		18.635	14.544	12.917	12.028	11.464	11.073	10.786	10.566	10.391	10.250
0.100	7	3.589	3.257	3.074	2.961	2.883	2.827	2.785	2.752	2.725	2.703
0.050		5.591	4.737	4.347	4.120	3.972	3.866	3.787	3.726	3.677	3.637
0.025		8.073	6.542	5.890	5.523	5.285	5.119	4.995	4.899	4.823	4.761
0.010		12.246	9.547	8.451	7.847	7.460	7.191	6.993	6.840	6.719	6.620
0.005		16.236	12.404	10.882	10.050	9.522	9.155	8.885	8.678	8.514	8.380
0.100	8	3.458	3.113	2.924	2.806	2.726	2.668	2.624	2.589	2.561	2.538
0.050		5.318	4.459	4.066	3.838	3.687	3.581	3.500	3.438	3.388	3.347
0.025		7.571	6.059	5.416	5.053	4.817	4.652	4.529	4.433	4.357	4.295
0.010		11.259	8.649	7.591	7.006	6.632	6.371	6.178	6.029	5.911	5.814
0.005		14.688	11.042	9.596	8.805	8.302	7.952	7.694	7.496	7.339	7.211
0.100	9	3.360	3.006	2.813	2.693	2.611	2.551	2.505	2.469	2.440	2.416
0.050		5.117	4.256	3.863	3.633	3.482	3.374	3.293	3.230	3.179	3.137
0.025		7.209	5.715	5.078	4.718	4.484	4.320	4.197	4.102	4.026	3.964
0.010		10.561	8.022	6.992	6.422	6.057	5.802	5.613	5.467	5.351	5.257
0.005		13.614	10.107	8.717	7.956	7.471	7.134	6.885	6.693	6.541	6.417
0.100	10	3.285	2.924	2.728	2.605	2.522	2.461	2.414	2.377	2.347	2.323
0.050		4.965	4.103	3.708	3.478	3.326	3.217	3.135	3.072	3.020	2.978
0.025		6.937	5.456	4.826	4.468	4.236	4.072	3.950	3.855	3.779	3.717
0.010		10.044	7.559	6.552	5.994	5.636	5.386	5.200	5.057	4.942	4.849
0.005		12.826	9.427	8.081	7.343	6.872	6.545	6.302	6.116	5.968	5.847

TABLE C.4 The F Distribution (Continued)

		df(n)									
Prob	df(d)	12	14	17	20	24	30	40	60	120	∞
0.100	1	60.705	61.073	61.464	61.740	62.002	62.265	62.529	62.794	63.061	63.328
0.050		243.906	245.364	246.918	248.013	249.052	250.095	251.143	252.196	253.253	254.314
0.025		976.708	982.528	988.733	993.103	997.249	1001.414	1005.598	1009.800	1014.020	1018.258
0.010		6106.321	6142.674	6181.435	6208.730	6234.631	6260.649	6286.782	6313.030	6339.391	6365.864
0.005		24426.366	24571.767	24726.798	24835.971	24939.565	25043.628	25148.153	25253.137	25358.573	25464.458
0.100	2	9.408	9.420	9.433	9.441	9.450	9.458	9.466	9.475	9.483	9.491
0.050		19.413	19.424	19.437	19.446	19.454	19.462	19.471	19.479	19.487	19.496
0.025		39.415	39.427	39.439	39.448	39.456	39.465	39.473	39.481	39.490	39.498
0.010		99.416	99.428	99.440	99.449	99.458	99.466	99.474	99.482	99.491	99.499
0.005		199.416	199.428	199.441	199.450	199.458	199.466	199.475	199.483	199.491	199.500
0.100	3	5.216	5.205	5.193	5.184	5.176	5.168	5.160	5.151	5.143	5.134
0.050		8.745	8.715	8.683	8.660	8.639	8.617	8.594	8.572	8.549	8.526
0.025		14.337	14.277	14.213	14.167	14.124	14.081	14.037	13.992	13.947	13.902
0.010		27.052	26.924	26.787	26.690	26.598	26.505	26.411	26.316	26.221	26.125
0.005		43.387	43.172	42.941	42.778	42.622	42.466	42.308	42.149	41.989	41.828
0.100	4	3.896	3.878	3.858	3.844	3.831	3.817	3.804	3.790	3.775	3.761
0.050		5.912	5.873	5.832	5.803	5.774	5.746	5.717	5.688	5.658	5.628
0.025		8.751	8.684	8.611	8.560	8.511	8.461	8.411	8.360	8.309	8.257
0.010		14.374	14.249	14.115	14.020	13.929	13.838	13.745	13.652	13.558	13.463
0.005		20.705	20.515	20.311	20.167	20.030	19.892	19.752	19.611	19.468	19.325
0.100	5	3.268	3.247	3.223	3.207	3.191	3.174	3.157	3.140	3.123	3.105
0.050		4.678	4.636	4.590	4.558	4.527	4.496	4.464	4.431	4.398	4.365
0.025		6.525	6.456	6.381	6.329	6.278	6.227	6.175	6.123	6.069	6.015
0.010		9.888	9.770	9.643	9.553	9.466	9.379	9.291	9.202	9.112	9.020
0.005		13.384	13.215	13.033	12.903	12.780	12.656	12.530	12.402	12.274	12.144
0.100	6	2.905	2.881	2.855	2.836	2.818	2.800	2.781	2.762	2.742	2.722
0.050		4.000	3.956	3.908	3.874	3.841	3.808	3.774	3.740	3.705	3.669
0.025		5.366	5.297	5.222	5.168	5.117	5.065	5.012	4.959	4.904	4.849
0.010		7.718	7.605	7.483	7.396	7.313	7.229	7.143	7.057	6.969	6.880
0.005		10.034	9.877	9.709	9.589	9.474	9.358	9.241	9.122	9.001	8.879
0.100	7	2.668	2.643	2.615	2.595	2.575	2.555	2.535	2.514	2.493	2.471
0.050		3.575	3.529	3.480	3.445	3.410	3.376	3.340	3.304	3.267	3.230
0.025		4.666	4.596	4.521	4.467	4.415	4.362	4.309	4.254	4.199	4.142
0.010		6.469	6.359	6.240	6.155	6.074	5.992	5.908	5.824	5.737	5.650
0.005		8.176	8.028	7.868	7.754	7.645	7.534	7.422	7.309	7.193	7.076
0.100	8	2.502	2.475	2.446	2.425	2.404	2.383	2.361	2.339	2.316	2.293
0.050		3.284	3.237	3.187	3.150	3.115	3.079	3.043	3.005	2.967	2.928
0.025		4.200	4.130	4.054	3.999	3.947	3.894	3.840	3.784	3.728	3.670
0.010		5.667	5.559	5.442	5.359	5.279	5.198	5.116	5.032	4.946	4.859
0.005		7.015	6.872	6.718	6.608	6.503	6.396	6.288	6.177	6.065	5.951
0.100	9	2.379	2.351	2.320	2.298	2.277	2.255	2.232	2.208	2.184	2.159
0.050		3.073	3.025	2.974	2.936	2.900	2.864	2.826	2.787	2.748	2.707
0.025		3.868	3.798	3.722	3.667	3.614	3.560	3.505	3.449	3.392	3.333
0.010		5.111	5.005	4.890	4.808	4.729	4.649	4.567	4.483	4.398	4.311
0.005		6.227	6.089	5.939	5.832	5.729	5.625	5.519	5.410	5.300	5.188
0.100	10	2.284	2.255	2.224	2.201	2.178	2.155	2.132	2.107	2.082	2.055
0.050		2.913	2.865	2.812	2.774	2.737	2.700	2.661	2.621	2.580	2.538
0.025		3.621	3.550	3.474	3.419	3.365	3.311	3.255	3.198	3.140	3.080
0.010		4.706	4.601	4.487	4.405	4.327	4.247	4.165	4.082	3.996	3.909
0.005		5.661	5.526	5.379	5.274	5.173	5.071	4.966	4.859	4.750	4.639

(Continued)

TABLE C.4 The *F* Distribution (Continued)

		df(*n*)									
Prob	**df(*d*)**	**1**	**2**	**3**	**4**	**5**	**6**	**7**	**8**	**9**	**10**
0.100	11	3.225	2.860	2.660	2.536	2.451	2.389	2.342	2.304	2.274	2.248
0.050		4.844	3.982	3.587	3.357	3.204	3.095	3.012	2.948	2.896	2.854
0.025		6.724	5.256	4.630	4.275	4.044	3.881	3.759	3.664	3.588	3.526
0.010		9.646	7.206	6.217	5.668	5.316	5.069	4.886	4.744	4.632	4.539
0.005		12.226	8.912	7.600	6.881	6.422	6.102	5.865	5.682	5.537	5.418
0.100	12	3.177	2.807	2.606	2.480	2.394	2.331	2.283	2.245	2.214	2.188
0.050		4.747	3.885	3.490	3.259	3.106	2.996	2.913	2.849	2.796	2.753
0.025		6.554	5.096	4.474	4.121	3.891	3.728	3.607	3.512	3.436	3.374
0.010		9.330	6.927	5.953	5.412	5.064	4.821	4.640	4.499	4.388	4.296
0.005		11.754	8.510	7.226	6.521	6.071	5.757	5.525	5.345	5.202	5.085
0.100	13	3.136	2.763	2.560	2.434	2.347	2.283	2.234	2.195	2.164	2.138
0.050		4.667	3.806	3.411	3.179	3.025	2.915	2.832	2.767	2.714	2.671
0.025		6.414	4.965	4.347	3.996	3.767	3.604	3.483	3.388	3.312	3.250
0.010		9.074	6.701	5.739	5.205	4.862	4.620	4.441	4.302	4.191	4.100
0.005		11.374	8.186	6.926	6.233	5.791	5.482	5.253	5.076	4.935	4.820
0.100	14	3.102	2.726	2.522	2.395	2.307	2.243	2.193	2.154	2.122	2.095
0.050		4.600	3.739	3.344	3.112	2.958	2.848	2.764	2.699	2.646	2.602
0.025		6.298	4.857	4.242	3.892	3.663	3.501	3.380	3.285	3.209	3.147
0.010		8.862	6.515	5.564	5.035	4.695	4.456	4.278	4.140	4.030	3.939
0.005		11.060	7.922	6.680	5.998	5.562	5.257	5.031	4.857	4.717	4.603
0.100	15	3.073	2.695	2.490	2.361	2.273	2.208	2.158	2.119	2.086	2.059
0.050		4.543	3.682	3.287	3.056	2.901	2.790	2.707	2.641	2.588	2.544
0.025		6.200	4.765	4.153	3.804	3.576	3.415	3.293	3.199	3.123	3.060
0.010		8.683	6.359	5.417	4.893	4.556	4.318	4.142	4.004	3.895	3.805
0.005		10.798	7.701	6.476	5.803	5.372	5.071	4.847	4.674	4.536	4.424
0.100	16	3.048	2.668	2.462	2.333	2.244	2.178	2.128	2.088	2.055	2.028
0.050		4.494	3.634	3.239	3.007	2.852	2.741	2.657	2.591	2.538	2.494
0.025		6.115	4.687	4.077	3.729	3.502	3.341	3.219	3.125	3.049	2.986
0.010		8.531	6.226	5.292	4.773	4.437	4.202	4.026	3.890	3.780	3.691
0.005		10.575	7.514	6.303	5.638	5.212	4.913	4.692	4.521	4.384	4.272
0.100	17	3.026	2.645	2.437	2.308	2.218	2.152	2.102	2.061	2.028	2.001
0.050		4.451	3.592	3.197	2.965	2.810	2.699	2.614	2.548	2.494	2.450
0.025		6.042	4.619	4.011	3.665	3.438	3.277	3.156	3.061	2.985	2.922
0.010		8.400	6.112	5.185	4.669	4.336	4.102	3.927	3.791	3.682	3.593
0.005		10.384	7.354	6.156	5.497	5.075	4.779	4.559	4.389	4.254	4.142
0.100	18	3.007	2.624	2.416	2.286	2.196	2.130	2.079	2.038	2.005	1.977
0.050		4.414	3.555	3.160	2.928	2.773	2.661	2.577	2.510	2.456	2.412
0.025		5.978	4.560	3.954	3.608	3.382	3.221	3.100	3.005	2.929	2.866
0.010		8.285	6.013	5.092	4.579	4.248	4.015	3.841	3.705	3.597	3.508
0.005		10.218	7.215	6.028	5.375	4.956	4.663	4.445	4.276	4.141	4.030
0.100	19	2.990	2.606	2.397	2.266	2.176	2.109	2.058	2.017	1.984	1.956
0.050		4.381	3.522	3.127	2.895	2.740	2.628	2.544	2.477	2.423	2.378
0.025		5.922	4.508	3.903	3.559	3.333	3.172	3.051	2.956	2.880	2.817
0.010		8.185	5.926	5.010	4.500	4.171	3.939	3.765	3.631	3.523	3.434
0.005		10.073	7.093	5.916	5.268	4.853	4.561	4.345	4.177	4.043	3.933
0.100	20	2.975	2.589	2.380	2.249	2.158	2.091	2.040	1.999	1.965	1.937
0.050		4.351	3.493	3.098	2.866	2.711	2.599	2.514	2.447	2.393	2.348
0.025		5.871	4.461	3.859	3.515	3.289	3.128	3.007	2.913	2.837	2.774
0.010		8.096	5.849	4.938	4.431	4.103	3.871	3.699	3.564	3.457	3.368
0.005		9.944	6.986	5.818	5.174	4.762	4.472	4.257	4.090	3.956	3.847
0.100	21	2.961	2.575	2.365	2.233	2.142	2.075	2.023	1.982	1.948	1.920
0.050		4.325	3.467	3.072	2.840	2.685	2.573	2.488	2.420	2.366	2.321
0.025		5.827	4.420	3.819	3.475	3.250	3.090	2.969	2.874	2.798	2.735
0.010		8.017	5.780	4.874	4.369	4.042	3.812	3.640	3.506	3.398	3.310
0.005		9.830	6.891	5.730	5.091	4.681	4.393	4.179	4.013	3.880	3.771
0.100	22	2.949	2.561	2.351	2.219	2.128	2.060	2.008	1.967	1.933	1.904
0.050		4.301	3.443	3.049	2.817	2.661	2.549	2.464	2.397	2.342	2.297
0.025		5.786	4.383	3.783	3.440	3.215	3.055	2.934	2.839	2.763	2.700
0.010		7.945	5.719	4.817	4.313	3.988	3.758	3.587	3.453	3.346	3.258
0.005		9.727	6.806	5.652	5.017	4.609	4.322	4.109	3.944	3.812	3.703

TABLE C.4 The F Distribution (Continued)

		df(n)									
Prob	df(d)	12	14	17	20	24	30	40	60	120	∞
0.100	11	2.209	2.179	2.147	2.123	2.100	2.076	2.052	2.026	2.000	1.972
0.050		2.788	2.739	2.685	2.646	2.609	2.570	2.531	2.490	2.448	2.404
0.025		3.430	3.359	3.282	3.226	3.173	3.118	3.061	3.004	2.944	2.883
0.010		4.397	4.293	4.180	4.099	4.021	3.941	3.860	3.776	3.690	3.602
0.005		5.236	5.103	4.959	4.855	4.756	4.654	4.551	4.445	4.337	4.226
0.100	12	2.147	2.117	2.084	2.060	2.036	2.011	1.986	1.960	1.932	1.904
0.050		2.687	2.637	2.583	2.544	2.505	2.466	2.426	2.384	2.341	2.296
0.025		3.277	3.206	3.129	3.073	3.019	2.963	2.906	2.848	2.787	2.725
0.010		4.155	4.052	3.939	3.858	3.780	3.701	3.619	3.535	3.449	3.361
0.005		4.906	4.775	4.632	4.530	4.431	4.331	4.228	4.123	4.015	3.904
0.100	13	2.097	2.066	2.032	2.007	1.983	1.958	1.931	1.904	1.876	1.846
0.050		2.604	2.554	2.499	2.459	2.420	2.380	2.339	2.297	2.252	2.206
0.025		3.153	3.082	3.004	2.948	2.893	2.837	2.780	2.720	2.659	2.595
0.010		3.960	3.857	3.745	3.665	3.587	3.507	3.425	3.341	3.255	3.165
0.005		4.643	4.513	4.372	4.270	4.173	4.073	3.970	3.866	3.758	3.647
0.100	14	2.054	2.022	1.988	1.962	1.938	1.912	1.885	1.857	1.828	1.797
0.050		2.534	2.484	2.428	2.388	2.349	2.308	2.266	2.223	2.178	2.131
0.025		3.050	2.979	2.900	2.844	2.789	2.732	2.674	2.614	2.552	2.487
0.010		3.800	3.698	3.586	3.505	3.427	3.348	3.266	3.181	3.094	3.004
0.005		4.428	4.299	4.159	4.059	3.961	3.862	3.760	3.655	3.547	3.436
0.100	15	2.017	1.985	1.950	1.924	1.899	1.873	1.845	1.817	1.787	1.755
0.050		2.475	2.424	2.368	2.328	2.288	2.247	2.204	2.160	2.114	2.066
0.025		2.963	2.891	2.813	2.756	2.701	2.644	2.585	2.524	2.461	2.395
0.010		3.666	3.564	3.452	3.372	3.294	3.214	3.132	3.047	2.959	2.868
0.005		4.250	4.122	3.983	3.883	3.786	3.687	3.585	3.480	3.372	3.260
0.100	16	1.985	1.953	1.917	1.891	1.866	1.839	1.811	1.782	1.751	1.718
0.050		2.425	2.373	2.317	2.276	2.235	2.194	2.151	2.106	2.059	2.010
0.025		2.889	2.817	2.738	2.681	2.625	2.568	2.509	2.447	2.383	2.316
0.010		3.553	3.451	3.339	3.259	3.181	3.101	3.018	2.933	2.845	2.753
0.005		4.099	3.972	3.834	3.734	3.638	3.539	3.437	3.332	3.224	3.112
0.100	17	1.958	1.925	1.889	1.862	1.836	1.809	1.781	1.751	1.719	1.686
0.050		2.381	2.329	2.272	2.230	2.190	2.148	2.104	2.058	2.011	1.960
0.025		2.825	2.753	2.673	2.616	2.560	2.502	2.442	2.380	2.315	2.247
0.010		3.455	3.353	3.242	3.162	3.084	3.003	2.920	2.835	2.746	2.653
0.005		3.971	3.844	3.707	3.607	3.511	3.412	3.311	3.206	3.097	2.984
0.100	18	1.933	1.900	1.864	1.837	1.810	1.783	1.754	1.723	1.691	1.657
0.050		2.342	2.290	2.233	2.191	2.150	2.107	2.063	2.017	1.968	1.917
0.025		2.769	2.696	2.617	2.559	2.503	2.445	2.384	2.321	2.256	2.187
0.010		3.371	3.269	3.158	3.077	2.999	2.919	2.835	2.749	2.660	2.566
0.005		3.860	3.734	3.597	3.498	3.402	3.303	3.201	3.096	2.987	2.873
0.100	19	1.912	1.878	1.841	1.814	1.787	1.759	1.730	1.699	1.666	1.631
0.050		2.308	2.256	2.198	2.155	2.114	2.071	2.026	1.980	1.930	1.878
0.025		2.720	2.647	2.567	2.509	2.452	2.394	2.333	2.270	2.203	2.133
0.010		3.297	3.195	3.084	3.003	2.925	2.844	2.761	2.674	2.584	2.489
0.005		3.763	3.638	3.501	3.402	3.306	3.208	3.106	3.000	2.891	2.776
0.100	20	1.892	1.859	1.821	1.794	1.767	1.738	1.708	1.677	1.643	1.607
0.050		2.278	2.225	2.167	2.124	2.082	2.039	1.994	1.946	1.896	1.843
0.025		2.676	2.603	2.523	2.464	2.408	2.349	2.287	2.223	2.156	2.085
0.010		3.231	3.130	3.018	2.938	2.859	2.778	2.695	2.608	2.517	2.421
0.005		3.678	3.553	3.416	3.318	3.222	3.123	3.022	2.916	2.806	2.690
0.100	21	1.875	1.841	1.803	1.776	1.748	1.719	1.689	1.657	1.623	1.586
0.050		2.250	2.197	2.139	2.096	2.054	2.010	1.965	1.916	1.866	1.812
0.025		2.637	2.564	2.483	2.425	2.368	2.308	2.246	2.182	2.114	2.042
0.010		3.173	3.072	2.960	2.880	2.801	2.720	2.636	2.548	2.457	2.360
0.005		3.602	3.478	3.342	3.243	3.147	3.049	2.947	2.841	2.730	2.614
0.100	22	1.859	1.825	1.787	1.759	1.731	1.702	1.671	1.639	1.604	1.567
0.050		2.226	2.173	2.114	2.071	2.028	1.984	1.938	1.889	1.838	1.783
0.025		2.602	2.528	2.448	2.389	2.331	2.272	2.210	2.145	2.076	2.003
0.010		3.121	3.019	2.908	2.827	2.749	2.667	2.583	2.495	2.403	2.305
0.005		3.535	3.411	3.275	3.176	3.081	2.982	2.880	2.774	2.663	2.545

(*Continued*)

TABLE C.4 The *F* Distribution (Continued)

		df(n)									
Prob	**df(d)**	**1**	**2**	**3**	**4**	**5**	**6**	**7**	**8**	**9**	**10**
0.100	23	2.937	2.549	2.339	2.207	2.115	2.047	1.995	1.953	1.919	1.890
0.050		4.279	3.422	3.028	2.796	2.640	2.528	2.442	2.375	2.320	2.275
0.025		5.750	4.349	3.750	3.408	3.183	3.023	2.902	2.808	2.731	2.668
0.010		7.881	5.664	4.765	4.264	3.939	3.710	3.539	3.406	3.299	3.211
0.005		9.635	6.730	5.582	4.950	4.544	4.259	4.047	3.882	3.750	3.642
0.100	24	2.927	2.538	2.327	2.195	2.103	2.035	1.983	1.941	1.906	1.877
0.050		4.260	3.403	3.009	2.776	2.621	2.508	2.423	2.355	2.300	2.255
0.025		5.717	4.319	3.721	3.379	3.155	2.995	2.874	2.779	2.703	2.640
0.010		7.823	5.614	4.718	4.218	3.895	3.667	3.496	3.363	3.256	3.168
0.005		9.551	6.661	5.519	4.890	4.486	4.202	3.991	3.826	3.695	3.587
0.100	25	2.918	2.528	2.317	2.184	2.092	2.024	1.971	1.929	1.895	1.866
0.050		4.242	3.385	2.991	2.759	2.603	2.490	2.405	2.337	2.282	2.236
0.025		5.686	4.291	3.694	3.353	3.129	2.969	2.848	2.753	2.677	2.613
0.010		7.770	5.568	4.675	4.177	3.855	3.627	3.457	3.324	3.217	3.129
0.005		9.475	6.598	5.462	4.835	4.433	4.150	3.939	3.776	3.645	3.537
0.100	26	2.909	2.519	2.307	2.174	2.082	2.014	1.961	1.919	1.884	1.855
0.050		4.225	3.369	2.975	2.743	2.587	2.474	2.388	2.321	2.265	2.220
0.025		5.659	4.265	3.670	3.329	3.105	2.945	2.824	2.729	2.653	2.590
0.010		7.721	5.526	4.637	4.140	3.818	3.591	3.421	3.288	3.182	3.094
0.005		9.406	6.541	5.409	4.785	4.384	4.103	3.893	3.730	3.599	3.492
0.100	27	2.901	2.511	2.299	2.165	2.073	2.005	1.952	1.909	1.874	1.845
0.050		4.210	3.354	2.960	2.728	2.572	2.459	2.373	2.305	2.250	2.204
0.025		5.633	4.242	3.647	3.307	3.083	2.923	2.802	2.707	2.631	2.568
0.010		7.677	5.488	4.601	4.106	3.785	3.558	3.388	3.256	3.149	3.062
0.005		9.342	6.489	5.361	4.740	4.340	4.059	3.850	3.687	3.557	3.450
0.100	28	2.894	2.503	2.291	2.157	2.064	1.996	1.943	1.900	1.865	1.836
0.050		4.196	3.340	2.947	2.714	2.558	2.445	2.359	2.291	2.236	2.190
0.025		5.610	4.221	3.626	3.286	3.063	2.903	2.782	2.687	2.611	2.547
0.010		7.636	5.453	4.568	4.074	3.754	3.528	3.358	3.226	3.120	3.032
0.005		9.284	6.440	5.317	4.698	4.300	4.020	3.811	3.649	3.519	3.412
0.100	29	2.887	2.495	2.283	2.149	2.057	1.988	1.935	1.892	1.857	1.827
0.050		4.183	3.328	2.934	2.701	2.545	2.432	2.346	2.278	2.223	2.177
0.025		5.588	4.201	3.607	3.267	3.044	2.884	2.763	2.669	2.592	2.529
0.010		7.598	5.420	4.538	4.045	3.725	3.499	3.330	3.198	3.092	3.005
0.005		9.230	6.396	5.276	4.659	4.262	3.983	3.775	3.613	3.483	3.377
0.100	30	2.881	2.489	2.276	2.142	2.049	1.980	1.927	1.884	1.849	1.819
0.050		4.171	3.316	2.922	2.690	2.534	2.421	2.334	2.266	2.211	2.165
0.025		5.568	4.182	3.589	3.250	3.026	2.867	2.746	2.651	2.575	2.511
0.010		7.562	5.390	4.510	4.018	3.699	3.473	3.304	3.173	3.067	2.979
0.005		9.180	6.355	5.239	4.623	4.228	3.949	3.742	3.580	3.450	3.344
0.100	31	2.875	2.482	2.270	2.136	2.042	1.973	1.920	1.877	1.842	1.812
0.050		4.160	3.305	2.911	2.679	2.523	2.409	2.323	2.255	2.199	2.153
0.025		5.549	4.165	3.573	3.234	3.010	2.851	2.730	2.635	2.558	2.495
0.010		7.530	5.362	4.484	3.993	3.675	3.449	3.281	3.149	3.043	2.955
0.005		9.133	6.317	5.204	4.590	4.196	3.918	3.711	3.549	3.420	3.314
0.100	32	2.869	2.477	2.263	2.129	2.036	1.967	1.913	1.870	1.835	1.805
0.050		4.149	3.295	2.901	2.668	2.512	2.399	2.313	2.244	2.189	2.142
0.025		5.531	4.149	3.557	3.218	2.995	2.836	2.715	2.620	2.543	2.480
0.010		7.499	5.336	4.459	3.969	3.652	3.427	3.258	3.127	3.021	2.934
0.005		9.090	6.281	5.171	4.559	4.166	3.889	3.682	3.521	3.392	3.286
0.100	33	2.864	2.471	2.258	2.123	2.030	1.961	1.907	1.864	1.828	1.799
0.050		4.139	3.285	2.892	2.659	2.503	2.389	2.303	2.235	2.179	2.133
0.025		5.515	4.134	3.543	3.204	2.981	2.822	2.701	2.606	2.529	2.466
0.010		7.471	5.312	4.437	3.948	3.630	3.406	3.238	3.106	3.000	2.913
0.005		9.050	6.248	5.141	4.531	4.138	3.861	3.655	3.495	3.366	3.260
0.100	34	2.859	2.466	2.252	2.118	2.024	1.955	1.901	1.858	1.822	1.793
0.050		4.130	3.276	2.883	2.650	2.494	2.380	2.294	2.225	2.170	2.123
0.025		5.499	4.120	3.529	3.191	2.968	2.808	2.688	2.593	2.516	2.453
0.010		7.444	5.289	4.416	3.927	3.611	3.386	3.218	3.087	2.981	2.894
0.005		9.012	6.217	5.113	4.504	4.112	3.836	3.630	3.470	3.341	3.235

TABLE C.4 The F Distribution (Continued)

		df(n)									
Prob	df(d)	1	2	3	4	5	6	7	8	9	10
0.100	23	1.845	1.811	1.772	1.744	1.716	1.686	1.655	1.622	1.587	1.549
0.050		2.204	2.150	2.091	2.048	2.005	1.961	1.914	1.865	1.813	1.757
0.025		2.570	2.497	2.416	2.357	2.299	2.239	2.176	2.111	2.041	1.968
0.010		3.074	2.973	2.861	2.781	2.702	2.620	2.535	2.447	2.354	2.256
0.005		3.475	3.351	3.215	3.116	3.021	2.922	2.820	2.713	2.602	2.484
0.100	24	1.832	1.797	1.759	1.730	1.702	1.672	1.641	1.607	1.571	1.533
0.050		2.183	2.130	2.070	2.027	1.984	1.939	1.892	1.842	1.790	1.733
0.025		2.541	2.468	2.386	2.327	2.269	2.209	2.146	2.080	2.010	1.935
0.010		3.032	2.930	2.819	2.738	2.659	2.577	2.492	2.403	2.310	2.211
0.005		3.420	3.296	3.161	3.062	2.967	2.868	2.765	2.658	2.546	2.428
0.100	25	1.820	1.785	1.746	1.718	1.689	1.659	1.627	1.593	1.557	1.518
0.050		2.165	2.111	2.051	2.007	1.964	1.919	1.872	1.822	1.768	1.711
0.025		2.515	2.441	2.360	2.300	2.242	2.182	2.118	2.052	1.981	1.906
0.010		2.993	2.892	2.780	2.699	2.620	2.538	2.453	2.364	2.270	2.169
0.005		3.370	3.247	3.111	3.013	2.918	2.819	2.716	2.609	2.496	2.377
0.100	26	1.809	1.774	1.735	1.706	1.677	1.647	1.615	1.581	1.544	1.504
0.050		2.148	2.094	2.034	1.990	1.946	1.901	1.853	1.803	1.749	1.691
0.025		2.491	2.417	2.335	2.276	2.217	2.157	2.093	2.026	1.954	1.878
0.010		2.958	2.857	2.745	2.664	2.585	2.503	2.417	2.327	2.233	2.131
0.005		3.325	3.202	3.067	2.968	2.873	2.774	2.671	2.563	2.450	2.330
0.100	27	1.799	1.764	1.724	1.695	1.666	1.636	1.603	1.569	1.531	1.491
0.050		2.132	2.078	2.018	1.974	1.930	1.884	1.836	1.785	1.731	1.672
0.025		2.469	2.395	2.313	2.253	2.195	2.133	2.069	2.002	1.930	1.853
0.010		2.926	2.824	2.713	2.632	2.552	2.470	2.384	2.294	2.198	2.097
0.005		3.284	3.161	3.026	2.928	2.832	2.733	2.630	2.522	2.408	2.287
0.100	28	1.790	1.754	1.715	1.685	1.656	1.625	1.592	1.558	1.520	1.478
0.050		2.118	2.064	2.003	1.959	1.915	1.869	1.820	1.769	1.714	1.654
0.025		2.448	2.374	2.292	2.232	2.174	2.112	2.048	1.980	1.907	1.829
0.010		2.896	2.795	2.683	2.602	2.522	2.440	2.354	2.263	2.167	2.064
0.005		3.246	3.123	2.988	2.890	2.794	2.695	2.592	2.483	2.369	2.247
0.100	29	1.781	1.745	1.705	1.676	1.647	1.616	1.583	1.547	1.509	1.467
0.050		2.104	2.050	1.989	1.945	1.901	1.854	1.806	1.754	1.698	1.638
0.025		2.430	2.355	2.273	2.213	2.154	2.092	2.028	1.959	1.886	1.807
0.010		2.868	2.767	2.656	2.574	2.495	2.412	2.325	2.234	2.138	2.034
0.005		3.211	3.088	2.953	2.855	2.759	2.660	2.557	2.448	2.333	2.210
0.100	30	1.773	1.737	1.697	1.667	1.638	1.606	1.573	1.538	1.499	1.456
0.050		2.092	2.037	1.976	1.932	1.887	1.841	1.792	1.740	1.683	1.622
0.025		2.412	2.338	2.255	2.195	2.136	2.074	2.009	1.940	1.866	1.787
0.010		2.843	2.742	2.630	2.549	2.469	2.386	2.299	2.208	2.111	2.006
0.005		3.179	3.056	2.921	2.823	2.727	2.628	2.524	2.415	2.300	2.176
0.100	31	1.765	1.729	1.689	1.659	1.630	1.598	1.565	1.529	1.489	1.446
0.050		2.080	2.026	1.965	1.920	1.875	1.828	1.779	1.726	1.670	1.608
0.025		2.396	2.321	2.239	2.178	2.119	2.057	1.991	1.922	1.848	1.768
0.010		2.820	2.718	2.606	2.525	2.445	2.362	2.275	2.183	2.086	1.980
0.005		3.149	3.026	2.891	2.793	2.697	2.598	2.494	2.385	2.269	2.144
0.100	32	1.758	1.722	1.682	1.652	1.622	1.590	1.556	1.520	1.481	1.437
0.050		2.070	2.015	1.953	1.908	1.864	1.817	1.767	1.714	1.657	1.594
0.025		2.381	2.306	2.223	2.163	2.103	2.041	1.975	1.905	1.831	1.750
0.010		2.798	2.696	2.584	2.503	2.423	2.340	2.252	2.160	2.062	1.956
0.005		3.121	2.998	2.864	2.766	2.670	2.570	2.466	2.356	2.240	2.114
0.100	33	1.751	1.715	1.675	1.645	1.615	1.583	1.549	1.512	1.472	1.428
0.050		2.060	2.004	1.943	1.898	1.853	1.806	1.756	1.702	1.645	1.581
0.025		2.366	2.292	2.209	2.148	2.088	2.026	1.960	1.890	1.815	1.733
0.010		2.777	2.676	2.564	2.482	2.402	2.319	2.231	2.139	2.040	1.933
0.005		3.095	2.973	2.838	2.740	2.644	2.544	2.440	2.330	2.213	2.087
0.100	34	1.745	1.709	1.668	1.638	1.608	1.576	1.541	1.505	1.464	1.419
0.050		2.050	1.995	1.933	1.888	1.843	1.795	1.745	1.691	1.633	1.569
0.025		2.353	2.278	2.195	2.135	2.075	2.012	1.946	1.875	1.799	1.717
0.010		2.758	2.657	2.545	2.463	2.383	2.299	2.211	2.118	2.019	1.911
0.005		3.071	2.948	2.814	2.716	2.620	2.520	2.415	2.305	2.188	2.060

(*Continued*)

TABLE C.4 The F Distribution (Continued)

		df(n)									
Prob	**df(d)**	**1**	**2**	**3**	**4**	**5**	**6**	**7**	**8**	**9**	**10**
0.100	35	2.855	2.461	2.247	2.113	2.019	1.950	1.896	1.852	1.817	1.787
0.050		4.121	3.267	2.874	2.641	2.485	2.372	2.285	2.217	2.161	2.114
0.025		5.485	4.106	3.517	3.179	2.956	2.796	2.676	2.581	2.504	2.440
0.010		7.419	5.268	4.396	3.908	3.592	3.368	3.200	3.069	2.963	2.876
0.005		8.976	6.188	5.086	4.479	4.088	3.812	3.607	3.447	3.318	3.212
0.100	40	2.835	2.440	2.226	2.091	1.997	1.927	1.873	1.829	1.793	1.763
0.050		4.085	3.232	2.839	2.606	2.449	2.336	2.249	2.180	2.124	2.077
0.025		5.424	4.051	3.463	3.126	2.904	2.744	2.624	2.529	2.452	2.388
0.010		7.314	5.179	4.313	3.828	3.514	3.291	3.124	2.993	2.888	2.801
0.005		8.828	6.066	4.976	4.374	3.986	3.713	3.509	3.350	3.222	3.117
0.100	45	2.820	2.425	2.210	2.074	1.980	1.909	1.855	1.811	1.774	1.744
0.050		4.057	3.204	2.812	2.579	2.422	2.308	2.221	2.152	2.096	2.049
0.025		5.377	4.009	3.422	3.086	2.864	2.705	2.584	2.489	2.412	2.348
0.010		7.234	5.110	4.249	3.767	3.454	3.232	3.066	2.935	2.830	2.743
0.005		8.715	5.974	4.892	4.294	3.909	3.638	3.435	3.276	3.149	3.044
0.100	50	2.809	2.412	2.197	2.061	1.966	1.895	1.840	1.796	1.760	1.729
0.050		4.034	3.183	2.790	2.557	2.400	2.286	2.199	2.130	2.073	2.026
0.025		5.340	3.975	3.390	3.054	2.833	2.674	2.553	2.458	2.381	2.317
0.010		7.171	5.057	4.199	3.720	3.408	3.186	3.020	2.890	2.785	2.698
0.005		8.626	5.902	4.826	4.232	3.849	3.579	3.376	3.219	3.092	2.988
0.100	55	2.799	2.402	2.186	2.050	1.955	1.884	1.829	1.785	1.748	1.717
0.050		4.016	3.165	2.773	2.540	2.383	2.269	2.181	2.112	2.055	2.008
0.025		5.310	3.948	3.364	3.029	2.807	2.648	2.528	2.433	2.355	2.291
0.010		7.119	5.013	4.159	3.681	3.370	3.149	2.983	2.853	2.748	2.662
0.005		8.554	5.843	4.773	4.181	3.800	3.531	3.330	3.173	3.046	2.942
0.100	60	2.791	2.393	2.177	2.041	1.946	1.875	1.819	1.775	1.738	1.707
0.050		4.001	3.150	2.758	2.525	2.368	2.254	2.167	2.097	2.040	1.993
0.025		5.286	3.925	3.343	3.008	2.786	2.627	2.507	2.412	2.334	2.270
0.010		7.077	4.977	4.126	3.649	3.339	3.119	2.953	2.823	2.718	2.632
0.005		8.495	5.795	4.729	4.140	3.760	3.492	3.291	3.134	3.008	2.904
0.100	70	2.779	2.380	2.164	2.027	1.931	1.860	1.804	1.760	1.723	1.691
0.050		3.978	3.128	2.736	2.503	2.346	2.231	2.143	2.074	2.017	1.969
0.025		5.247	3.890	3.309	2.975	2.754	2.595	2.474	2.379	2.302	2.237
0.010		7.011	4.922	4.074	3.600	3.291	3.071	2.906	2.777	2.672	2.585
0.005		8.403	5.720	4.661	4.076	3.698	3.431	3.232	3.076	2.950	2.846
0.100	80	2.769	2.370	2.154	2.016	1.921	1.849	1.793	1.748	1.711	1.680
0.050		3.960	3.111	2.719	2.486	2.329	2.214	2.126	2.056	1.999	1.951
0.025		5.218	3.864	3.284	2.950	2.730	2.571	2.450	2.355	2.277	2.213
0.010		6.963	4.881	4.036	3.563	3.255	3.036	2.871	2.742	2.637	2.551
0.005		8.335	5.665	4.611	4.029	3.652	3.387	3.188	3.032	2.907	2.803
0.100	100	2.756	2.356	2.139	2.002	1.906	1.834	1.778	1.732	1.695	1.663
0.050		3.936	3.087	2.696	2.463	2.305	2.191	2.103	2.032	1.975	1.927
0.025		5.179	3.828	3.250	2.917	2.696	2.537	2.417	2.321	2.244	2.179
0.010		6.895	4.824	3.984	3.513	3.206	2.988	2.823	2.694	2.590	2.503
0.005		8.241	5.589	4.542	3.963	3.589	3.325	3.127	2.972	2.847	2.744
0.100	120	2.748	2.347	2.130	1.992	1.896	1.824	1.767	1.722	1.684	1.652
0.050		3.920	3.072	2.680	2.447	2.290	2.175	2.087	2.016	1.959	1.910
0.025		5.152	3.805	3.227	2.894	2.674	2.515	2.395	2.299	2.222	2.157
0.010		6.851	4.787	3.949	3.480	3.174	2.956	2.792	2.663	2.559	2.472
0.005		8.179	5.539	4.497	3.921	3.548	3.285	3.087	2.933	2.808	2.705
0.100	150	2.739	2.338	2.121	1.983	1.886	1.814	1.757	1.712	1.674	1.642
0.050		3.904	3.056	2.665	2.432	2.274	2.160	2.071	2.001	1.943	1.894
0.025		5.126	3.781	3.204	2.872	2.652	2.494	2.373	2.278	2.200	2.135
0.010		6.807	4.749	3.915	3.447	3.142	2.924	2.761	2.632	2.528	2.441
0.005		8.118	5.490	4.453	3.878	3.508	3.245	3.048	2.894	2.770	2.667
0.100	∞	2.706	2.303	2.084	1.945	1.847	1.774	1.717	1.670	1.632	1.599
0.050		3.841	2.996	2.605	2.372	2.214	2.099	2.010	1.938	1.880	1.831
0.025		5.024	3.689	3.116	2.786	2.567	2.408	2.288	2.192	2.114	2.048
0.010		6.635	4.605	3.782	3.319	3.017	2.802	2.639	2.511	2.407	2.321
0.005		7.879	5.298	4.279	3.715	3.350	3.091	2.897	2.744	2.621	2.519

TABLE C.4 The F Distribution (Continued)

		df(*n*)									
Prob	df(*d*)	12	14	17	20	24	30	40	60	120	∞
0.100	35	1.739	1.703	1.662	1.632	1.601	1.569	1.535	1.497	1.457	1.411
0.050		2.041	1.986	1.924	1.878	1.833	1.786	1.735	1.681	1.623	1.558
0.025		2.341	2.266	2.183	2.122	2.062	1.999	1.932	1.861	1.785	1.702
0.010		2.740	2.639	2.527	2.445	2.364	2.281	2.193	2.099	2.000	1.891
0.005		3.048	2.926	2.791	2.693	2.597	2.497	2.392	2.282	2.164	2.036
0.100	40	1.715	1.678	1.636	1.605	1.574	1.541	1.506	1.467	1.425	1.377
0.050		2.003	1.948	1.885	1.839	1.793	1.744	1.693	1.637	1.577	1.509
0.025		2.288	2.213	2.129	2.068	2.007	1.943	1.875	1.803	1.724	1.637
0.010		2.665	2.563	2.451	2.369	2.288	2.203	2.114	2.019	1.917	1.805
0.005		2.953	2.831	2.697	2.598	2.502	2.401	2.296	2.184	2.064	1.932
0.100	45	1.695	1.658	1.616	1.585	1.553	1.519	1.483	1.443	1.399	1.349
0.050		1.974	1.918	1.855	1.808	1.762	1.713	1.660	1.603	1.541	1.470
0.025		2.248	2.172	2.088	2.026	1.965	1.900	1.831	1.757	1.677	1.586
0.010		2.608	2.506	2.393	2.311	2.230	2.144	2.054	1.958	1.853	1.737
0.005		2.881	2.759	2.625	2.527	2.430	2.329	2.222	2.109	1.987	1.851
0.100	50	1.680	1.643	1.600	1.568	1.536	1.502	1.465	1.424	1.379	1.327
0.050		1.952	1.895	1.831	1.784	1.737	1.687	1.634	1.576	1.511	1.438
0.025		2.216	2.140	2.056	1.993	1.931	1.866	1.796	1.721	1.639	1.545
0.010		2.562	2.461	2.348	2.265	2.183	2.098	2.007	1.909	1.803	1.683
0.005		2.825	2.703	2.569	2.470	2.373	2.272	2.164	2.050	1.925	1.786
0.100	55	1.668	1.630	1.587	1.555	1.522	1.487	1.450	1.408	1.362	1.308
0.050		1.933	1.876	1.812	1.764	1.717	1.666	1.612	1.553	1.487	1.412
0.025		2.190	2.114	2.029	1.967	1.904	1.838	1.768	1.692	1.607	1.511
0.010		2.526	2.424	2.311	2.228	2.146	2.060	1.968	1.869	1.761	1.638
0.005		2.779	2.658	2.523	2.425	2.327	2.226	2.118	2.002	1.876	1.733
0.100	60	1.657	1.619	1.576	1.543	1.511	1.476	1.437	1.395	1.348	1.291
0.050		1.917	1.860	1.796	1.748	1.700	1.649	1.594	1.534	1.467	1.389
0.025		2.169	2.093	2.008	1.944	1.882	1.815	1.744	1.667	1.581	1.482
0.010		2.496	2.394	2.281	2.198	2.115	2.028	1.936	1.836	1.726	1.601
0.005		2.742	2.620	2.486	2.387	2.290	2.187	2.079	1.962	1.834	1.689
0.100	70	1.641	1.603	1.559	1.526	1.493	1.457	1.418	1.374	1.325	1.265
0.050		1.893	1.836	1.771	1.722	1.674	1.622	1.566	1.505	1.435	1.353
0.025		2.136	2.059	1.974	1.910	1.847	1.779	1.707	1.628	1.539	1.436
0.010		2.450	2.348	2.234	2.150	2.067	1.980	1.886	1.785	1.672	1.540
0.005		2.684	2.563	2.428	2.329	2.231	2.128	2.019	.900	1.769	1.618
0.100	80	1.629	1.590	1.546	1.513	1.479	1.443	1.403	.358	1.307	1.245
0.050		1.875	1.817	1.752	1.703	1.654	1.602	1.545	1.482	1.411	1.325
0.025		2.111	2.035	1.948	1.884	1.820	1.752	1.679	1.599	1.508	1.400
0.010		2.415	2.313	2.199	2.115	2.032	1.944	1.849	1.746	1.630	1.494
0.005		2.641	2.520	2.385	2.286	2.188	2.084	1.974	.854	1.720	1.563
0.100	100	1.612	1.573	1.528	1.494	1.460	1.423	1.382	.336	1.282	1.214
0.050		1.850	1.792	1.726	1.676	1.627	1.573	1.515	.450	1.376	1.283
0.025		2.077	2.000	1.913	1.849	1.784	1.715	1.640	.558	1.463	1.347
0.010		2.368	2.265	2.151	2.067	1.983	1.893	1.797	.692	1.572	1.427
0.005		2.583	2.461	2.326	2.227	2.128	2.024	1.912	.790	1.652	1.485
0.100	120	1.601	1.562	1.516	1.482	1.447	1.409	.368	.320	1.265	1.193
0.050		1.834	1.775	1.709	1.659	1.608	1.554	.495	.429	1.352	1.254
0.025		2.055	1.977	1.890	1.825	1.760	1.690	.614	.530	1.433	1.310
0.010		2.336	2.234	2.119	2.035	1.950	1.860	.763	.656	1.533	1.381
0.005		2.544	2.423	2.288	2.188	2.089	1.984	.871	.747	1.606	1.431
0.100	150	1.590	1.550	1.504	1.470	1.434	1.396	.353	.305	1.247	1.169
0.050		1.817	1.758	1.691	1.641	1.590	1.535	.475	.407	1.327	1.223
0.025		2.032	1.955	1.867	1.801	1.736	1.665	.588	.502	1.402	1.271
0.010		2.305	2.203	2.088	2.003	1.918	1.827	.729	.620	1.493	1.331
0.005		2.506	2.385	2.250	2.150	2.050	1.944	.830	.704	1.559	1.374
0.100	∞	1.546	1.505	1.457	1.421	1.383	1.342	.295	.240	1.169	1.000
0.050		1.752	1.692	1.623	1.571	1.517	1.459	.394	.318	1.221	1.000
0.025		1.945	1.866	1.776	1.708	1.640	1.566	.484	.388	1.268	1.000
0.010		2.185	2.082	1.965	1.878	1.791	1.696	.592	.473	1.325	1.000
0.005		2.358	2.237	2.101	2.000	1.898	1.789	.669	.533	1.364	1.000

TABLE C.5 The Chi-Square Distribution

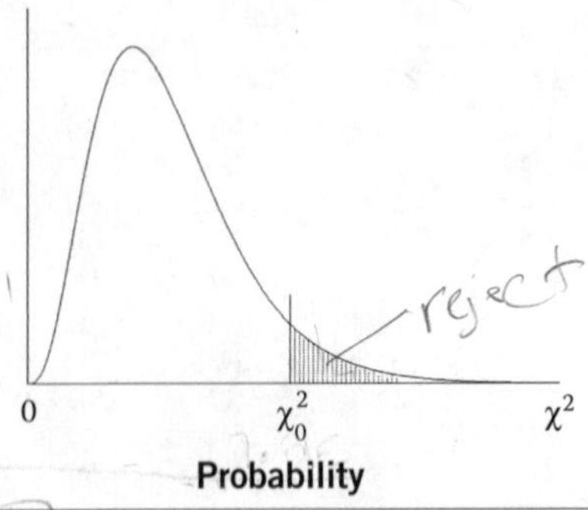

Probability

df	0.2500	0.1000	0.0500	0.0250	0.0100	0.0050	0.0025	0.0010	0.0005
1	1.323	2.706	3.841	5.024	6.635	7.879	9.141	10.828	12.116
2	2.773	4.605	5.991	7.378	9.210	10.597	11.983	13.816	15.202
3	4.108	6.251	7.815	9.348	11.345	12.838	14.320	16.266	17.730
4	5.385	7.779	9.488	11.143	13.277	14.860	16.424	18.467	19.997
5	6.626	9.236	11.070	12.833	15.086	16.750	18.386	20.515	22.105
6	7.841	10.645	12.592	14.449	16.812	18.548	20.249	22.458	24.103
7	9.037	12.017	14.067	16.013	18.475	20.278	22.040	24.322	26.018
8	10.219	13.362	15.507	17.535	20.090	21.955	23.774	26.124	27.868
9	11.389	14.684	16.919	19.023	21.666	23.589	25.462	27.877	29.666
10	12.549	15.987	18.307	20.483	23.209	25.188	27.112	29.588	31.420
11	13.701	17.275	19.675	21.920	24.725	26.757	28.729	31.264	33.137
12	14.845	18.549	21.026	23.337	26.217	28.300	30.318	32.909	34.821
13	15.984	19.812	22.362	24.736	27.688	29.819	31.883	34.528	36.478
14	17.117	21.064	23.685	26.119	29.141	31.319	33.426	36.123	38.109
15	18.245	22.307	24.996	27.488	30.578	32.801	34.950	37.697	39.719
16	19.369	23.542	26.296	28.845	32.000	34.267	36.456	39.252	41.308
17	20.489	24.769	27.587	30.191	33.409	35.718	37.946	40.790	42.879
18	21.605	25.989	28.869	31.526	34.805	37.156	39.422	42.312	44.434
19	22.718	27.204	30.144	32.852	36.191	38.582	40.885	43.820	45.973
20	23.828	28.412	31.410	34.170	37.566	39.997	42.336	45.315	47.498
21	24.935	29.615	32.671	35.479	38.932	41.401	43.775	46.797	49.011
22	26.039	30.813	33.924	36.781	40.289	42.796	45.204	48.268	50.511
23	27.141	32.007	35.172	38.076	41.638	44.181	46.623	49.728	52.000
24	28.241	33.196	36.415	39.364	42.980	45.559	48.034	51.179	53.479
25	29.339	34.382	37.652	40.646	44.314	46.928	49.435	52.620	54.947
26	30.435	35.563	38.885	41.923	45.642	48.290	50.829	54.052	56.407
27	31.528	36.741	40.113	43.195	46.963	49.645	52.215	55.476	57.858
28	32.620	37.916	41.337	44.461	48.278	50.993	53.594	56.892	59.300
29	33.711	39.087	42.557	45.722	49.588	52.336	54.967	58.301	60.735
30	34.800	40.256	43.773	46.979	50.892	53.672	56.332	59.703	62.162
31	35.887	41.422	44.985	48.232	52.191	55.003	57.692	61.098	63.582
32	36.973	42.585	46.194	49.480	53.486	56.328	59.046	62.487	64.995
33	38.058	43.745	47.400	50.725	54.776	57.648	60.395	63.870	66.403
34	39.141	44.903	48.602	51.966	56.061	58.964	61.738	65.247	67.803
35	40.223	46.059	49.802	53.203	57.342	60.275	63.076	66.619	69.199
40	45.616	51.805	55.758	59.342	63.691	66.766	69.699	73.402	76.095
45	50.985	57.505	61.656	65.410	69.957	73.166	76.223	80.077	82.876
50	56.334	63.167	67.505	71.420	76.154	79.490	82.664	86.661	89.561
55	61.665	68.796	73.311	77.380	82.292	85.749	89.035	93.168	96.163
60	66.981	74.397	79.082	83.298	88.379	91.952	95.344	99.607	102.695
70	77.577	85.527	90.531	95.023	100.425	104.215	107.808	112.317	115.578
80	88.130	96.578	101.879	106.629	112.329	116.321	120.102	124.839	128.261
100	109.141	118.498	124.342	129.561	135.807	140.169	144.293	149.449	153.167
120	130.055	140.233	146.567	152.211	158.950	163.648	168.082	173.617	177.603
150	161.291	172.581	179.581	185.800	193.208	198.360	203.214	209.265	213.613

Index

D

E

F

G

Q

R

S

T

U